October 29–30, 2012
Austin, Texas, USA

I0054767

**Association for
Computing Machinery**

Advancing Computing as a Science & Profession

ANCS'12

Proceedings of the Eighth ACM/IEEE Symposium on

Architectures for Networking
and Communications Systems

Sponsored by:

ACM SIGARCH, ACM SIGCOMM & IEEE COMPUTER SOCIETY

Supported by:

Intel & NETRONOME

Association for Computing Machinery

Advancing Computing as a Science & Profession

The Association for Computing Machinery
2 Penn Plaza, Suite 701
New York, New York 10121-0701

Notice to Past Authors of ACM-Published Articles

ISBN: 978-1-4503-1685-9 (Digital)

ISBN: 978-1-4503-1931-7 (Print)

Additional copies may be ordered prepaid from:

ACM Order Department
PO Box 30777
New York, NY 10087-0777, USA

Phone: 1-800-342-6626 (USA and Canada)
+1-212-626-0500 (Global)
Fax: +1-212-944-1318
E-mail: acmhelp@acm.org
Hours of Operation: 8:30 am – 4:30 pm ET

Printed in the USA

ANCS 2012 General Chair's Welcome

It is my great pleasure to welcome you to the Eighth ACM/IEEE Symposium on Architectures for Networking and Communications Systems (ANCS 2012). This year's conference will be held at the AT&T Executive Education and Conference Center on the campus of the University of Texas in Austin, TX, USA, on October 29–30, 2012.

This year's conference continues the tradition of ANCS to provide an opportunity for participants to exchange first-hand the cutting-edge system-oriented research results at the intersection of computer and network systems architecture from both academia and industry. We have an exciting program with 19 high-quality papers, 14 posters, and two keynote presentations by Gordon Brebner and Teemu Koponen.

In addition to the technical program, I hope you enjoy the conference banquet on Monday evening. I also hope you have a chance to explore the restaurants, nightlife, and sights of Austin. This year, ANCS is co-located with ICNP 2012, which will be held right after ANCS, to reduce travel expenses for those who attend both conferences.

I would like to thank all the members of the organizing committee for their great efforts to make this conference a success. I would like to thank Andrew Moore and Viktor Prasanna, who served as program chairs and, together with the technical program committee, put together an exciting technical program. I also thank Charlie Wiseman and his poster selection committee for organizing two poster sessions. My thanks go to all the authors who submitted their papers and posters and who present their work at the conference. I also want to thank Michela Becchi, who served both as finance chair and as student travel grant chair, for her great help in organizing this conference.

I also want to acknowledge and thank our sponsors and supporters. ANCS 2012 is co-sponsored by ACM SIGARCH, ACM SIGCOMM, and IEEE TCCA. Our industry supporters are Intel and Netronome. ACM SIGARCH and ACM SIGCOMM provided student travel support.

Finally, I want to thank all of the attendees of ANCS 2012 for coming to the conference. I hope you enjoy the conference program and find this conference an opportunity to exchange ideas, make new contacts, and renew old ones.

Tilman Wolf
ANCS 2012 General Chair
University of Massachusetts Amherst, USA

Message from the Program Co-Chairs

Welcome to the Eighth Symposium on Architectures for Networking and Communications Systems (ANCS 2012). It is our pleasure to present an excellent conference program for the meeting consisting of nineteen technical papers. The program also consists of two keynote addresses by leading researchers in the community, as well as two poster sessions. In addition, we are pleased to present a tutorial on NetFPGA.

A total of 64 papers were submitted from across the world to the conference. The 19 accepted papers represent a competitive acceptance rate of 29%. Each paper was assigned to at least 3, and in many cases 4, TPC members. The 23 members of the technical program committee and external reviewers completed an impressive 205 reviews! The program committee subsequently participated in an extensive on-line discussion to select the set of papers to be accepted.

A tremendous amount of dedicated hard work has gone into putting this program together. We are most grateful and fortunate for such an outstanding TPC team who helped tackle the challenges in this process in a professional, responsive, and friendly fashion. Several papers received shepherding and we extend our additional gratitude to those TPC members who acted as shepherds. We want to acknowledge the contributions of General Chair Tilman Wolf who offered his insights and "mentored us" throughout the process. We appreciate the opportunity to serve as program co- chairs and thank the steering committee for selecting us to participate in this important role.

These are exciting times for networking research as we move forward in the software and multicore era, and we are delighted to bring you the latest in networking architecture innovation. We hope that you will enjoy the exceptional technical program while exploring the Austin area.

Andrew W. Moore and Viktor K. Prasanna
ANCS 2012 Program Co-Chairs

Table of Contents

ANCS 2012 Conference Organization ix

ANCS 2012 Sponsors & Supporters x

Keynote Address 1

- **Softly Defined Networking** 1
  Gordon Brebner *(Xilinx, Inc.)*

Session 1: New Architectures

- **On the Feasibility of Completely Wireless Datacenters** 3
  Ji-Yong Shin, Emin Gün Sirer, Hakim Weatherspoon *(Cornell University)*, Darko Kirovski *(Microsoft Research)*

- **Popularity-Driven Coordinated Caching in Named Data Networking** 15
  Jun Li, Hao Wu, Bin Liu, Jianyuan Lu, Yi Wang *(Tsinghua University)*, Xin Wang *(State University of New York)*,
  Yanyong Zhang, Lijun Dong *(Rutgers University)*

Session 2: Multicore Networking

- **NetSlices: Scalable Multi-Core Packet Processing in User-Space** 27
  Tudor Marian *(Google)*, Ki Suh Lee, Hakim Weatherspoon *(Cornell University)*

- **Cache-Aware Affinitization on Commodity Multicores for High-Speed Network Flows** 39
  Vishal Ahuja, Matthew Farrens, Dipak Ghosal *(University of California, Davis)*

Session3: High-Performance Middlesystems

- **xOMB: Extensible Open Middleboxes with Commodity Servers** 49
  James W. Anderson, Ryan Braud, Rishi Kapoor, George Porter, Amin Vahdat
  (University of California, San Diego)

- **NetBump: User-Extensible Active Queue Management with Bumps on the Wire** 61
  Mohammad Al-Fares, Rishi Kapoor, George Porter, Sambit Das *(University of California, San Diego)*,
  Hakim Weatherspoon *(Cornell University)*, Balaji Prabhakar *(Stanford University)*,
  Amin Vahdat *(University of California, San Diego & Google, Inc.)*

Poster Session 1

- **Fast Longest Prefix Name Lookup for Content-Centric Network Forwarding** 73
  Fu Li, Fuyu Chen, Jianming Wu *(Huawei Technologies)*,
  Haiyong Xie *(Huawei Technologies & University of Science and Technology of China)*

- **Hardware Support for Dynamic Protocol Stacks** 75
  Ariane Keller, Daniel Borkmann, Stephan Neuhaus *(ETH Zurich)*

- **Low-Latency Modular Packet Header Parser for FPGA** 77
  Viktor Puš, Lukas Kekely *(CESNET a.l.e.)*, Jan Kořenek *(Brno University of Technology)*

- **M-DFA (Multithreaded DFA): An Algorithm for Reduction of State Transitions
  and Acceleration of REGEXP Matching** 79
  Cheng-Hung Lin *(National Taiwan Normal University)*, Jyh-Charn Liu *(Texas A&M University)*

- **A New Embedded Platform for Rapid Development of Network Applications** 81
  Jan Korenek, Pavol Korcek, Vlastimil Kosar, Martin Zadnik, Jan Viktorin *(Brno University of Technology)*

- **Structural Compression of Packet Classification Trees** 83
  Xiang Wang, Zhi Liu, Yaxuan Qi, Jun Li *(Tsinghua University)*

- **Toward Fast NDN Software Forwarding Lookup Engine Based on Hash Tables** 85
  Won So, Ashok Narayanan, Dave Oran, Yaogong Wang *(Cisco Systems, Inc.)*

Session 4: System Design for Data Center Networks

- **Host-Based Multi-Tenant Technology for Scalable Data Center Networks** 87
  Koichi Onoue, Naoki Matsuoka, Jun Tanaka *(Fujitsu Laboratories Ltd.)*
- **Distributed Adaptive Routing for Big-Data Applications Running on Data Center Networks** 99
  Eitan Zahavi *(Mellanox Technologies & Technion)*, Isaac Keslassy, Avinoam Kolodny *(Technion)*

Session 5: Efficiency

- **Efficient Buffering and Scheduling for a Single-Chip Crosspoint-Queued Switch** 111
  Zizhong Cao, Shivendra S. Panwar *(Polytechnic Institute of New York University)*
- **Efficient Traffic Aware Power Management in Multicore Communications Processors** ...123
  Muhammad Faisal Iqbal, Lizy K. John *(University of Texas at Austin)*

Keynote Address 2

- **Software Is the Future of Networking** 135
  Teemu Koponen *(VMware, Inc.)*

Poster Session 2

- **Attendre: Mitigating Ill Effects of Race Conditions in OpenFlow via Queueing Mechanism** 137
  Xiaoye Sun, Apoorv Agarwal, Tze Sing Eugene Ng *(Rice University)*
- **Dynamic Frequency Scaling Architecture for Energy Efficient Router** 139
  Wenliang Fu, Tian Song, Shian Wang *(Beijing Institute of Technology)*, Xiaojun Wang *(Dublin City University)*
- **Extensible Hierarchical Simulation of Network Systems** 141
  Xinming Chen, Tilman Wolf *(University of Massachusetts, Amherst)*
- **FlowOS: A Pure Flow-Based Vision of Network Traffic** 143
  Abdul Alim, Mehdi Bezahaf *(Lancaster University)*, Laurent Mathy *(University of Liege)*
- **Modular Design of Data Vortex Switch Network** 145
  Brett Burley, Qimin Yang *(Harvey Mudd College)*
- **PVNs: Making Virtualized Network Infrastructure Usable** 147
  Shufeng Huang, James Griffioen, Ken Calvert *(University of Kentucky)*
- **Securing Multi-Core Multi-Threaded Packet Processors** 149
  Danai Chasaki *(Villanova University)*

Session 6: Algorithms Improving Performance

- **A Hardware Spinal Decoder** 151
  Peter A. Iannucci, Kermin Elliott Fleming, Jonathan Perry, Hari Balakrishnan, Devavrat Shah *(Massachusetts Institute of Technology)*
- **Fast Submatch Extraction Using OBDDs** 163
  Liu Yang *(Rutgers University)*, Pratyusa Manadhata, William Horne, Prasad Rao *(HP Laboratories)*, Vinod Ganapathy *(Rutgers University)*
- **LEAP: Latency- Energy- and Area-optimized Lookup Pipeline** 175
  Eric N. Harris, Samuel L. Wasmundt, Lorenzo De Carli, Karthikeyan Sankaralingam *(University of Wisconsin-Madison)*, Cristian Estan *(Broadcom Corporation)*

Session 7: Mobility Enabling Architectures

- **Floating Ground Architecture: Overcoming the One-Hop Boundary of Current Mobile Internet** 187
  Hajime Tazaki *(NICT)*, Rodney Van Meter *(Keio University)*,
  Ryuji Wakikawa *(Toyota InfoTechnology Center, U.S.A., Inc.)*, Noriyuki Shigechika *(RCA Co., Ltd.)*,
  Keisuke Uehara, Jun Murai *(Keio University)*

- **ECOS: Leveraging Software-Defined Networks to Support Mobile Application Offloading** 199
  Aaron Gember, Christopher Dragga, Aditya Akella *(University of Wisconsin-Madison)*

Session 8: Content-Centric Architectures, Platforms, and Mechanisms

- **On Pending Interest Table in Named Data Networking** 211
  Huichen Dai, Bin Liu *(Tsinghua University)*, Yan Chen *(Northwestern University)*,
  Yi Wang *(Tsinghua University)*

- **Coexist: Integrating Content Oriented Publish/Subscribe Systems with IP** 223
  Jiachen Chen *(University of Goettingen)*,
  Mayutan Arumaithurai *(University of Goettingen & NEC Laboratories Europe)*,
  Xiaoming Fu *(University of Goettingen)*, K. K. Ramakrishnan *(AT&T Labs-Research)*

Session 9: Enhancing Security

- **MCA2: Multi-Core Architecture for Mitigating Complexity Attacks** 235
  Yehuda Afek *(Tel Aviv University)*, Anat Bremler-Barr *(The Interdisciplinary Center)*,
  Yotam Harchol, David Hay *(The Hebrew University)*, Yaron Koral *(Tel Aviv University)*

- **Malacoda: Towards High-Level Compilation of Network Security Applications on Reconfigurable Hardware** 247
  Sascha Muehlbach *(Center for Advanced Security Research Darmstadt)*,
  Andreas Koch *(Technische Universitaet Darmstadt)*

Author Index 258

ANCS 2012 Conference Organization

General Chair: Tilman Wolf *(University of Massachusetts Amherst, USA)*

Program Chairs: Andrew W. Moore *(University of Cambridge, UK)*
Viktor Prasanna *(University of Southern California, USA)*

Finance Chair: Michela Becchi *(University of Missouri, Columbia, USA)*

Poster Chair: Charlie Wiseman *(Wentworth Institute of Technology, USA)*

Student Travel Grant Chair: Michela Becchi *(University of Missouri, Columbia, USA)*

Web Chair: Xinming Chen *(University of Massachusetts Amherst, USA)*

Local Arrangements Chair: EJ Kim *(Texas A&M University, USA)*

Program Committee: Michela Becchi *(University of Missouri, USA)*
Gordon Brebner *(Xilinx Labs, USA)*
Greg Byrd *(North Carolina State University, USA)*
Qunfeng Dong *(University of Science and Technology of China, China)*
Hans Eberle *(Oracle Labs, USA)*
Holger Fröning *(University of Heidelberg, Germany)*
Euan Harris *(Arista Networks, USA)*
Hoang Le *(ISC8, USA)*
Jun Li *(Tsinghua University, China)*
Bill Lin *(University of California, San Diego, USA)*
Derek McAuley *(Nottingham University, UK)*
Kieran Mansley *(Solarflare, USA)*
David Meyer *(Cisco Systems, USA)*
Mario Nemirovsky *(Barcelona Supercomputing Center, Spain)*
George Neville-Neil *(FreeBSD Foundation, USA)*
Luigi Rizzo *(Università di Pisa, Italy)*
Tom Rodeheffer *(Microsoft Research, USA)*
Dimitrios Serpanos *(ISI/RC ATHENA & University of Patras, Greece)*
Frederico Silla *(Universitat Politècnica de València, Spain)*
Satnam Singh *(Google, USA)*
Ripduman Sohan *(University of Cambridge, UK)*
Russ Tessier *(University of Massachusetts Amherst, USA)*
Ola Tørudbakken *(Oracle Systems, Norway)*
Fang Yu *(Microsoft Research, USA)*

Soft*ly* Defined Networking

Gordon Brebner
Xilinx Labs
2100 Logic Drive
San Jose, CA 95124, USA
gordon.brebner@xilinx.com

ABSTRACT

Software Defined Networking (SDN) has been described as the hope and hype for the future of networking. Definitions vary, but one research direction has been to separate the control plane from the data plane, introducing abstractions that can provide a global network view, a description of required behavior, and a model of packet forwarding. This is intended as a way to open up the closed-box and vertically-integrated, proprietary packet switches and routers, and allow applications greater scope for accessing networking capabilities. A somewhat less disruptive direction for SDN is to see it as the provision of programmatic interfaces (APIs) into existing networking equipment, whether or not there is an explicit separation of control and data planes.

Regardless of precise SDN definitions, while the worthy goal is to address ossification of the Internet, the "S" for "software" in SDN perhaps unintentionally ossifies views of the respective roles of hardware and software. Specifically, it introduces an inbuilt assumption that there is relatively dumb switching hardware for high-speed packet forwarding, and relatively intelligent software running on processors for lesser-speed packet forwarding and networking control. This reflects the tradition that hardware is expensive and hard to build, whereas software is cheaper and easier to design and extend.

The viewpoint is exemplified to a large extent by OpenFlow, a pioneering inter-plane communication interface, sometimes wrongly equated with the broader aspirations of SDN. In OpenFlow 1.0, the version mainly implemented to date, the model of packet forwarding was a single lookup table for matching certain pre-defined packet fields, collecting statistics, and carrying out simple actions. In later versions, the model has been extended to allow multiple sequential lookup tables, plus a group table, with extended ranges of pre-defined packet fields and actions. The OpenFlow protocol provides access to this simple model of packet forwarding.

Research work, such as the Frenetic language and implementation, has civilized the OpenFlow interface at a higher level of abstraction, but it is questionable whether the underlying OpenFlow abstraction itself needs to offer such a limited view of forwarding architecture and its programmability, including being tied to existing ossified protocols, given the rather richer programmable capabilities that are actually possible.

To address increasing requirements for both high throughput and low latency in networking, together with more complex functions within core networks, real switching hardware can be very complex, with numerous highly-programmable features. In practice, it can
encompass multiple line cards connected by a high-speed backplane switching fabric, with the line cards carrying out physical interfacing, packet classification, packet editing, and traffic management functions. These are implemented by a mixture of ASIC, FPGA, NPU, and multi-core technologies on the line cards. In other words, it can be much more elaborate than just the simple "switching ASIC" that is often postulated in SDN discussions, and so is deserving of more sophisticated abstractions.

In particular, the use of FPGA (programmable logic) technology in switching hardware offers scope for 'soft hardware', with the potential to blur the distinctions between traditional capabilities and roles of hardware and software, and hence open the door to Soft*ly* Defined Networking that involves a less restrictive view of hardware-software partitioning. However, for this vision to become real, such technology must prove both its ability to deliver the necessary high performance and its ability to be programmed in a high-level manner. This involves going beyond the well-known NetFPGA research vehicle, which supports line rates of just 4x1G initially and 4x10G in its latest version, together with a Verilog hardware design programming experience.

Recent research has been addressing these issues successfully. Contemporary FPGA devices can perform network processing functions at up to 400 Gbit/sec line rates, with 1 Tbit/sec rates on the horizon for the next generation. This indicates that performance levels normally associated with fixed ASIC devices can be achieved using this completely programmable technology. Moreover, it is possible to compile high-level network processing descriptions directly and automatically to soft architectures formed from programmable logic that deliver this performance.

Given these results, there are interesting research questions around investigating the implications for SDN. One aspect involves enlarging the scope perceived for programmability in the data plane, and hence its capabilities relative to those of the control plane. Another aspect involves the descriptions of required behavior, and the nature of programmatic interfaces, to ensure that these are not overly constrained by conventional software programming assumptions.

Categories and Subject Descriptors

C.2.6 COMPUTER-COMMUNICATION NETWORKS: Internetworking -- *Routers*.

Keywords: Domain-specific languages, High-speed network processing, Soft hardware, Softly Defined Networking

On the Feasibility of Completely Wireless Datacenters

Ji-Yong Shin, Emin Gün Sirer, and
Hakim Weatherspoon
Dept. of Computer Science, Cornell University
Ithaca, NY, USA
{jyshin, egs, hweather}@cs.cornell.edu

Darko Kirovski
Microsoft Research
Redmond, WA, USA
darkok@microsoft.com

ABSTRACT

Conventional datacenters, based on wired networks, entail high wiring costs, suffer from performance bottlenecks, and have low resilience to network failures. In this paper, we investigate a radically new methodology for building wire-free datacenters based on emerging 60GHz RF technology. We propose a novel rack design and a resulting network topology inspired by Cayley graphs that provide a dense interconnect. Our exploration of the resulting design space shows that wireless datacenters built with this methodology can potentially attain higher aggregate bandwidth, lower latency, and substantially higher fault tolerance than a conventional wired datacenter while improving ease of construction and maintenance.

Categories and Subject Descriptors

C.2.1 [**Computer-Communication Networks**]: Network Architecture and Design—*Network topology, Wireless communication*

General Terms

Design, Experimentation, Performance

Keywords

60GHz RF, Wireless data center

1. INTRODUCTION

Performance, reliability, cost of the switching fabric, power consumption, and maintenance are some of the issues that plague conventional wired datacenters [2, 16, 17]. Current trends in cloud computing and high-performance datacenter applications indicate that these issues are likely to be exacerbated in the future [1, 4].

In this paper, we explore a radical change to the construction of datacenters that involves the removal of all but power supply wires. The workhorses of communication in

this new design are the newly emerging directional, beam-formed 60GHz RF communication channels characterized by high bandwidth (4-15Gbps) and short range (≤ 10 meters). New 60GHz transceivers [40, 43] based on standard 90nm CMOS technology make it possible to realize such channels with low cost and high power efficiency (< 1W). Directional ($25° - 60°$ wide) short-range beams employed by these radios enable a large number of transmitters to simultaneously communicate with multiple receivers in tight confined spaces.

The unique characteristics of 60GHz RF modems pose new challenges and tradeoffs. The most critical questions are those of feasibility and structure: can a large number of transceivers operate without signal interference in a densely populated datacenter? How should the transceivers be placed and how should the racks be oriented to build practical, robust and maintainable networks? How should the network be architected to achieve high aggregate bandwidth, low cost and high fault tolerance? And can such networks compete with conventional wired networks?

To answer these questions, we propose a novel datacenter design—because its network connectivity subgraphs belong to a class of Cayley graphs [6], we call our design a Cayley datacenter. The key insight behind our approach is to arrange servers into a densely connected, low-stretch, failure-resilient topology. Specifically, we arrange servers in cylindrical racks such that inter- and intra-rack communication channels can be established and form a densely connected mesh. To achieve this, we replace the network interface card (NIC) of a server with a Y-switch that connects a server's system bus with two transceivers positioned at opposite ends of the server box. This topology leads to full disappearance of the classic network switching fabric (e.g., no top-of-rack switches, access routers, copper and optical interconnects) and has far-reaching ramifications on performance.

Overall, this paper makes three contributions. First, we present the first constructive proposal for a fully wireless datacenter. We show that it is possible for 60GHz technology to serve as the sole and central means of communication in the demanding datacenter setting. Second, we propose a novel system-level architecture that incorporates a practical and efficient rack-level hardware topology and a corresponding geographic routing protocol. Finally, we examine the performance and system characteristics of Cayley datacenters. Using a set of 60GHz transceivers, we demonstrate that signals in Cayley datacenters do not interfere with each other. We also show that, compared to a fat-tree [37, 38] and a conventional datacenter, our proposal exhibits higher bandwidth, substantially improved latency due to the switching fabric being integrated into server nodes, and lower power consumption. Cayley datacenters exhibit strong fault tolerance due to a routing scheme that can fully explore the mesh: Cayley datacenters can maintain connectivity to over 99% of live nodes until up to 55% of total nodes fail.

2. 60GHZ WIRELESS TECHNOLOGY

In this section, we briefly introduce the communication characteristics of the newly emerging 60GHz wireless technology, which is the foundation of our datacenter.

Propagation of RF (radio frequency) signals in the 57-64GHz sub-band is severely attenuated because of the resonance of oxygen molecules, which limits the use of this sub-band to relatively short distances [34]. Consequently 57-64GHz is unlicensed under FCC rules and open to short-range point-to-point applications. Several efforts are aiming to standardize the technology, with most of them tailored to home entertainment: two IEEE initiatives, IEEE 802.15.3c and 802.11.ad [26,55], WiGig 7Gbps standard with beam-forming [56], and ECMA-387/ISO DS13156 6.4Gbps spec [14] based upon Georgia Tech's design [43].

In this paper, we focus on a recent integrated implementation from Georgia Tech whose characteristics are summarized in Table 1:

Category	Characteristic
Technology	Standard 90nm CMOS
Packaging	Single chip Tx/Rx in QFN
Compliance	ECMA TC48
Power	0.2W (at output power of 3dBm)
Range	$\leq 10m$
Bandwidth	4-15Gbps

Table 1: 60GHz wireless transceiver characteristics [43].

More details about 60GHz transceiver characteristics can be found from a link margin, which models communication between a transmitter (Tx) and a receiver (Rx). The link margin, M, is the difference between the received power at which the receiver stops working and the actual received power, and can be expressed as follows:

$$M = P_{TX} + G_{TX+RX} - L_{Fade} - L_{Implementation}$$
$$-FSL - NF - SNR, \qquad (1)$$

where P_{TX} and G_{TX+RX} represent transmitted power and overall joint transmitter and receiver gain which is dependent upon the geometric alignment of the Tx↔Rx antennae [57]. Free-space loss equals $FSL = 20 \log_{10}(4\pi D/\lambda)$, where D is the line-of-sight Tx↔Rx distance and λ wavelength. The noise floor $NF \sim 10 \log_{10}(R)$ is dependent upon R, the occupied bandwidth. SNR is the signal to noise ratio in dBs which links a dependency to the bit error rate as $BER = \frac{1}{2}\mathrm{erfc}(\sqrt{SNR})$ for binary phase-shift keying (BPSK) modulation for example. Loss to fading and implementation are constants given a specific system. From Equation 1, one can compute the effects of constraining different communication parameters.

Figure 1 illustrates a planar slice of the geometric communication model we consider in this paper. A transmitter antenna radiates RF signals within a lobe—the surface of the lobe is a level-set whose signal power is equal to one half of the maximum signal power within the lobe. Because the

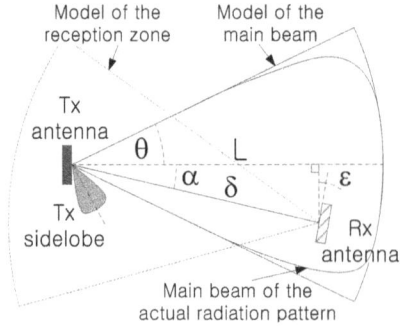

Figure 1: Geometric communication model.

attenuation is very sharp in the 60GHz frequency range, a receiver antenna should be within the bound of a transmitter's beam for communication. The beam is modeled as a cone with an angle θ and length L. Using a spherical coordinate system centered at transmitter's antenna, one can define the position of the receiver antenna with its radius, δ, elevation α, and azimuth β. The plane of the receiver antenna can then be misaligned from the plane of the transmitter antenna by an angle ε along the elevation and γ along the azimuth. We use a modeling tool developed at Georgia Tech to convert $\{\alpha, \beta, \gamma, \varepsilon, \delta, L, \theta\}$ into G_{TX+RX}. Through personal communication with Georgia Tech's design team, we reduced our space of interest to $25° \leq \theta \leq 45°$ as a constraint to suppress side lobes. Based on design parameters from the antenna prototypes developed by the same team, we model a reception zone of the receiver that is in identical shape to the main transmitter beam. We limit ε and γ to be smaller than θ such that the transmitter is located within the reception zone and assume a BER of 10^{-9} at 10Gbps bandwidth within $L < 3$ meters range. We do not utilize beam-steering[1] and assume that the bandwidth can be multiplexed using both time (TDD) and frequency division duplexing (FDD).

The design parameters of the transceiver are optimized for our datacenter design and lead to a higher bandwidth and less noisier transceiver design compared to off-the-shelf 60GHz transceivers for HDMI [49]. More research in 60GHz RF design with a focus on Cayley datacenters can further improve performance.

3. CAYLEY DATACENTER DESIGN

This section introduces Cayley datacenter architecture, the positioning of the 60GHz transceivers in a wireless datacenter, and the resulting network topology. We also introduce a geographical routing protocol for this unique topology and adopt a MAC layer protocol to address the hidden terminal and the masked node problem.

3.1 Component Design

In order to maximize opportunities for resource multiplexing in a wireless datacenter, it is important to use open spaces efficiently, because the maximum number of live connections in the network is proportional to the volume of the datacenter divided by that of a single antenna beam. We

---

[1]Typically, reconnection after beam-steering involves training of communication codebooks involving delays on the order of microseconds, which is not tolerable in datacenters.

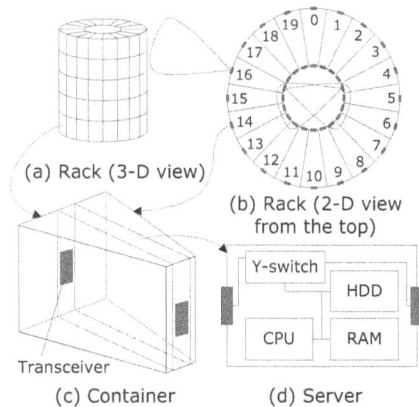

(a) Rack (3-D view)

(b) Rack (2-D view from the top)

Transceiver

(c) Container

(d) Server

Figure 2: Rack and server design.

focus on the network topology that would optimize key performance characteristics, namely latency and bandwidth.

To separate the wireless signals for communications within a rack and among different racks, we propose cylindrical racks (Figure 2.a) that store servers in prism-shaped containers (Figure 2.c). This choice is appealing, because it partitions the datacenter volume into two regions: intra- and inter-rack free space. A single server can be positioned so that one of its transceivers connects to its rack's inner-space and another to the inter-rack space as the rack illustrated in Figure 2.b. A rack consists of S stories and each story holds C containers; we constrain $S = 5$ and $C = 20$ for brevity of analysis and label servers in the same story from 0 to 19 starting from the 12 o'clock position in a clockwise order.

The prism containers can hold commodity half-height blade servers. A custom built Y-switch connects the transceivers located on opposite sides of the server (Figure 2.d). The Y-switch, whose design is discussed at the end of this section, multiplexes incoming packets to one of the outputs.

3.2 Topology

The cylindrical racks we propose utilize space and spectrum efficiently and generalize to a topology that can be modeled as a mesh of Cayley graphs.

A Cayley graph [6] is a graph generated from a group of elements G and a generator set $S \subseteq G$. Set S excludes the identity element $e = g \cdot g^{-1}$, where $g \in G$, and $h \in S$ iff $h^{-1} \in S$. Each vertex $v \in V$ of a Cayley graph (V, E) corresponds to each element $g \in G$ and edge $(v_1, v_2) \in E$ iff $g_1 \cdot g_2^{-1} \in S$. This graph is vertex-transitive, which facilitates the design of a simple distributed routing protocol and is generally densely connected, which adds fault tolerance to the network [50].

When viewed from the top, connections within a story of the rack form a 20-node, degree-k Cayley graph, where k depends on the signal's radiation angle (Figure 3.a). This densely connected graph provides numerous redundant paths from one server to multiple servers in the same rack and ensures strong connectivity.

The transceivers on the exterior of the rack stitch together Cayley subgraphs in different racks. There is a great flexibility in how a datacenter can be constructed out of these racks, but we pick the simplest topology by placing the racks in rows and columns for ease of maintenance. Figure 3.b illustrates an example of the 2-dimensional connectivity of 9 racks in 3 by 3 grids: a Cayley graph sits in the center of each rack and the transceivers on the exterior of the racks

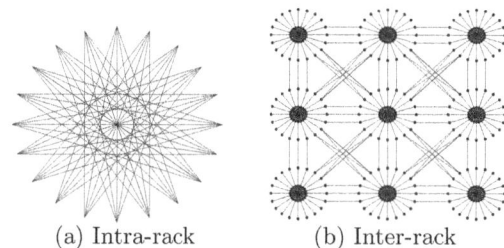

(a) Intra-rack   (b) Inter-rack

Figure 3: Cayley datacenter topology when $\theta = 25°$

connect the subgraphs together. Further, since the wireless signal spreads in a cone shape, a transceiver is able to reach servers in different stories, both within and across racks.

3.3 Routing Protocol

A routing protocol for datacenters should enable quick routing decisions, utilize a small amount of memory, and find efficient routes involving few network hops. A geographic routing technique for our topology can fulfill these conditions.

3.3.1 Diagonal XYZ Routing

The uniform structure of Cayley datacenters lends itself to a geographical routing protocol. The routing protocol that we investigate in this paper is called diagonal XYZ routing.

Similar to XY routing [21], diagonal XYZ routing finds an efficient route to the destination at a low computational and storage cost using geographical information. We define the geographical identity g_k of a server k as (rx, ry, s, i), where rx and ry are the x and y coordinates of the rack, s corresponds to the ordinal number for the story, and i is the index of the server within a story. Cayley datacenters use this identity to address the servers.

The geographical identity facilitates finding a path in the Cayley datacenter network. The routing protocol determines the next hop by comparing the destination of a packet to the identity of the sever holding the packet. Based on rx and ry values, the protocol finds an adjacent rack of the server that is closest to the destination. The s value is then used to reach the story height of the destination that the packet should arrive. Finally, the i value is used to forward the packet using the shortest path to the destination server within the same story. Algorithm 1 describes the details about the routing algorithm.

Because the topology has a constant fanout, diagonal XYZ routing requires very little state to be maintained on each host. Every host keeps and consults only three tables to determine the next destination for a packet.

- **Inter-rack routing table**: Maps 8 horizontal directions towards adjacent racks to directly reachable servers on the shortest path to the racks.

- **Inter-story routing table**: Maps 2 vertical directions to directly reachable servers in the same rack of the table owner leading to the desired story.

- **Intra-story routing table**: Maps 20 server index i's to directly reachable servers in the same story in the same rack of the table owner. The servers in the table are on the precomputed shortest path leading to server i.

Inter-rack and inter-story routing tables maintain story s as the secondary index for lookup. Using this index,

5

Algorithm 1 Diagonal XYZ routing

Require: g_{curr}: geographical identity of the server, where the packet is currently at
$\quad g_{dst}$: geographical identity of the packet's final destination
$\quad r_{curr}$: rack of g_{curr}
$\quad r_{dst}$: rack of g_{dst}
$\quad R_{adj}$: set of adjacent racks of r_{curr}
$\quad T_{InterRack}$: inter-rack routing table of curr
$\quad T_{InterStory}$: inter-story routing table of curr
$\quad T_{IntraStory}$: intra-story routing table of curr
Ensure: g_{next}: geographical identity of next destination
$\quad$ **if** IsInDifferentRack(g_{curr}, g_{dst}) **then**
$\quad\quad r_{next} \leftarrow r_{dst}$.GetMinDistanceRack($R_{adj}$)
$\quad\quad dir \leftarrow r_{curr}$.GetHorizontalDirection($r_{next}$)
$\quad\quad G \leftarrow T_{InterRack}$.LookupGeoIDs($dir, g_{dst}.s$)
$\quad$ **else if** IsInDifferentStory(g_{curr}, g_{dst}) **then**
$\quad\quad dir \leftarrow g_{curr}$.GetHorizontalDirection($g_{dst}$)
$\quad\quad G \leftarrow T_{InterStory}$.LookupGeoIDs($dir, g_{dst}.s$)
$\quad$ **else if** IsDifferentServer(g_{curr}, g_{dst}) **then**
$\quad\quad G \leftarrow T_{IntraStory}$.LookupGeoIDs($g_{dst}.i$)
$\quad$ **else**
$\quad\quad G \leftarrow g_{dst}$
$\quad$ **end if**
$\quad g_{next} \leftarrow$ RandomSelect(G)

LookupGeoIDs($dir, g_{dst}.s$) returns the identities with the closest s value to $g_{dst}.s$ among the ones leading to dir.

For all three tables, LookupGeoIDs returns multiple values, because a transceiver can communicate with multiple others. The servers found from the table lookup all lead to the same number of hops to the final destination. Thus, the routing protocol pseudo-randomly selects one of the choices to evenly distribute the traffic and to allow a TCP flow to follow the same path. We use a pseudo-random hashing of the packet header like the Toepliz Hash function [28].

The directionality of the radio beam, the presence of multiple transceivers per node and the low latency of the Y-switch makes it possible for Cayley datacenters to deploy cut-through switching [30], which starts routing a packet immediately after receiving and reading the packet header. While this is generally not usable in wireless communication based on omni-directional antennae, unless special methodologies, such as signal cancellation is employed [8,20]— Cayley datacenter servers employ this optimization.

3.3.2 *Adaptive Routing in Case of Failure*

Compared to a conventional datacenter, a Cayley datacenter has a distinct failure profile. Conventional datacenters are dependent on switches for network connectivity and consequently a switch failure can disconnect many servers. Cayley datacenters, on the other hand, can compensate for the failure of nodes and racks by utilizing some of the many alternative paths in their rich topology. We employ an adaptive routing scheme such as a variant of face routing [27] with the diagonal XYZ routing. Due to space constraints, we do not detail our adaptive routing scheme, but our previous work [47] shows that the routing scheme can circumvent randomly failed racks with less than 5us latency overhead.

3.4 MAC Layer Arbitration

A transceiver in a Cayley datacenter can communicate with approximately 7 to over 30 transceivers depending on its configuration. As a result, communication needs to be coordinated. However, due to the directionality of the signal, all transceivers that can communicate with the same

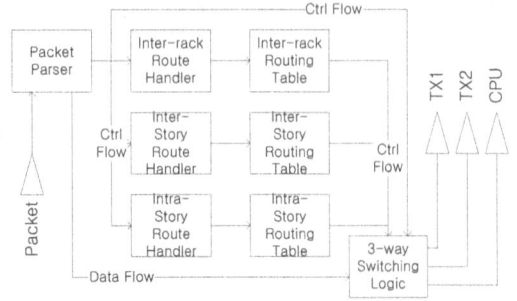

Figure 4: Y-switch schematic.

transceiver act as hidden terminals for each other. Such multiple hidden terminals can lead to a masked node problem [46] that causes collisions if a regular ready-to-send (RTS)/clear-to-send (CTS) based MAC protocol [31] is used.

Therefore, we adopt a dual busy tone multiple access (DBTMA) [23, 24] channel arbitration/reservation scheme. DBTMA is based on an RTS/CTS protocol, but it employs an additional out of band tone to indicate whether the transceivers are transmitting or receiving data. This tone resolves the masked node problem by enabling nodes both at the sending and receiving end to know whether other nodes are already using the wireless channel.

We use a fraction of the dedicated frequency channel for this tone and control messages using FDD so that they do not interfere with the data channel.

3.5 Y-Switch Implementation

The Y-switch is a simple customized piece of hardware that plays an important role in a Cayley datacenter. High-level schematic of this switch is shown in Figure 4. When the Y-switch receives a packet, it parses the packet header and forwards the packet to the local machine or one of the transceivers. The decisions are made by searching through one of the three routing tables described in Section 3.3.1. To analyze the feasibility of the proposed Y-switch design, we implemented the Y-switch design for Xilinx FPGA in Simulink [39] and verified that, for an FPGA running at 270MHz, its switching delay is less than 4ns.

4. PHYSICAL VALIDATION

Before evaluating the performance of Cayley datacenters, we validate the assumptions behind the Cayley design with physical 60GHz hardware. Specifically, we quantify communication characteristics and investigate the possibility of interference problems that may interfere with realizing the Cayley datacenter.

We conduct our experiments using Terabeam/HXI 60GHz transceivers [25] (Figure 5.a). While the Terabeam/HXI transceivers are older and therefore not identical to the Georgia Tech's transceiver described in Section 2, they provide a good baseline for characterizing 60GHz RF signals. This is a conservative platform, previously used in [22], over which modern hardware would provide further improvements. For instance, the Terabeam antennae are large and emit relatively broad side lobes and the signal-guiding horns catch some unwanted signals. In contrast, recently proposed CMOS-based designs can be smaller than a dime, effectively suppress side lobes, and do not use signal-guiding horns at all [36, 43]. To compensate for the noise stemming from the

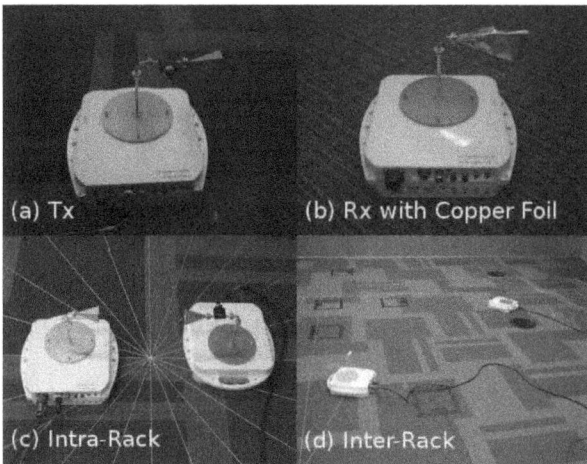

Figure 5: 60GHz Tx, Rx, and measurements on a Cayley datacenter floor plan

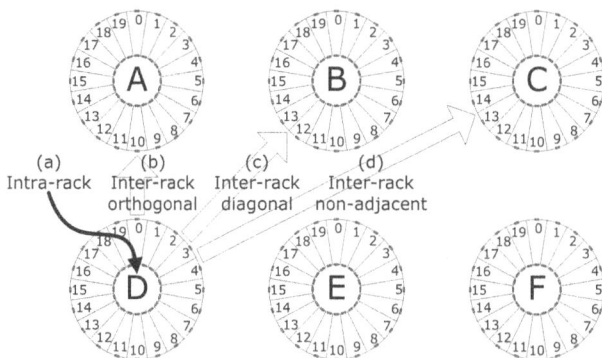

Figure 6: Interference measurement summary

older horn design, we augment one side of the receiver's horn with a copper foil (Figure 5.b). The devices are statically configured to emit signals in a $\theta = 15°$ arc, which is narrower than the Georgia Tech's transceiver.

We validate our model with physical hardware by first measuring how the received signal strength (RSS) varies as a function of the angle between the transmitter and receiver. We then build a real-size floor plan of a Cayley datacenter with a 2 by 3 grid of racks based on Table 2, place transmitter-receiver pairs in their physical locations, and examine whether the signal strength is sufficient for communication (Figure 5.c and d). Finally, we quantify the amount of interference for all possible receiver and transmitter pairs in intra-rack space, in inter-rack space both between adjacent and non-adjacent racks, and in different rack stories. Due to the symmetric circular structure of racks on a regular grid, evaluating a subset of transceiver pairs on the 2 by 3 grid is sufficient to cover all cases.

In the following experiments, we primarily examine RSS as a measure of signal quality in relationship to a vendor-defined base. We configure the transmission power of the Terabeam transmitter for all experiments such that a receiver directly facing the transmitter receives signal at -46dB. This is a conservative level, as the minimum error-free RSS for this hardware is −53dB in a noisy environment [52], and the typical default noise level we measured in a datacenter-like environment was approximately -69dB.

4.1 Received Signal Strength and Facing Directions

The most basic assumption that the Cayley datacenter design makes of the underlying hardware is that a transmitter and a receiver pair can communicate when they are within each other's signal zone. To validate this assumption, we examine the signal strength of a transmitter-receiver pair, placed one meter apart, as a function of the facing angle ε (i.e. $\alpha, \beta = 0°$ and $\delta = 1$ meter in Figure 1). In an ideal scenario with no interference, a receiver would not read any signals when ε exceeds θ.

Figure 7 shows that the received signal strength is significantly above the error-free threshold when $\varepsilon \leq \theta = 15°$ and is negligible when $\varepsilon > 15°$. This confirms that the pair can communicate when oriented in the prescribed manner, and more importantly, that there is negligible interference from a transmitter on an unintended receiver whose reception zone does not cover the transmitter.

4.2 Intra-Rack Space

The cylindrical rack structure we propose divides free space into intra- and inter-rack spaces in order to achieve high free space utilization. Such cylindrical racks would not be feasible if there was high interference within the dense intra-rack space (Figure 6.a). To evaluate if this is the case, we measure the interference within a rack by measuring the signal strength at all receivers during a transmission.

Figure 8 demonstrates that only the receivers within the 15° main signal lobe of the transmitter (receivers at positions 9 and 10 for transmitter 0) receive a signal at a reliable level. The rest of the servers do not receive any signal interference. In part, this is not surprising given the previous experiment. But it confirms that any potential side lobes and other leaked signals from the transmitter do not affect the adjacent receivers.

4.3 Orthogonal Inter-Rack Space

Eliminating all wires from a datacenter requires the use of wireless communication between racks. Such communication requires that the signals from nodes on a given rack can successfully traverse the free space between racks. We first examine the simple case of communication between racks placed at 90° to each other (Figure 6.b).

Figure 9 shows that a transmitter-receiver pair can communicate between racks only when their signal zones are correctly aligned. For clarity, the graph omits symmetrically equivalent servers and plots the received signal strength of servers 6 to 10 on rack A. Other servers on rack A at positions less than 6 or greater than 14 show no signal from servers 0 to 2 on rack D. The graph shows that server 0 on rack D can transmit effectively to server 10 on rack A without any interference to any other servers, as expected.

4.4 Diagonal Inter-Rack Space

Cayley datacenters take advantage of diagonal links between racks in order to provide link diversity and increase bandwidth. We next validate whether the transceivers in our cylindrical racks can effectively utilize such diagonal paths (Figure 6.c).

Figure 10 shows the received signal strength between diagonally oriented racks, and demonstrates that the intended transmitter-receiver pairs can communicate successfully. Once again, the figure omits the symmetrical cases

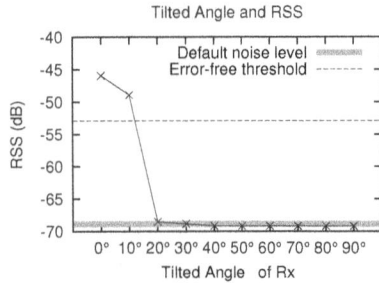

Figure 7: Facing direction of Rx and RSS.

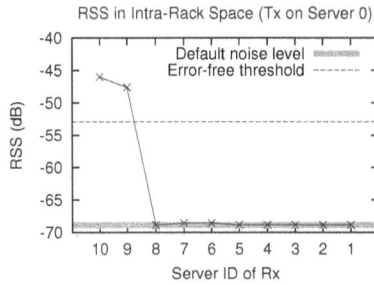

Figure 8: RSS in intra-rack space.

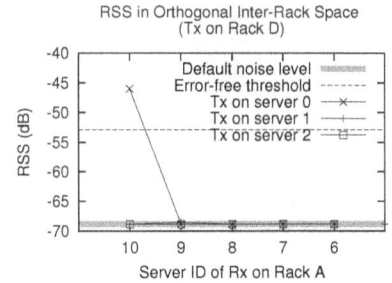

Figure 9: RSS in inter-rack space between racks in orthogonal positions.

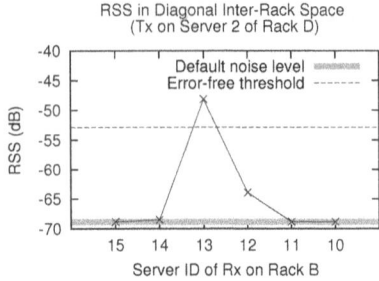

Figure 10: RSS in inter-rack space between racks in diagonal positions.

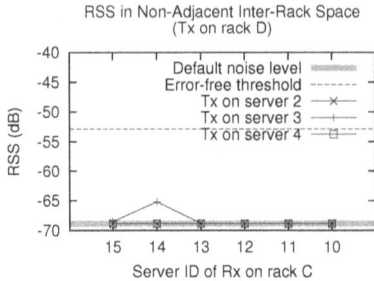

Figure 11: RSS in inter-rack space between non-adjacent racks.

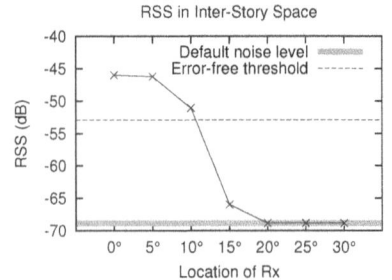

Figure 12: RSS in inter-story space.

(e.g. transmitter on server 3 of rack D), and no signal from far away servers (e.g. 0, 1, 4, 5 of rack D) reaches rack B at all. The signal strength in this experiment is as high as the orthogonal case despite the increased distance due to transmit power adjustment. The case of receiver on server 12 represents an edge case in our model: the signal strength is slightly above the background level because the node is located right at the boundary of the transmission cone. This signal level, while not sufficient to enable reliable communication, can potentially pose an interference problem. To avoid this problem, one can slightly increase the transmitter's signal's angle so that it sends a stronger signal. Alternatively, one can narrow the transmitter's signal angle to eliminate the signal spillover.

4.5 Non-Adjacent Racks

While Cayley datacenters utilize only the wireless links between adjacent racks, it is possible for signals from non-adjacent racks to interfere with each other (Figure 6.d). This experiment examines the attenuation of the signal between non-adjacent racks and quantifies the impact of such interference.

Figure 11 shows the impact of three transmitters on rack D and the receivers on non-adjacent rack C. The transmitters are calibrated to communicate with their adjacent racks B and E. The measurements show that receivers on rack C receive no signal or weak signal not strong enough for communication, but when multiple non-adjacent transmitters send the weak signal (i.e. transmitter on server 3 and receiver on server 14), the noise rate could potentially become too great. For this reason, we propose placing non-reflective curtains, made of conductors such as aluminum or copper foil, that block the unwanted signal. Such curtains can be

placed in the empty triangles in Figure 3.b without impeding access.

4.6 Inter-Story Space

Finally, we examine the feasibility of routing packets along the z-axis, between the different stories on racks. To do so, we orient a transmitter-receiver pair exactly as they would be oriented when mounted on prism-shaped servers placed on different stories of a rack, and examine signal strength as the receivers are displaced from $0°$ to $30°$ following the z-axis.

Figure 12 shows that the signal is the strongest at the center of the main lobe and drops quickly towards the edge of the signal zone. When the receiver reaches the borderline ($15°$) of the signal, it only picks up a very weak signal. Once the receiver moves beyond the $15°$ point, it receives no signal. Overall, the signal strength drops very sharply towards the edge of the signal, and except for the $15°$ borderline case, transceivers on different stories can reliably communicate.

4.7 Summary

In summary, we have evaluated transceiver pairs in a Cayley datacenter and demonstrated that the signal between pairs that should communicate is strong and reliable, with little interference to unintended receivers. Calibrating the antenna or using conductor curtains can address the few borderline cases when the signal is weaker than expected or where there is potential interference. Although not described in detail, we also tested for potential constructive interference. We verified with two transmitters that even when multiple nodes transmit simultaneously, the signals do not interfere with the unintended receivers, namely the receivers in positions that received negligible or no signal in Figures 7 through 12. Overall, these physical experiments demonstrate that extant 60GHz transceivers achieve the sharp attenuation and well-formed beam that can enable the directed communication topology of a Cayley datacenter, while controlling interference.

5. PERFORMANCE AND COST ANALYSIS

In this section, we quantify the performance, failure resilience, and cost of Cayley datacenters in comparison to a fat-tree and a conventional wired datacenter (CDC).

5.1 Objectives

We seek to answer the following questions about the feasibility of wireless datacenters:

○ **Performance**: How well does a Cayley datacenter perform and scale?

By measuring the maximum aggregate bandwidth and packet delivery latency using a fine-grain packet level simulation model with different benchmarks, we compare the performance with fat-trees and CDCs.

○ **Failure resilience**: How well can a Cayley datacenter handle failures?

Unlike wired datacenters, server failures can affect routing reliability in Cayley datacenters because each server functions as a router. Thus, we measure the number of node pairs that can connect to each other under an increasing number of server failures.

○ **Cost**: How cost effective is a Cayley datacenter compared to wired datacenters?

The wireless transceivers and Y-switches are not yet available in the market. We estimate and parameterize costs based on the technologies that wireless transceivers use and compare the price of a Cayley datacenter with a CDC based on the expected price range of 60GHz transceivers.

5.2 Test Environments

Because datacenters involve tens of thousands of servers and 60GHz transceivers are not yet massively produced, it is impossible to build a full Cayley datacenter at the moment. Therefore, we built a fine-grained packet level simulation to evaluate the performance of different datacenter designs.

We model, simulate, and evaluate the MAC layer protocol including busy tones, routing protocol, and relevant delays in the switches and communication links both for Cayley datacenters and CDCs. From the simulation, we can measure packet delivery latency, packet hops, number of packet collisions, number of packet drops from buffer overflow or timeout and so on. The simulator can construct the 3-dimensional wireless topology depending on the parameters such as the transceiver configurations, the distance between racks, and the size of servers. We also model, simulate, and evaluate the hierarchical topology of a fat-tree and a CDC given the number of ports and oversubscription rate of switches in each hierarchy.

5.3 Base Configurations

Throughout this section, we evaluate Cayley datacenters along with fat-trees and CDCs with 10K server nodes. Racks are positioned in a 10 by 10 grid for Cayley datacenters. We

Cayley datacenter parameter	Value
Inner radius	0.25 (meter)
Outer radius	0.89 (meter)
Distance between racks	1 (meter)
Height of each story	0.2 (meter)
# of servers per story	20
# of stories per rack	5
# of servers per rack	100
Bandwidth per wireless data link	10 Gbps
Bandwidth per wireless control link	2.5 Gbps
Switching delay in Y-switch	4 ns

Table 2: Cayley datacenter configurations

Conventional datacenter parameter	Value
# of servers per rack	40
# of 1 GigE ports per TOR	40
# of 10 GigE port per TOR	2 to 4
# of 10 GigE port per AS	24
# of 10 GigE port per CS sub-unit	32
Buffer per port	16MB
Switching delay in TOR	6 μs
Switching delay in AS	3.2 μs
Switching delay in CS	5 μs

Table 3: Conventional datacenter configurations

use the smallest configurable signal angle of $25°$ to maximize the number of concurrent wireless links in the Cayley datacenter and distance of one meter between racks for ergonomic reasons [47].

For CDCs and fat-trees, we simulate a conservative topology consisting of three levels of switches, top of rack switches (TOR), aggregation switches (AS), and core switches (CS) in a commonly encountered oversubscribed hierarchical tree [13]. Oversubscription rate x indicates that among the total bandwidth, the rate of the bandwidth connecting the lower hierarchy to that connecting the upper hierarchy is $x : 1$. The oversubscription rates in a real datacenter are often larger than 10 and can increase to over several hundred [5, 17]. To be conservative, we configure CDCs to have oversubscription rates between 1 and 10, where the rate 1 represents the fat-tree.

The basic configurations for Cayley datacenters and CDCs are described in Tables 2 and 3 respectively. The number of switches used for CDC varies depending on the oversubscription rate in each switch. The configuration and delays for the switches are based on the data sheets of Cisco products [9, 10, 12].

We focus exclusively on traffic within the datacenter, which account for more than 80% of the traffic even in client-facing web clouds [17]. Traffic in and out of the Cayley datacenter can be accommodated without hot spots through transceivers on the walls and ceiling as well as wired injection points.

5.4 Performance

In this subsection, we measure the key performance characteristics, maximum aggregate bandwidth and average and maximum packet delivery latency of Cayley datacenters, fat-trees and CDCs, using a detailed packet level simulator. The evaluation involves four benchmarks varying the packet injection rates and packet sizes:

○ **Local Random**: A source node sends packets to a random destination node within the same pod. The pod of a CDC is set to be the servers and switches connected under the same AS. The pod of a Cayley

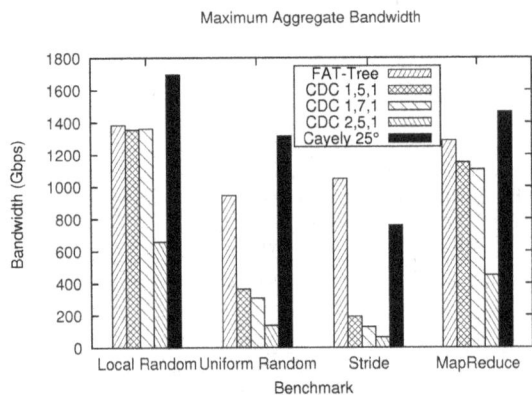

Figure 13: Maximum aggregate bandwidth.

datacenter is set to be the servers in a 3 by 3 grid of racks.

○ **Uniform random**: Source and destination nodes for a packet are randomly selected among all nodes with uniform probability.

○ **Stride**: Source node with a global ID x sends packets to the destination node with ID $mod(x +$ (total # of servers)/2, total # of servers).

○ **MapReduce**: (1) A source node sends messages to the nodes in the same row of its rack. (2) The nodes that receive the messages send messages to the nodes in the same columns of their racks. (3) All the nodes that receive the messages exchange data with the servers in the same pod and outside the pod with 50% probability each. This benchmark resembles the MapReduce application used in Octant [54].

We use different oversubscription rates in each level of switch in the CDC and use three numbers to indicate them: each number represents the rate in TOR, AS, and CS in order. For example, (2,5,1) means the oversubscription rate of TOR is 2, that of AS is 5, and that of CS is 1 and a fat-tree is equivalent to (1,1,1).

5.4.1 Bandwidth

We measure the maximum aggregate bandwidth while every node pair is sending a burst of 500 packets. The results are summarized in Figure 13.

For all cases, the Cayley datacenter shows higher maximum aggregate bandwidth than any CDC. A Cayley datacenter takes advantage of high bandwidth, oversubscription-free wireless channels. The figure clearly shows the disadvantage of having oversubscribed switches in CDCs: when the majority of packets travel outside of a rack or above a AS, as in uniform random and stride, the bandwidth falls below 50% of Cayley datacenter's bandwidth.

Fat-trees perform noticeably better than all CDCs except for local random, where no packet travels above AS's. However, Cayley datacenters outperform fat-trees for all cases except the stride benchmark. Packets from the stride benchmark travel through the largest amount of hop counts, thus it penalizes the performance of the Cayley datacenter.

5.4.2 Packet Delivery Latency

We measure packet delivery latencies by varying the packet injection rate and packet size. Figure 14 and 15

show the average and maximum latencies, respectively. The columns separate the type of benchmarks and the rows divide the packet sizes that we use. Packets per server per second injection rates ranged from 100 to 500.

Local random is the most favorable and stride is the least favorable traffic for all datacenters from a latency point of view: packets travel a longer distance in order of local random, MapReduce, uniform random, and stride.

Overall, the average packet delivery latencies of Cayley datacenters are an order of magnitude smaller (17 to 23 times) than those of fat-trees and all CDCs when the traffic load is small. This is because datacenter switches have relatively larger switching delay than the custom designed Y-switch and Cayley datacenters use wider communication channels. For local random and MapReduce benchmarks that generate packets with relatively small network hops (Figure 14.a and d), Cayley datacenters outperform fat-trees and CDCs for almost all cases.

For all other benchmarks, CDC (2,5,1) performs noticeably worse than all others, especially when traffic load is large, because the TOR is oversubscribed. The latency of CDC (2,5,1) skyrockets once uniform random and stride traffic overloads the oversubscribed switches and packets start to drop due to buffer overflow (Figures 14.b and c). Besides CDC (2,5,1), fat-tree and other CDCs maintain relatively stable average latencies except for during the peak load. The amount of traffic increases up to 8MBps per server: 8MBps per server is approximately the same amount of traffic generated per server as the peak traffic measured in an existing datacenter [32].

Cayley datacenters generally maintain lower latency than fat-trees and CDCs. The only case when the Cayley datacenters' latency is worse, is near the peak load. When running uniform random and stride benchmarks under the peak load, Cayley datacenters deliver packets slower than fat-tree, CDC (1,5,1), and CDC (1,7,1) (the last row of Figures 14.b and c). The numbers of average network hops for a Cayley datacenter are 11.5 and 12.4 whereas those of the tree-based datacenters are 5.9 and 6 for uniform random and stride benchmarks. Competing for a data channel at each hop with relatively large packets significantly degrades the performance of Cayley datacenters compared to fat-trees and CDC (1,5,1) and (1,7,1).

The maximum packet delivery latency shows the potential challenge in a Cayley datacenter (Figure 15). Although the average latencies are better than CDCs, Cayley datacenters show a relatively steep increase in maximum latency as traffic load increases. Therefore, the gap between average and maximum latency for packet delivery becomes larger depending on the amount of traffic. However, except for under the peak traffic load, the maximum latency of the Cayley datacenter is less than 3.04 times as large as the latency of a fat-tree, and is smaller than CDCs for most cases. Therefore, Cayley datacenters are expected to show significantly better latency on average than fat-tree and CDCs, except under peak load for applications similar to stride.

In summary, except for handling the peak traffic for uniform random and stride benchmark, the Cayley datacenter performance is better than or comparable to fat-tree and CDC. As the average number of hops per packet increases, the performance of Cayley datacenters quickly decreases. This shows that Cayley datacenters may not also be as scalable as CDC, which has stable wired links with

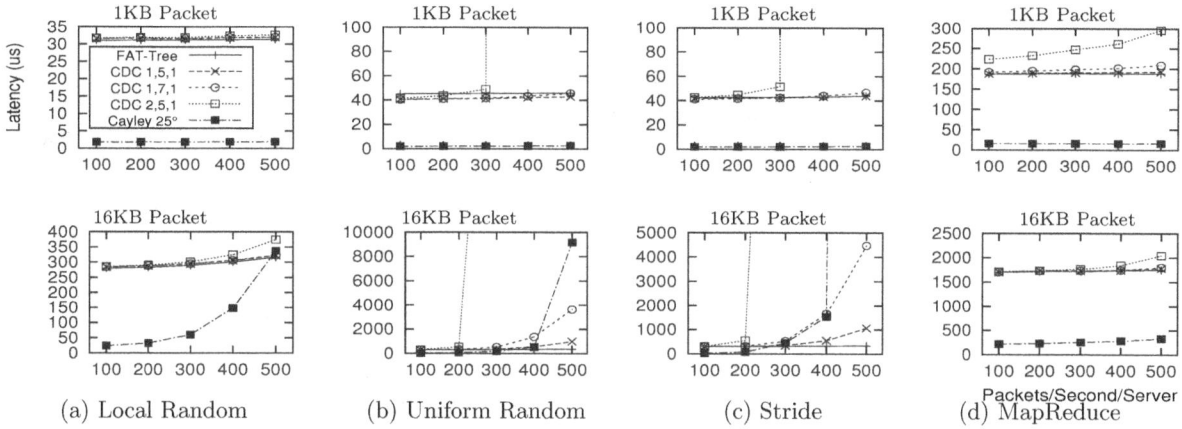

Figure 14: Average packet delivery latency.

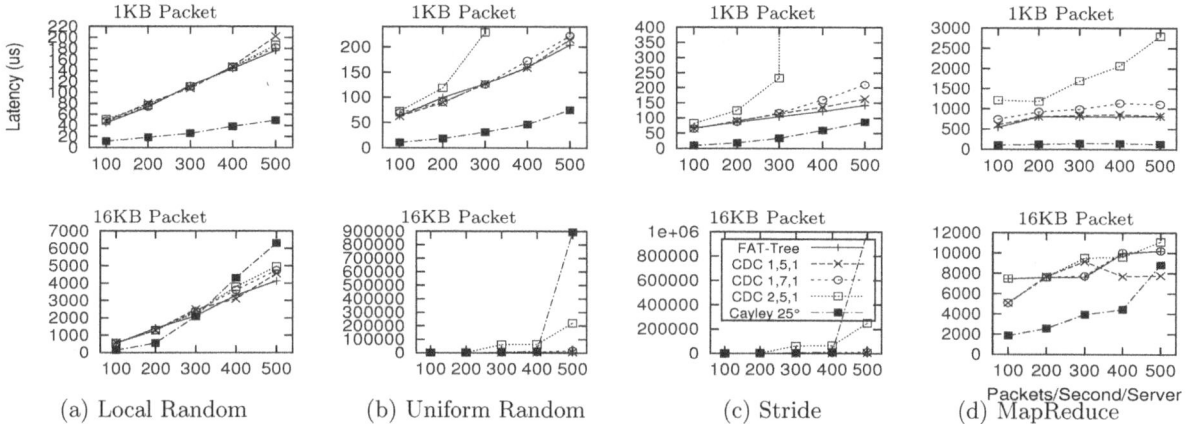

Figure 15: Maximum packet delivery latency.

smaller number of network hops. Cayley datacenters may not be suitable to handle applications requiring large number of network hops per packet, but this type of applications also penalizes the CDC performance as we observed for CDC (2,5,1). In reality, datacenter applications such as MapReduce usually resembles the local random benchmark, which does not saturate oversubscribed (aggregate) switches [5, 32]. Further, the experimental results demonstrate that Cayley datacenters perform the best for MapReduce. Consequently, Cayley datacenters may be able to speed up a great portion of datacenter applications. Even for larger scale datacenters, engineering the application's traffic pattern as in [3] will enable applications to run in Cayley datacenters more efficiently than in fat-trees and CDCs.

5.5 Failure Resilience

We evaluate how tolerant Cayley datacenters are to failures by investigating the impact of server failures on connections between live nodes (Figure 16). We select the failing nodes randomly in units of individual node, story, and rack. We run 20 tests for each configuration and average the results. The average of standard deviation for the 20 run is less than 6.5%.

Server nodes start to disconnect when 20%, 59%, and 14% of the nodes, stories, and racks fail, respectively. However, over 99% of the network connections are preserved until more than 55% of individual nodes or stories fail. Over 90% of the connections are preserved until 45% of racks fail.

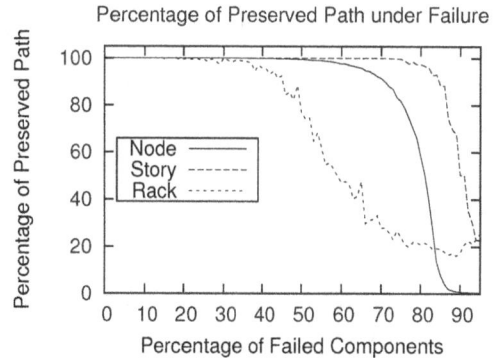

Figure 16: Percentage of preserved path under failure.

Assuming failure rates of servers are the same in wireless datacenters as fat-tree based datacenters and CDCs, then a Cayley datacenter can be more resilient to network failures. This is mainly because wireless datacenters do not have conventional switches which can be critical points of failure and the failures catastrophic enough to partition a Cayley datacenter is very rare [17].

5.6 Cost Comparison

It is complicated to compare two technologies when one is commercially mature and the other is yet to be commer-

11

Config	#TOR	#AS	#CS	#CS chassis	Cost ($)
2,5,1	250	26	8	1	1,818,500
1,7,1	250	48	12	2	2,229,000
1,5,1	250	52	16	2	2,437,000
fat-tree	250	88	96	10	6,337,000

Table 4: CDC networking equipment cost for 10K nodes

cialized. We can easily measure the cost of a fat-tree and a CDC, but the cost of a Cayley datacenter is not accurately measurable. However, we parameterize the costs of Cayley datacenters and compare the cost for different values of 60GHz transceiver cost.

Hardware cost: We compare the cost of the wireless and the wired datacenters based on the network configurations that we used so far. The price comparison can start from the NIC—typically priced at several tens of dollars [42]—and the Y-switch. In our system, we replace the NIC with the proposed simple Y-switch and at least two transceivers. Y-switches consist of simple core logic, host interface, such as a PCI express bus, and interface controllers. Thus, we expect the price of a Y-switch to be comparable to a NIC.

The price differences between wireless and wired datacenters stem from the wireless transceivers and the switches. The cost required for CDC and fat-tree to connect 10K servers based on the price of TOR, AS, and CS [44] are summarized in Table 4. The total price ranges from US$1.8M to US$2.4M for CDCs and US$6.3M for a fat-tree. Since the cost of a fat-tree can be very high, it should be able to use commodity switches [38] and the cost can vary much depending on the switch configuration. Thus, we mainly focus on the comparison between CDCs and Cayley datacenters.

60GHz transceivers are expected to be inexpensive, due to their level of integration, usage of mature silicon technologies (90nm CMOS), and low power consumption which implies low-cost packaging. We cannot exactly predict the market price, but the total cost of network infrastructure excluding the Y-switch in Cayley datacenters can be expressed as a function,

$$Cost_{Cayley}(cost_t, N_{server}) = 2 \times cost_t \times N_{server}, \quad (2)$$

where $cost_t$ is the price for a transceiver and N_{server} is the number of servers in a datacenter. From this function, we can find out that as long as $cost_t$ is less than US$90, Cayley datacenters can connect 10K servers with lower price than a CDC. Similarly, if $cost_t$ becomes US$10, the cost of transceivers in Cayley datacenters can be 1/9 of CDC switches. Considering the rapidly dropping price of silicon chips [18] we expect the transceiver's price to quickly drop to less than US$90 even if it starts with a high cost. This comparison excludes the wire price for CDC, so there is an additional margin, where $cost_t$ can grow higher to achieve lower cost than CDC.

Power consumption: The maximum power consumption of a 60GHz transceiver is less than 0.3 watts [43]. If all 20K transceivers on 10K servers are operating at their peak power, the collective power consumption becomes 6 kilowatts. TOR, AS, and a subunit of CS typically consume 176 watts, 350 watts, and 611 watts, respectively [9–11]. In total, wired switches typically consumes 58 kilowatts to 72 kilowatts depending on the oversubscription rate for datacenter with 10K servers. Thus, a Cayley datacenter can

consume less than 1/12 to 1/10 of power to switch packets compared to a CDC.

Besides the lower price and power, lower maintenance costs stemming from the absence of wires and substantially increased tolerance to failure can be a strong point for wireless datacenters. In summary, we argue that 60GHz could revolutionize datacenter construction and maintenance.

6. PUTTING IT ALL TOGETHER

The summary of our findings throughout the evaluation of Cayley datacenters are as follows. The merits of completely wireless Cayley datacenters over fat-trees and conventional datacenters are:

o **Ease of maintenance through inherent fault tolerance**: Densely connected wireless datacenters have significantly greater resilience to failures than wired datacenters, in part because they do not have switches which can cause correlated loss of connectivity and in part because the wireless links provide great path diversity. Additionally, installing new or replacing failed components can be easier than in a CDC, since only rewiring power cables is necessary.

o **Performance**: Cayley datacenters can perform better than or comparable to fat-trees and CDCs. Cayley datacenters achieve the highest maximum aggregate bandwidth for most benchmarks and deliver packets at a significantly lower latency, especially for MapReduce-like benchmarks and when traffic load is moderate.

o **Cost**: The price of networking components in a Cayley datacenter is expected to be less than those in CDC depending on the market price of wireless transceivers for comparable performance. Power consumption and expected maintenance costs are significantly lower than CDC.

Characteristics and limitations of Cayley datacenters are:

o **Interference**: Orientation of transceivers on the cylindrical racks and characteristics of 60GHz signals limit the interference and enable reliable communication.

o **MAC layer contention**: Sharing of the wireless channel followed by MAC layer contention greatly influence the overall performance: the lower the contention, the greater the performance.

o **Hop count**: Performance depends on the number of network hops, because each hop entails MAC layer arbitration.

o **Scalability**: Due to the multi hop nature of the topology, scalability is not as good as CDC. Yet, this limitation can be overcome by tuning applications to exhibit spatial locality when possible.

These points summarize the challenges, open problems, opportunities, benefits, and feasibility for designing a wireless datacenter.

7. RELATED WORK

Ramachandran et al. [45] outlined the benefits and challenges for removing wires and introducing 60GHz communication within a datacenter and Vardhan et al. [53] explored the potentials of 60GHz antennae emulating an existing tree-based topology. We share many of their insights and also conclude that 60GHz wireless networks can improve conventional datacenters. Further, we address some of the problems identified by the authors. We propose a novel rack-level architecture, use real 60GHz transceivers and realistic parameters, and provide an extensive evaluation of the performance of the proposed wireless datacenters.

Although we focused on Georgia Tech's transmitter design [43], other research groups are also developing CMOS-based 60GHz transceivers [15, 51]. While the technology was developed initially for home entertainment and mobile devices, other groups are looking at deploying it more broadly [41]. Our work on building completely wireless datacenters extends this line of research and tests the limits of 60GHz technology.

Flyways [22] and [35] are wireless networks based on 60GHz or 802.11n organized on top of wired datacenter racks. They provide supplementary networks for relieving congested wired links or for replacing some of the wired switches. In contrast, wireless links are the main communication channels in Cayley datacenters.

Zhang et al. [58] proposed using 3D beamformation and ceiling reflection of 60GHz signals in datacenters using networks like Flyways to reduce interference. Cayley datacenters use cone-shape 3D beams, but use a novel cylindrical rack design to isolate signals and avoid interference.

A scalable datacenter network architecture by Al-Fares et al. [2] and Portland [38] employ commodity switches in lieu of expensive high-performance switches in datacenters and provide a scalable oversubscription-free network architecture. They achieve high performance at a lower cost, but significantly increase the number of wires.

CamCube consists of a 3-dimensional wired torus network and APIs to support application specific routing [3]. Although the motivation and goal of our paper is different from those of CamCube, combining their approach of application specific routing is expected to enhance the performance of our Cayley datacenter design.

The MAC layer protocol that we used [23,24] is not developed specifically for Cayley datacenters; as a result, there may be inefficiencies that arise. Alternatively, there are other MAC layer protocols developed specifically for 60GHz technology and directional antennae [7,33,48], but they require global arbitrators or multiple directional antennae collectively pointing to all directions. These are not suitable for datacenters. Designing a specialized MAC layer protocol for wireless datacenters is an open problem.

While our design adopted XY routing for Cayley datacenters, other variations of routing protocols for interconnecting networks, such as [19,21,29], can be adapted to our design.

8. CONCLUSION

In this paper, we proposed a radically novel methodology for building datacenters which displaces the existing massive wired switching fabric, with wireless transceivers integrated within server nodes.

For brevity and simplicity of presentation, we explore the design space under the assumption that certain parameters such as topology and antenna performance are constant. Even in this reduced search space, we identify the strong potential of Cayley datacenters: while maintaining higher bandwidth, Cayley datacenters substantially outper-

form conventional datacenters and fat-trees with respect to latency, reliability, power consumption, and ease of maintenance. Issues that need further improvements are extreme scalability and performance under peak traffic regimes.

Cayley datacenters open up many avenues for future work. One could focus on each aspect of systems research related to datacenters and their applications and try to understand the ramifications of the new architecture. We feel that we have hardly scratched the surface of this new paradigm and that numerous improvements are attainable. Some interesting design considerations involve understanding the cost structure of individual nodes and how it scales with applications: is it beneficial to parallelize the system into a substantially larger number of low-power low-cost less-powerful processors and support hardware? What data replications models yield best reliability vs. traffic overhead balance? Could an additional global wireless network help with local congestion and MAC-layer issues such as the hidden terminal problem? What topology of nodes resolves the max-min degree of connectivity across the network? How should software components be placed within the unique topology offered by a Cayley datacenter? How does performance scale as the communication sub-band shifts higher in frequency? Would some degree of wired connectivity among servers internal to a single rack benefit performance? As the 60GHz technology matures, we expect many of the issues mentioned here to be resolved and novel wireless networking architectures to be realized in datacenters.

9. ACKNOWLEDGMENTS

We'd like to thank the Georgia Tech team for providing us with the specifications and the communication model of the 60GHz wireless transceivers; Han Wang for helping implement the Y-switch for FPGAs; Srikanth Kandula, Jitendra Padhye, Victor Bahl, Dave Harper, and Dave Maltz for providing us with the 60GHz antennae for physical validations; Daniel Halperin for his help in experimenting with the 60GHz antennae; Bobby Kleinberg for helping name the project; and Deniz Altinbuken and Tudor Marian for insightful feedback on an earlier version of this paper. This work was supported in part by National Science Foundation grants No. 0424422, 1040689, 1053757, 1111698, and SA4897-10808PG.

10. REFERENCES

[1] M. Armbrust, A. Fox, R. Griffith, A. D. Joseph, R. Katz, A. Konwinski, G. Lee, D. Patterson, A. Rabkin, I. Stoica, and M. Zaharia. A view of cloud computing. *Communications of the ACM*, 53(4), 2010.

[2] M. Al-Fares, A. Loukissas, and A. Vahdat. A scalable, commodity data center network architecture. *SIGCOMM*, 2008.

[3] H. Abu-Libdeh, P. Costa, A. Rowstron, G. O'Shea, and A. Donnelly. Symbiotic routing in future data centers. *SIGCOMM*, 2010.

[4] R. Buyya, C. S. Yeo, and S. Venugopal. Market-oriented cloud computing: vision, hype, and reality for delivering IT services as computing utilities. *HPCC*, 2008.

[5] T. Benson, A. Akella, and D. A. Maltz. Network traffic characteristics of data centers in the wild. *IMC*, 2010.

[6] A. Cayley. On the theory of groups. *American Journal of Mathematics*, 11(2), 1889.

[7] X. Chen, J. Lu, and Z. Zhou. An enhanced high-rate WPAN MAC for mesh networks with dynamic bandwidth management. *GLOBECOM*, 2005.

[8] J. I. Choi, M. Jain, K. Srinivasan, P. Levis, and S. Katti. Achieving single channel, full duplex wireless communication. *MOBICOM*, 2010.

[9] Cisco. Cisco Nexus 5000 series architecture: the building blocks of the unified fabric. http://www.cisco.com/en/US/prod/collateral /switches/ps9441/ps9670/white_paper_c11-462176.pdf, 2009.

[10] Cisco. Cisco Catalyst 4948 switch. http://www.cisco.com/en/US/prod/collateral/switches/ps5718/ ps6021/product_data_sheet0900aecd8017a72e.pdf, 2010.

[11] Cisco. Cisco Nexus 7000 series environment. http://www.cisco.com/en/US/prod/collateral/switches/ps9441/ ps9402/ps9512/Data_Sheet_C78-437759.html, 2010.

[12] Cisco. Cisco Nexus 7000 F-series modules. http://www.cisco.com/en/US/prod/collateralswitches/ps9441/ ps9402/at_a_glance_c25-612979.pdf, 2010.

[13] Cisco. Cisco data center infrastructure 2.5 design guide. http://www.cisco.com/en/US/docs/solutions/Enterprise/ Data_Center/DC_Infra2_5/DCI_SRND_2_5_book.html, 2010.

[14] Ecma International. Standard ECMA-387: high rate 60GHz PHY, MAC and HDMI PAL. http://www.ecma-international.org/publications/standards/Ecma-387.htm, 2008.

[15] B. Floyd, S. Reynolds, U. Pfeiffer, T. Beukema, J. Grzyb, and C. Haymes. A silicon 60GHz receiver and transmitter chipset for broadband communications. *ISSCC*, 2006.

[16] A. Greenberg, J. Hamilton, D. A. Maltz, and P. Patel. The cost of a cloud: research problems in data center networks. *SIGCOMM Compututer Communication Review*, 39(1), 2008.

[17] A. Greenberg, J. R. Hamilton, N. Jain, S. Kandula, C. Kim, P. Lahiri, D. A. Maltz, P. Patel, and S. Sengupta. VL2: a scalable and flexible data center network. *SIGCOMM*, 2009.

[18] C. Gianpaolo, D. M. Xavier, S. Regine, V. W. L. N., and W. Linda. Inventory-driven costs. *Harvard Business Review*, 83(3), 2005.

[19] P. Gratz, B. Grot, and S.W. Keckler. Regional congestion awareness for load balance in networks-on-chip. *HPCA*, 2008.

[20] S. Gollakota, S. D. Perli, and D. Katabi. Interference alignment and cancellation. *SIGCOMM*, 2009.

[21] C. J. Glass, L. M. Ni, and L. M. Ni. The turn model for adaptive routing. *ISCA*, 1992.

[22] D. Halperin, S. Kandula, J. Padhye, P. Bahl, and D. Wetherall. Augmenting data center networks with multi-gigabit wireless links. *SIGCOMM*, 2011.

[23] Z.J. Hass and J. Deng. Dual busy tone multiple access (DBTMA)-a multiple access control scheme for ad hoc networks. *IEEE Transactions on Communications*, 50(6), 2002.

[24] Z. Huang, C.-C. Shen, C. Srisathapornphat, and C. Jaikaeo. A busy-tone based directional MAC protocol for ad hoc networks. *MILCOM*, volume 2, 2002.

[25] HXI. http://www.hxi.com, 2012.

[26] IEEE 802.15 Working Group for WPAN. http://www.ieee802.org/15/.

[27] E. Kranakis, H. Singh, and J. Urrutia. Compass routing on geometric networks. *Canadian Conference on Computational Geometry*, 1999.

[28] H. Krawczyk. LFSR-based Hashing and Authentication. *CRYPTO*, 1994.

[29] J. Kim, D. Park, T. Theocharides, N. Vijaykrishnan, and C. R. Das. A low latency router supporting adaptivity for on-chip interconnects. *DAC*, 2005.

[30] P. Kermani and L. Kleinrock. Virtual cut-through: a new computer communication switching technique. *Computer Networks*, 3, 1979.

[31] P. Karn. MACA - a new channel access method for packet radio. *ARRL Computer Networking Conference*, 1990.

[32] S. Kandula, S. Sengupta, A. Greenberg, P. Patel, and R. Chaiken. The nature of data center traffic: measurements & analysis. *IMC*, 2009.

[33] T. Korakis, G. Jakllari, and L. Tassiulas. A MAC protocol for full exploitation of directional antennas in ad-hoc wireless networks. *MobiHoc*, 2003.

[34] V. Kvicera and M. Grabner. Rain attenuation at 58 GHz: prediction versus long-term trial results. *EURASIP Journal on Wireless Communications and Networking*, 2007(1), 2007.

[35] Y. Katayama, K. Takano, N. Ohba, and D. Nakano. Wireless data center networking with steered-beam mmWave links. *WCNC*, 2011.

[36] M. M. Khodier and C. G. Christodoulou. Linear array geometry synthesis with minimum sidelobe level and null control using particle swarm optimization. *IEEE Transactions on Antennas and Propagation*, 53(8), 2005.

[37] C. E. Leiserson. Fat-trees: universal networks for hardware-efficient supercomputing. *IEEE Transaction on Computers*, 34(10), 1985.

[38] R. N. Mysore, A. Pamboris, N. Farrington, N. Huang, P. Miri, S. Radhakrishnan, V. Subramanya, and A. Vahdat. PortLand: a scalable fault-tolerant layer 2 data center network fabric. *SIGCOMM*, 2009.

[39] Mathworks. Simulink—simulation and model-based design. http://www.mathworks.com/products/simulink/.

[40] J. Nsenga, W. V. Thillo, F. Horlin, A. Bourdoux, and R. Lauwereins. Comparison of OQPSK and CPM for communications at 60 GHz with a nonideal front end. *EURASIP Journal on Wireless Communications and Networking*, 2007(1), 2007.

[41] A. M. Niknejad. Siliconization of 60GHz. *IEEE Microwave Magazine*, 11(1), 2010.

[42] newegg.com. Intel PWLA8391GT 10/100/1000 Mbps PCI PRO/1000 GT desktop adapter 1 x RJ45. http://www.newegg.com/Product/Product.aspx? Item=N82E16833106121, 2012.

[43] S. Pinel, P. Sen, S. Sarkar, B. Perumana, D. Dawn, D. Yeh, F. Barale, M. Leung, E. Juntunen, P. Vadivelu, K. Chuang, P. Melet, G. Iyer, and J. Laskar. 60GHz single-chip CMOS digital radios and phased array solutions for gaming and connectivity. *IEEE Journal on Selected Areas in Communications*, 27(8), 2009.

[44] PEPPM. Cisco Current Price List. http://www.peppm.org/Products/cisco/price.pdf, 2012.

[45] K. Ramachandran, R. Kokku, R. Mahindra, and S. Rangarajan. 60 GHz data-center networking: wireless => worry less? *NEC Technical Report*, 2008.

[46] S. Ray, J.B. Carruthers, and D. Starobinski. Evaluation of the masked node problem in ad hoc wireless LANs. *IEEE Transactions on Mobile Computing*, 4(5), 2005.

[47] J.-Y. Shin, E. G. Sirer, H. Weatherspoon, and D. Kirovski. On the feasibility of completely wireless data centers. *Cornell CIS Tech Report*, 2011.

[48] S. Singh, F. Ziliotto, U. Madhow, E.M. Belding, and M.J.W. Rodwell. Millimeter wave WPAN: cross-layer modeling and multi-hop architecture. *INFOCOM*, 2007.

[49] SiBeam White Paper. Designing for high definition video with multi-gigabit wireless technologies. http://www.sibeam.com/whtpapers/ Designing_for_HD_11_05.pdf, 2005.

[50] K.W. Tang and R. Kamoua. Cayley pseudo-random (CPR) protocol: a novel MAC protocol for dense wireless sensor networks. *WCNC*, 2007.

[51] M. Tanomura, Y. Hamada, S. Kishimoto, M. Ito, N. Orihashi, K. Maruhashi, and H. Shimawaki. TX and RX front-ends for 60GHz band in 90nm standard bulk CMOS. *ISSCC*, 2008.

[52] Terabeam. Terabeam Gigalink field installation and service manual. 040-1203-0000 Rev B, 2003.

[53] H. Vardhan, N. Thomas, S.-R. Ryu, B. Banerjee, and R. Prakash. Wireless data center with millimeter wave network. *GLOBECOM*, 2010.

[54] B. Wong, I. Stoyanov, and E. G. Sirer. Octant: a comprehensive framework for the geolocalization of internet hosts. *NSDI*, 2007.

[55] WG802.11 - Wireless LAN Working Group. http://standards.ieee.org/develop/ project/802.11ad.html.

[56] Wireless Gigabit Alliance. http://wirelessgigabitalliance.org, 2010.

[57] S. K. Yong and C.-C. Chong. An overview of multigigabit wireless through millimeter wave technology: potentials and technical challenges. *EURASIP Journal on Wireless Communications and Networking*, 2007(1), 2007.

[58] W. Zhang, X. Zhou, L. Yang, Z. Zhang, B. Y. Zhao, and H. Zheng. 3D Beamforming for Wireless Data Centers. *HotNets*, 2011.

Popularity-driven Coordinated Caching in Named Data Networking

Jun Li, Hao Wu, Bin Liu,
Jianyuan Lu, Yi Wang
Dept. of Computer Science
Tsinghua University, Beijing

Xin Wang
Dept. of Electrical & Computer
Engineering, State Univ. of New York
Stony Brook, New York

Yanyong Zhang, Lijun Dong
WINLAB, Rutgers University
North Brunswick, NJ

ABSTRACT

The built-in caching capability of future Named Data Networking (NDN) promises to enable effective content distribution at a global scale without requiring special infrastructure. The aim of this work is to design efficient caching schemes in NDN to achieve better performance at both the network layer and application layer. With the specific objective of minimizing the inter-ISP (Internet Service Provider) traffic and average access latency, we first formulate the optimization problems for different objectives and then solve them to obtain the optimal replica placement. Then we develop popularity-driven caching schemes which dynamically place the replicas in the caches on the en-route path in a coordination fashion. Simulation results show that the performances of our caching algorithms are much closer to the optimum and outperform the widely used schemes in terms of the inter-ISP traffic and the average number of access hops. Finally, we thoroughly evaluate the impact of several important design issues such as network topology, cache size, access pattern and content popularity on the caching performance and demonstrate that the proposed schemes are effective, stable, scalable and with reasonably light overhead.

Categories and Subject Descriptors

C.2.1 [**Computer Systems Organization**]: Network Architecture and Design

General Terms

Algorithms, Performance, Design.

Keywords

Named Data Networking; Modeling; Dynamic caching; Coordinated caching; Popularity-based.

1. INTRODUCTION

The modern usage of Internet has become largely content-oriented, i.e. users tend not to care where (from which host) and how (via which protocol) to obtain a piece of content, but are more interested in fast and reliable content retrieval. Meanwhile, driven by increasing content sizes and content types, Internet traffic has been rapidly growing at an unprecedented rate. This explosive growth in traffic poses a

significant challenge to the underlying network, as network capacity cannot satisfy the exponentially growing demand. Content-centric overlay networks such as Content Delivery Network (CDN) and Peer-to-Peer (P2P) are then introduced to effectively improve the content distribution efficiency. However, these incremental designs have to deploy extra application-oriented overlay mechanisms and need dedicated components for the architecture, which leads to unscalable solutions. To meet the huge demand of content dissemination in the Internet, it is necessary to rethink the future Internet architecture which can bridge the gap between name-based content delivery and the underlying host-to-host communication infrastructure.

The clean slate Named Data Networking (NDN) [1], also called Content-Centric networking (CCN) [1], is recently proposed for this purpose and widely regarded as one of the most promising architectures for future networks. Quite different from the current IP-based network, this new paradigm features name-based routing and systematic in-network caching. To be specific, in-network caching can directly cache content at each node (say router) on the forwarding path. By typically caching the popular contents at the router, in-network caching can reduce both the overall network load and the access delay. Subsequent requests no longer need to be served directly by the content source which may be far away, but can be served by a closer NDN router along the routing path. Though Internet caching has already been extensively studied, caching in NDN faces a different set of challenges.

In today's Internet (like CDNs), caches are located in specific severs and replicas can be placed in any of these caches. In NDN, however, replicas of the objects are cached along the en-route paths so the requested objects can be obtained with much shorter latencies. This design significantly differs from traditional Internet caching, and can seamlessly integrate routing and content retrieval without introducing much overhead. In addition, NDN caching is universal as it not only applies to the content carried by any protocol, but also applies to all the content from users other than the content providers (e.g. CDNs). Since there is a tremendous amount of content in the Internet, line-speed packet processing is required by NDN

---

[1] We use NDN and CCN interchangeably in the paper.

to support name-based data forwarding and caching. Therefore, the storage of each router in NDN is technologically limited by memory access speed. Due to limited cache capacity on each node, careful cache placement is critical to maximizing the benefit. Our goal in this paper is to find suitable caching locations according to specific objectives, given network topology, content access pattern and various caching constraints.

Among many metrics that can benefit from caching, the main objective of our work is to minimize the inter-ISP traffic at the administrative boundaries in NDN. Our motivations for this particular objective are as follows:

- Intra-ISP links are usually over-provisioned, while inter-ISP links tend to be the bottlenecks which often suffer from congestion [2]. Reducing inter-ISP traffic will significantly alleviate congestion and thus improve the network performance at a global scale.

- Since the inter-ISP links are much more costly than internal links, the reduction of Inter-ISP traffic will greatly reduce the deployment cost for ISPs and thus cut down the inter-ISP charging [2].

- By investigating popularity-based in-network caching strategies in NDN with the special objective, we intend to thoroughly remove the caching redundancy and accommodate as many diverse content items as possible in caching system, which yields highest cache hit rate as well as minimizing inter-ISP traffic. Meanwhile, since a fraction of requests are satisfied within the ISP, caching draws the most popular content closer to the end users and helps to reduce the number of access hops, which will in turn alleviate the traffic burden within an ISP.

The other objective of the work is to explore better caching algorithms to further reduce the access delay without increasing inter-ISP traffic. In addition, a fewer number of access hops can then result in light traffic load within the ISP. To summarize, our ultimate objective is thus to improve the overall network performance in terms of inter-ISP and intra-ISP bandwidth consumption as well as access latency, with effective in-network caching.

Intuitively, coordinated caching among the routers is a promising approach to achieving reduced inter-ISP traffic, but several important issues need to be addressed: 1) Caching principle. Although NDN suggests a multi-path usage to enhance the network performance, it is a non-deterministic variation depending upon the future protocol. It is difficult to model such kind of non-determinacy. For simplicity, at least at the beginning of NDN, we restrict content caching following the en-route principle as the first step towards a full-fledged one. 2) Caching consistency. Practically, an ISP has several gateways to interconnect with provider-ISPs or peer-ISPs. Obviously, maintaining caching consistency among multiple gateways is a very challenging problem. In this paper, we assume an ISP only has a single gateway and plan to extend our solution to multiple gateways in the future work.

The main problem addressed in the paper is to provide effective caching strategies that enable NDN routers within an ISP to coordinate their caching decisions. Unlike most of the earlier work in conventional CDN caching with a focus on minimizing the access latency without considering the resulting bandwidth consumption, we make the following contributions:

- We formulate the problem-solving models with the aim of concurrently minimizing inter-ISP traffic and minimizing the average number of access hops, in order to obtain an optimal solution to the replica placement.

- Guided by optimal replica placement, we present two popularity-based caching algorithms, named TopDown and AsympOpt, where caching is coordinated implicitly among the routers on the path and the routers can make online decision independently. The proposed algorithms can significantly reduce inter-ISP traffic as well as decrease access latencies. Particularly, AsympOpt can achieve the best overall performance, which is very close to the results of optimal solutions.

- We evaluate the performance of the proposed caching algorithms with optimum solutions and study the impact of a variety of factors such as network topology, request pattern, object popularity and cache capacity etc. Simulation results demonstrate that the proposed algorithms exhibit stability and scalability under a wide range of workloads without introducing much overhead.

The rest of the paper is organized as follows: Section 2 surveys the related work. The system model and problem statement are presented in Section 3. Section 4 describes the coordinated dynamic caching scheme. The simulation model, impact factors and the simulation results are discussed in Section 5. Finally, we conclude this paper in Section 6.

2. RELATED WORK

Internet Caching plays an important role in enhancing content delivery, as caching can reduce network traffic and alleviate server load, thereby decreasing access latencies and improving user-perceived Quality of Experience. A large body of research has been done in this field which has led to great successes, such as web proxies [3], object caches [4] and CDN [5], while caching in NDN is a new area and very little investigative work has been published, to our best knowledge. The similar idea of having Internet routers cache passing data as suggested in NDN has been studied in en-route caching, which equips each node in a network with a cache and enables the nodes along the routing path to cache formerly requested objects in the network for future reuse [6]. Dong et al. [7] presented independent content caching and replacement algorithms for intermediate nodes with limited storage, but the work can only reach local optimality with the mathematical model. Due to the complexity of solving the optimization problem, this scheme is limited to be used for small-scaled network. Walter et al. [8] presented an in-network caching architecture based on content routers, which discovers resources in the network proximity. However, cooperation is limited to neighborhood and cannot reach the optimum in the network. In contrast, we strive to thoroughly remove the redundancy with implicit cooperation among caches in order to efficiently utilize the available cache space and minimize inter-ISP traffic.

The most recent works [9] [10] [11]have dived into the study on CCN caching. D.Rossi et al. [9] presented a quite thorough simulation study of CCN caching performance. However, they assumed that named content in CCN can in principle take any path in the network, while we argue that routing to the content sources is determined by the CCN routing protocol and content objects are cached along the en-route path. Kideok et al. [11] proposed a lightweight caching scheme named WAVE, in which the popular contents can be pushed closer to the end users. However, WAVE cannot eliminate the caching redundancy from perspective of the whole network and is limited to achieve good caching performance locally. Psaras et al. [10] focused on modeling caching trees of content-centric networking, which is complementary to our work.

In order to achieve better caching performance, we propose dynamic caching schemes based on content popularity. Traditional approaches towards network caching have placed large caches at specific points in the network, with little or no coordination between the caches. In contrast, routers with limited storage in our schemes independently make the caching decision based on the recent content requests of its subordinates; thus nodes implicitly share their caching information and coordinate in minimizing the redundancy. Besides, the proposed caching strategy can work online to adapt to various network changes.

3. SYSTEM MODEL AND PROBLEM STATEMENT

In this section, we first briefly introduce the NDN architecture and then present a system model of the routers/caches within a single ISP, which are followed by formulating our caching problem.

3.1 System Model

NDN architecture is featured with the availability of built-in network storage and receiver-driven chunk level transport. That is, each router on the Internet is equipped with a cache and can replicate passing contents to serve the subsequent requests without the need of forwarding them to their source servers. In addition, the unit used for transmission is the segment of content, named chunk.

NDN uses a globally unique identifier (e.g. a hash function of a URL) to recognize a content object. The content retrieval procedure is as follows: (1) The content names are published into network by different Internet applications. (2) An end user who is interested in a particular type of content sends out an *interest packet* with the name of the requested content. The *interest packet* propagates along the

routing path towards the content source. (3) Each router receiving an *interest* checks whether the requested content is present in its local cache by looking up a *Content Store* (CS) table. If there is a hit, the router sends the matched data piece to the requester along the reverse path. Otherwise, it forwards the request to the interfaces determined by the *Forwarding Information Base* (FIB). Ongoing requests are recorded in a *Pending Interest Table* (PIT) for later sending back the requested data through the reverse path towards the sources of the interests. (4) When the target content travels from the source (or cache) downward to the requester, the router on the forwarding path will determine whether to replicate the content according to the caching strategy.

The requested content objects are distributed at the repositories outside the ISP in the model, due to the fact that the majority of the content requests can not be satisfied within an ISP without caching. Generally it takes fewer hops to retrieve a content object inside the ISP, so we do not consider caching those objects whose sources happen to reside in the investigated ISP, for fully exploiting cache capacity for better content retrieval performance. Hence, we simplify the model as illustrated in Figure 1. When an edge router (say R_{14}) receives an *interest* for content object (say O_1), it will forward the *interest* towards the content source guided by FIB. When the target content objects are sent in reply to *interest packets* and travel along the way back to the requester following the chain of PIT entries, the NDN router on the path determines whether to replicate the passing content according to the caching strategy. As in the example, node R_4 replicates the content object O_1 and serves the later requests for O_1 from the edge routers within its subtree, that is, R_{12}, R_{13}, R_{14}.

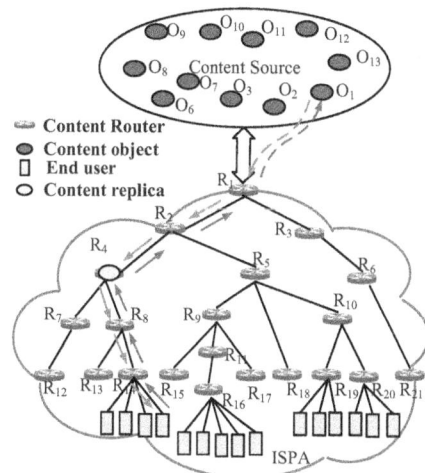

Figure 1. An illustrated caching model

From the above, all the content resources retrieved from the sources outside the given ISP pass through its gateway and are cached at some inner nodes on the path to the requesters. Thus, our caching infrastructure is hierarchical, where the requests are only interfaced at the end nodes of the hierarchy and routed towards a single root cache (i.e., ISP gateway). Accordingly, the flat graph of ISP topology can be simplified as a tree model, with gateway being the root node. Figure 1 shows the example tree abstracted from the router-level topology of an ISP.

The purpose of our work is to design and evaluate the NDN caching strategies in order to achieve a system goal. To make full use of in-networking caching, we focus on minimizing the inter-ISP traffic and reducing the number of access hops by caching the requested contents within an ISP using the available caching space of routers. To be specific, we address the problem of online coordinated caching decision in the following environment: once a request for a piece of chunk fails to be satisfied by the caches inside an ISP, the requested object item will be fetched from the external source and traveled via the gateway of the investigated ISP. We aim to design appropriate caching strategies which can work online to achieve the best gain. Each router in the ISP independently determines whether to replicate the passing items in its associated cache. The objective of our work is to minimize the inter-ISP traffic and also minimize the average number of access hops by caching frequently requested objects at selected routers inside the ISP.

3.2 Problem Statement

It is impracticable to find the solution to an optimization problem including all the optimization objectives. In our case, the objective of minimizing inter-ISP traffic and the objective of minimizing average access hops are not always consistent. In this subsection, we first give out the definition and notations, and then we formulate individual problem-solving model for each objective and obtain the optimal solution to the replica placement respectively. The solutions are used to guide the caching design and taken as bounds to the performance.

3.2.1 Definition and Notations

According to the tree topology of our system model in Figure 1, we consider a caching tree (routing tree), whose node set N is partitioned into two parts: a set of end nodes U and a set of intermediate nodes ($N-U$), and the gateway node R_1 being the root of the tree. End nodes are responsible for managing the interest requests from their users. Each node j ($j \in N$) is equipped with a cache, whose storage capacity is C_j.

Let $[i \rightarrow j]$ be the unique routing path from node i up to node j, $i, j \in N$. We say j is the upstream node (or ancestor), while i is the downstream node (or descendant), if node j is closer to the root R_1 than node i. Let $Ancestors(j)$ denote the set of ancestors of node j, i.e., the nodes in the unique en-route path $[j \rightarrow R_1]$ (Node j is included).

Let O be the set of interested objects in the Internet. For the requested content object o_k ($o_k \in O$), we define:

s^k: The size of the object o_k.

q_i^k: Request rate for object o_k at node i ($i \in U$).

$Cache^k$: The node set which replicates the object o_k and serves the requests for object o_k within an ISP.

A_i^j: The hop counts between end node i ($i \in U$) and node j ($j \in Ancestors(i)$).

X_j^k is a Boolean variable which equals to 1 if node j is a cache of o_k, that is,

$$X_j^k = \begin{cases} 1 & j \in Cache^k \\ 0 & j \notin Cache^k \end{cases} \quad j \in N \quad (1)$$

3.2.2 Optimization Problem for Minimizing inter-ISP Traffic

Given a set of caches and request rates, our optimization problem is to decide which objects should be stored in the caching system and where to cache them, in order to satisfy the objective of minimizing inter-ISP traffic. This optimization objective can be interpreted as maximizing the inter-ISP traffic savings by the caching system, with the limited cache storage at each router. That is,

$$\max\left(\sum_{k \in O}\sum_{i \in U}\sum_{j \in Ancestors(i)} q_i^k s^k X_j^k\right) \quad (2)$$

$$\text{s.t.} \quad \sum_{k \in O} s^k X_j^k \leq C_j, \text{ for all } j \in N \quad (3)$$

$$\sum_{j \in Ancestors(i)} X_j^k \leq 1, \text{ for all } i \in U \quad (4)$$

The first constraint is a capacity constraint, which requires the amount of storage space occupied by objects cached at a router not exceed its capacity. The second constraint means that the number of the caches which serve the requests of object o_k from end node i is no more than 1, i.e., there is at most one replica for object o_k in the unique en-route path $[i \rightarrow R_1]$ from this end node i to the root R_1.

3.2.3 Optimization Problem for Minimizing Average Access Hops

The objective of minimizing the average number of access hops for requesting contents can be formulated as follows,

$$\min(H_{avg})$$

$$\text{s.t.} \quad \sum_{k \in O} s^k X_j^k \leq C_j, \text{ for all } j \in N$$

With the objective, our optimization problem is to decide which objects should be stored in the caching system and where to cache them, given a set of caches and requests rates. More specific, we try to find the optimal replica placement, i.e., the values of X_j^k ($o_k \in O$, $j \in N$), which leads to the smallest number of access hops on average, with the constraint of cache capacity. Thus, the problem turns out to be a Mixed Integer Problem. In addition, for ease of mathematical expression, we change the objective of minimizing average response hops to the equivalent objective of maximizing the saved average hops by caching in network. The optimization problem is therefore given as,

$$\max\left(\sum_{o_k \in O}\sum_{i \in U}\sum_{j \in Ancestors(i)} q_i^k s^k (HOP - A_i^j) X_j^k\right) \quad (5)$$

$$\text{s.t.} \quad \sum_{o_k \in O} s^k X_j^k \leq C_j, \text{ for all } j \in N \quad (6)$$

$$\sum_{j \in Ancestors(i)} X_j^k \leq 1, \text{ for all } i \in U \quad (7)$$

Here, *HOP* is the average number of router traverse hops without caching in network. In today's Internet, packets traverse an average of around 12 to 14 hops. We remove 2 hops which account for the hops between the router and the user/content source and take 11 hops for *HOP* in our evaluation section [12] [13]. The first constraint is a capacity constraint, which requires the objects cached at a router could not exceed its storage capacity. The second constraint means that the number of the caches which serve the requests of object o_k from end node i is no more than 1, i.e., there is at most one replica of object o_k caching at the nodes falling in the set of the *Ancestors(i)*, in order to make the benefit of accommodating as many diverse objects as possible in caching system.

Each of the above linear programming problems is a Mixed Integer Problem (MIP). We obtain the optimal solution using GLPK [14], given the caching tree and request rate of each object at the end nodes. The complexity of solving the MIP problem mainly depends on the network size, content population and cache capacity of each router. For the given topology with 50 nodes (each node can accommodate 10 content items) and 1000 object items in the network, it takes several seconds to determine the cached objects and their optimal caching locations. However, as the tree topology enlarges to model the real ISP scale, say with the scale of 200 nodes (each accommodating 20 content items) and 5000 content items (which is still much fewer than the actual number of contents in Internet), it will take more than 3 weeks to solve the optimization problem on a high-end server. Obviously, the huge computational power prohibits the real-time online decision making. Again, considering the fact that the object request rates are not priori-known and typically difficult to be predicted, the optimal caching decision is impractical to be achieved. Thus instead, we will turn to developing online caching algorithms in Section 4, while taking the optimization solutions as the guide to caching algorithm design and the bounds for performance evaluation.

4. SYSTEM DESIGN AND CACHING ALGORITHMS

As mentioned earlier, the system can be modeled as a tree-like routing topology shown in Figure 1. We first clarify the two notations *Level* and *Tier* for the tree topology, as illustrated in Figure 2. *Tier* is denoted as the distance from the intermediate node to the closest end node (each end node being Tier 1), measured by the number of hops. Obviously, lower tier caches are closer to the end users. In contrast, *Level* is denoted as the distance from each node to the root R_1 (R_1 being Level 1), also measured by the number of hops. End nodes (nodes in Tier 1) correspond to the highest level caches and are responsible for monitoring the requests from the end users and the root node corresponds to the lowest level cache. The objects contained in the caches at the lower level can be shared and accessed by sub-tree nodes at the higher level. A user request travels from an end node (requester) towards the root, until the requested object is found. If the requested object cannot be found even at the root level, the request is redirected to the source content server which contains the interested object. Once the object is found, it is sent along the reverse path towards the requester. Each cache along the forwarding path independently decides the content replication according to a chosen caching strategy, which is determined by dynamic caching algorithms proposed later in this section.

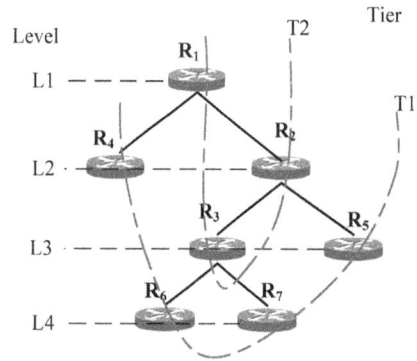

Figure 2. Toy model of caching topology

Our primary goal is to maximize the gain in cross-domain traffic from the ISP's perspective and in access latency from the user's perspective by dynamically creating replicas for popular contents. In this section, we first introduce the procedure of dynamic caching. We then propose two coordinated dynamic caching strategies to achieve our design goal. The strategies are also expected to run online.

4.1 Dynamic Caching

The access pattern at end nodes changes over time, so the caching strategy has to track the object request rate at end nodes, and adapts the caching decision to better achieve the design goal. The popularity of an object is measured by the request rate for the object and is generally stable during a short time period. The caching algorithm is invoked at regular time intervals to determine the replica placement positions based on the most recent histories of request statistics as well as the available cache capacity. As the object popularity varies over time, only the most recent access histories are kept to reduce the memory occupation. The invocation interval for a caching decision is chosen according to the total arrival rates at the end nodes and the change frequency of the content popularity. A shorter interval is preferred for a higher request rate in order to adapt the caching decision quickly to the changing access patterns. However, a shorter interval incurs a higher overhead. On the contrary, a longer interval is suited better to the stable access patterns.

We go a step further into the details of the popularity-based dynamic caching process. Each access node maintains a set of request counters for selected objects and dynamically created object items and calculates the historical request statistics periodically to form the access profile. Herein, the selected objects refer to the most popular objects presented in the last caching interval, based on the observation that recently popular files will tend to be accessed more frequently than others in the near future. Those sustained in dynamic items region are the emerging popular contents in the caching round. Dynamic items are maintained according to the object arrival rate and those with longer arrival interval of requests for a particular object will make room for the subsequent requested objects. As a result, unpopular objects are screened out and the profile of content catalog is dramatically cut down.

Each node spreads the request information along the determined routing paths. In this way, each node gets the needed request profile and individually makes the placement decision according to the caching scheme. When an object is fetched after being requested, each router on the paths determines whether to replicate it or not based on the object tags which are set by the recent caching decision made in this round according to the caching algorithms presented in the next subsections. An object can be tagged by an *update* mark or a *replication* mark. The *replication* mark indicates that the fetched object can be cached at the router upon its arrival and its replica creation time is then set to the current time. Meanwhile, the *replication* mark turns to an *update* one. The *update* mark indicates that there is a replica of the object in the current cache, so a request for the object can be served by the cache until the lifetime of the replica exceeds a preset value. Once an object is obsolete, the request for this object has to be forwarded to the source instead of being served by the cache and the *update* mark of the object turns to a *replication* mark. Here the threshold of object lifetime is introduced for the purpose of keeping the cached object refreshed.

4.2 TopDown Caching Algorithm

TopDown caching algorithm consists of two procedures: information aggregation and decision making. In this algorithm, each node makes its caching decision for each object according to its popularity measured by the aggregated request statistics of its subtree.

The algorithm is invoked at the commencement of a new interval and starts the process of information aggregation from end nodes upwards to the root. To illustrate the algorithm, we use node R_j as an example, be it the end node or the intermediate node. An end node obtains the most recent request history covering the latest interval by calling *GetReqHistory*, while an intermediate node aggregates request records sent from all of its children by calling *Aggregate*. The obtained request records are then sorted in the descending order of the number of requests (#request) and those whose #request is less than the threshold are removed. The threshold can be used to screen out the unpopular objects for reducing the computing complexity. The result is stored in A_j. In this way, TopDown gets sorted request records at each node by aggregating request records from the bottom *level* (the highest *level*) up to the top *level* (the root). Here, the *level* is defined as the distance from the nodes in this layer to the root in terms of hops, as above mentioned.

	Algorithm 1 TopDown Caching Algorithm
	Information Aggregation (ReqHistory)
1	for layer ← bottom level to root do
2	for each node R_j in the layer do
3	if $R_j \in U$ then
4	ReqRec[]$_j$ ← Get ReqHistory( )
5	Else
6	ReqRec[]$_j$ ← Aggregate (Children(R_j), Records)
7	end if
8	A_j ← Sort-Dec (ReqRec[]$_j$, threshold)
9	end for
10	end for
11	Return (A_j)
	TopDown decision making (A_j)
12	for layer ← root to bottom level do
13	for each node R_j in the layer do
14	for each record $r \in A_j$ do
15	if r.ObjectID Exist-In R_j 's CachedTable then
16	MarkUpdate (r.ObjectID)
17	Append ($DelTable_j$, r.ObjectID)
18	else if Available-Space ≥ Size (r.ObjectID) then
19	MarkReplicate (r.ObjectID)
20	Append ($DelTable_j$, r.ObjectID)
21	end if
22	end if
23	end for
24	Delete ($A_{Children(R_j)}$, $DelTable_j$)
25	end for
26	end for

With the request information, the decision making process is from the top to the bottom. Line 15 starts with the record from the top of A_j and fetches the record in turn from the top till reaching the number of objects (or chunks) which the node can cache. If the *ObjectID* of the record is found in the R_j 's *CachedTable* which is a list of cached object items determined by the previous interval, an *update* mark is set to the *ObjectID*. Otherwise, a *replication* mark is set to the *ObjectID*. Thus, the corresponding router storage is assigned for the object item tagged by *update* and *replication* in this caching round and then actually implemented upon the arrival of the retrieved content.

For eliminating redundancy, Topdown creates an additional deletion tracking table called *Deltable*. The current node should append *ObjectID* of those objects marked to be cached locally (including *update* and *replication*) to a deletion tracking table and send the table to all its children. Each child will delete the corresponding records existing in the deletion tracking table from its request record table, and then makes the caching decision based on the table containing the local request records.

According to the TopDown algorithm, a caching example is illustrated in Figure 3 (a) with the request profile in Table 1. We assume each router can only accommodate one object in the example. In Figure 3, R_i stands for a content router, and O_j stands for an object. The circles with O_j indicate where the replicas of O_j are placed according to TopDown caching algorithm.

Table 1. Content request profile

Object	Request at R4	Request at R5	Request at R6	Request at R7
O_1	20	20	20	20
O_2	10	10	10	10
O_3	8	8	8	8
O_4	6	6	6	6
O_5	5	5	5	5
O_6	4	4	4	4

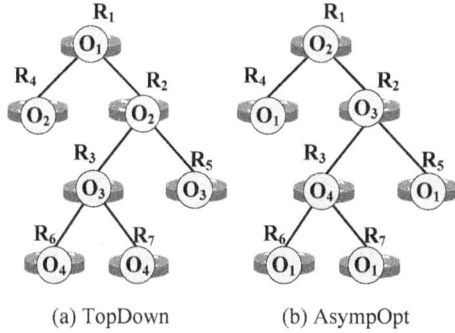

(a) TopDown   (b) AsympOpt

Figure 3. Examples for Caching Algorithms

From the caching example, we can see that TopDown algorithm makes caching decision from root down towards end nodes and places the most popular content objects at the lower *levels*. We can also observe that TopDown can thoroughly eliminate the caching redundancy with global coordination, which can achieve the objective of minimizing inter-ISP traffic and maximizing cache hit rate.

4.3 AsympOpt Caching Algorithm

TopDown caching algorithm can eliminate caching redundancy and accommodate more diverse content objects, which increases the gain in reducing user access latency. From the observation on Figure 3(a) and the solution to the optimization for minimizing user access latency, we can draw the most popular content objects closer to the end nodes and further reduce the access delay, without sacrificing performance of hit rate or inter-ISP traffic. An example of caching placement determined by AsympOpt caching algorithm, which is an improvement of TopDown, is presented in Figure 3(b). AsympOpt caches the globally most popular content objects from the lowest *tier* to the highest *tier*. Each node in T1 (*tier* 1, end nodes layer) is tagged by first priority P_1 in T1. In other tiers, those having no parent in the same tier are tagged as P_1 in this tier. Otherwise, the node should be tagged after its parent's priority. Take the model in Figure 2 for instance. R_1, R_2, R_3 are in the same tier T2. R_1 has no parent in the this tier, while R_2's parent is R_1 and R_3's parent is R_2. Therefore, R_1 is tagged by priority P_1, R_2 is tagged by priority P_2, and R_3 is tagged by priority P_3. The values of Tier and priority are determined by the routing topology.

At starting time, all cache stores are empty and caching system will begin with *Information Aggregation* procedure of the TopDown algorithm in order to get the global popularity rank *GloRank* and distribute the result to each node. The subscript of Object O_j stands for its *GloRank*. Information Aggregation procedure is iterated periodically during the execution of AsympOpt caching. The iterative period is determined by the popularity change.

Given the routing topology and caching capacity C_j of router R_j, we have *StartValue*(j) and *RangeValue*(j) of R_j. Herein, *StartValue*(j) is determined by the R_j's Tier and its priority, and *RangeValue*(j) is the total cache capacity on R_j's forwarding route. With the value, router R_j picks out C_j global popular contents to be cached locally. If the local popularity of some object is not consistent with the global popularity and exceeds the range of popularity difference, those which have relatively low *GloRank* but are locally popular will be cached. In this way, we try to compensate for the difference between the global popularity rank and the local one (Line 10~Line 23).

Aggregate procedure and Mark procedure are similar to the corresponding procedures in Algorithm 1 except that the records of cached contents will be removed from the request profile which is sent to the parents. For brevity, we don't describe the procedures in detail in the paper.

We take an example to explain AsympOpt caching algorithm when the local and global popularity rank are not consistent. Let's consider the end node R_6. Its *StartValue*(6) =1 and *RangeValue*(6) = 4, provided each router can only cache one content object. If the request number for O_1 ranks lower than 5 in popularity and thus local popularity of O_1 exceeds the range value, O_5 which is not in the expected caching list for being globally popular but relatively locally popular is cached at R_6; if the request number for O_1 ranks the first four in popularity, O_1 is still chosen to be cached.

Algorithm 2 AsympOpt Caching Algorithm
1   for layer ← lowest tier to highest tier do
2   for each node R_j in the layer sorted by ascending priority do
3   if $R_j \in U$ then
4   ReqRec[]$_j$ ← Get ReqHistory( )
5   Else
6   ReqRec[]$_j$ ← **Aggregate** (Children(R_j), Records)
7   end if
8   end for
9   for each node R_j in the layer sorted by descending priority do
10   A_j ← Sort-Dec (ReqRec[]$_j$, threshold)
11   st ← StartValue(j) obtained from child in the closest tier
12   while k<= C_j do
13   if each GloRank(st) Exist-in first C_j records of A_j then
14   **Mark**(GloRank(st))
15   else
16   count ++
17   end if
18   st ++
19   end while
20   for k=1 to count do
21   re ← top record of A_j
22   if re.ObjectID is out of the RangeValue(j) then **Mark**(re)
23   end for
24   SendtoParent (A_j)
25
26   **Mark(r)**
27   if r.ObjectID Exist-In R_j 's CachedTable then
28   MarkUpdate (r.ObjectID)
29   Remove (A_j , r)
30   else
31   MarkReplicate (r.ObjectID)
32   Remove (A_j , r)
33   end if
34
35   **Aggregate (Children(R_j), Records)**
36   Initialize (ReqRec[]$_j$)
37   for each child R_k in Children (R_j) do
38   for each record $r \in A_k$ do
39   if r.objectID Exist-In ReqRec[]$_j$ then
40   r$_j$. ReqCount= r$_j$. ReqCount + r. ReqCount
41   else
42   Insert the record r
43   end if
44   end for
45   end for
46   return (ReqRec[]$_j$)

5. PERFORMANCE EVALUATION

In this section, the experimental results of the caching algorithms are presented and analyzed.

5.1 Simulation Settings

Since NS2 has some limitation to simulate the new NDN paradigm [10], we establish our own simulation to evaluate the presented caching schemes.

5.1.1 Simulation environment

Our algorithms are tested and evaluated using both synthetic and real network topologies that have different structural properties.

We employ the Georgia Tech Internetwork Topology Model (GT-ITM) toolkit [15] to generate the router-level network topology using the Transit-Stub model. A shortest path tree for rooting level topology is abstracted from the generated graph. Each node is equipped with a cache. User requests are sent to the end nodes of the tree. Each node, including the end nodes, checks the requested object in its local cache before forwarding the request. If the object is not found, the request will be forwarded to the next-hop node along the routing path towards the root until it reaches a node that caches the requested object or out of the ISP via a gateway node towards the source of the requested object. In either case, an object copy is sent along the reverse path to the requesting end node. Each node on the path can replicate the passing object based on the caching criteria.

We employ different network topologies in the evaluation, including the topology of University of Wisconsin AS59, UUNet Alternet AS701and UNINETT AS224 as listed in Table 2. These topologies vary in size and ISP type. AS701 is a tier-1 ISP, while AS59 and AS224 are Stub ISPs. We abstract the routing topologies from the listed network by ospf (open shortest path first) routing algorithm.

Table 2. Real Network Topology

Network	AS number	Nodes	Number of Tiers	Number of Layers
Uni. of Wisconsin	AS59	41	3	5
UUNet Alternet	AS701	75	3	8
UNINETT	AS224	208	5	9

5.1.2 Input data

We mainly use the synthetic input data in the caching performance evaluation, and also use the actual trace collected from Tsinghua University for validation purpose.

5.1.2.1 Synthetic input data

Let $O = \{o_1, o_2 ... o_N\}$ denote a set of cacheable objects. We assume that requests are identical and independently distributed (i.i.d.) within the set O in a considered time frame such that each request refers to an object o_k with a probability p_k without memorizing previous requests. The objects are ordered according to the decreasing access probabilities $p_1 \geq p_2 \geq ... \geq p_N$.

The requests at each end node follow the Poisson arrival $P_n(t) = (\lambda t)^n e^{-\lambda t} / n!$. We assume the average request rate λ

at edge routers follows a uniform distribution and the objects' popularity is governed by the Zipf distribution $f(i) = i^{-\alpha} / \sum i^{-\alpha}$ ($0.5 \leq \alpha \leq 2$) [16], where α is the skewness factor indicating the concentration degree of object access. Zipf distribution defines the probability of accessing an object at rank i out of N available objects.

5.1.2.2 Real trace
The real trace is collected from Tsinghua University campus network. The duration of the trace is one hour and the trace accounts for around 100G bytes. We apply DPI (Deep Packet Inspection) to parse the trace and use the urls in HTTP as the input data.

5.1.3 Performance metrics for evaluation
Our goal is to find the optimum caching locations to maximize the benefits. The most typical metrics are inter-ISP traffic or hit rate (The difference of the two metrics mainly lie in whether content size is involved), and the access delay measured by the number of hops that a given request travels in the network. So, the two metrics are tested.

- Saving Rate of inter-ISP Traffic (SR-CDT): the ratio of the Inter-ISP traffic saved by caching with respect to the total Inter-ISP traffic incurred without caching.
- Saving Rate of Hops (SR-Hops): we define the response hops as the number of the routers traveled by the response packets from the source (or cache) to the requester. SR-Hops is the ratio of the average number of response hops reduced by caching over the number of response hops without caching (11 hops as above mentioned).

5.1.4 Caching schemes for comparisons
We compare our caching schemes with other caching schemes: Leaving Copies Everywhere (LCE) [17], Leaving Copies with Probability (LCProb), and Leaving Copies with Uniform Probability (LCUniP). LCE is currently used in most hierarchical caches and the same caching algorithm was applied in the most influential article for NDN [1]. In LCE, each cache on the delivery path replicates the copy of the object with the Least Recently Used (LRU) replacement algorithm. LCE is widely used due to its high performance and ease of implementation. LCUniP and LCProb are similar to LCE except that the retrieved content is not blindly cache at each passing node, but selectively cached by probability to eliminate redundancy. LCUniP caches the passing content with uniform probability at each router, while LCProb is with caching probability 1/(hop count along the path).

In addition, we compare our caching algorithms with the optimization solutions.

5.2 Comparison with Optimization Solutions
As mentioned above, it takes an extremely long time to solve the optimization problem with a 200-node network and it is thus impractical to compare the solution at such a scale. Therefore, we compare our algorithms with the optimal solutions in the 50-node topology with each router

caching 10 objects (or chunks), serving for 1000 object (or chunk) interests in the network. The object request arrivals follow the Poisson distribution and the popularity of requested objects follows the Zipf distribution with skewness parameter $\alpha = 0.9$.

In Table 3, Optimal Solution 1 stands for the solution to the objective of minimizing inter-ISP traffic, while Optimal Solution 2 stands for the solution to the objective of minimizing average number of access hops. We can observe that both AsympOpt and TopDown closely approach the bound of SR-CDT performance, so they both work well on achieving the objective of minimizing inter-ISP traffic and maximizing cache hit rate. As for SR-Hops which measures the access latency, AsympOpt is much closer to the bound and achieves the optimum performance from the user perspective. In contrast, though TopDown slightly outperforms AsympOpt in terms of SR-CDT, it is much inferior to AsympOpt in terms of SR-Hops. In summary, the proposed algorithms can achieve nearly optimal performance in small-scale networks. Particularly, the performance of AsympOpt is very close to the bounds obtained from the optimization solutions.

Table 3. Comparison with Optimal Solution

Metric	Optimal Solution 1	Optimal Solution 2	AsympOpt	TopDown
SR-CDT	0.4834	/	0.4775	0.4834
SR-Hops	/	0.4500	0.4462	0.3713

5.3 Performance Impact Factors
The efficiency of caching depends on factors such as network topology, request pattern, cache capacity and object popularity. In this section, we compare our algorithms with baseline algorithms LCE, LCProb and LCUniP (10% probability) considering these factors

The default settings are listed as follows. The ISP routing topology is a tree with 200 nodes including 103 end nodes, where each node is equipped with a cache that can serve for 100,000 object items. The number of average requests at each end node (which follow the Poisson arrival with parameter λt) follows a uniform distribution U(20000,40000), where λ is the request arrival rate at the end node and t is the observation interval for caching decision. Object popularity follows a Zipf distribution with $\alpha = 0.9$.

5.3.1 Impact of the cache size on performance
The cache size at each node is described as the value relative to the total size of all objects available in the network and called relative cache size. We compare the effectiveness of different caching algorithms across a range of cache sizes, from 0.01% percent to 0.12% percent, with the total object size of 100,000 chunks. (The default relative cache size 0.04% will be used in the following simulations.)

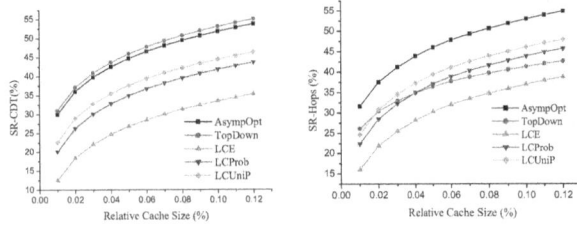

(a) Synthetic input

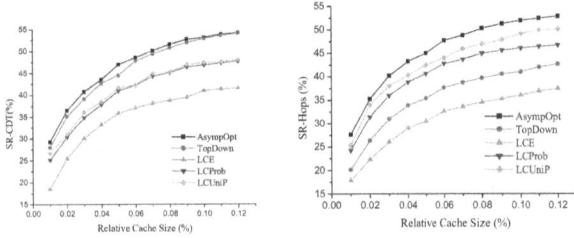

(b) Real trace

Figure 4. Caching performance vs. cache size

Figure 4 compares the saving rate of inter-ISP traffic (SR-CDT) and saving rate of average response Hops (SR-Hops) with the synthetic load and real trace, respectively. The simulation results show that all the algorithms provide steady performance improvement as the cache size increases. In general, AsympOpt significantly outperform LCE, LCProb and LCUniP both in SR-CDT and SR-Hops. TopDown performs best, but achieves a marginal improvement in terms of SR-CDT compared with AsympOpt. However, AsympOpt performs much better in terms of SR-Hops than TopDown. Therefore, AsympOpt is preferable for achieving better whole performance. Among the three baseline algorithms, LCUnip (10%) performs the best, while LCE performs the worst as expected in terms of both metrics. We also test the naïve random en-route replica placement, that is LCUnip with 50% probability and find that LCUnip (10%) outperforms LCUnip (50%) in both of the performance metrics.

We also observe that as the relative cache size increases, the slope of the curves turns to be flatter, that is, the performance gain decreases with the increased capacity. Considering the tradeoff between the cost and performance gain, we can deploy suitable cache capacity with the curves.

Although the result curves are largely consistent with the curves created by the synthetic input, the advantage of our algorithms over baseline algorithms seem to decline in case of real trace input. It may be because the trace we've got is not general enough and we only analyze the web requests. The comprehensive evaluation on real traces will be our future work.

5.3.2 Impact of the request pattern on performance

5.3.2.1 Popularity Skewness
We assume that the request pattern follows the Zipf distribution, that is, the frequency of a request is the inverse of its rank in the request popularity. The Zipf skewness

parameter α indicates the degree of concentration of object requests. When the values of α are close to 1, it indicates that a few objects (also known as "hot spots") attract the majority of the requests; while when values are close to 0, it means the object popularity is almost uniform. We examine the impact of request frequency distribution on the effectiveness of our caching schemes.

Figure 5 shows the performance curves as a function of Zipf parameter α over a range from 0.5 to 2, with the relative cache size of 0.04%. The curves remain approximately constant with the skewness factor. Because of the large base of content population, the most popular contents are almost cached in different cases.

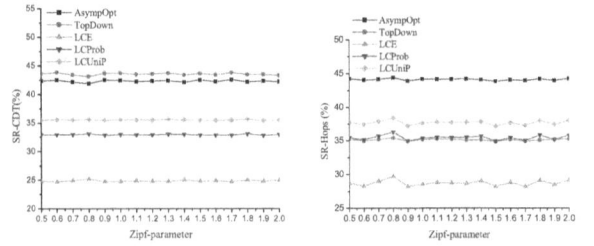

Figure 5. Caching performance vs. Zipf skewness

5.3.2.2 Object popularity fluctuation
As the object popularity changes with time owing to the variation in user interests, we further study the impact of varying object popularity on the caching performance. The X axis shows the popularity variation range, with maximum change of 60 popularity rank for each round. Though there is some fluctuation because of the random generation for dataset, in general slight variation on object popularity has advantage over sharp variation. It is more difficult for the caching algorithms to adapt quickly in response to severe and quick popularity alteration. Both SR-CDT and SR-Hops are the worst in case of sharp variation in popularity. As shown in Figure 6, all the algorithms can keep good efficiency and stability regardless of the popularity change. The insensitiveness of the proposed algorithms provides better adaptation to different network environment and makes the assumption of unchanged popularity reasonable.

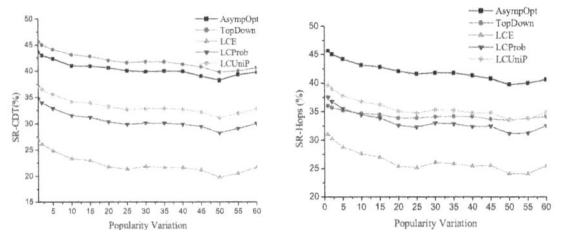

Figure 6. Caching performance vs. Popularity variation

5.3.3 Impact of content population on performance
Internet contents are anticipated to dramatically increase due to the explosive growth in user-generated contents and

many emerging applications. To shed light on this issue, we have conducted experiments to gain a deeper insight into the impact of object population increase on the performance of caching algorithms, with a special focus on its scalability and robustness.

Figure 7 shows that the proposed algorithms still gain much better performance than the baseline algorithms, when increasing the number of object items while keeping the caching capacity fixed. Our simulation also shows that when enlarging the cache capacity with the increase in #objects, the caching performance in terms of SR-CDT and SR-Hops remains good. Besides, as the content population exceeds 100,000, caching performance tends to be convergent. The stable property is very favorable, as the content items increase rapidly nowadays.

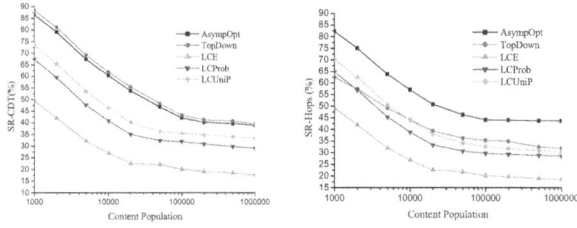

Figure 7. Caching performance vs. content population

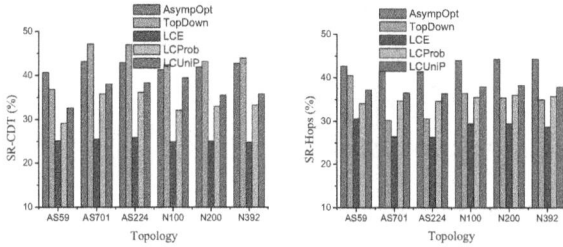

Figure 8. Caching performance vs. topology

5.3.4 *Impact of the ISP topology on performance*
We generate three different router-level topologies by GT-ITM and extract the routing trees rooting from the gateway: 100 nodes with 53 end nodes, 200 nodes with 103 end nodes, and 392 nodes with 262 end nodes, respectively. And we test the algorithms over real networks as well. Figure 8 illustrates that the performances of the presented caching schemes are insensitive to the topology change, which ensures the scalability of our proposed caching algorithms and the ease of deployment.

5.3.5 *Discussion of simulation results*
From the simulation results presented above, we reach some conclusions about our caching schemes.

1) *Effectiveness*
From Figure 4-8, we can see that AsympOpt outperforms the baseline algorithms both in SR-CDT and SR-Hops, and TopDown performs well in saving inter-ISP traffic, which is our first order objective. In all the cases, the algorithms

are effective and seem to achieve better performance in presence of more randomness in access pattern.

2) *Scalability*
The proposed caching algorithms show great scalability as demonstrated in Figure 7, 8. The caching performance of the schemes increases with the increasing cache size. Moreover, enlarging ISP's network scale will not impose a negative impact on algorithms' performance, possibly because caching favors popular contents, which are not pertinent to object population. Further, when object population is large enough, caching performance is convergent to a reasonable value, which is good news for the content explosive era.

3) *Stability*
The algorithms are insensitive to the variation in user behaviors on object requests. The relative performance remains stable regardless of the distribution of requested objects and popularity alteration, even in the worst case where there is extremely dynamic change of object popularity. Besides, AsympOpt is superior to the baseline algorithms in almost all cases with various parameters.

5.4 Preliminary measurement of overheads
We analyzed the caching scheme and conducted preliminary experiments mainly on the default 200-node topology for measuring the cost on three folded: storage use, communication overhead and execution time.

5.4.1 *Communication overhead*
In order to characterize communication overhead, we first list the related notations as follows,

D_i : the number of nodes who is i hops apart from the root.

L: the longest distance between end node and root

C: homogeneous cache capacity

M: the number of content profile sent for information aggregation

N: the number of all nodes. $N = \sum_{i=0}^{L} D_i$

F: content population

H_{ij} : hop count to satisfy the request for O_j at end node i

Then, the average communication overhead is as follows,

AsympOpt $4\sum_{i=0}^{L} D_i \cdot [M - (L - i + 1) \cdot C] / N$

TopDown: $4 \cdot [\sum_{i=0}^{L-1} \sum_{j=i+1}^{L} D_j \cdot C + M \cdot (N-1)] / N$

LCProb: $4 \cdot \sum_{i=1}^{N} \sum_{j=1}^{F} E[H_{ij}] / N$

LCProb: $\sum_{i=1}^{N} \sum_{j=1}^{F} E[H_{ij}] / N$

LCE: 0

Take 200-node topology for instance, the communication overhead for all the algorithms is : AsympOpt 11.3 Kbytes ,

TopDown 12.464Kbytes, LCProb 424M bytes, LCUniP 105 Mbytes.

5.4.2 *Storage overhead*

Storage cost for LCE, LCProb and LCUniP is negligible, while AsympOpt needs $(4F-2/N+8)$ bytes and TopDown needs $(4F-2/N+8)$ bytes.

5.4.3 *Execution time*

The job execution time during each caching round is listed as follows: 1304ms for AsympOpt, 2577ms for TopDown, 402540ms for LEC, 400784ms for LCProb, and 398672 for LCUniP. Our algorithms save hundreds of seconds of running time and account for limited communication overhead mostly because the algorithms only deal with access statistics of selected objects and are well scaled with the increasing object population, while the baseline algorithms have to cope with each arriving object. Therefore, we can expect our algorithms to gain favorable result under the large-scale network serving numerous objects.

6. CONCLUSION AND FUTURE WORK

We have developed coordinated caching schemes to reduce the redundant traffic going through the NDN networks, trying to minimize both inter-ISP traffic and average number of access hops. The main goal of this paper is to propose efficient caching algorithms that can make dynamic caching decisions on the fly. The proposed algorithms achieve the performance that is close to the optimum (especially for AsympOpt) with the favorable saving rate in inter-ISP traffic and considerably improve the performance of access delay and intra-ISP link consumption measured by the number of hops traveled. A variety of factors that can impact the caching performance are considered in the simulations and our algorithms are demonstrated to be effective, stable and scalable with varying network topology, cache capacity, objects request pattern, popularity variation and population covered by end routers. As part of our future work, we plan to extend our dynamic caching solution by considering multiple gateways, as well as the use of multiple paths in NDN.

7. ACKNOWLEDGMENTS

This paper is partially supported by NSFC (61073171), China Postdoctoral Science Foundation (Grant No. 023230012), Tsinghua University Initiative Scientific Research Program (20121080068), Specialized Research Fund for the Doctoral Program of Higher Education of China (20100002110051), and Ningbo Natural Science Foundation (Grant No. 2010A610121). We would like to thank Greg Byrd and the anonymous reviewers for their helpful comments and suggestions.

8. REFERENCES

[1]. Jacobson, V., Smetters, D. K., and Thornton, J. D. et al., "Networking named content", Proceedings of the 5th international conference on Emerging networking experiments and technologies, ACM, 2009, pp.1-12.

[2]. Hefeeda, M. and Noorizadeh, B., "On the Benefits of Cooperative Proxy Caching for Peer-to-Peer Traffic," Parallel and Distributed Systems, IEEE Transactions on, vol. 21, no. 7, pp.998-1010, 2010.

[3]. Rodriguez, P., Spanner, C., and Biersack, E. W., "Analysis of web caching architectures: hierarchical and distributed caching," Networking, IEEE/ACM Transactions on, vol. 9, no. 4, pp.404-418, 2001.

[4]. Hefeeda, M., Hsu, C. H., and Mokhtarian, K., "Design and Evaluation of a Proxy Cache for Peer to Peer Traffic," IEEE Transactions on Computers, vol. 60, no.7, pp.964-977, 2011.

[5]. Pallis, G. and Vakali, A., "Insight and perspectives for content delivery networks," Communications of the ACM, vol. 49, no. 1, pp.101-106, 2006.

[6]. Bhattacharjee, S., Calvert, K. L., and Zegura, E. W., "Self-organizing wide-area network caches", Seventeenth Annual Joint Conference of the IEEE Computer and Communications Societies (Infocom'98), pp.600-608, 1998.

[7]. Lijun Dong, Dan Zhang, Yanyong Zhang, and Dipankar Raychaudhuri, "Optimal Caching with Content Broadcast in Cache-and-Forward Networks", ICC' 2011, 2011.

[8]. Walter Wong, Marcus Giraldi, Mauricio F.Magalhaes, and Jussi Kangasharj, "Content Routers: Fetching Data on Network Path", ICC'2011, pp.1-6, 2011.

[9]. D.Rossi and G.Rossini, "Caching performance of content centric networks under multi-path routing (and more)," Telecom ParisTech, Technical Report, 2011.

[10]. Psaras, I., Clegg, R., Landa, R., Chai, W., and Pavlou, G., "Modelling and Evaluation of CCN-Caching Trees," NETWORKING 2011, pp.78-91, 2011.

[11]. Cho, K., Lee, M., Park, K., Kwon, T. T., Choi, Y., and Pack, S., "WAVE: Popularity-based and Collaborative In-network Caching for Content-Oriented Networks", Infocom'2012 workshops, pp.316-321, 2012.

[12]. Schwartz, Y., Shavitt, Y., and Weinsberg, U., "A measurement study of the origins of end-to-end delay variations", Passive and Active Measurement, Lecture notes in Computer Science, vol. 6032/2010, pp.21-30, 2010.

[13]. P.V.Mieghem, "Performance analysis of communications networks and systems",Cambridge Univ. Press, 2006, pp.358.

[14]. "GLPK: GNU Linear Programming Kit," http://www.gnu.org/s/glpk/, 2012.

[15]. Calvert, K. I., Doar, M. B., and Zegura, E. W., "Modeling internet topology," Communications Magazine, IEEE, vol. 35, no. 6, pp.160-163, 1997.

[16]. Hefeeda, M. and Saleh, O., "Traffic modeling and proportional partial caching for peer-to-peer systems," IEEE/ACM Transactions on Networking (TON), vol. 16, no. 6, pp.1447-1460, 2008.

[17].Laoutaris, N., Syntila, S., and Stavrakakis, I., "Meta algorithms for hierarchical web caches", IEEE International Conference on Performance, Computing and Communications, pp. 445-452, 2004.

NetSlices: Scalable Multi-Core Packet Processing in User-Space

Tudor Marian[*]
Google
Mountain View, CA
tudorm@google.com

Ki Suh Lee
Cornell University
Ithaca, NY
kslee@cs.cornell.edu

Hakim Weatherspoon
Cornell University
Ithaca, NY
hweather@cs.cornell.edu

ABSTRACT

Modern commodity operating systems do not provide developers with *user-space* abstractions for building high-speed packet processing applications. The conventional raw socket is inefficient and unable to take advantage of the emerging hardware, like multi-core processors and multi-queue network adapters. In this paper we present the NetSlice operating system abstraction. Unlike the conventional raw socket, NetSlice tightly couples the hardware and software packet processing resources, and provides the application with control over these resources. To reduce shared resource contention, NetSlice performs domain specific, coarse-grained, spatial partitioning of CPU cores, memory, and NICs. Moreover, it provides a streamlined communication channel between NICs and user-space. Although backward compatible with the conventional socket API, the NetSlice API also provides batched (multi-) send / receive operations to amortize the cost of protection domain crossings. We show that complex user-space packet processors—like a protocol accelerator and an IPsec gateway—built from commodity components can scale linearly with the number of cores and operate at 10Gbps network line speeds.

Categories and Subject Descriptors

C.2.6 [**Computer-Communication Networks**]: Internetworking; D.4.4 [**Operating Systems**]: Communications Management; D.4.7 [**Operating Systems**]: Organization and Design; D.4.8 [**Operating Systems**]: Performance

Keywords

Software routers, software packet processors, software router performance, operating systems support

1. INTRODUCTION

Extensible and programmable router support is becoming more important within today's experimental networks [1,2,6,40,51]. Indeed, general purpose packet processors enable the rapid prototyp-

---

[*]Work performed while at Cornell University

ing, testing, and validation of novel protocols. For example, Open-Flow [47] evolved quickly into a mature specification, and was able to do so by leveraging highly extensible NetFPGA [44] forwarding elements. Moreover, the OpenFlow specification is currently being incorporated into silicon fabric by enterprise grade router manufacturers. Such extensible router support seamlessly enables the deployment of functionality that is currently implemented by network providers through special purpose network middleboxes, like protocol accelerators and performance enhancement proxies [4, 7, 8].

Traditionally, the tradeoff between specialized hardware packet processors and software packet processors running on general purpose commodity hardware has been, and remains still, one of high performance versus ease of programmability. The currency for packet processors is performance. More recently, several significant efforts strived to render networking hardware more extensible [3, 10, 44]. Conversely, software routers have successfully harnessed the raw horsepower of modern hardware to achieve considerably high data rates [27, 40, 45]. However, for the sake of performance, such software routers were devised to run within the kernel, at a low level immediately on top of the hardware.

Writing a packet processor on domain specific, albeit extensible, hardware is hard since the developer needs to be aware of low level issues, intricacies, and limitations. We argue that building packet processors in the kernel, even when taking advantage of elegant frameworks such as Click [41], is equally difficult. In particular, the developer does not simply learn a new "programming paradigm." She needs to be aware of the idiosyncrasies of the memory allocator (e.g. small virtual address spaces, the limit on physically contiguous memory chunks, the inability to swap out pages), understand various execution contexts and their preemptive precedence (e.g. interrupt context, bottom half, task / user context), understand synchronization primitives and how they are intimately intertwined with the execution contexts (e.g. when an execution context is not allowed to block), deal with the lack of standard development tools like debuggers, and handle the lack of fault isolation. A bug in a conventional monolithic kernel brings the system into an inconsistent state and is typically lethal—leading at best to a crash, or worse, may corrupt data on persistent storage or cause permanent hardware component failure.

Although user-space packet processing applications could ease the development burden and provide fault isolation, the premium on performance has rendered such an option largely invalid for all but modest data rates. Packet processors running in user-space on modern operating systems (OSes) are rarely able to saturate modern networks [24,27,48], given that 10 Gigabit Ethernet (GbE) Network Interface Controllers (NICs) are currently a commodity. Yet the opportunity to achieve both performance, fault isolation, and programmability rests in taking advantage of multi-core processors and multi-queue NICs. However, to scale linearly with the number of cores, contention must be kept to a minimum. Conventional wisdom, and Amdahl's law [13,39], states that when adding processors to a system the benefit grows at most linearly while the costs (cache coherency, memory / bus contention, serialization) grow quadratically. Unfortunately, operating systems fail to provide general-purpose abstractions for packet processing in user-space that take advantage of modern hardware transparently. For example, packet processors built with the *raw socket*—the de-facto packet processing mechanism—are unable to sustain high rates.

In this paper we report on the design and implementation of NetSlice—a new operating system abstraction that enables linear performance scaling with the number of cores while processing packets in user-space. We achieves this through an efficient raw communication channel akin to the raw socket, that leverages modern hardware. NetSlice performs *spatial partitioning* (i.e. exclusive assignment instead of time sharing) of the CPU cores, memory, and multi-queue NIC resources at coarse granularity, to aggressively reduce overall memory and interconnect contention.

NetSlice provides high performance and multi-core scalability. It tightly couples the hardware and software resources involved in packet processing. The spatial partitioning effectively offers the illusion of a battery of independent, isolated SMP machines working in parallel with little contention. At the same time, each individual NetSlice partition is designed to provide a fast, lightweight, streamlined path for packets between the NICs and the user-space raw endpoint. Moreover, the NetSlice application programming interface (API) exposes fine-grained control over the hardware resources, and provides efficient batched send / receive operations.

NetSlice is practical; it works out-of-the-box with vanilla Linux kernels running stock NIC device drivers (simply build and load NetSlice at runtime), achieving high-performance without requiring any invasive patches (e.g. it requires no new system-calls or modified zero-copy drivers). Unlike NetTap [17] and the more recent netmap [55], NetSlice does not rely on zero-copy techniques, though it could benefit from them; consequently, NetSlice is able to trivially enforce strict address space isolation, as well as provide seamless portability and usability. NetSlice is self-contained, as portable as any device driver, and easy to deploy, requiring only a simple kernel extension that can be loaded at runtime.

We show that complex user-space packet processors built with NetSlice—a protocol accelerator and an IPsec gateway—closely match the performance of state-of-the art in-kernel Route-Bricks [27] variants. Moreover, NetSlice packet processors scale linearly with the number of cores and operate at nominal 10Gbps network line speeds, vastly exceeding alternative user-space implementations that rely on the conventional raw socket. NetSlice is fundamentally different than high-speed in-kernel variants like RouteBricks since the latter does not provide fault isolation. Further, RouteBricks does not enable high-speed packet processing in user-space any more than the conventional raw socket does.

The contributions of this paper are as follows:

- We argue that the conventional abstractions (e.g. the raw socket) are ill-suited for packet processing applications.

- We propose NetSlice—a new operating system abstraction for developing packet processors in user-space that can leverage modern hardware.

- We show that the throughput of NetSlice applications scales linearly with the number of cores, closely following the performance of state-of-the-art, in-kernel variants. NetSlice also provides fault isolation.

- NetSlice requires only a simple kernel extension which can be loaded at runtime, providing hardware independent drop-in replacement for conventional raw sockets.

The rest of the paper is structured as follows. Section 2 expands on the motivation behind user-mode packet processors. Section 3 details the NetSlice design and implementation while Section 4 presents our evaluation. Section 5 contains the related work and Section 6 concludes.

2. RAW SOCKETS AND MANY CORES: WHERE HAVE ALL MY CYCLES GONE?

The new reality is that software packet processors must scale with the number of cores (e.g. routing throughput should increase as the number of cores increase). This is true even for user-space packet processors. Currently, packet processors and packet capture libraries rely on the raw socket (PF_PACKET and SOCK_RAW) and BSD Packet Filter [46] (BPF/LSF). Unfortunately, these abstractions were designed when the ratio between single CPU performance (expressed in cycles/MIPS/MFLOPS) and network speed remained the same over time. By shifting the focus from single CPU scaling to placing many cores on the same silicon chip, the semiconductor industry has ushered in a new world in which fast networks are driven in unison by many slow cores. For example, modern 10 Gbps commodity network adaptors are commonplace, and while the number of cores per chip has been steadily increasing, single core performance has been stagnant for years.

The raw socket and other sister operating systems abstractions for packet processing in user-space are overly general, and in need of an overhaul. The issues stem from the fact that the entire network stack handles the raw socket in the same fashion it handles a regular endpoint (e.g. TCP or UDP) socket—essentially taking the least common denominator between the two. However, unlike TCP or UDP sockets, a raw socket is different in that it manipulates the entire traffic seen by the host. Given today's network capabilities and the relatively slow cores, such traffic is sufficient to overwhelm a host that uses raw sockets. We argue that applications are unable to take advantage of modern hardware since:

1. The raw socket abstraction is too general and provides the user-mode application with no control over the physical resources involved in packet reception and transmission.

2. Although simple and common to all types of sockets, the socket API is largely inefficient.

3. The conventional network stack is loosely coupled. In particular, the hardware and software resources that are involved in packet processing are loosely coupled, which results in increased contention.

4. Likewise, the conventional network stack was built for the general, most common case. As a result, the path taken by a packet between the NIC and the user-space raw endpoint is unnecessarily expensive.

Engler et al. [30] have similarly argued for an "end-to-end" approach against the high cost introduced by high-level abstractions. A fixed set of high level abstractions has been known to **i)** hurt application performance, **ii)** hide information from applications, and **iii)** limit the functionality of applications. The conventional (raw) socket is such an example: it offers a single, arguably ossified, API which abstracts away the path taken by a packet between the NIC and the application, thus providing no control over the hardware resources utilized, which is why applications fail to perform.

In what follows, we will expand on the four above claims. First, the socket API does not provide tight control over the physical resources involved in packet processing. For example, the user-mode application has no control over the path taken by a packet between some NICs queue and the raw endpoint. Second, although providing a simplified interface, the socket API is largely inefficient. For example, it requires a system call for every packet send / receive operation (the asynchronous I/O interface is currently only used for file operations, since it does not support ordering—equally important for both TCP send/receive and UDP send operations).

Third, the network stack is loosely coupled. For example, the raw socket endpoint is loosely coupled with the network stack by virtue of the user-mode task it belongs to. Since processing is performed in a separate protection domain, an additional cost is incurred due to packet copies between address spaces, cache pollution, context switches, and scheduling overheads. The cost depends on the CPU affinity of the user-mode task relative to the corresponding in-kernel network stack that processed the packets in the first place. In general, there are several choices where the user-mode task may run with respect to the in-kernel network stack:

- **Same-core:** in lockstep on the same CPU with the in-kernel network stack;

- **Hyperthread:** concurrently on a peer hyperthread of the CPU that runs the in-kernel network stack;

- **Same-chip:** concurrently on a CPU that shares the Last Level Cache (LLC), e.g. L3 for Nehalem;

- **Different-chip:** concurrently on a CPU that belongs to a different packaging socket / silicon die.

The first option, i.e. same-core, is ideal in terms of cache performance, however one has to consider the cost of frequent context switches and the impedance mismatch between the in-kernel network stack running in softirq context (a type of bottom half), at a strictly higher priority than user-mode tasks, and the user-mode task. If the user-mode task is not allocated sufficient CPU cycles to clear the socket buffers in a timely fashion, packets will be dropped.

If hyperthreads are available, the second option may be ideal. However, hyperthreads need to be simultaneous—the CPU can fetch instructions from multiple threads in a single cycle. Hyperthreads are not ideal if they work on separate data (i.e. at different physical locations in memory), since they would split all shared cache levels into half. However, if hyperthreads work on shared data, like the packets passed between a user-mode task and the in-kernel network stack, then this scenario has the potential of also reducing cache misses beyond the LLC. Alternatively, two CPUs may only share the LLC—the third option—and still reduce the number of cache misses. The final option is sub-optimal, since every packet would induce an additional LLC cache miss.

However, the kernel scheduler defaults to dynamically selecting on which CPU to run the user-mode task, constantly re-evaluating its past decision, and potentially migrating the task onto a different CPU. Although the user-space application is able to choose a CPU affinity to request on which CPUs to run, the socket interface provides no insight into what the placement should be. The traffic may have been handled by the in-kernel network stack on any of the available CPUs. Worse, the raw socket was designed to receive traffic from all queues of every NIC, traffic that is handled by all (interrupt receiving) CPUs, thus increasing the cross-core contention overhead (e.g. cache coherency, cache pollution).

Fourth, and final, the in-kernel network stack is overly general, bulky, and unnecessarily expensive. To illustrate this, consider a user-space application processing the entire traffic by means of raw sockets. For the system depicted in Figure 1(a), in order to utilize the available CPU cores, boilerplate solutions either use several raw sockets in parallel, one per process / thread, or a single raw socket and load balance traffic to several worker threads.

If several raw sockets are used in parallel, each received packet is processed by protocol handlers as many times as there are raw sockets, and a copy of the packet is delivered to each of the raw sockets. Moreover, the original packet is also passed to the default in-kernel IP layer. To implement a packet processor in user-space, an additional firewall rule is needed that instructs the kernel to needlessly drop the packet. Berkeley Packet Filters (BPF) can be installed on each raw socket in an attempt to disjointly split the traffic, however:

- BPF filters are expensive, and they scale poorly with an increase in the number of sockets [62].

- Writing non-overlapping filters for all possible traffic patterns is hard at best, and requires a priori knowledge of traffic characteristics, not to mention the complexity of handling traffic imbalances. Filters may be installed at runtime, by reacting to the traffic patterns, however, installing filters on the fly at rates around 10Gbps is not feasible [61].

- Without understanding the NICs opaque hash function that classifies flows to queues we are unable to predict the CPU that will be executing the kernel network stack, hence filters may exacerbate interference (e.g. cache misses). Such predictions are only possible if the interrupts from queues are issued in a deterministic fashion, and if the classification function is itself deterministic. The issue is further aggravated by using NICs from different vendors, which implement different classification functions. For example, the Intel 82598/82599 NICs do not map flows to queues deterministically unless Flow Director is enabled, yet Flow Director is not supported by the 82598 NIC; Myricom 10GbE NICs use a different, albeit deterministic, hash function that is configurable to some degree.

Alternatively, a single raw socket may be used to load balance and quickly dispatch traffic to several worker threads. In this scenario, there are two potential contention spots. First, between the in-kernel network stacks running on all (interrupt receiving) CPUs and the dispatch task, and second, between the dispatch task and the worker threads (we evaluate this scenario in Section 4).

3. NetSlice

We argue that user-mode processes need complete control over the entire path taken by packets, all the way from the NICs to the applications and back. NetSlice relies on a four pronged approach to provide an efficient OS abstraction for packet processing in user-space. First, NetSlice spatially partitions the hardware resources at coarse granularity to reduce interference / contention. Second, the NetSlice API provides the application with fine-grained control over the hardware resources. Third, NetSlice provides a streamlined path for packets between the NICs and user-space. Fourth, NetSlice exports an efficient API.

The core of the NetSlice design consists of spatial partitioning of the hardware resources involved in packet processing. In particular, we provide an array of independent packet processing execution contexts that "slice" the network traffic to exploit parallelism and minimize contention. We call such an execution context a NetSlice.

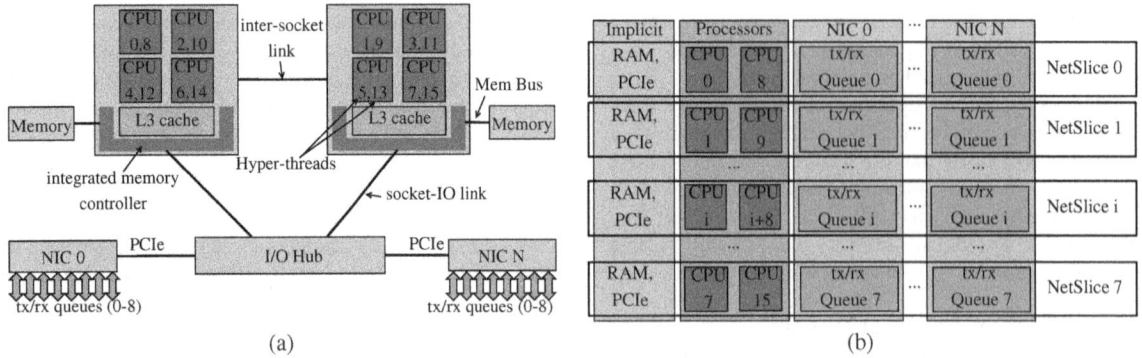

Figure 1: Nehalem cores and cache layout (a) and corresponding NetSlice spatial partitioning example (b).

Each NetSlice tightly couples all software and hardware components like NICs and CPUs—executing the in-kernel network stack and the user-mode task tightly coupled with each-other.

As network speeds have continued to increase and vendors have switched focus from individual CPU performance scaling to increasing the number of cores per chip, a single core handling traffic at line rate from a single network interface has few, if any, cycles to spare. Modern NICs attempt to address the issue by supporting in hardware multiple transmit (tx)/receive (rx) queues that can be handled in parallel by different cores. A NetSlice packet processing execution context consists of one such tx and one rx queue per attached NIC, and two (or more) *tandem* CPUs. Importantly, a tx/rx NIC queue belongs to a single context, hence each NetSlice context can perform any interface-to-interface forwarding independently. While the NIC queues and CPU cores are resources explicitly partitioned by NetSlice, each execution context also consists of implicit resources, like a share of the physical memory, PCIe bus bandwidth, etc. The tandem CPUs share at the very least the LLC; NetSlice defaults to using hyperthreads if available. NetSlice automatically binds the tx/rx queues of each context to issue interrupts exclusively to one of the peer CPUs in the context—we call this the *k-peer* CPU; we call the other CPU(s) the *u-peer* CPU(s). The in-kernel (NetSlice) network stack executes on the k-peer CPU, while the user-mode (NetSlice) task runs simultaneously on the u-peer CPU. A NetSlice may have more than two CPUs: several threads execute in user-mode to balance the processing load between user- and kernel-space.

There are as many NetSlices as there are CPU tandems. For our experimental setup depicted in Figure 1(a), NetSlice partitions resources as depicted in Figure 1(b). Every NIC is configured with eight tx/rx queues, associating the ith tx/rx queue of every NIC (e.g. NICs 0 and 1 in Figure 1(a)) with tandem pairs consisting of CPUs i (k-peer) and $i + 8$ (u-peer). Each NIC issues interrupts signaling events pertaining to the ith queue to the ith CPU exclusively. Through this technique, no two k-peer CPUs will handle packets on the same NIC queue, thus eliminating the costs of contention like locking, cache coherency, and cache misses. This scheme that binds NIC queues to CPUs was previously evaluated for 1Gbps NICs [18] and is the keystone to RouteBricks' individual forwarding element scaling. (RouteBricks relies on Click [41] which uses a polling driver—the same functionality provided by the "New API" (NAPI) [56] hybrid polling in conjunction with NIC device interrupt coalescence.)

NetSlice exposes fine-grained control over the hardware resources of the entire packet processing execution context to the user-mode application. For example, it provides control over which CPU the in-kernel (NetSlice) network stack is executing with respect to the user-mode application to take advantage of the physical cache layout. The added control is key to minimizing inter-CPU contention in general, and cache misses and cache coherency penalties in particular.

Importantly, the path a packet takes through each NetSlice execution context is streamlined, bypassing the default, bulky, in-kernel general purpose network stack. NetSlice hijacks the packets at an early stage subsequent to DMA reception and before it would have been handed off to the network stack. Next it performs minimal processing while in kernel context executing on the k-peer CPU, and then passes the packets to the user-space application to be processed in overlapped (pipelined) fashion, on the u-peer CPU. Notably, on an entire NetSlice path there is a single spinlock being used per send / receive direction. The spinlock is specialized for synchronization between the communicating execution contexts, namely between a bottom half and a task context.

While the NetSlice API provides tight control over physical resources, it also supersedes and extends the socket API while maintaining a level of backwards compatibility. In addition to single-packet send and receive system calls, the NetSlice API also supports batched operations to amortize the cost associated with protection domain crossings. Similarly, the Linux kernel has recently begun to provide support for receiving and sending multiple messages on a socket through two new system calls (`recvmmsg` and `sendmmsg`).

3.1 NetSlice Implementation and API

The raw NetSlice API extends the device-file interface and leverages the flexibility of the `ioctl` mechanism. User-mode libraries may use it to create a more elegant API, in the same fashion the Packet CAPture (pcap) library [9] is layered on top of the raw socket. These user-mode applications perform conventional file operations using the familiar API (`read`/`write`/`poll`) over each slice, which map to corresponding operations over the per-NetSlice data flows. For example, a conventional read operation will return the next available packet, block if no packet is available, or return `-EAGAIN` if there are no packets available and the device was opened with the `O_NONBLOCK` flag set. We implemented the NetSlice abstraction as a set of character devices with the same major number and N minor numbers—one minor number for each of the N slices. By overloading the device-file interface we gained portability since NetSlice could reside in a kernel runtime loadable module, whereas new system calls cannot.

The `ioctl` mechanism was sufficient to provide the NetSlice additional control and API extensions. Consider, for example, the `NETSLICE_CPUMASK_GET ioctl` request—it returns the affin-

```
1:  #include "netslice.h"                              25:  for (i = 0; i < IOVS; i++) {
2:                                                     26:     iov.iov_base = malloc(MTU_LARGE);
3:  struct netslice_iov {                              27:     iov.iov_len = MTU_LARGE;
4:     void *iov_base;                                 28:  }
5:     size_t iov_len; /* capacity */                  29:  if (mlockall(MCL_CURRENT) < 0)
6:     size_t iov_rlen;/* returned length */           30:     EXIT_FAIL_MSG("mlockall");
7:     int flags; /* selective per-packet operations */ 31:
8:  } iov[IOVS];                                       32:  for (;;) {
9:                                                     33:     ssize_t cnt, wcnt = 0;
10: struct netslice_rw_multi {                         34:     if ((cnt = read(fd, iov, IOVS)) < 0)
11:    int flags;                                      35:         EXIT_FAIL_MSG("read");
12: } rw_multi;                                         36:
13:                                                     37:     for (i = 0; i < cnt; i++)
14: struct netslice_cpu_mask {                         38:         /* iov_rlen marks bytes read */
15:    cpu_set_t k_peer, u_peer;                        39:         scan_pkg(iov[i].iov_base, iov[i].iov_rlen);
16: } mask;                                             40:     do {
17:                                                     41:         size_t wr_iovs;
18: fd = open("/dev/netslice-1", O_RDWR);               42:         /* write iov_rlen bytes */
19:                                                     43:         wr_iovs = write(fd, &iov[wcnt], cnt-wcnt);
20: rw_multi.flags = MULTI_READ | MULTI_WRITE;          44:         if (wr_iovs < 0)
21: ioctl(fd, NETSLICE_RW_MULTI_SET, &rw_multi);        45:             EXIT_FAIL_MSG("write");
22: ioctl(fd, NETSLICE_CPUMASK_GET, &mask);             46:         wcnt += wr_iovs;
23: sched_setaffinity(getpid(), sizeof(cpu_set_t),      47:     } while (wcnt < cnt);
24:    &mask.u_peer);                                   48:  }
```

Figure 2: One NetSlice batched read/write example.

ity mask of the tandem CPUs, providing the current user-mode task with fine control over the CPU it runs atop. The NETSLICE_TX_CSUM_SET ioctl allows the user-mode application to offload TCP, IP, both or no checksum computation (alternatively, one may set a per-packet flag in the netslice_iov). The in-kernel Net-Slice stack has the knowledge to enable hardware specific offload computation to spare CPUs from unnecessarily spending cycles.

Once the NETSLICE_RW_MULTI_SET ioctl is issued, the user-mode application may overload the read/write calls to send and receive an array of datagrams encoded into the parameters. Note that this is fundamentally different than the readv / writev calls which can only perform scatter-gather of a *single* datagram (or packet) per call. Batched packet receive and send operations are instrumental in mitigating the overheads of issuing a system call per operation. At the same time, batching reduces per packet locking overheads, e.g. spinlock induced cycle waste and cache coherency overheads, between the user-mode task while executing system calls and the in-kernel NetSlice network stack.

Figure 2 shows an example of application code using NetSlice batched read / write for a naïve deep packet inspection tool. Commenting out lines 37 and 39, the application forwards packets behaving as a regular router. The array of buffers are passed to the read and write functions encoded in netslice_iov structures. The example consists of a single NetSlice (namely the first Net-Slice) hence the application will only receive packets classified to be handled by the first queue of each NIC. To handle the entire traffic, the example can be easily extended to accommodate all available queues using either an equal number of threads or processes.

For outgoing packets, the routing decision is performed by default within the in-kernel NetSlice stack. However, there is an ioctl request (NETSLICE_NOROUTE_SET) that allows applications to instruct NetSlice that routing will be performed in userspace (by encoding the chosen output interface within the parameters of the write call). If the hardware decides which NIC rx queue to place the received packets onto, the software is responsible for selecting an outbound NIC queue to transmit packets on. For the conventional network stack, the core kernel or the device driver is responsible with implementing this functionality. NetSlice provides a specialized classification "virtual func-

tion" that overrides driver or kernel provided hash functions (by updating the select_queue function pointer of net_device structures). The NetSlice classification function ensures that the packets which belong to a particular NetSlice context are placed solely on the tx queues associated with the context. Unlike the classification functions provided by the device drivers (e.g. the Myricom myri10ge driver provides the myri10ge_select_queue function) or the kernel's default simple_tx_hash, the NetSlice classification function is cheaper, consisting only of three load operations, one arithmetic, one bitwise mask operation, and no conditional branches (by contrast, simple_tx_hash has three conditional statements).

Instead of a character device, we could have implemented Net-Slice by extending the socket interface with a new socket type (e.g. SOCK_RAW). However, the current approach enabled us to seamlessly commandeer received packets immediately after reception. A new PF_PACKET socket does not curtail the default network stack, nor does it prevent the kernel from performing additional processing per packet (e.g. pass packets through all relevant protocol handlers).

3.2 Discussion

NetSlice does not rely on zero-copy techniques, unlike prior work for which zero-copy was the keystone in boosting the performance of I/O and network paths [28, 52]. Instead, NetSlice copies each packet once between user- and kernel-space, trading off CPU cycles in exchange for flexibility and portability. The reason we can afford to make this tradeoff is because modern NUMA (non-uniform memory access) architectures that replaced the Front-Side-Bus with point-to-point interconnects can be CPU bound when processing packets [27]; with the caveat that cache coherency overhead and cache misses can be kept sufficiently low [21]. As we will show in Section 4, NetSlice minimizes these overheads and achieves linear scalability with the number of cores, while maintaining the cost of packet copies constant per CPU. The cost is roughly a cache miss for the first load plus the time it takes to copy the remaining bytes which the hardware prefetching already brought in the LLC. This works to our advantage, since CPU cycles and even entire cores are easier to scale than interconnect and bus capacities: Moore's Law

31

currently results in an increase in cores, and the goal is to fully utilize each core. Further, commodity NICs are expected to support an increasing number of queues, since they are also used to support virtualization.

NetSlice leverages modern hardware to render zero-copy an orthogonal issue, less pivotal for performance, which can be supported in the future if so desired. By avoiding zero-copy, NetSlice gains added portability and usability. Currently, NetSlice comprises of a single runtime loadable module that works out-of-the-box with vanilla Linux kernels running stock NIC device drivers.

By contrast, zero-copy techniques, like memory mapping NIC DMA rings or the NIC's entire address space into user-space [17, 55] are not only invasive, they also break isolation. New device drivers may have to be written and supported, and kernels may have to be extended or modified accordingly. For example, the kernel scheduler will have to ensure that the user-mode task controlling a NIC is scheduled sufficient CPU time and is not preempted for long continuous periods [19, 25] of time, or packets may be dropped (when the DMA rings fill up). Moreover, current OSes lack support for delivering events to user-space in a timely fashion—mechanisms like epoll or kqueue / kevent do not currently feature interrupt delivery. Myricom's MX-10G is one such technology that provides zero-copy drivers. However, the MX-10G drivers are intended for TCP/IP and UDP/IP communication endpoints, and do not support interface-to-interface forwarding, the most basic functionality of a router or general purpose packet processor. In fact, the MX-10G drivers do not provide support out-of-the-box for two different NICs at the same time on the same machine.

Likewise, NetSlice is also orthogonal to the large body of past work that relocated the networking stack into user-space [17,33,35, 53,59]. For example, user-space networking may very well be built on top of NetSlice, although we did not yet implement the network stack encapsulation for replacing endpoint sockets. In our experience, conventional TCP and UDP sockets still perform sufficiently well, to date. Moreover, given that a typical host may have an arbitrarily large number of concurrent TCP and UDP connections, it is not clear that user-space networking, even built over NetSlice, would perform better than the current network stack since it would require efficient inter-process-communication for de-multiplexing and multiplexing traffic between user-mode applications. Nevertheless, we plan to investigate NetSlice support for endpoint sockets in the future.

4. EVALUATION

We evaluated software packet processors running NetSlice and compared them against the state-of-the-art user-space and in-kernel equivalent implementations. We have ported packet processors to run over RouteBricks [27] forwarding elements, as well as to run in user-space using the pcap library [9]. Pcap is implemented on top of the conventional raw (PF_PACKET) sockets. We also linked the pcap applications with Phil Wood's libpcap-mmap library [60], which uses the memory mapping functionality of PF_PACKET sockets, known as PACKET_MMAP. (PF_RING [26] sockets are roughly an alternative implementation of the PACKET_MMAP approach. Further, PF_RING sockets require an invasive patch that alters the core-kernel codebase, unlike the readily available PACKET_MMAP sockets or NetSlice.) A kernel that is built with the PACKET_MMAP support copies each packet onto a circular buffer mapped into user-space before optionally adding it to the socket's queue. The user-space application can poll the arrival of new packets and receive them without the cost of issuing additional system calls. (The PACKET_MMAP support does not implement

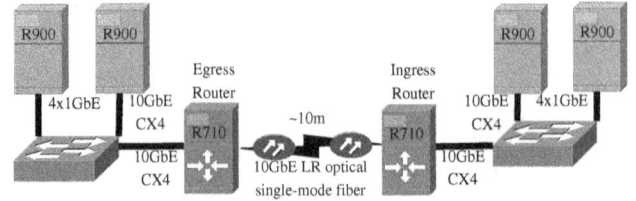

Figure 3: Experimental evaluation physical topology.

a zero-copy receive technique, a packet is copied the same number of times as with a traditional socket.) The NetSlice batched receive operation achieves the same net effect, however, unlike the NetSlice batched transmit, PACKET_MMAP sockets do not offer equivalent support for outbound packets. During our experiments, we set the circular buffer size to the value that yielded the best performance, incidentally it was the maximum value. (For our experimental machines, the PCAP_MEMORY=max request yielded a 1.93GB circular buffer.) We did not compare with the libnetfilter_queue libipq packet redirection mechanism since it consistently crashed at high data rates (the netlink sockets it relies on proved to be inadequate for high data rates). Further, we did not use the TPACKET_V3 pcap support since it was only recently introduced with Linux kernel versions 3.2, whereas Net-Slice currently only supports Linux kernel versions between 2.6.20 and 2.6.36.

NetSlice consists of 1814 lines of kernel module code and 2981 lines of user-space applications—a router, an IPsec gateway (839 lines of AES ports), and an implementation of the Maelstrom [15] protocol accelerator.

Our evaluation answers the following questions:

- What is the performance of NetSlice with respect to the state-of-the-art, for both routing and IPsec?

- What is the performance breakdown of the NetSlice techniques? To quantify this scenario, we funnel all traffic to be handled by a single NIC queue: there is no interference from extraneous CPUs and NIC resources, and we are able to quantify, in isolation:

  - The benefit of streamlining packet paths;
  - The NetSlice performance with respect to possible peer CPUs placement;
  - The benefit of NetSlice batched operations;
  - NetSlice added latency and CPU usage.

- How does NetSlice scale with the number of cores?

- Can complex packet processors built with NetSlice deliver the advertised performance increase?

4.1 Experimental Setup

We deployed a testbed topology as depicted in Figure 3, with four Dell PowerEdge R900 machines serving as end-hosts that generate and receive traffic. The traffic is aggregated by two Cisco Catalyst 4948 series switches before being routed through a pair of identical Dell PowerEdge R710 machines, which we refer to as the egress and ingress routers. The egress and the ingress routers run various packet processor variants, like NetSlice, RouteBricks, or pcap.

Each R900 machine is a four socket 2.40GHz quad core Xeon E7330 (Penryn) with 6MB of L2 cache and 32GB of RAM—the

E7330 is effectively a pair of two dual core CPUs packaged on the same chip, each with 3MB of L2 cache. By contrast, the R710 machines are dual socket 2.93GHz Xeon X5570 (Nehalem) with 8MB of shared L3 cache and 12GB of RAM, 6GB connected to each of the two CPU sockets. The Nehalem CPUs support hardware threads, or hyperthreads, hence the operating system manages a total of 16 processors. Each R710 machine is equipped with two Myri-10G NICs, one CX4 10G-PCIE-8B-C+E NIC and one 10G-PCIE-8B-S+E NIC with a 10G-SFP-LR transceiver. Figure 1(a) depicts the R710 internal structure with two NICs.

The egress router is connected to the ingress router via a 10 meter single-mode fiber optic patch cable, and each router is connected to the corresponding switch through a 6 meter CX4 cable. Two of the R900 machines are each equipped with an Intel 82598EB 10-Gigabit CX4 NIC, while the other two R900 machines are connected to the switches through all of their four Broadcom NetXtreme II BCM5708 Gigabit Ethernet NICs. We use the additional R900 machines, although the egress and ingress routers only have one 10GbE connection on each side, since a single R900 machine with a 10GbE interface is unable to receive (in the best configuration) more than roughly 5Gbps worth of MTU size (1500 byte packets) traffic. The packet rate (pps) for the R710 router with the Myricom 10GbE NIC was the same for both a kernel-level router and NetSlice: 1215k pps using minimum size packets (64 bytes; or 174 Mbps) and 822k pps with MTU size packets (1500 bytes; or 9.68 Gbps). The same observation applies for the R900 client with the Intel 10GbE NIC. RouteBricks altered the NIC driver to increase the packet rate by performing DMA transfers of small packets in a single transaction on the PCIe bus. Our evaluation does not include such batching since it is not clear it is possible on our Myricom NICs. Further, we did not use the Intel 82598 NIC for the evaluation comparisons since it does not support the Intel Flow Director and does not provide a deterministic mapping from a flow to a queue. We used the Myricom 10GbE NIC for all evaluation comparison.

Unless specified otherwise, we generate traffic between the R900 machines with Netperf [5] that consists of MTU size UDP packets at line rate (10Gbps). The machines run the Linux kernel version 2.6.28-17; we use the myri10ge version 1.5.1 driver for the Myri-10G NICs and the ixgbe version 2.0.44.13 driver for the Intel NICs. Both drivers support NAPI and are configured with factory default interrupt coalescence parameters. To enable RouteBricks, we modified the myri10ge driver to work in polling mode with Click (we used Linux kernel version 2.6.24.7 with Click, the most recent version supported).

All values presented are averaged over multiple independent runs, between as low as eight and as high as 32 runs; the error bars denote standard error of the mean and are always present, although most of the time they are sufficiently small to be virtually invisible.

4.2 Forwarding / Routing

Figure 4 shows the UDP payload throughput for the most basic functionality—packet routing with MTU size packets. We compare the NetSlice implementation with the default in-kernel routing, a RouteBricks implementation, and with the best configurations of pcap user-space solutions. Utilizing all NIC queues and all CPUs, NetSlice forwards packets at nominal line rate (roughly 9.7Gbps for MTU packet size and MAC layer overhead), as do the kernel and RouteBricks routing. However, the best pcap variants top off at about 2.25Gbps. There is no difference between pcap and pcap-mmap, while Click user-space does in fact perform worse.

For each case, the Figure shows the additional scenario in which all traffic is handled by a single NIC queue (per available NIC). In

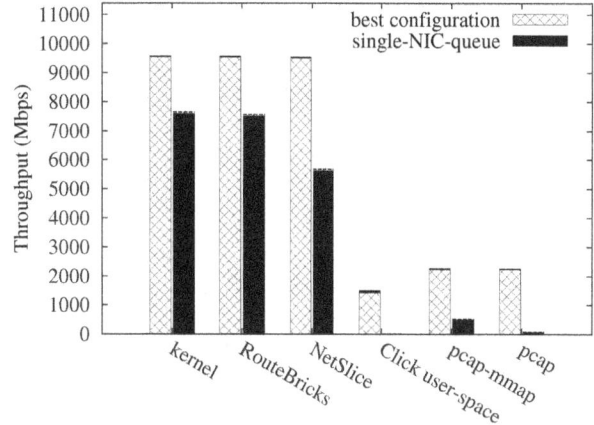

Figure 4: Packet routing throughput.

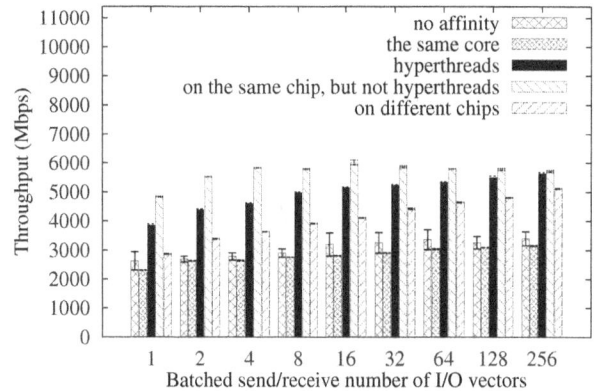

Figure 5: Routing throughput for single a NIC queue, a single NetSlice, and various u-peer CPU placements.

this case, the kernel achieves 7.59Gbps, while NetSlice achieves 74% of the kernel throughput, while pcap-mmap achieves $\frac{1}{11}$ of the throughput achieved by NetSlice, and about 7.6 times better than regular pcap. As expected, in-kernel variants perform better since routing is performed at an early stage, and less CPU work (zero-copy forwarding) is expended per dropped packet [49].

The take-away is that the NetSlice kernel to user-space communication channel is highly efficient, even when a single channel is used (a single NetSlice is two CPUs and one NIC queue per NIC). Moreover, using more than a single NetSlice easily sustains line rate—currently, our clients are not able to generate more than 10Gbps worth of MTU-size packet traffic.

Next, we evaluate the importance of the u-peer CPU placement. User-space processing takes place on the u-peer CPU as part of the spatial partitioning that isolates individual NetSlices. Here we used a single NetSlice to stress one communication channel that handles all traffic in isolation. Since only two tandem CPU cores and one NIC queue per NIC are utilized, the experiment only accounts for direct interference (like cache coherency, cache misses due to pollution) within a single NetSlice. Additional indirect interference is expected in the general case, however, the NetSlice spatial partitioning was designed precisely to keep such interference to a minimum. Figure 5 shows the throughput given various core placement

Configuration	Packets / μs	RTT (μs)
Linux kernel	1/100	242.24 ± 42.14
Linux kernel	10/100	279.48 ± 42.74
NetSlice (no batching)	1/100	255.38 ± 39.98
NetSlice (no batching)	10/100	308.10 ± 44.51
NetSlice (128-batch)	1/100	255.67 ± 40.18
NetSlice (128-batch)	10/100	301.33 ± 42.16

Table 1: Round-trip-time (RTT) between the end-hosts.

Datarate (Mbps)	CPU usage (% of single CPU)		
	Total	k-peer	u-peer
1000.9	31.34 ± 1.00	1.90 ± 0.30	29.44 ± 0.73
2001.8	63.49 ± 1.39	12.41 ± 0.72	51.09 ± 0.68
3002.7	100.38 ± 0.88	24.56 ± 1.61	75.82 ± 0.92
4003.2	102.28 ± 2.69	14.29 ± 3.41	87.99 ± 0.72
5003.1	103.40 ± 2.78	17.94 ± 2.90	85.46 ± 1.36

Table 2: CPU usage: One NetSlice (2 CPUs) forwarding.

choices and the number of I/O vectors used for batched operations. There are several key observations. First, if the user-mode task does not use the CPU affinity as instructed by NetSlice, the default choice made by the OS scheduler is suboptimal. Moreover, the high error bars imply that the kernel does not attempt to perform smart task placement. The Linux scheduler is primitive in that it typically moves a task on the runqueue of a different CPU only if the current CPU is deemed congested.

The second observation is that using the same CPU for both in-kernel and user-space processing performs the worst—there are simply not enough cycles to counter the excessive overheads introduced by the context switches. Additionally, there is an impedance mismatch between the task context and the in-kernel processing that is performed in a softirq context and is of strictly higher priority than the task, i.e. the task is not scheduled enough cycles. This setting is complicated further by the kernel's per-CPU `ksoftirqd` threads that are spawned to act as rate-limiters during receive-livelock scenarios [49].

The third observation is that same-chip and hyperthread placement outperform the scenario in which the user-space processing happens on a different chip. This is consistent with the memory hierarchy—i.e. accessing the shared L3 cache is faster than accessing data over the QuickPath inter-socket link. However, note that the gap between same-chip and different-chip data access decreases considerably with the increase in the number of I/O vectors. This is likely because batching increases code and data locality, and hardware optimizations like pre-fetching and pipelined processing are in effect. Batched processing also improves the performance of user-space processing on the hyperthread, however to a lesser extent than same-chip placement, presumably because the hardware threads still contend for functional units (e.g. ALUs) within the (shared) physical core.

The best case is when the peer CPUs are on the same chip yet are not hyperthreads. However, the Figure shows the scenario in which a single NetSlice is used, hence only the peer CPUs are utilized, all the remaining cores are idle. In the general case, such a placement choice is only viable when there is a lower number of NetSlices than there are available CPUs. By default, NetSlice performs userspace processing on the sibling hyperthread, if one is available—having two sibling hyperthreads work on different NetSlices would split the cache levels (higher than the LLC) into half.

Figure 5 also shows the performance increase due to NetSlice batching. For the default peer CPU placement (i.e. sibling hyperthreads) we observe a 46.2% increase in aggregate throughput from singleton send / receive to 256 batched I/O vectors shuttled between user-space and the kernel in a single operation, even though the kernel uses the fast system call processor instructions (e.g. SYSENTER).

In summary, the forwarding throughput of a single NetSlice is eleven times larger than the best pcap. Of that, batching provides a 46.2% throughput increase, peer CPU placement provides a 78.3% throughput increase when batching is used and 66.6% increase with no batching, while the streamlined path of packets (with no batch-

ing or peer CPU placement) provides a 4.5 times throughput increase over pcap-mmap. This coarse break-down does not reveal the subtle interaction between NIC queues nor the cross core, cross PCIe bus, and cross QuickPath interconnect interference.

Table 1 shows the additional latency introduced by a single NetSlice forwarding element. The experiment shows the roud-triptime (RTT) between an R900 end-host and the Ingress Router while traffic flows through the Egress Router (Figure 3). The Table depicts the Egress Router performing standard in-kernel forwarding, and forwarding through NetSlice with batching both disabled and enabled (128 I/O vectors). The table shows two scenarios, one in which the sender issues packets at a steady rate of one every $100\mu s$ and a second in which the sender issues 10 packets in rapid succession every $100\mu s$. The two-way latency introduced by NetSlice is $19\mu s$ on average (at most $28\mu s$), which is half the standard deviation of the reported RTTs. The latency introduced by NetSlice is in fact smaller than the effects of NIC interrupt coalescence (IC) and NAPI, two ubiquitous techniques that have been universally adopted to the detriment of latency (e.g. the myri10ge driver defaults the `rx-usecs` IC parameter to $75\mu s$ and does not compile without NAPI).

Table 2 shows the NetSlice CPU utilization while forwarding traffic through a single NIC queue, for increasing input data rates. As expected, the CPU utilization increases with the data rate, however, the increase is less sharp for rates greater than 3Gbps. This is due to batching which, while seldom used for rates less than 3Gbps, is required to forward packets at increasingly higher data rates, as we previously showed in Figure 5.

4.3 IPsec

Next we evaluate a CPU intensive packet processing task, namely IPsec encryption with 128 bit key (typically used by VPNs). We implemented AES encryption in Cipher-block Chaining (CBC) [31] mode of operation. Our experiments focused on steady-state performance, hence the key establishment protocol is not evaluated. We use IPsec to evaluate how NetSlice scales with the number of cores. IPsec accelerators typically need all the CPU cycles they can spare and two NetSlices proved sufficient to forward all the 10Gbps MTU-size traffic that our testbed was able to generate.

Ideally, NetSlice should only trail RouteBricks by a constant factor (per CPU) due to the cycles spent performing an additional copy per packet and the overhead of protection domain crossings (system calls). Figure 6 shows that NetSlice does scale linearly with the number of CPUs, closely following RouteBricks. The NetSlice throughput is roughly 8% less than that of RouteBricks. RouteBricks tops off at 9157Mbps, about 600Mbps shy of nominal line rate. NetSlice tops off at about 8513Mbps. We expect both NetSlice and RouteBricks to continue to scale linearly given more cores. By contrast, the user-space variants using pcap scale poorly and are unable to take advantage of the current technology trend towards placing many independent cores on the same silicon die. (The Figure reports on the best user-space pcap variants with

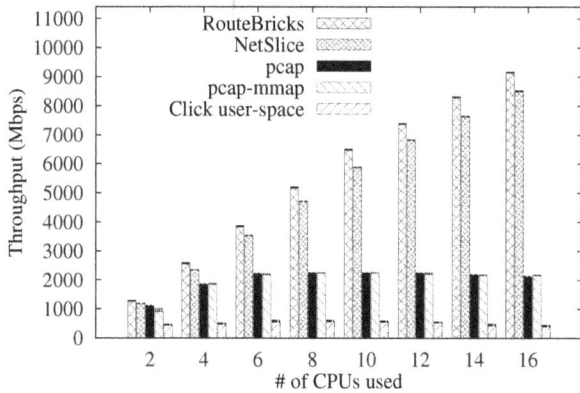

Figure 6: IPsec throughput scaling with cores.

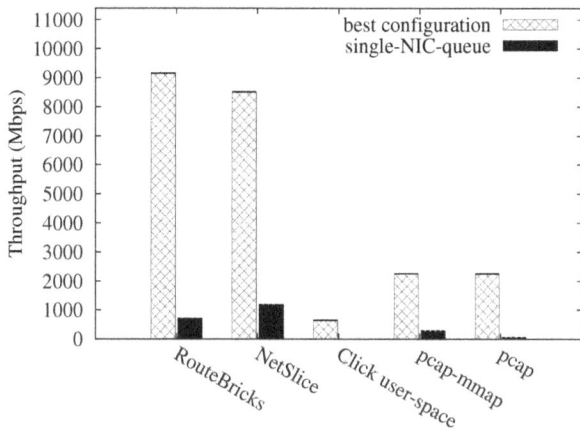

Figure 7: IPsec throughput, all vs. single NIC queue.

a dispatch thread load balancing packets to threads bound to CPUs exclusively.)

Figure 7 shows IPsec throughput results for the best configurations of NetSlice, RouteBricks, and pcap user-space solutions along with the additional scenario in which all traffic is handled by a single NIC queue. First, notice that the pcap variants top off at about 2258Mbps in the common case and perform poorly when traffic is handled by a single NIC queue. As with routing, the pcap-mmap outperforms conventional pcap in the latter scenario and NetSlice, like RouteBricks, vastly outperforms the user-space variants. Net-Slice also achieves better throughput than RouteBricks since all traffic that is routed to a single NIC queue is handled by Route-Bricks with a single CPU in kernel mode, whereas NetSlice handles it with a pair of CPUs, one running in kernel-mode and one running the user-mode task.

The take-away is that NetSlice scales with the number of available cores as good as the in-kernel RouteBricks implementation does. By contrast, user-space variants that use conventional raw sockets scale poorly and are comprehensively outperformed by NetSlice and RouteBricks during a CPU intensive task like IPsec.

4.4 The Maelstrom Protocol Accelerator

To get more insight into the feasibility of building highly complex protocol accelerators atop NetSlice, we ported and evaluated the Maelstrom [15] appliance. Maelstrom is a performance enhancement proxy developed to overcome the poor performance of TCP when loss occurs on high bandwidth / high latency network links. Maelstrom appliances work in tandem, each appliance located at the perimeter of the network and facing a LAN on one side and a high bandwidth / high latency WAN link on the opposite side. The appliances perform forward error correction (FEC) encoding over the egress traffic on one side and decoding over the ingress traffic on the opposite side. In Figure 3, the egress and the ingress routers are running Maelstrom appliances, with the egress router encoding over all traffic originating from the clients on the same LAN and destined for the clients on the LAN behind the ingress router. The ingress router receives both the original IP traffic and the additional FEC traffic and forwards the original traffic and potentially any recovered traffic to original destination nodes. In reality, each Maelstrom appliance works both as an encoder and as an decoder at the same time.

The existing hand-tuned, in-kernel version of Maelstrom is about 8432 lines of C code. It is self contained with few calls into the exported kernel base symbols, and it was not yet retrofitted to take advantage of multi-queue NICs and multiple cores. By contrast, the NetSlice implementation required 934 lines of user-space C code, not counting the NetSlice kernel module or the 263 lines of hash-table implementation. For MTU size packets, NetSlice achieves a goodput of 6993.69 ± 35.7Mbps for a throughput of 8952.04 ± 37.25Mbps. For the nominal FEC parameters we used (namely, for every r=8 packets, an additional c=3 FEC packets are sent), there is a 27.27% overhead. NetSlice achieves close to maximum goodput: $6993.69\text{Mbps} \times \left(1 + \frac{c}{r+c}\right) = 8901\text{Mbps}$.

5. RELATED WORK

It has been well known that large scale cache coherent, possibly NUMA, multiprocessors require careful operating system design, or else various bottlenecks prevent the systems from realizing their performance potential. Indeed, operating systems like Tornado/K42 [32, 42, 58] have been carefully designed to minimize contention by clustering and replicating key kernel data structures, and by employing intricate scheduling algorithms that, for example, take NUMA locality into account.

More recently, there have been several research efforts that aimed at redesigning the OS from the ground up in order to effectively exploit the emerging and now ubiquitous multi-core architectures. Corey [20] is an ExoKernel-like OS within which shared kernel data structures and kernel intervention are kept to a minimum, while applications are given explicit control over the sharing of resources. This allows the Corey kernel to perform finer grained locking of highly accessed data structures, like process memory regions. The Barrelfish research operating system [16] explores how to structure the OS as a distributed system in order to best utilize future multi- and many-core, potentially heterogeneous systems. Similarly, the Helios [50] operating system tackles building and tuning applications for heterogeneous systems through satellite kernels. Satellite kernels export a uniform set of OS abstractions across all CPUs and communicate one with another by means of explicit message passing instead of relying on a cache coherent memory system. The Tessellation OS [43] introduces a "nanovisor" to enforce strict spatial and temporal resource multiplexing between library OSes. To ensure resource isolation, the Tessellation OS envisions hardware support for resources that have been traditionally hard to share, like caches and memory bandwidth.

Like the Tessellation OS, NetSlice performs spatial partitioning of the CPU, memory, and multi-queue NIC resources at coarse granularity. However, the NetSlice partitioning is domain specific, and the performance isolation need not be strongly enforced, instead it is implicit by the design of the NetSlice abstraction itself.

Historically, there have been a large number of zero-copy user-space network stacks proposed [17, 26, 28, 35, 52, 53, 55, 59]. Their general approach was to eliminate the OS involvement on the communication path, and virtualize the NIC while providing direct, low-level access to the network. Some of these approaches relied on hardware support, for example, U-Net [59] and its commercial successor VIA [14, 29] required a communications co-processor capable of demultiplexing packets into user-space buffers, and an on-board MMU (Memory Management Unit) to perform RDMA (Remote DMA) [12]. The former is not available on typical Ethernet NICs, moreover, IOMMUs are currently unable to handle page faults [34]. Other systems [22, 23, 28] rely on virtual memory and page protection techniques, however, on demand memory mapping of shared buffers is tricky and can be unnecessarily expensive; such techniques are yet to be adopted by mainstream kernels.

NetSlice is orthogonal to prior work that placed the network stack in user-space. Further, it does not use zero-copy techniques since they were not necessary (see Section 3.2) and they would have prevented portability.

Software routers Achilles' heel has been, and continues to be, the low performance with respect to their hardware counterparts. Nevertheless, recent efforts, like RouteBricks [27], have shown that modern multi-core architectures and multi-queue NICs are well suited for building low-range software routers, albeit in kernel-space. RouteBricks relies on a cluster of PCs fitted with Nehalem multi-core CPUs and multi-queue NICs, connected through a k-degree butterfly interconnect. Packets are forwarded / routed at aggregate rates of 24.6Gbps per PC, however, the interconnect routing algorithm introduces packet re-ordering. The PacketShader [37] software router takes RouteBricks further by providing an entire framework for general-purpose packet processing that utilizes the Graphics Processing Unit (GPU).

Internally, RouteBricks uses the Click [41] modular router, an elegant framework for building functionality from smaller building blocks that can be arranged in an acyclic control flow graph. However, Click is aimed at building routers and does not easily express general packet processing; e.g. it cannot support global state that extends across building blocks.

In general, software routers are implemented within the kernel [36, 57], early in the network stack and below the (raw) socket interface. Full blown software routers like RouteBricks [27] may require distributed coordination algorithms to decide interconnect forwarding paths [11]. By contrast, NetSlice provides support for user-space implementation of individual packet processing units, independent of interconnects. Complex packet processing logic, like rule-based forwarding [54], or the distributed coordination in RouteBricks may be seamlessly built using NetSlice (NetSlice does not re-order packets).

NetSlice can be used to implement the XORP [38] open source routing platform, or to provide rapid prototyping of OpenFlow [47] forwarding elements. For example, the current NetFPGA [44] reference implementation is limited to four 1GbE interfaces (the recently launched NetFPGA-10G supports four 10GbE interfaces), whereas NetSlice is only limited by the number of CPUs and PCIe connections a commodity server can support. Moreover, developers need not have intimate Verilog knowledge, or worry about details such as gateware real-estate.

The new Threaded NAPI (TNAPI) PF_RING [26] support improves on the raw socket by creating one virtual NIC per receive queue and capturing traffic from each virtual NIC with a different user-space thread (somewhat similar to NetSlice's spatial partitioning). In general, such packet capture techniques are only optimized for the receive path, ignoring the transmission, which means that they provide no support for efficient interface-to-interface forwarding, which is the most basic software router / packet processor functionality.

6. CONCLUSION

The end of CPU frequency scaling is ushering in a world of slow cores and fast networks. This paper introduced the NetSlice operating system abstraction that enables building scalable packet processors in user-space. NetSlice tightly couples the hardware and software packet processing resources by performing domain specific, coarse-grained, spatial partitioning of CPU cores, memory, and NIC resources. NetSlice also provides the application with control over these resources. On top of each resource partition, NetSlice superimposes independent streamlined communication channels to shuttle packets between NICs and user-space and bypass the default network stack. While it is backward compatible with the conventional socket API, the NetSlice API also provides batched send / receive operations to amortize the cost of protection domain crossings. Further, NetSlice is portable, working with existing device drivers. We demonstrate NetSlice by showing that complex user-space packet processors can scale linearly with the number of cores and operate at nominal 10Gbps line speeds.

7. AVAILABILITY

The NetSlice source code is published under BSD license and is freely available for download at http://netslice.cs. cornell.edu.

8. ACKNOWLEDGEMENTS

We would like to thank our shepherd, Ripduman Sohan, and the anonymous reviewers for their comments. This work was partially funded and supported by an IBM Faculty Award, NetApp Faculty Fellowship, and Alfred P. Sloan award received by Hakim Weatherspoon. Further partial support includes NSF TRUST (No. 0424422), NSF Future Internet Architecture (No. 1040689), NSF CAREER (No. 1053757), and DARPA Computer Science Study Panel (No. D11AP00266).

9. REFERENCES

[1] GENI: Global Environment for Network Innovations. http://www.geni.net.

[2] Internet2. http://www.internet2.edu.

[3] Juniper Networks: Open IP Service Creation Program (OSCP). http://www.ictnetworks.com.au/pdf/1000167-en.pdf.

[4] Netequalizer. http://netequalizer.com.

[5] Netperf. http://netperf.org.

[6] NLR: National LambdaRail. http://www.nlr.net.

[7] PacketLogic. http://proceranetworks.com.

[8] Riverbed. http://www.riverbed.com.

[9] tcpdump / libpcap. http://www.tcpdump.org.

[10] Cisco opening up IOS. http://www.networkworld.com/news/2007/121207-cisco-ios.html, 2007.

[11] M. Al-Fares, A. Loukissas, and A. Vahdat. A scalable, commodity data center network architecture. In *Proceedings of the ACM SIGCOMM 2008 conference on Data communication*, SIGCOMM '08, pages 63–74, New York, NY, USA, 2008. ACM.

[12] AMD. I/O Virtualization Specification, 2007.

[13] G. M. Amdahl. Validity of the single processor approach to achieving large scale computing capabilities. In *Proceedings of the April 18-20, 1967, spring joint computer conference*, AFIPS '67 (Spring), pages 483–485, New York, NY, USA, 1967. ACM.

[14] P. Balaji, P. Shivam, P. Wyckoff, and D. Panda. High performance user level sockets over gigabit ethernet. In *Proceedings of the IEEE International Conference on Cluster Computing*, CLUSTER '02, pages 179–, Washington, DC, USA, 2002. IEEE Computer Society.

[15] M. Balakrishnan, T. Marian, K. Birman, H. Weatherspoon, and E. Vollset. Maelstrom: transparent error correction for lambda networks. In *Proceedings of the 5th USENIX Symposium on Networked Systems Design and Implementation*, NSDI'08, pages 263–278, Berkeley, CA, USA, 2008. USENIX Association.

[16] A. Baumann, P. Barham, P.-E. Dagand, T. Harris, R. Isaacs, S. Peter, T. Roscoe, A. Schüpbach, and A. Singhania. The multikernel: a new os architecture for scalable multicore systems. In *Proceedings of the ACM SIGOPS 22nd symposium on Operating systems principles*, SOSP '09, pages 29–44, New York, NY, USA, 2009. ACM.

[17] S. Blott, J. Brustoloni, and C. Martin. NetTap: An Efficient and Reliable PC-Based Platform for Network Programming. In *Proceedings of OPENARCH*, 2000.

[18] R. Bolla and R. Bruschi. Pc-based software routers: high performance and application service support. In *Proceedings of the ACM workshop on Programmable routers for extensible services of tomorrow*, PRESTO '08, pages 27–32, New York, NY, USA, 2008. ACM.

[19] H. Bos, W. de Bruijn, M. Cristea, T. Nguyen, and G. Portokalidis. FFPF: fairly fast packet filters. In *Proceedings of the 6th conference on Symposium on Opearting Systems Design & Implementation - Volume 6*, OSDI'04, pages 24–24, Berkeley, CA, USA, 2004. USENIX Association.

[20] S. Boyd-Wickizer, H. Chen, R. Chen, Y. Mao, F. Kaashoek, R. Morris, A. Pesterev, L. Stein, M. Wu, Y. Dai, Y. Zhang, and Z. Zhang. Corey: an operating system for many cores. In *Proceedings of the 8th USENIX conference on Operating systems design and implementation*, OSDI'08, pages 43–57, Berkeley, CA, USA, 2008. USENIX Association.

[21] E. Bröse. ZeroCopy: Techniques, Benefits and Pitfalls.

[22] J. C. Brustoloni and P. Steenkiste. Effects of buffering semantics on i/o performance. In *Proceedings of the second USENIX symposium on Operating systems design and implementation*, OSDI '96, pages 277–291, New York, NY, USA, 1996. ACM.

[23] J. C. Brustoloni, P. Steenkiste, and C. Brustoloni. User-Level Protocol Servers with Kernel-Level Performance. In *Proc. of IEEE Infocom Conference*, pages 463–471, 1998.

[24] P. Crowley, M. E. Fluczynski, J.-L. Baer, and B. N. Bershad. Characterizing processor architectures for programmable network interfaces. In *Proceedings of the 14th international conference on Supercomputing*, ICS '00, pages 54–65, New York, NY, USA, 2000. ACM.

[25] W. de Bruijn and H. Bos. Beltway buffers: Avoiding the os traffic jam. In *INFOCOM*, 2008.

[26] L. Deri and S. Suin. Effective traffic measurement using ntop. *Comm. Mag.*, 38(5):138–143, May 2000.

[27] M. Dobrescu, N. Egi, K. Argyraki, B.-G. Chun, K. Fall,

G. Iannaccone, A. Knies, M. Manesh, and S. Ratnasamy. RouteBricks: exploiting parallelism to scale software routers. In *Proceedings of the ACM SIGOPS 22nd symposium on Operating systems principles*, SOSP '09, pages 15–28, New York, NY, USA, 2009. ACM.

[28] P. Druschel and L. L. Peterson. Fbufs: a high-bandwidth cross-domain transfer facility. In *Proceedings of the fourteenth ACM symposium on Operating systems principles*, SOSP '93, pages 189–202, New York, NY, USA, 1993. ACM.

[29] D. Dunning, G. Regnier, G. McAlpine, D. Cameron, B. Shubert, F. Berry, A. M. Merritt, E. Gronke, and C. Dodd. The Virtual Interface Architecture. *IEEE Micro*, 18(2):66–76, Mar. 1998.

[30] D. R. Engler, M. F. Kaashoek, and J. O'Toole, Jr. Exokernel: an operating system architecture for application-level resource management. In *Proceedings of the fifteenth ACM symposium on Operating systems principles*, SOSP '95, pages 251–266, New York, NY, USA, 1995. ACM.

[31] S. Frankel, R. Glenn, and S. Kelly. The AES-CBC Cipher Algorithm and Its Use with IPsec (RFC 3602), 2003.

[32] B. Gamsa, O. Krieger, J. Appavoo, and M. Stumm. Tornado: maximizing locality and concurrency in a shared memory multiprocessor operating system. In *Proceedings of the third symposium on Operating systems design and implementation*, OSDI '99, pages 87–100, Berkeley, CA, USA, 1999. USENIX Association.

[33] G. R. Ganger, D. R. Engler, M. F. Kaashoek, H. M. Briceño, R. Hunt, and T. Pinckney. Fast and flexible application-level networking on exokernel systems. *ACM Trans. Comput. Syst.*, 20(1):49–83, Feb. 2002.

[34] P. Geoffray. A Critique of RDMA. High-Perf. Comp. '06.

[35] P. Geoffray, L. Prylli, and B. Tourancheau. BIP-SMP: high performance message passing over a cluster of commodity SMPs. In *Proceedings of the 1999 ACM/IEEE conference on Supercomputing (CDROM)*, Supercomputing '99, New York, NY, USA, 1999. ACM.

[36] D. Guo, G. Liao, L. N. Bhuyan, B. Liu, and J. J. Ding. A scalable multithreaded L7-filter design for multi-core servers. In *Proceedings of the 4th ACM/IEEE Symposium on Architectures for Networking and Communications Systems*, ANCS '08, pages 60–68, New York, NY, USA, 2008. ACM.

[37] S. Han, K. Jang, K. Park, and S. Moon. PacketShader: a GPU-accelerated software router. In *Proceedings of the ACM SIGCOMM 2010 conference*, SIGCOMM '10, pages 195–206, New York, NY, USA, 2010. ACM.

[38] M. Handley, E. Kohler, A. Ghosh, O. Hodson, and P. Radoslavov. Designing extensible IP router software. In *Proceedings of the 2nd conference on Symposium on Networked Systems Design & Implementation - Volume 2*, NSDI'05, pages 189–202, Berkeley, CA, USA, 2005. USENIX Association.

[39] M. D. Hill and M. R. Marty. Amdahl's law in the multicore era. *Computer*, 41(7):33–38, July 2008.

[40] J.-C. Huang, M. Monchiero, Y. Turner, and H.-H. S. Lee. Ally: OS-Transparent Packet Inspection Using Sequestered Cores. In *Proceedings of the 2011 ACM/IEEE Seventh Symposium on Architectures for Networking and Communications Systems*, ANCS '11, pages 1–11, Washington, DC, USA, 2011. IEEE Computer Society.

[41] E. Kohler, R. Morris, B. Chen, J. Jannotti, and M. F.

Kaashoek. The click modular router. *ACM Trans. Comput. Syst.*, 18(3):263–297, Aug. 2000.

[42] O. Krieger, M. Auslander, B. Rosenburg, R. W. Wisniewski, J. Xenidis, D. Da Silva, M. Ostrowski, J. Appavoo, M. Butrico, M. Mergen, A. Waterland, and V. Uhlig. K42: building a complete operating system. In *Proceedings of the 1st ACM SIGOPS/EuroSys European Conference on Computer Systems 2006*, EuroSys '06, pages 133–145, New York, NY, USA, 2006. ACM.

[43] R. Liu, K. Klues, S. Bird, S. Hofmeyr, K. Asanović, and J. Kubiatowicz. Tessellation: space-time partitioning in a manycore client OS. In *Proceedings of the First USENIX conference on Hot topics in parallelism*, HotPar'09, pages 10–10, Berkeley, CA, USA, 2009. USENIX Association.

[44] J. W. Lockwood, N. McKeown, G. Watson, G. Gibb, P. Hartke, J. Naous, R. Raghuraman, and J. Luo. NetFPGA–An Open Platform for Gigabit-Rate Network Switching and Routing. In *Proceedings of the 2007 IEEE International Conference on Microelectronic Systems Education*, MSE '07, pages 160–161, Washington, DC, USA, 2007. IEEE Computer Society.

[45] Y. Ma, S. Banerjee, S. Lu, and C. Estan. Leveraging parallelism for multi-dimensional packet classification on software routers. In *Proceedings of the ACM SIGMETRICS international conference on Measurement and modeling of computer systems*, SIGMETRICS '10, pages 227–238, New York, NY, USA, 2010. ACM.

[46] S. McCanne and V. Jacobson. The bsd packet filter: a new architecture for user-level packet capture. In *Proceedings of the USENIX Winter 1993 Conference Proceedings on USENIX Winter 1993 Conference Proceedings*, USENIX'93, pages 2–2, Berkeley, CA, USA, 1993. USENIX Association.

[47] N. McKeown, T. Anderson, H. Balakrishnan, G. Parulkar, L. Peterson, J. Rexford, S. Shenker, and J. Turner. Openflow: enabling innovation in campus networks. *SIGCOMM Comput. Commun. Rev.*, 38(2):69–74, Mar. 2008.

[48] A. G. Miklas, S. Saroiu, A. Wolman, and A. D. Brown. Bunker: a privacy-oriented platform for network tracing. In *Proceedings of the 6th USENIX symposium on Networked systems design and implementation*, NSDI'09, pages 29–42, Berkeley, CA, USA, 2009. USENIX Association.

[49] J. C. Mogul and K. K. Ramakrishnan. Eliminating receive livelock in an interrupt-driven kernel. In *Proceedings of the 1996 annual conference on USENIX Annual Technical Conference*, ATEC '96, pages 9–9, Berkeley, CA, USA, 1996. USENIX Association.

[50] E. B. Nightingale, O. Hodson, R. McIlroy, C. Hawblitzel, and G. Hunt. Helios: heterogeneous multiprocessing with satellite kernels. In *Proceedings of the ACM SIGOPS 22nd symposium on Operating systems principles*, SOSP '09, pages 221–234, New York, NY, USA, 2009. ACM.

[51] R. Niranjan Mysore, A. Pamboris, N. Farrington, N. Huang, P. Miri, S. Radhakrishnan, V. Subramanya, and A. Vahdat. PortLand: a scalable fault-tolerant layer 2 data center

network fabric. In *Proceedings of the ACM SIGCOMM 2009 conference on Data communication*, SIGCOMM '09, pages 39–50, New York, NY, USA, 2009. ACM.

[52] V. S. Pai, P. Druschel, and W. Zwaenepoel. IO-Lite: a unified I/O buffering and caching system. *ACM Trans. Comput. Syst.*, 18(1):37–66, Feb. 2000.

[53] S. Pakin, M. Lauria, and A. Chien. High performance messaging on workstations: Illinois fast messages (FM) for Myrinet. In *Proceedings of the 1995 ACM/IEEE conference on Supercomputing (CDROM)*, Supercomputing '95, New York, NY, USA, 1995. ACM.

[54] L. Popa, N. Egi, S. Ratnasamy, and I. Stoica. Building extensible networks with rule-based forwarding. In *Proceedings of the 9th USENIX conference on Operating systems design and implementation*, OSDI'10, pages 1–6, Berkeley, CA, USA, 2010. USENIX Association.

[55] L. Rizzo. Netmap: a novel framework for fast packet I/O. In *Proceedings of the 2012 USENIX conference on Annual Technical Conference*, USENIX ATC'12, pages 9–9, Berkeley, CA, USA, 2012. USENIX Association.

[56] J. H. Salim, R. Olsson, and A. Kuznetsov. Beyond softnet. In *Proceedings of the 5th annual Linux Showcase & Conference - Volume 5*, ALS '01, pages 18–18, Berkeley, CA, USA, 2001. USENIX Association.

[57] M. J. Schultz, B. Wun, and P. Crowley. A Passive Network Appliance for Real-Time Network Monitoring. In *Proceedings of the 2011 ACM/IEEE Seventh Symposium on Architectures for Networking and Communications Systems*, ANCS '11, pages 239–249, Washington, DC, USA, 2011. IEEE Computer Society.

[58] R. C. Unrau, O. Krieger, B. Gamsa, and M. Stumm. Experiences with locking in a NUMA multiprocessor operating system kernel. In *Proceedings of the 1st USENIX conference on Operating Systems Design and Implementation*, OSDI '94, Berkeley, CA, USA, 1994. USENIX Association.

[59] T. von Eicken, A. Basu, V. Buch, and W. Vogels. U-Net: a user-level network interface for parallel and distributed computing. In *Proceedings of the fifteenth ACM symposium on Operating systems principles*, SOSP '95, pages 40–53, New York, NY, USA, 1995. ACM.

[60] P. Wood. http://public.lanl.gov/cpw, 2008.

[61] Z. Wu, M. Xie, and H. Wang. Swift: a fast dynamic packet filter. In *Proceedings of the 5th USENIX Symposium on Networked Systems Design and Implementation*, NSDI'08, pages 279–292, Berkeley, CA, USA, 2008. USENIX Association.

[62] M. Yuhara, B. N. Bershad, C. Maeda, and J. E. B. Moss. Efficient packet demultiplexing for multiple endpoints and large messages. In *Proceedings of the USENIX Winter 1994 Technical Conference on USENIX Winter 1994 Technical Conference*, WTEC'94, pages 13–13, Berkeley, CA, USA, 1994. USENIX Association.

Cache-Aware Affinitization on Commodity Multicores for High-Speed Network Flows

Vishal Ahuja, Matthew Farrens and Dipak Ghosal
University of California, Davis, CA 95616, USA
{vahuja, mfarrens, dghosal}@ucdavis.edu

ABSTRACT

For a given TCP or UDP flow, protocol processing of incoming packets is performed on the core that receives the interrupt, while the user-space application which consumes the data may run on the same or a different core. If the cores are not the same, additional costs due to context switches, cache misses, and the movement of data between the caches of the cores may occur. The magnitude of this cost depends upon the processor affinity of the user-space process relative to the network stack. In this paper we present a prototype implementation of a tool which enables the application processing and protocol processing to occur on cores which share the lowest cache level. The Cache-Aware Affinity Deamon (CAAD) analyzes the topology of the die and the NIC characteristics and conveys information to the sender which allows the entire end-to-end path for each new flow to be be managed and controlled. This is done in a light-weight manner for both uni- and bi-directional flows. Measurements show that for bulk data transfers using commodity multicore machines, the use of CAAD improves the overall TCP throughput by as much as 31%, and reduces the cache miss rate as much as 37.5%. GridFTP combined with CAAD improves the download time for big file transfers by up to 18%.

Categories and Subject Descriptors

C.2.5 [**Local and Wide-Area Networks**]: High-speed; C.2.2 [**Network Protocols**]: Applications

General Terms

Performance

Keywords

Cache Affinity, Processor Affinity, Receive Livelock, High-Speed Networks.

1. INTRODUCTION

The era of single-core processors is over, a point made clear at the Computer Science and Telecommunications Board (CTSB) symposium "The Future of Computing Performance: Game Over or Next Level?" [9, 20]. For a variety of reasons CPU cycle times have reached a plateau, and designers have turned to increasing the number of CPUs per die in order to improve throughput. At the same time, network speeds continue to climb, providing ever higher bandwidth and lower latency - the network is now capable of providing data faster than it can be consumed by a single processor.

Network processing has long been performed in a multi-threaded manner, with different portions of the network stack being executed at different times (some during the socket system calls, some in the context of receive packet processing, etc.). Therefore, the problem of data arriving faster than it can be processed is currently being dealt with by exploiting the multicore nature of modern processors and having different parts of the network stack computations execute on different cores.

Unfortunately, as the number of processors increases, the cost of sharing the network control structures between processors grows as well; in fact, if not done carefully such partitioning may actually have a negative impact on performance. For example, the execution of a few hundred network processing instructions on a single TCP segment can end up requiring tens of thousands of CPU cycles, because many cycles are lost while waiting for the cache lines that are shared by all the processors to be synchronized by the cache coherence logic. These lost cycles often limit the overall system performance, and in the case of end-to-end data transfer, lower throughput.

What is needed is an approach which provides a more informed usage of cores within a mulitcore system when doing network processing. Cores should not be chosen arbitrarily, but rather cores should be chosen that share the lowest possible level of the cache structure[1]. For example, when a given core (e.g., core A) is selected to do the protocol/interrupt processing, the core that shares the L2 cache with core A should execute the corresponding user-level application. Doing so will lead to fewer context switches, improved cache performance, and ultimately higher overall throughput.

In this paper we describe our tool which provides this capability, the Cache-Aware Affinity Deamon (CAAD), and show its impact on file download times. We show that current state of the art techniques (Receive Side Scaling

---

[1]In this document we consider the L1 cache to be at a lower level than the L2 cache, L2 lower than L3, etc.

[3], irqbalance [1], Receive Flow Steering [10], and Receive Packet Steering [11]) provide throughput which is lower than the line rate even when using multiple flows. Next, we describe CAAD, and show that a cache aware placement of protocol processing and application processing provides an overall improvement in throughput of 26% for a single TCP flow and 31% for multiple TCP flows. It also increases throughput by 45% for a single UDP flow and 28% for multiple UDP flows. The main reason for the improvement is a reduction in cache misses, which leads to a reduction in stall cycles. We also show that using CAAD can improve the performance of two different bulk data transfer protocols, GridFTP[5] and Flow Bifurcation Manager[4].

The remainder of this paper is organized as follows. In Section 2, we provide an overview of the typical packet receiving process in a modern operating system, and present various research efforts aimed at helping multicore systems handle high speed network data in an efficient manner. In Section 3, we describe our experimental setup and show (using two different commodity multicore machines) that the LINUX kernel using a number of different of optimizations (RFS, RPS, RSS, and irqbalance) achieves throughput lower than the line rate. In Section 4, we use pseudo-code to outline the algorithm that CCAD uses, and Section 5 gives more details of the algorithm. Section 6 presents the results of our experiments measuring the performance of CAAD. Finally, we discuss our conclusions in Section 7.

2. BACKGROUND

In most systems today, when a packet is received the NIC performs a sequence of steps: it consults the (on-board) RSS tables, DMA-transfers the packet into the appropriate queue in the host memory, and then interrupts the core(s) based on the mask set found in the RSS tables. The interrupted core will run the top-half (the interrupt handler), which acknowledges the NIC and does the bare minimum processing (with interrupts disabled) necessary to prepare and schedule a bottom half context (softirq) for deferred execution. It then re-enables interrupts and the hardware context that was interrupted resumes.

At some later time, the softirq will run on the same core that scheduled it. The softirq will do the bulk of the packet processing (routing table validity checking, netfilter processing, IP processing, TCP/UDP processing, ack handling, etc.), and based on an internal map place the packet into the appropriate socket's queue. It then issues a wake up call to the socket's owner, to notify the pending **recvmsg** system calls that packets are available.

As pointed out in [21] and [17], in the LINUX kernel there is a large amount of state associated with each TCP flow, which must be updated as packets arrive or depart. The core which is responsible for making these updates is chosen based on where the NIC deposits packets, while the user-level application which consumes (or generates) the associated data is generally placed based on overall system load, and may execute on a completely different core. If different cores end up copying connection data to and from the user-level application and dealing with incoming and outgoing packets, the cache lines holding a connection's state may have to move between cores. Since the cache coherency hardware must keep all caches consistent and coherent, cache lines moving may necessitate the invalidation of lines in the caches of other cores. These invalidations may cause a core

to experience read misses and force it to access the data from another core's cache, or even from external memory. Clearly both of these types of accesses are slower than accessing local memory, and negatively impact performance.

If a single core is processing a flow, there are no memory allocation problems, because both allocation and freeing of buffers are fast since the buffers are in the local pool. On the other hand, when multiple cores are involved in the processing of a flow, the kernel allocates a buffer on the core that first receives the packet from the rx ring buffer, and frees a buffer on the core which calls **recvmsg()**. Clearly freeing buffers on a remote core is a time-consuming operation.

In order to parallelize network processing across multiple cores, present-day NICs support multiple receive and transmit descriptor queues. When such a NIC receives a packet, it applies a flow classification to the packet and assigns it to one of the receive queues. In this way, different packets are assigned to different queues, which in turn are affinitized to different CPUs. This mechanism is generally known as **Receive-side Scaling** (RSS) [3].

Receive Packet Steering (RPS) [11] is essentially a simplified RSS, done in software (i.e., what the NIC does with the on-board RSS tables, interrupts, and DMA queues is done in RPS using extra linked-lists in the host memory). Since it is a software technique, it does not control which CPU will handle the interrupts. It only selects the CPU which will perform the protocol processing above the interrupt handler. Unlike RPS, RSS directs an incoming packet to a receive queue, and whichever CPU this queue is affinitized to will run the hardware interrupt handler.

While both RSS and RPS distribute the network processing load uniformly across cores, neither takes into account application locality. In order to provide this capability, **Receive Flow Steering** (RFS) [10] was developed. RFS steers the softirq processing (kernel processing) of the packets to the core/CPU where the corresponding application process is running. This technique does increase the data cache hit rate, but we show in Section 3 that it does not work well for a high throughput workload.

Irqbalance [1] scatters interrupts (of all types) across cores based on the load statistics, and in a sense is a variant of round-robin scheduling. In [7, 8, 22, 19] the effect of processor affinity on networking performance in multicore systems has been analyzed, and in [13] the authors propose a cache aware scheduling mechanism. However, their focus is on processor utilization rather than performance. Furthermore, their scheme works only for TCP and requires modification of the kernel.

With the advent of techniques like Direct Cache Access (DCA) [12, 15], processor affinity has become even more important. Using the LINUX system call interface, a user-space process can affinitize itself to a certain CPU, unfortunately, the socket interface does not provide any guidance in terms of which CPU to choose. The default operation of the kernel scheduler is to dynamically choose the CPU on which the user-space process will run based on the system load, which means it may migrate the process to a different CPU altogether.

In [6], a different design for a network stack has been proposed which offloads the network stack processing to a dedicated core. In this way, it eliminates the unnecessary sharing of network state among multiple cores.

One problem with the LINUX kernel is that there is only

Table 1: Configuration of the sender and receiver for the experimental setup.

Component	Receiver	Sender
Machine	Dell XPS	Custom built
Processor	Core 2 Duo Intel Q6600 @2.40GHz	Quadcore Intel corei7 @2.67 GHz
Memory	6 GByte DDR2	6 GByte DDR2
Bus	PCIe 16x	PCIe 16x
OS	LINUX 3.0.0-19	LINUX 3.0.0-19

a single **accept()** queue for all the new connections on a given TCP port. This means that the processing of new connections will be serialized. In [21], this processing is parallelized by adding per-core **accept()** queues, and ensuring that a server process that is waiting for a connection is assigned one from its own core's queue. Specifically, they have designed and implemented a scheme which enables the incoming connection processing on multicore machines to scale as the number of incoming requests grows. In our work we are not dealing with a large number of flows, but are instead focussed on maximizing the throughput when dealing with a small number of high speed flows.

In [24], a NIC design is proposed which steers incoming flows to the core on which the application process resides. However, their evaluation testbed only used a 1 Gbps link, so they were not able to observe that when multiple high speed flows are mapped to the same core, it becomes overwhelmed providing the protocol processing. Our experiments demonstrate this problem, and show that when dealing with high speed flows, one should avoid having the protocol processing and application processing occur on the same core.

In this paper we present a prototype implementation of an end-to-end tool which enables the application and protocol processing to occur on cores which share the lowest level cache. It takes into account the on-chip system topology (the number of cores and which cores share a cache) and NIC characteristics (number of queues, the flow hash function, and the hash function type) to determine a set of port numbers to convey to the sender, so that the entire end-to-end path for each new connection can be managed and controlled. This is done in a light-weight manner for both uni- and bi-directional flows. Our work differs from previous studies in that we are proposing a general purpose mechanism to control the entire end-to-end path of flows by providing affinity masks to sender and client applications in order to achieve improved network throughput. Our mechanism is not doing so in an application transparent manner, for which an API would be required - CAAD does require the application to be modified slightly. However, our results demonstrate that a significant improvement in throughput can be obtained by having a deamon running which automatically gathers all the system information necessary to allow the application to map to the correct core.

3. MOTIVATIONAL RESULTS

The testbed for our experiments was composed of two PCs connected back-to-back via a 10 Gbps link. Both the PCs are multicore machines (configured as shown in Table 1), and the 10 GbE NICs were mounted on 16-lane PCI-E slots. We used netperf [14] to generate UDP and TCP traffic between the two end-systems.The ring buffer's default size on the sender and the receiver is 512 entries, and also by default

NAPI and interrupt coalescing is enabled. Unless specified, we did not use jumbo frames. Since the end-systems are connected back to back, there was no network loss. We performed 30 second runs, and the throughput for each experimental run was recorded. All the reported values were averaged over ten runs. We performed the experiments using 20 runs as well, but saw no difference in the average throughput. Although our NICs support Direct Cache Access, we did not observe any improvement when it was enabled.

Irqbalance is a dynamic interrupt assignment daemon that is part of LINUX. By default, it does a simple interrupt load analysis on the CPUs every ten seconds (information available in the /etc/sysconfig/irqbalance file), and based on the analysis dynamically changes which CPU will handle which interrupt. The LINUX kernel uses irqbalance and RSS when RFS and RPS are not enabled.

In Figure 1 we see that high throughput is not achieved, even when using RFS and RPS. This is in part because of the following issues with RFS:

1. The application is woken up on the same CPU that ran the softirq. When a single high speed flow (10 Gbps or more) is being received, the overhead due to context switches significantly degrades the performance.

2. Our experiments indicate that for high speed flows, application processing and protocol processing is best done on cores which share the lowest level of cache. However, since LINUX is a general purpose operating system, such a strategy cannot be generalized - it may not hold true for architectures which do not have many levels of caches, for example. Since RFS is a part of the LINUX kernel, and dealing with high speed network data is not a universal problem, RFS chooses to adopt a generic approach. While we find that the generic approach does not work well for high throughput workloads, it may work well in environments where latency is the primary concern.

3. Using RFS, it is not possible to control the entire end-to-end path. Multiple flows may map to the same queue and therefore be processed by the same core, overwhelming it.

4. RFS does not take into account bi-directional flows.

Looking at Figure 1 we see that irqbalance used with RSS enabled NICs provides an end to end throughput of between 6 and 7.5 Gbps on a 10 Gbps link, even when using multiple flows. This is because irqbalance has a few drawbacks. For example, the (relatively) frequent analysis and changing of which CPU is handling which interrupt means that each time an interrupt is reassigned to a new CPU it may incur cold cache misses. In addition, irqbalance does not take into account application affinity, i.e. which CPU the application receiving the network data is running on. These factors lead to a limit on the throughput that is attainable.

4. ALGORITHM OVERVIEW

In this section we will describe the design of our algorithm. It proceeds in the following manner (the pseudo-code for the algorithm is presented in Algorithm 1, Algorithm 2, and Algorithm 3):

1. The sender application, before it connects to the receiver application, collects information about the sending machine's system features (the number of cores, which cores share a cache, the RSS hash function, the number of NIC queues, and the NIC interrupt affinity map).

2. The sender application sends this information to the CAAD running on the receiver machine.

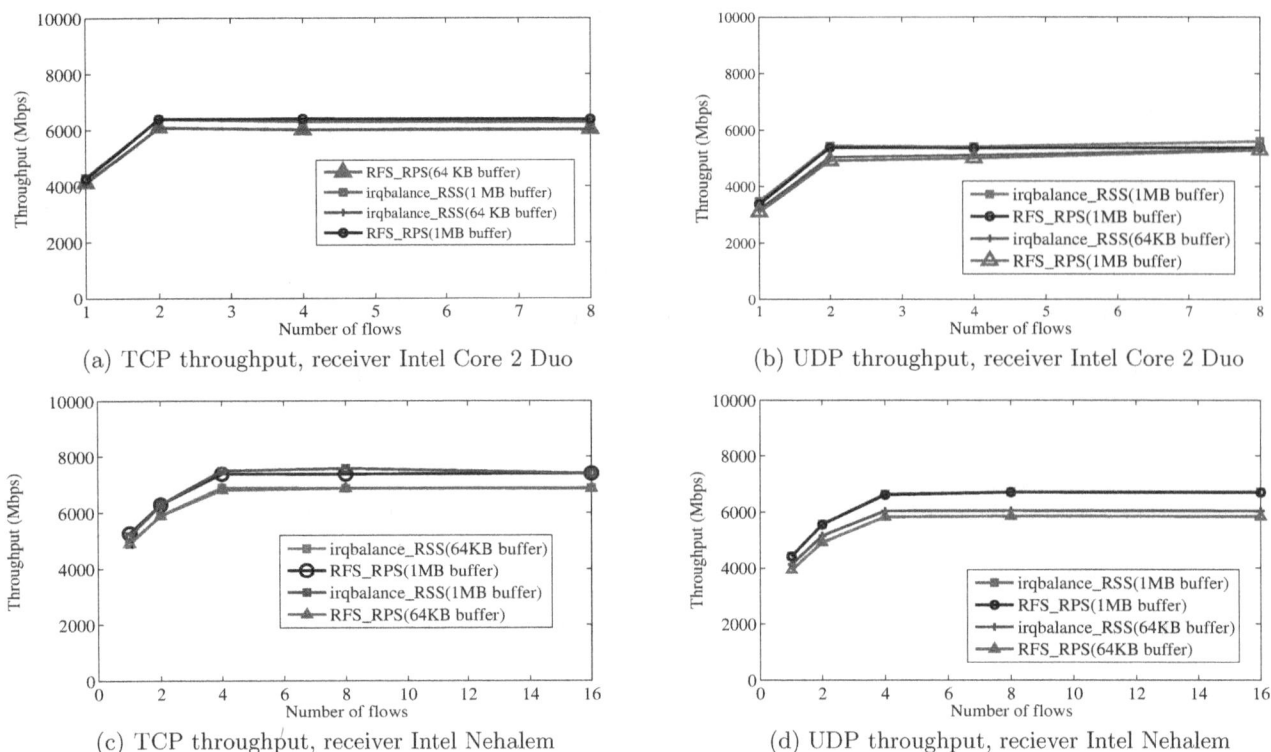

(a) TCP throughput, receiver Intel Core 2 Duo

(b) UDP throughput, receiver Intel Core 2 Duo

(c) TCP throughput, receiver Intel Nehalem

(d) UDP throughput, reciever Intel Nehalem

Figure 1: netperf stream tests. *RFS_RPS(64KB buffer)* and *RFS_RPS(1MB buffer)* denote the use of RFS and RPS with the socket buffer size set to 64KB (default) and 1MB, respectively. *irqbalance_RSS(64KB buffer)* and *irqbalance_RSS(1MB buffer)* denote the use of irqbalance and RSS with the socket buffer size set to 64KB (default) and 1MB, respectively. **RFS and RPS do not show any improvement over irqbalance.**

3. CAAD, which is already aware of the system features of the receiver machine, does the following:

1. Finds the core which is the least loaded.

2. Decides on a pair of ports (using the knowledge of the RSS hash function) so that the incoming flow is mapped to this least loaded core.

3. Finds the best core for application processing by taking into account the cache layout of the receiver machine.

4. Informs the sender application of the ports (sender and destination ports) and the core affinity mask.

5. Informs the receiver application of the destination port (so that the receiver binds itself to this port).

6. Informs the receiver application of the core affinity mask found in c), so that the receiver application can affinitize itself.

7. The sender application begins the data transfer.

In order to provide cache aware packet processing, two approaches can be adopted. One is to migrate the application-level thread (which consumes the incoming data) to the core where the NIC delivers the packets. This takes into account application locality, but it is a time-consuming process because migrating a thread involves acquiring scheduler locks and leads to cache misses when the thread runs on a different core. This is certainly not feasible for short lived connections.

The second approach would be to have the NIC steer the packets for a flow to the correct core. Unfortunately this would require the NIC to maintain a flow steering table, which would need to be extremely large since a single server may be simultaneously handling many thousands of TCP connections. Instead of using thread migration or forcing flows to a particular core, CAAD uses system information in such a way that the application processing and protocol processing can take place in a cache aware manner.

4.1 System Information

System information such as cache configuration, processor load, and interrupt affinity plays a critical role in deciding how to map the applications. Our results (presented in Section 6) show that when dealing with high speed network processing it is best to take into account the processor affinity, so that the network process can run on a core which shares a low level cache with the core that handles network interrupts. If this is not possible (due to an overload of the optimal core, for example) we can instead choose the core that shares a higher level cache - this may lead to suboptimal performance, but still provide much better network throughput than irqbalance does.

In order to achieve the desired mapping, we need to know the cache configuration of the multi-core processor. In LINUX systems, this information can be obtained from the *shared_cpu_map* file under the sysfs file system. CAAD reads this file during the initialization phase and stores the layout information in an internal data structure. CAAD also needs

42

```
sNumOfNICQs ← procNICQs ()
sCPUTopology ← procCPUTopology ()
sIntAff ← procInterruptAffinity() ()
sRSS ← ethtoolGetRssHashFields ()
mappingHash ← init()
sSock ← openSockConn ()
while True do
    cSock ← accept ()
    cSysInfo ← recvMsg ()
    /* cNumOfNICQs,cCPUTopology,cIntAff,cRSS,
       cIPAdd,sIPAdd,sPort all of the above are
       parsed from cSysInfo                    */
    if clientIPAdd exists in mappingHash then
        mlCore ← kstatCpu ()
        mappings ← mappingHash(clientIPAdd)
        mappings ← getMapping(mappings, mlCore)
    end
    else
        setOfMappings ← computeMapping ()
    end
end
```

Algorithm 1: *Main loop of CAAD. Gathers system information from the local machine. Accepts incoming connections and parses through the information sent by the client. If the client IP exists in the mappingHash table, it returns the (ports,core) mapping. If the client IP does not exist in the table, new mappings must be computed as shown in Algorithm 2.*

```
listOfPorts ← generateRandomPorts ()
mlCore ← kstatCpu ()
for port in listOfPorts do
    sRSSVal ← Hash(sRSS, port, sIPAdd, cIPAdd)
    sNICQueue ← sRSSVal mod sNumOfNICQs
    mappings ← mappingHash(clientIPAdd)
    mappings ← addMap(port, sIntAff[sNICQueue])
    if sIntAff[sNICQueue] == mlCore then
        cRSSVal ← Hash(cRSS, port, sIPAdd, cIPAdd)
        cNICQueue ← cRSSVal mod
        cNumOfNICQs
    end
end
appCore ← getAppCore(mlCore)
```

Algorithm 2: *Compute Mapping. New mappings are determined based on the knowledge of the RSS hash function and the interrupt affinity.*

```
/* sCPUTopology is a hash table with core
   numbers as keys and an ordered set of cores
   as values, which are stored in the order:
   hyperthreaded (or logical core sharing L1
   cache), cores sharing L2 cache and cores
   sharing L3 cache. To find the best core for
   application processing, we lookup this hash
   table using mlCore as the key. mlCore is
   the core on which the protocol processing
   will be done.                              */
```

Algorithm 3: *Selecting the best core for application processing based on the cache configuration of the receiver and sender machines.*

to know the interrupt affinity of network interfaces in order to direct the flow to the right core. For CAAD to work correctly, we must disable the ability of irqbalance to handle network interrupts, so that their affinity doesn't change dynamically. In LINUX, the interrupt affinity information can be obtained from the *smp_affinity* file of each device in the proc file system. Using the *rx-flow-hash* option in *ethtool*, we can obtain the hash options for a given network traffic type.

5. CAAD DETAILS

As stated previously, CAAD first obtains the system information, and then waits for incoming connections. A client wanting to transfer data to an application running on the receiving end-system will first connect to CAAD and exchange the client machine's system information. Upon receiving this information, CAAD performs a lookup in *mappingHash*, which is a hash table storing the clientIPs as keys, and a list of (ports,cores) combinations as values.

If the client exists in the hash table, CAAD finds the minimum loaded core (*mlCore*) using the *kstat_cpu* macro function. Once the core to use is identified, the incoming flow must be mapped to this core only. The mapping of flow to core cannot be controlled directly, because the hash function in Intel NICs is computed in the hardware and cannot be changed. However, we can accomplish the desired goal by exploiting an understanding of how Intel does the hashing.

The NIC hash function operates on a four tuple which consists of the source IP address, destination IP address, source port, and destination port. At the time of the connection, the source and destination IP addresses are known. Keeping these attributes fixed, the NIC's hash function can be precomputed in software to determine combinations of

source and destination ports such that the incoming flows map to specific cores. The hash computation is done only at the time of the connection, so it does not lie on the critical path. We verified that our hash computation is correct using the Receive Side Scaling (RSS) verification suite given in Intel's Developer Manual [2].

For a given client request, CAAD precomputes the RSS hash (*Hash function shown in Algorithm 2*) for multiple ports to ascertain the corresponding NIC queues. Since irqbalance is turned off for network interrupts, NIC queues are statically mapped to cores. (*sIntAff* is a hash table with NIC queues as keys and the corresponding cores as values, and a simple lookup in this hash table can be performed to find the cores corresponding to the NIC queues that we want to use.

CAAD stores in hash table *mappingHash* the combination of sender port, destination port, and the corresponding core. Having found the *mlCore*, it accesses *mappingHash* using the client IP address as the key to obtain the list of (ports, core) combinations that we had precomputed. Among the combinations, the one which contains *mlCore* gives us the ports that have to be used by the sender and receiver applications. The next step is to find the best core for the application processing to take place (*Algorithm 3*).

We now present an example using actual values, to help make this process clear. When a client application connects to CAAD, it provides all its system information. CAAD uses the information to figure out multiple (ports, cores) combinations for future use as well. For a client IP *192.168.0.1* for instance, the corresponding values will be stored in *map-*

pingHash as a list: ((5001,5001,1), (5002,5002,3) ...), which is based on the RSS hashing function. Thus, we know that if the client application uses port 5001 and the destination application uses port 5001, then the incoming flow will go to core 1, and so on.

The next step is to find out the least loaded core - for purposes of illustration, we will say that is core 3. CAAD accesses *mappingHash* using the clientIP *192.168.0.1* as the key to find that the port numbers to be used are 5002 for the client application and 5002 for the destination application. For future requests from this particular client, the RSS hash need not be performed, since we have what we require in *mappingHash*.

Considering modern CPU topologies, there are a number of choices where the user-space process may run with respect to the network stack:

Same core: The user-space process and the network stack process run on the same core.

Hyperthread: The user-space process and the network stack process run on peer hyperthreads of the same core.

Peer core: Different cores that share the lowest level cache (LLC).

Different chip: Cores that belong to different chips.

As far as cache performance is concerned, the first option is ideal, but if there is a high incoming data rate, the cost of frequent context switching will be excessive. Since the network stack process runs at a higher priority than the user-space process, the user-space process will not have enough CPU cycles to read the packets from its socket buffer, resulting in packet loss.

If hyperthreads are available, and the CPU can fetch instructions from multiple threads simultaneously, then this option may be preferable. Hyperthreads can further reduce cache misses because in the case of packet processing they will be accessing shared data, i.e. packets passed between the kernel process and the user process.

Using Peer cores can also lead to reduced cache misses for the same reasons given above, but the last option should be avoided since it will lead to additional cache misses instead of reducing them. The logic necessary to decide which approach to use is described in *Algorithm 3*.

To take into account bi-directional flows, CAAD computes the RSS hash function provided by the client machine, but this time all the parameters are already fixed. The receiver and sender IP are already known, and CAAD determines a set of ports for the sender and receiver threads in the previous step. Computing the RSS hash identifies the NIC queue for an incoming flow on the sender side. Once again, the NIC queues are statically mapped to cores (*cIntAff*), so we can easily determine the core that will be receiving (softirq processing) the incoming flow. Using the same logic as above, we can identify the core on which application processing should be done.

We did not explore dynamically changing the assignment of the flow to the core or the application to the core. This would correspond to solving a dynamic on-line scheduling problem. Instead, we focused on a greedy approach, mapping the flow to the least loaded core. Since we are hard-affinitizing the tasks to cores, the LINUX scheduler will automatically balance the load by moving other processes to the remaining least loaded cores.

As mentioned earlier, CAAD is not an application transparent technique, since applications must communicate with the daemon which is running on the receiver machine. However, the necessary modifications are not complicated, and using a control channel to establish the parameters of the data transfer channel has been used in other client-server applications. One of the most common is the original FTP application, which uses a control channel between the client and server to exchange applications commands and indicate client-side port numbers to describe where the server should connect for the data transfer [23].

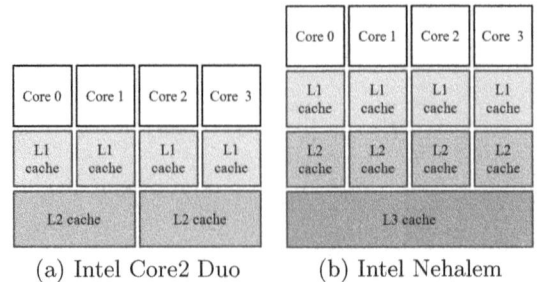

(a) Intel Core2 Duo   (b) Intel Nehalem

Figure 2: System/Cache Layout

6. RESULTS

In order to show the effectiveness of CAAD, we modified netperf, a widely used network performance benchmark which uses the Berkeley Sockets interface to measure different aspects of networking performance. The tests performed using netperf focus mainly on bulk data transfer using TCP and UDP.

The socket buffer for both the sender and the receiver was set to 8 MB (the maximum possible). We ran the tests with up to 16 parallel flows, but the throughput does not change beyond 4 flows, so for clarity of presentation we only show results for up to 8 flows. Since RFS and RPS did not perform any better than irqbalance, we have instead compared CAAD's performance to irqbalance and other static flow-to-core mappings. We performed ten 30 second runs, and the throughput for each experimental run was recorded and an average calculated.

For both TCP and UDP, our results demonstrate that CAAD is able to provide significantly better throughput than any of the other techniques. This is because using CAAD we are able to compute the optimal processor affinity based on the cache configuration of the receiving machine. We can see in Figure 3c and Figure 3d that the performance difference is due to a reduction in the cache misses when using CAAD. In the *same core* case, both the application and the protocol processing is performed on the same core, which means that the application will be context switched frequently (since protocol processing occurs at a strictly higher priority). Although the two processes share the L1 cache, the incoming data rate is very high, so there are many cache evictions. In the case of *different cpu*, the cache miss rate is high because the two processes do *not* share any caches.

Cache misses by themselves do not tell the complete story, however. The penalty paid when incurring a cache miss will also have a substantial impact on performance. Therefore, we used oprofile [16] to measure the CPU Stall cycles, which we can see from Figure 4 correlate well with the cache miss

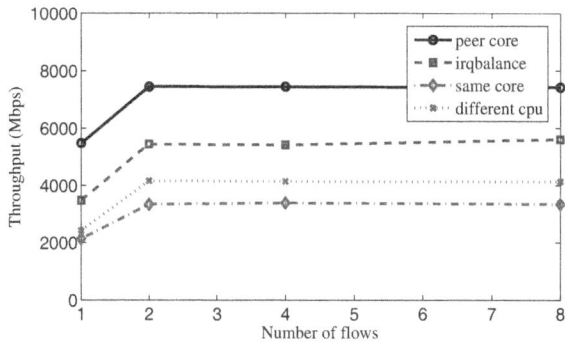

(a) UDP throughput using netperf

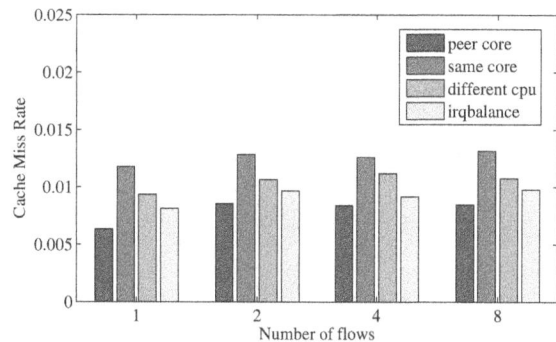

(b) Cache Miss Rates for UDP

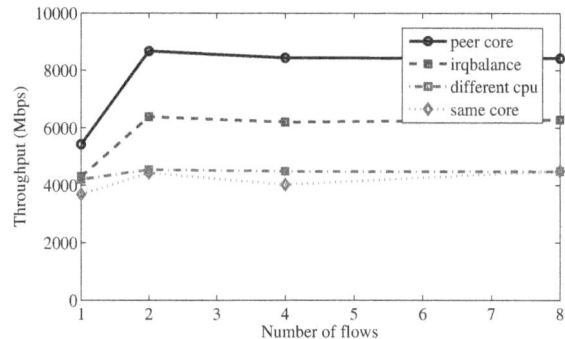

(a) TCP throughput using netperf

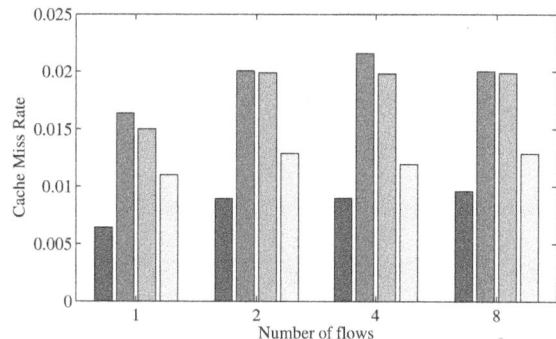

(b) Cache Miss Rates for TCP

Figure 3: Results using netperf with different affinity mappings. Receiver machine is an Intel Q660(Core2 Duo) with 2 cores in 2 cpus. In the figures, *peer core* refers to the core chosen by our tool, *same core* means that application and softirq processing is done on the same core, *different cpu* means that the application and softirq processing are done on cores on separate CPU dies, and *irqbalance* means irqbalance is running.

rate. As the cache miss rate goes up, so does the number of stall cycles.

For all the cases, the best throughput is achieved using 2 flows, since the total number of cores is 4. Referring to Figure 2, CAAD would assign 2 flows in the following manner: one flow would be mapped to *core 0* and the corresponding application would run on *core 1*, while the second flow would be mapped to *core 2* and the application would run on *core 3*. This leads to the best possible usage of the available resources. In the case of four flows, the same mapping would be used, so there would be two flows on each of the even numbered cores, and two applications each on the odd numbered ones. While this does lead to contention for resources, the throughput does not drop, because the overall rate that a core is processing remains the same. In essence each core is doing twice as much work, but the incoming flows are arriving half as fast. This holds for the case where there are 8 and 16 flows as well.

Figure 5 shows the results when the Intel Nehalem machine was used as the receiver. We performed the same experiments using netperf (TCP and UDP). This machine supports hyperthreads, so the operating system manages a total of eight logical cores[2]. Since hyperthreads are enabled, the affinity daemon arranges for the application and protocol processing to happen on logical cores which share all the cache levels. As pointed out in Section 5, if hyperthreads are available and the CPU can fetch instructions from mul-

tiple threads simultaneously, then this option is ideal. In the case of network data, hyperthreads can further reduce cache misses because the application and protocol processing will be accessing shared data and not splitting the cache.

We can see that using CAAD we can achieve better throughput than the *same core* and *irqbalance* approaches. Since this is a single chip machine, we could not test the *different cpu* case. The L2 cache miss rates shown in Figure 5b and Figure 5d are lower than what we see in Figure 3 because all the cores share an L3 cache.

6.1 Bulk Data Transfers

Private optical networks (Lambda Networks) are used by geographically distributed data centers to transmit and receive large amounts of data. On a single optical fiber strand, such networks provide multiple independent channels which operate over different wavelengths. These networks are for dedicated use, and suffer from very little congestion [18]. As a result, they are able to provide the high bandwidths required by data center applications. Currently an independent channel can support up to 40 Gbps, while the standard for 100 Gbps has been approved.

We decided to evaluate the effectiveness of our approach for bulk data transfers using our 10 Gbps testbed. Our target applications were GridFTP[5], a data transfer protocol commonly used in computational grids which is an extension of FTP and comes with many optimizations (parallel flows, pipelining, etc.), and Flow Bifurcation Manager

---

[2]LINUX treats each hyperthread as a separate core.

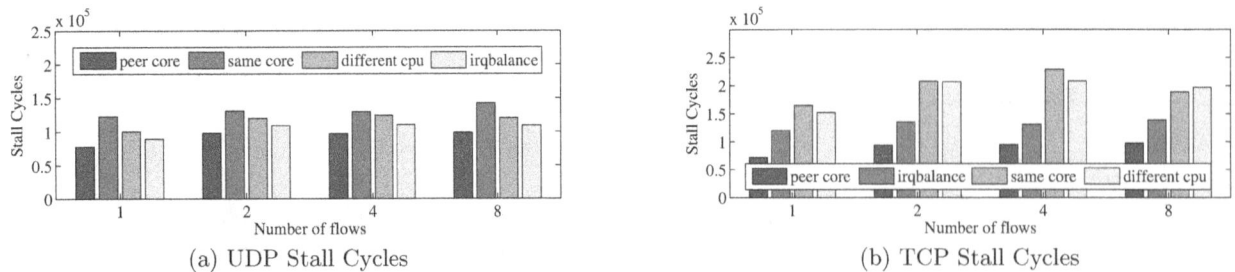

(a) UDP Stall Cycles

(b) TCP Stall Cycles

Figure 4: Stall Cycles due to cache misses. Receiver: Intel Core2 Duo.

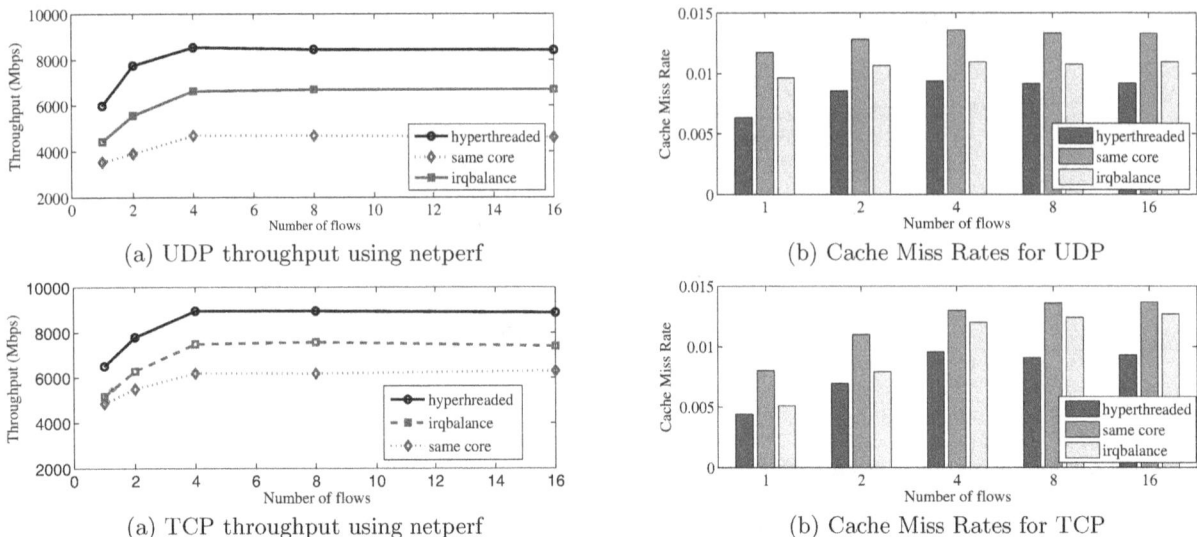

(a) UDP throughput using netperf

(b) Cache Miss Rates for UDP

(a) TCP throughput using netperf

(b) Cache Miss Rates for TCP

Figure 5: Results using netperf with different affinity mappings. Receiver machine is an Intel Nehalem Quad core. In the figures, *hyperthreaded* refers to the hyperthreaded core chosen by our tool, *same core* means that application and softirq processing is done on the same core, and *irqbalance* means irqbalance is running.

(FBM)[4], a rate-based technique which has been integrated into RBUDP and is used to determine the optimal number of parallel flows based on the state of the receiving end-system. The GridFTP protocol implementation includes the Globus-FTP-Control Library and the Globus-FTP-Client Library. We had to add around 100 lines of code to the control library to communicate with the affinity daemon before the client could actually begin the data transfer. Modifying FBM to communicate with the affinity daemon was even easier, requiring just an additional 30 lines of code.

The download time for each file transfer was recorded and all the reported values were averaged over ten runs. Figure 6 shows the performance improvement for both the architectures while transferring a 2 GB file. We can see that the download time for *gridftp-affinity* and *fbm-affinity* is approximate equal, and lower than when not using CAAD.

6.2 Bidirectional TCP flows

The network physical medium works in full duplex, so a 10Gbps connection means 10Gbps one way - it can, in fact, support 20Gbps at the same time when bi-directional flows are taken into consideration. Generally, TCP flows are bidirectional. Popular applications like ftp, ssh, and scp all use bidirectional TCP flows.

Techniques like RFS and the NIC design mentioned in [24] focus only on the receiving end, while CAAD uses the client

information that is exchanged in the beginning to determine the core binding for the application which initiates the connection as well. We used netperf to demonstrate the effectiveness of this approach. Although netperf does not fully support bidirectional data transfers, it provides a test called **TCP_MAERTS** to transmit data in the opposite direction.

In the experiments described in Section 6, the sender application is not receiving any data. Figure 7 shows the performance improvement that we can get when the sender application is also receiving data at a high speed, and is using the affinity mask suggested by CAAD. Using a single bi-directional flow, we achieve 11.1 Gbps as compared to 8.5 Gbps when CAAD is not used. The performance improvement is similar when using two bi-directional flows.

6.3 Single-core Network mismatch

Figure 8 shows that even when using advanced load balancing and flow steering techniques, a commodity multicore machine can experience something called *receive livelock*. Receive livelock occurs when the CPU spends all its time processing hard and soft interrupt requests, leading to starvation of user applications (which run at a lower priority). To avoid this problem, upon detecting receive livelock the kernel starts a ksoftirqd kernel thread (which runs at a user level priority) to do the packet processing, instead of the softirq context. However, this also means it cannot keep up

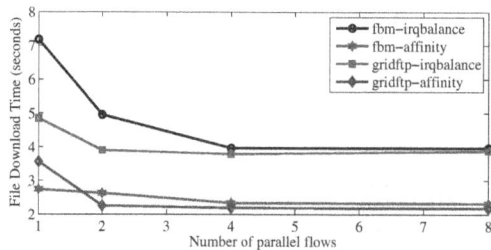

(a) Intel Core2 Duo receiver

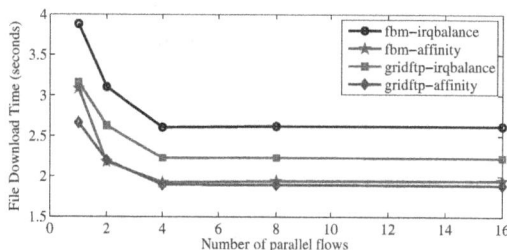

(b) Intel Nehalem receiver

Figure 6: Bulk File Transfer Using Parallel Flows and a file size of 2 GB. *gridftp_affinity* and *fbm_affinity* are gridFTP and FBM (respectively) used in conjunction with CAAD, while *gridftp_irqbalance* and *fbm_irqbalance* are gridFTP and FBM (respectively) used with irqbalance turned on. **Note the two figures have different vertical axis values.**

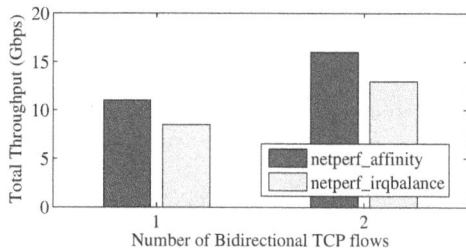

Figure 7: Bidirectional TCP flows: netperf_affinity shows bidirectional performance when CCAD was used along with netperf.

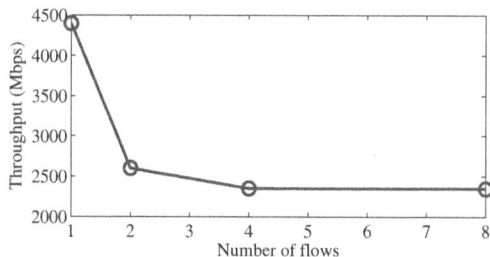

Figure 8: Receive Livelock can occur while using RFS

with the high incoming data rate, leading to packet losses within the receiving end system (socket buffer losses andor ring buffer losses).

In modern operating systems (e.g., LINUX), an instance of the network stack runs on each of the cores in parallel, and the kernel ensures that a single NIC is driven only by a single core (in order to avoid packet reordering.) Since each of the cores execute concurrently and receive interrupts arbitrarily, the order in which they process packets belonging to the same flow cannot be fixed. In the same vein, a multi-queue NIC ensures that packets belonging to the same flow are mapped to the same queue, and is viewed by the operating system as if it were multiple NICs (one per queue). The operating system avoids packet reordering by ensuring that the interrupts issued by a single NIC queue are processed only on a single core.

Multiple flows could also map to the same queue, because the NIC classification function takes into account only the

four tuple. For example, in our experiments, keeping the other three parameters fixed, if we send two flows using sender ports 5001 and 5009, both of these would go to the same NIC queue, and thus the same core. Figure 8 shows that the throughput degrades when multiple flows are to be processed by a single core even when RFS is enabled. This can happen for two reasons: 1) The protocol processing and application processing happen on the same core. If the protocol processing load is high, the applications become starved, or 2) If there are parallel flows, all of them may map to a single core, since RFS does not control the entire path. All RFS does is wake up the application on the core that does the protocol processing. This problem will not arise when using CAAD, because we choose the port numbers in a way that ensures the incoming flows are mapped to the least loaded core, and we also ensure that the application is processed on a separate core which shares a cache.

While a single core running at 3.5 GHz may be able to handle multiple flows, it would certainly be overwhelmed if there were multiple 10 Gbps flows. For the experiments shown in Figure 8 we modeled this by scaling down the core frequency from 2.4 to 1.6 GHz, to create a larger mismatch between core and network speed.

7. CONCLUSION

In this paper we have presented the Cache-Aware Affinity Deamon (CAAD), a tool which enables the application processing and protocol processing to occur on cores which share the lowest cache level. Doing so reduces the number of cache misses, which leads to a reduction in stall cycles and an improvement in performance. CAAD takes into account the on-chip system topology (the number of cores and which cores share a cache) and NIC characteristics (number of queues, the flow hash function, and hash function type) to determine a set of port numbers to convey to the sender, so that the entire end-to-end path for each new connection can be managed and controlled. This is done in a light-weight manner for both uni- and bi-directional flows.

Experimental results show that for bulk data transfers using commodity multicore machines, the use of CAAD provides an overall improvement in throughput of as much as 26% for a single TCP flow and 31% for multiple TCP flows. It also increases throughput by 45% for a single UDP flow and 28% for multiple UDP flows. In addition, we show that using CAAD can improve the performance of two different bulk data transfer protocols, GridFTP[5] and Flow Bifurcation Manager[4]. GridFTP combined with CAAD improves the download time for big file transfers by up to 18%.

Our work differs from previous studies in that we are proposing a general purpose mechanism to control the entire end-to-end path of flows. For future work, we plan to develop this into an API, so that no modifications are required for client-server applications.

8. ACKNOWLEDGMENTS

1. I am grateful to Tudor Marian(Google) for his invaluable comments, and spending a lot of time to help me understand different aspects of LINUX Networking.

2. This proposal was funded by a NSF grant CSR-0917315.

9. REFERENCES

[1] irqbalance. http://www.irqbalance.org/.

[2] Rss verification. http://www.intel.com/content/www/us/en/ethernet-controllers/82598-10-gbe-controller-datasheet.html.

[3] Microsoft corporation. scalable networking with rss, 2005.

[4] AHUJA, V., GHOSAL, D., AND FARRENS, M. Minimizing the data transfer time using multicore end-system aware flow bifurcation. In CCGrid, 2012.12th IEEE ACM International Symposium on Cluster, Cloud and Grid Computing (2012), IEEE.

[5] ALLCOCK, W., BRESNAHAN, J., KETTIMUTHU, R., LINK, M., DUMITRESCU, C., RAICU, I., AND FOSTER, I. The globus striped gridftp framework and server. In Proceedings of the 2005 ACM/IEEE conference on Supercomputing (2005), IEEE Computer Society, p. 54.

[6] BEN-YEHUDA, M., GOLDSHMIDT, O., KOLODNER, E. K., MACHULSKY, Z., MAKHERVAKS, V., SATRAN, J., SEGAL, M., SHALEV, L., AND SHIMONY, I. Ip only server. In USENIX Annual Technical Conference, General Track (2006), pp. 381–386.

[7] FOONG, A., FUNG, J., AND NEWELL, D. An in-depth analysis of the impact of processor affinity on network performance. In Networks, 2004.(ICON 2004). Proceedings. 12th IEEE International Conference on (2004), vol. 1, IEEE, pp. 244–250.

[8] FOONG, A., FUNG, J., NEWELL, D., ABRAHAM, S., IRELAN, P., AND LOPEZ-ESTRADA, A. Architectural characterization of processor affinity in network processing. In Performance Analysis of Systems and Software, 2005. ISPASS 2005. IEEE International Symposium on (2005), IEEE, pp. 207–218.

[9] FULLER, S., AND MILLETT, L. Computing performance: Game over or next level? Computer 44, 1 (2011), 31–38.

[10] HERBERT, T. rfs: receive flow steering, september 2010. http://lwn.net/Articles/381955/.

[11] HERBERT, T. rps: receive packet steering, september 2010. http://lwn.net/Articles/361440/.

[12] HUGGAHALLI, R., IYER, R., AND TETRICK, S. Direct cache access for high bandwidth network i/o. In ACM SIGARCH Computer Architecture News (2005), vol. 33, IEEE Computer Society, pp. 50–59.

[13] JANG, H., AND JIN, H. Miami: Multi-core aware processor affinity for tcp/ip over multiple network interfaces. In High Performance Interconnects, 2009. HOTI 2009. 17th IEEE Symposium on (2009), IEEE, pp. 73–82.

[14] JONES, R., ET AL. Netperf: a network performance benchmark. Information Networks Division, Hewlett-Packard Company (1996).

[15] KUMAR, A., HUGGAHALLI, R., AND MAKINENI, S. Characterization of direct cache access on multi-core systems and 10gbe. In High Performance Computer Architecture, 2009. HPCA 2009. IEEE 15th International Symposium on (2009), Ieee, pp. 341–352.

[16] LEVON, J., AND ELIE, P. Oprofile: A system profiler for linux. http://oprofile.sf.net, 2004.

[17] MARIAN, T. Operating systems abstractions for software packet processing in datacenters. PhD thesis, Cornell University, 2011.

[18] MARIAN, T., FREEDMAN, D., BIRMAN, K., AND WEATHERSPOON, H. Empirical characterization of uncongested optical lambda networks and 10gbe commodity endpoints. In Dependable Systems and Networks (DSN), 2010 IEEE/IFIP International Conference on (2010), IEEE, pp. 575–584.

[19] NARAYANASWAMY, G., BALAJI, P., AND FENG, W. Impact of network sharing in multi-core architectures. In Computer Communications and Networks, 2008. ICCCN'08. Proceedings of 17th International Conference on (2008), IEEE, pp. 1–6.

[20] PANDE, A., AND ZAMBRENO, J. Efficient translation of algorithmic kernels on large-scale multi-cores. In Intl. Work. Reconfigurable and Multicore Embedded Systems (WoRMES), IEEE Intl. Conf. Computational Science and Engineering (2009), IEEE Computer Society, pp. 915–920.

[21] PESTEREV, A., STRAUSS, J., ZELDOVICH, N., AND MORRIS, R. Improving network connection locality on multicore systems. In Proceedings of the EuroSys 2012 Conference, EuroSys 2012 (2012), EuroSys.

[22] SCOGLAND, T., BALAJI, P., FENG, W., AND NARAYANASWAMY, G. Asymmetric interactions in symmetric multi-core systems: analysis, enhancements and evaluation. In High Performance Computing, Networking, Storage and Analysis, 2008. SC 2008. International Conference for (2008), IEEE, pp. 1–12.

[23] STEVENS, W. TCP/IP Illustrated: the protocols, vol. 1. Addison-Wesley Professional, 1994.

[24] WU, W., DEMAR, P., AND CRAWFORD, M. A transport-friendly nic for multicore/multiprocessor systems. Parallel and Distributed Systems, IEEE Transactions on, 99 (2011), 1–1.

xOMB: Extensible Open Middleboxes
with Commodity Servers

James W. Anderson, Ryan Braud, Rishi Kapoor, George Porter, and Amin Vahdat

University of California, San Diego

{jwanderson,rbraud,rkapoor,gmporter,vahdat}@cs.ucsd.edu

ABSTRACT

This paper presents the design and implementation of an incrementally scalable architecture for middleboxes based on commodity servers and operating systems. xOMB, the eXtensible Open MiddleBox, employs general programmable network processing pipelines, with user-defined C++ modules responsible for parsing, transforming, and forwarding network flows. We implement three processing pipelines in xOMB, demonstrating good performance for load balancing, protocol acceleration, and application integration. In particular, our xOMB load balancing switch is able to match or outperform a commercial programmable switch and popular open-source reverse proxy while still providing a more flexible programming model.

Categories and Subject Descriptors

C.2.6 [**Computer-Communication Networks**]: Internetworking; D.2.11 [**Software Engineering**]: Software Architectures—*Patterns*

Keywords

middlebox, application-layer switch, network processing pipeline

1. INTRODUCTION

Network appliances and middleboxes performing forwarding, filtering, and transformation based on traffic contents have proliferated in the Internet architecture. Examples include load balancing switches [6–8] and reverse proxies [3, 10, 16–18], firewalls [15, 31], and protocol accelerators [13, 16, 18]. These middleboxes typically accept connections from potentially tens of thousands of clients, read messages from the connections, perform processing based on the message contents, and then forward the (potentially modified) messages to destination servers.

Middleboxes form the basis for scale-out architectures in modern data centers, being used in three main roles. First,

they perform *static load balancing* and possibly *filtering*, whereby they distribute (or drop) messages to server pools based on a fixed configuration. For example, a load balancing switch might forward requests to different server pools based on URL. Second, middleboxes perform *dynamic request routing and application integration*, where they execute service logic and use dynamic service state to forward requests to specific application servers and often compose replies from many application servers for the response. For example, front-end servers use object ids to direct requests to the back-end servers storing the objects. Third, middleboxes perform *protocol acceleration* by caching/compressing data and responding to requests directly from their caches. For example, services deployed across multiple data centers use middleboxes to cache content from remote data centers.

Unfortunately, the architecture of commercial hardware middleboxes consists of a mixture of custom ASICs, embedded processors, and software with, at best, limited extensibility. While firewalls, NATs, load balancing switches, VPN gateways, protocol accelerators, and other middleboxes perform logically similar functionality, they are individually designed by niche providers with non-uniform programming models. These boxes often command a significant price premium because of the need for custom hardware and software and their limited production volumes. Extending functionality to new protocols may require new custom hardware. Worse, expanding the processing or bandwidth capacity of a given middlebox may require replacing it with a higher-end model. Similarly, software reverse proxy middleboxes are specialized for specific protocols and, like their hardware counterparts, offer limited scalability and extensibility.

At first blush, software routers such as Click [28] or Route Bricks [21] may be employed to achieve extensible middlebox functionality. In fact, recent work [33] demonstrates the feasibility of such an approach with a middlebox architecture based on Click. However, these pioneering efforts are focused on per-packet processing, making them less applicable to the stream or flow-based processing common to the class of middleboxes we target in this work. For example, flow-based processing requires operating on a byte stream rather than individual packets and may require communication among multiple network elements to perform dynamic forwarding and rewriting. Further, they provide no specific support for managing and rewriting a large number of concurrent flows, instead focusing on high-speed pipeline processing of individual packets.

Hence, this paper presents xOMB (pronounced *zombie*), an eXtensible Open MiddleBox software architecture for building flexible, programmable, and incrementally scalable middleboxes based on commodity servers and operating systems. xOMB employs a general *programmable pipeline* for network processing, composed of xOMB-provided and user-defined C++ modules responsible for arbitrary parsing, transforming, and forwarding messages and streams. Modules can store state and dynamically choose different processing paths within a pipeline based on message content. A control plane automatically configures middleboxes and monitors both middleboxes and destination servers for fault tolerance and availability. xOMB provides a single, unified platform for implementing the various functions of static load balancing/filtering, dynamic request routing, and protocol acceleration.

Several additional xOMB features add power and ease of programming to the simple pipeline model. *Asynchronous processing modules* and *independent, per-connection processing* efficiently support network communication (e.g., to retrieve dynamic state) as part of message processing. *Arbitrary per-message metadata* allows modules to store and pass state associated with each message to other modules in the pipeline. Finally, xOMB automatically manages client and server connections, socket I/O, message data buffers, and message buffering, pairing, or reordering.

We implement and evaluate three sample pipelines: an HTTP load balancing switch (static load balancing), a front end to a distributed storage service based on Eucalyptus [29] implementing the S3 interface [1] (dynamic request routing), and an NFS [32] protocol accelerator. Forwarding through xOMB presents little overhead relative to direct access to back-end servers. More importantly, xOMB scales its network and processing performance with additional commodity servers and provides transparent support for dynamically growing the pool of available middleboxes. We also compare the performance of our xOMB load balancing switch against a commercially available hardware load balancing switch and the leading open source reverse proxy. xOMB matches or outperforms the commercial switch and open source proxy in most cases, while providing a more flexible and powerful programming model.

2. OVERVIEW

Middleboxes can be defined as network elements that process, forward, and potentially modify traffic on the path between a source and a destination. With this generic definition, routers and switches can also be classified as middleboxes. In this paper, we focus on an *active middle box* ("middlebox" for short), a network device that performs programmable traffic processing based on entire packet contents rather than just on headers. We identify three key features differentiating middleboxes from traditional routers and switches: middleboxes i) understand the application semantics of network data, ii) may modify or even completely replace the contents of that data, and iii) despite being "in the middle," middleboxes may terminate connections and initiate new connections.

Although the xOMB design is general to a variety of middlebox applications, we primarily focus on their use in data centers in this paper as this deployment scenario stresses all aspects of our design and presents a particularly challenging use case, with strict requirements for performance and reliability. We begin by examining *load balancing switches*

(LBSs), specialized middleboxes widely used to distribute requests among dynamic server pools in data centers.

2.1 Load Balancing Switches

Data center services rely on specialized hardware LBSs that serve as the access point for services at a particular data center, abstracting back-end service topology and server membership and enabling incremental scalability and fault tolerance of server pools. When a client initiates a request, rather than returning the IP address of an application server, the service instead returns the IP address of a LBS. Client packets must then pass through the LBS on the way to their ultimate destination.

LBSs may operate at the packet-header or packet-payload granularity. In the first case, they forward packets to back-end servers by transparently rewriting IP destination addresses in packet headers. Care must be taken to ensure that all packets belonging to the same flow are mapped to the same server. Many commodity switches provide basic hardware support for appropriate flow hashing to deliver such functionality. A straightforward design might employ OpenFlow [11] coupled with commercially available switches to map flows to back-end servers by monitoring group membership and load information. Hence, we focus on the more challenging (and commercially relevant) instance where LBSs must operate at the granularity of packet-payloads, performing arbitrary processing on application-level messages before forwarding them to appropriate back-end servers.

Performance optimizations, used either independently or in addition to protocol acceleration, are another example of LBS functionality. For example, an LBS will maintain persistent TCP connections to back-end webservers, re-writing HTTP 1.0 requests as 1.1 if necessary to use one of the existing connections. Switches may also implement connection collapsing, in which many incoming client TCP connections are multiplexed onto a small number of persistent TCP connections at the application servers. These functions provide several benefits. First, they eliminate the overhead of establishing new TCP connections and the delay for it to ramp up to the available bandwidth, reducing overall latency. Second, servers incur some degree of per-connection processing overhead, so if every client connection requires a connection from the middlebox to the back-end, then both the middleboxes and backends incur this cost. By "absorbing" the client connections through connection collapsing, the middlebox reduces back-end load and also reduces its own load by having fewer back-end connections to manage. Connection collapsing and request rewriting for persistent connections is one of the most popular uses for commercial LBSs [9,30].

While some commercial LBSs can process byte streams, this capability is typically limited to a small set of protocols (e.g., HTTP) with restricted programming models. Further, as running at line speeds often requires specialized hardware employing custom ASICs, it is typically impossible for LBSs to perform message forwarding or rewriting based on state maintained across connections or to initiate RPCs to look up non-local state for dynamic request routing, requiring large-scale service providers to employ proprietary software solutions. The goal of our work is to address these challenges with a scalable, easy-to-program architecture built entirely from commodity server components.

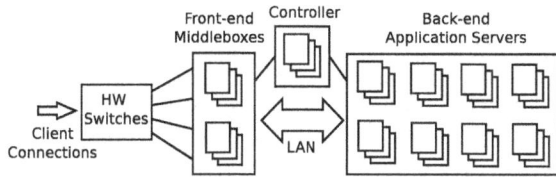

Figure 1: System Architecture

2.2 Architecture

Figure 1 shows the xOMB architecture. The major components include commodity hardware switches, our front-end software middleboxes, back-end application servers, and a controller for coordination. The software middleboxes communicate with each other, with the controller, and with *agent processes*—used for collecting statistics such as machine load on each back-end server—using an RPC framework.

Hardware switches. While our software middleboxes can perform application-level packet inspection to aid forwarding decisions, we can optionally leverage existing commodity hardware switches to act as the single point of contact for client requests. These commodity switches employ line-rate hashing to map flows to an array of our programmable software middleboxes [35].

Software middleboxes. Commodity servers with software middleboxes function as the front-end switches for xOMB. They parse, process, and forward streams of requests and responses between clients and the back-end application servers. xOMB flexibly supports arbitrary protocol and application logic through user-defined processing modules. Deployments can scale processing capacity by "stacking" xOMB servers either vertically (every server runs the same modules) or horizontally (servers run different modules and form a processing chain). Additionally, the middleboxes provide distributed failure detection and load monitoring.

Application servers. Application servers (e.g., webservers) form the back end of our system and process and respond to client requests according to the protocol(s) they are serving. These servers may be grouped into logical *pools* with related resources or functionality.

Controller. The controller provides a central rendezvous point for managing front-end and back-end configuration and membership. In addition to coordinating and storing server membership, the controller also stores a limited amount of service hard state. The controller may be implemented as a replicated state machine for fault tolerance.

3. DESIGN

The middleboxes form the core of the xOMB framework, with the primary goals of flexibility, programmability, and performance. They provide complete control over data processing, allowing them to work with any protocol, including proprietary, institution-specific ones. Our high-level approach to arbitrary byte stream processing is to terminate client TCP connections at the middlebox, execute the appropriate *modular processing pipeline* (pipeline) containing user-defined processing logic on an incoming byte stream, and then transmit the resulting byte stream over a new TCP connection to the appropriate back-end server. We leave extensions to message-oriented protocols such as UDP to

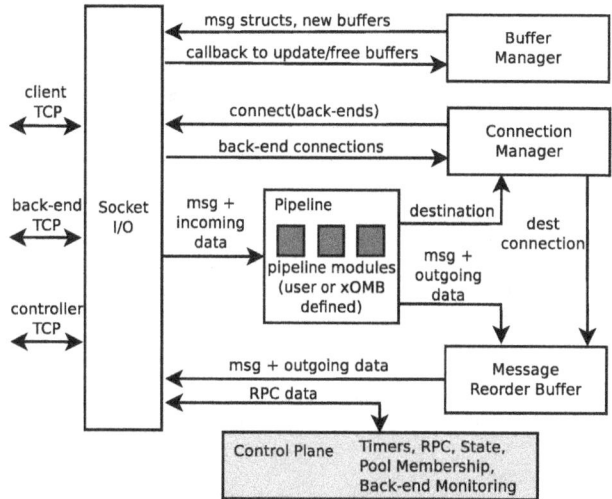

Figure 2: Anatomy of a xOMB Server

future work. A separate control plane configures the middleboxes, manages membership, performs monitoring, and schedules and executes timers.

xOMB automatically handles low-level functionality necessary for high-performance processing, allowing the programmer to concentrate on the application logic. xOMB abstracts connection management, socket I/O, and data buffering. Additionally, xOMB targets *request-oriented* protocols, which comprise most of today's Internet services. In these protocols, request *messages* have exactly one logical response message, and responses should be returned in the order requested. xOMB tracks requests and buffers and reorders responses to meet this requirement. Although our design allows arbitrary data processing, our aim is to make handling common processing patterns and standard protocols as simple as possible (see §4.3.2).

Figure 2 shows an overview of the anatomy of a xOMB server. The user simply defines the processing modules (the dark gray boxes) that plug into the xOMB pipeline framework. In this section, we discuss the design of the xOMB core architecture, and we delve into the key implementation details in §4.

3.1 Pipelines and Modules

xOMB divides data stream processing into three logical stages: protocol parsing, filtering/transformation, and forwarding. Each stage is composed of an arbitrary collection of *modules* that dynamically determine the processing path of a message. As each module can process different messages independently and concurrently, we refer to the complete DAG of modules and any additional control-flow logic as the pipeline. Because requests and responses require different handling, middleboxes use separate pipelines for each direction.

3.1.1 API

Each pipeline module represents a single processing task, and modules may be composed of other modules. Modules may be either synchronous or asynchronous; asynchronous modules allow processing tasks to retrieve state over the network without blocking. A simple API consists of methods for initializing state, receiving failure/membership notifica-

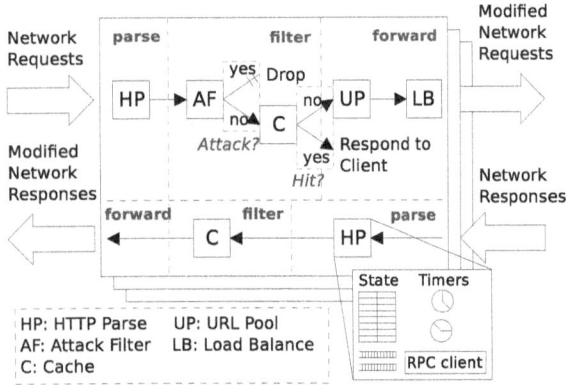

Figure 3: Example Pipeline

tions, and a `process()` method to execute the module's task. Middlebox programmers must only define a set of modules and the pipeline to link them together.

When a middlebox reads a chunk of data from a socket into a message buffer, it passes the buffer to the pipeline, which successively invokes module `process()` methods until either a module halts processing or every module has been executed. Modules may store soft state in memory, schedule timers, and, for asynchronous modules, make RPCs to retrieve or store shared persistent state. Modules pass processing results to other modules, or between requests and responses, through arbitrary *metadata* pointers associated with each message. For example, a parser module may set a data structure representing the parsed message as a metadata value.

`xOMB` assigns every connection its own pipeline object to increase throughput and simplify programming. The `xOMB` concurrency model uses *strands* [5] based on each connection, meaning that at most one thread will execute pipeline or I/O instructions for a given connection at a time. The advantage of this model over explicit locking is that it elegantly allows multiple cores to execute processing or I/O for different connections in parallel. Moreover, if one connection's pipeline makes an RPC (which is asynchronous), control will immediately transfer to another connection's pipeline until receiving the reply, further increasing throughput and utilization. Because many modules need to track per-message or per-connection state (e.g., the number of bytes parsed), per-connection pipelines have the additional advantage of simplifying programming by automatically giving each module private state. `xOMB` also provides support for modules storing state across connections (§3.1.3).

3.1.2 Example Pipeline: HTTP

Figure 3 shows an example request and response pipeline for processing HTTP traffic, broken into parse, filter, and forward stages.

Protocol Parsing. The protocol parsing stage transforms the raw byte stream into discrete application-defined messages and sets application-specific metadata in the message structure. On a parse error, the module sets a message flag to close the connection. The parser indicates when it has parsed a complete message along with the total bytes parsed. The middlebox reads these fields to create a new buffer pointing to any remaining bytes for the next message.

Parsers can be chained together to simplify the logic for different subsets of a complicated protocol. For example, parsing XML-RPC might use an HTTP parser followed by an XML parser. Our sample HTTP parser sets a metadata structure representing an HTTP request, including the protocol version, request method, path, headers, etc.

Filtering and Transformation. Filter modules perform arbitrary transformations on messages. Our example request pipeline has three filters illustrating common uses. First, an *AttackDetector* checks the request against a set of attack signatures using string matching expressions loaded from Snort [15] rules. If the message matches a rule, the filter sets the "drop connection" flag.

Next, the pipeline uses a *Cache* module for protocol acceleration. In the request pipeline, the module checks whether a cache entry exists for the path set in the HTTP metadata. If so, the module sets the message buffer pointers to the cached response and sets the message destination to the client. In the response pipeline, the cache module checks the response headers in the HTTP metadata and stores the response if permitted.

The final filtering step (omitted in Figure 3) performs HTTP version rewriting to allow middleboxes to maintain persistent connections to the back-end webservers, even when clients do not support them. If the HTTP metadata version is 1.0, the request pipeline module rewrites the headers to version 1.1 and adds the appropriate "Host" field. The response pipeline also uses a version filter that rewrites the response back to HTTP/1.0 for requests that were transformed. Similarly, because cached responses will always be version 1.1, the request pipeline sends these through a response version filter as well.

Forwarding. The forwarding stage sets the message destination based on metadata set by the previous stages. Forwarding can be as simple as selecting a back-end server from a pool to as complicated as computing the destination from a dynamically populated forwarding table. Response pipelines do not have forwarding modules; the middlebox automatically sends responses to the requesting client.

Our example request pipeline uses two forwarding modules. The *URLPool* module partitions back-end server pools based on the paths from the URL that they may serve. This module periodically reads configuration state from the controller that maps URL path prefixes to server pools. For example, paths beginning with "/image" go to one set of servers and paths beginning with "/video" go to another set. By using the HTTP metadata path, the module sets a metadata field with the pool for the longest prefix match. The *LoadBalance* module selects a destination server from the designated pool using a specified load balancing policy. We have implemented simple load balancers that use round-robin or random selection.

Figure 4 shows a sample module implementing random forwarding. The module sets the message destination to a random server in the specified pool. If the specified pool does not contain any servers, the module returns an error status that signals the pipeline logic to initiate alternate processing or close the connection.

3.1.3 Module State and Configuration

Integrating programmable middleboxes into complex distributed protocols requires that the middleboxes can access potentially large amounts of service state necessary for making forwarding decisions. We distinguish two kinds of

```
class RandomForwardModule : public Module {
 public:
  MessageStatus process(MessagePtr m) {
    // get membership from control plane
    MembershipSet members =
      Membership::getMembers(m->getPool());

    if (members.empty())
      // tell xomb to close connection
      return MessageStatus::Error;

    // set destination on message structure
    m->setDest(random(members)->addr());

    // tell xomb to start forwarding data
    return MessageStatus::Complete;

    // (Modules that need more data can
    //  return MessageStatus::ReadMore)
  }
};
```

Figure 4: A forwarding module implementing random forwarding to a pool of servers specified in the message metadata.

module state: configuration state and dynamic state. Configuration state specifies parameters such as rates, cache sizes, numbers of connections, etc., and any global state that changes infrequently. Examples of configuration state include the set of fingerprints used by the *AttackDetector* module or the path prefix to server pool mapping used by the *URLPool* module. xOMB uses the controller to manage all global configuration state, stored as a map from state name to opaque binary value. This map can be queried through an RPC interface by a middlebox when it starts, allowing modules to retrieve necessary configuration parameters. Optional metadata includes a version number and time duration, which tells modules how long they should use the current value before checking for a new version.

Dynamic state consists of any unique state that a module references or retrieves for each message. Consider an object store directory that maps billions of object ids to back-end servers. Modules typically cannot prefetch and store a complete copy of such forwarding state because the total state is too large, keeping a consistent copy would be too expensive, or both. xOMB modules may dynamically construct forwarding tables during message processing and may control the rate at which dynamic state is updated. Modules can retrieve required state on demand by making asynchronous RPCs to application services such as a back-end metadata server. The ability to build dynamic forwarding tables is a significant advantage afforded by the general programmability of xOMB middleboxes relative to less flexible callback-based models (§3.3).

Modules store global state—including both configuration and dynamic state—in memory that persists across connections. To allow modules to manage memory effectively, xOMB passes membership and failure notifications to every module so that they may discard unneeded state. Additionally, modules may set timers to perform periodic state maintenance to optimize storage or purge stale state.

3.2 Control Plane

Although pipelines and modules form the core of xOMB, a number of other components complete the system functionality and convenience.

3.2.1 Membership

xOMB middleboxes and back-end servers require the current server membership for various pools. As middleboxes and servers join and leave, updated membership must be disseminated efficiently. In addition to basic pool membership, xOMB must assign both back-end and middlebox servers to be monitored by one or more middleboxes. These assignments should remain balanced as middleboxes and servers are added and removed. Finally, we also require that there be no manual configuration for adding or removing servers—all membership pools and monitoring assignments must be updated automatically.

To achieve these requirements, the xOMB controller manages pool membership and monitors assignments. Although using a centralized controller may not scale to the largest systems, it is a simple solution and should be sufficient for thousands to tens-of-thousands of servers [22]. The controller may be implemented as a replicated state machine for high availability, or replaced with a coordination system such as [26].

When a new server comes online, its agent process makes a `join` RPC to the controller, registering itself as either a middlebox or an application server, and specifies the sets of pools to which it should be added and any pools for which it requires membership updates. The controller records this request and informs all servers who have registered interest in membership updates for that pool. When failure detectors inform the controller of a server failure, it similarly notifies all registrants.

3.2.2 Monitoring

Services must respond to failures and changing server loads. Front-ends such as xOMB implement this functionality as they direct requests to back-end servers: middleboxes can avoid sending requests to failed machines and shift traffic to less loaded servers. To effectively minimize service time for requests, the middleboxes need the current liveness and load status of all servers, generally including other middleboxes.

Each middlebox collects load information from a set of servers assigned by the controller at a configurable interval. The xOMB agent on each server reports machine-level information, such as load, CPU, network, and memory utilization, but application server agents may report more detailed application information, such as the number of active connections or operations per second.

3.2.3 Failure Detector

xOMB employs an active failure detector to quickly detect unresponsive servers. The controller assigns every middlebox a set of servers to monitor. Middleboxes ping each of their monitored servers at a configurable period. For more reliable failure detection, xOMB supports assigning multiple middlebox failure detectors for every server.

3.3 Design Discussion

While a general modular/pipeline approach is common in system design [28, 37], it represents a novel architecture for programmable middleboxes, which typically use a layered approach with protocol-specific callbacks [8, 17]. For example, a conventional middlebox may provide callbacks for processing a new TCP flow, part of an HTTP request (such as the URL), or a complete HTTP request.

The primary advantage of callbacks is that, as long as the

product supports your protocol, they make it straightforward to implement simple protocol-specific handling for a particular set of events. Because vendors tailor the set of callbacks to only specific supported protocols, they can provide a high level of integration for switch programmers, abstracting details such as protocol parsing and loading shared libraries—the programmer only needs to provide bodies of the desired event handlers. Additionally, the callbacks take as arguments the relevant fields for the event, eliminating the need for metadata objects attached to messages.

In contrast, xOMB modular pipelines provide four important advantages over callbacks. First, asynchronous modules allow message processing to perform RPCs to retrieve or store state over the network. The programmable middleboxes that we surveyed have the limitation that callbacks must run to completion and must not block, thus precluding this critical functionality for implementing dynamic request routing. Second, xOMB pipelines are more flexible because they are not limited to a fixed set of protocols or callbacks. Third, xOMB pipelines elegantly allow modules to pass arbitrary per-message state to other modules through the message metadata, enabling cross-module processing logic. While callbacks could provide similar functionality, this must be supported by the framework; current systems do not allow this and instead require setting global variables, a much more complicated and error-prone approach, and may limit processing parallelism if accessing these global variables requires locking. Finally, xOMB pipelines are potentially more efficient, because parsing modules only need to parse the minimal amount of bytes necessary to complete the desired processing. Furthermore, the pipeline can be programmed to immediately begin processing message fragments rather than waiting for the complete message, potentially reducing latency and overhead for large messages that can be processed with streaming logic.

xOMB pipelines can be structured to provide all of the convenience of callbacks while maintaining the above advantages. For example, we envision including parsers for popular protocols as part of the xOMB distribution. Further, pipelines can emulate a set of callbacks by using a series of modules with methods for each callback. These modules can wait for a desired event to occur and invoke the respective handler method with arguments from the message metadata.

4. IMPLEMENTATION

We now discuss xOMB's implementation in more detail.

4.1 Pipeline Libraries

To simplify pipeline programming, xOMB separates module and pipeline implementation from the xOMB C++ implementation. The user provides pipelines written in C or C++ as shared-object libraries linked with the required modules, also written in C/C++. xOMB allows users to define simple pipelines by just adding modules in the desired order to a pipeline module list. A xOMB server can load multiple pipelines, each associated with a separate port.

4.2 Control Module

The middlebox implements its portion of the control plane in a global control module. Upon startup, the control module first initializes the middlebox by creating threads for asynchronous I/O dispatch [5], and starts an RPC server to

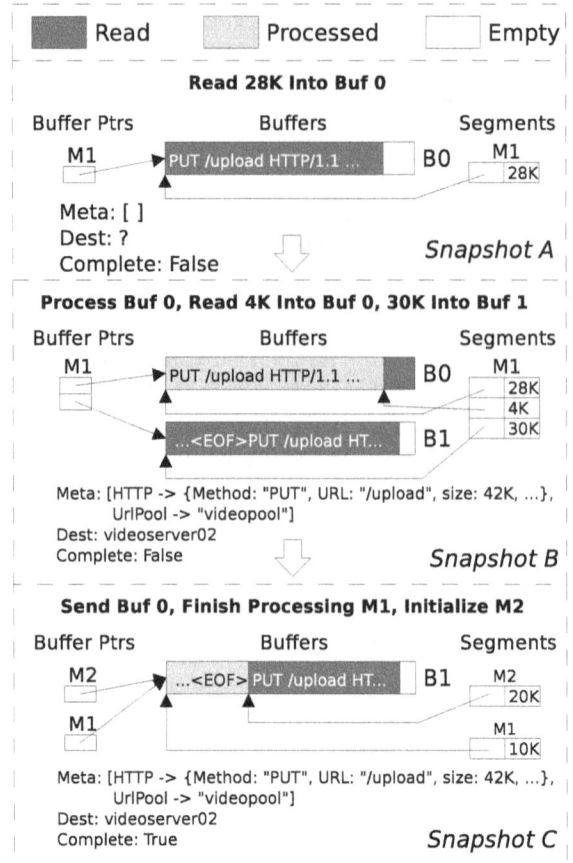

Figure 5: Message and Buffer State Snapshots

receive calls from the controller. Next, it dynamically loads specified pipeline shared-object libraries and begins listening for client connections on the data plane. Finally, it joins the controller and retrieves any global configuration parameters.

As described in §3.2, the middlebox receives server pool membership and monitoring assignments from the controller. The control module stores the pool assignments in a shared membership module and schedules timers for its monitoring tasks. A shared load monitor module queries servers for load information and stores the results. When the control module receives a failure notification from the controller, it notifies all pipelines so that the modules can update their state and respond appropriately. However, services such as health and load monitoring are optional. While we expect such functionality in many production environments, simpler deployments may not have this requirement.

4.3 Data Plane

The data plane listens for client connections on one or more ports, each of which has its own request and response pipeline. Although the xOMB architecture is general to any protocol, we focus on TCP-based protocols in this paper. Upon accepting a client TCP connection, the middlebox creates a client connection object that holds the client socket, the request pipeline, and a data structure to buffer and reorder responses. xOMB then creates a new message structure and buffer and reads data from the socket. We will illustrate pipeline processing with the state snapshots in Figure 5.

4.3.1 Messages and Buffers

The middlebox creates a message data structure for each

request and response to buffer message data during processing. Buffer management—how buffers are allocated, accessed, and copied—is a critical design detail for building a high performance middlebox. xOMB uses memory efficiently by avoiding user-level copying and freeing buffered data once it has been sent, even if the message is not complete. We use reference-counted buffer pointers to simplify memory management for data that persists across multiple messages.

Each message structure maintains a list of pointers to fixed-size buffers. Because buffers may not be full or may contain data for multiple messages, the message also has a list of *segments*—contiguous substrings in a buffer—each holding a pointer to the start of the segment (an offset into the buffer) and the segment length. The structure also holds fields for the total buffered data size and the number of bytes parsed, queued, and sent. The segment list and byte counts represent shifting windows of data to be processed and sent.

To maximize memory efficiency, the middlebox always fills every buffer by using scatter/gather I/O (reading into multiple buffers with one system call) and passing allocated but unused buffers to the next message. When a read completes, the middlebox adds segments to the message pointing to the newly read bytes. Snapshot A of Figure 5 shows message $M1$ after reading 28K into buffer $B0$.

Message processing proceeds one segment at a time. The pipeline returns one of three results: either the message is complete, an error occurred, or the pipeline needs more data. If the message is complete, the middlebox constructs a new message with any remaining unparsed buffer pointers and segments. On an error, the middlebox discards the message and closes the connection. If the pipeline returns incomplete, the middlebox performs another read.

Snapshot B of Figure 5 shows $M1$ after the middlebox has processed $B0$. The HTTP parser has set metadata representing the request, including the total request size of $42K$, *URLPool* has set the destination pool, and *LoadBalance* has set the destination server. Because $M1$ was not complete, the pipeline returned that it needed more, and the middlebox read another 30K into buffer $B1$.

Once the pipeline has determined the destination for the message, the middlebox sends any processed segments to the destination. As the middlebox reads and processes new buffers and segments, it simultaneously sends previous ones. By discarding buffers after sending them, the middlebox uses only a small amount of memory for each message, regardless of the total message size. The middlebox limits the amount of data buffered for any message by not reading on a connection while the total buffered size exceeds a threshold; the middlebox eventually closes connections for reads that take too long. Because message data typically will not fall on buffer boundaries, when the pipeline has processed a complete message, xOMB copies pointers for any buffered but unprocessed bytes from the completed message into a new message and invokes the pipeline with the new message before attempting another read.

Figure 5 C shows the state after processing $B1$ through the end of $M1$. While the pipeline processed $B1$, the middlebox concurrently sent $B0$ and then freed it. Because $M1$ is complete but $B1$ is not empty, the middlebox creates a new message $M2$ with initial buffers and segments as shown and empty metadata (not shown). The middlebox will process $M2$ before attempting to read more from the socket because

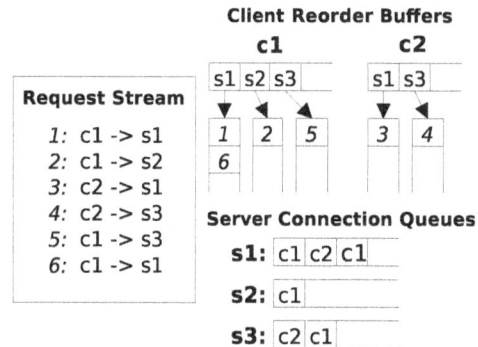

Figure 6: Reorder Buffer Example

$M2$ may be complete and, if so, processed before blocking on further data.

4.3.2 Connection Pool and Message Reordering

Most Internet protocols can be classified as either *request-oriented*, for which each logical request has one logical response, or *streaming*, where the byte stream cannot be separated into logical message boundaries. Although many request-oriented protocols, such as HTTP, have a one-to-one mapping between request and response messages, this is not always the case: for instance, NFS (see §6.4) may send multiple data fragment messages in response to a read request. xOMB relies on the parsing module to denote the logical message boundaries to when they span multiple protocol messages.

For streaming protocols, xOMB uses a unique server connection for each client connection and proxies the transformed pipeline data. For request-oriented protocols, xOMB maintains a pool of connections to back-end servers, with a configurable number of reusable connections per server. The connection collapsing performed by xOMB middleboxes can significantly increase back-end server efficiency by multiplexing a large number of client connections onto a small number of server connections (§6.2).

Because of connection collapsing, xOMB may interleave requests from different clients on server connections. Additionally, because xOMB may distribute pipelined client requests among different servers, the responses may arrive out-of-order. xOMB automatically demultiplexes, buffers, and reorders the responses to the clients, simplifying pipeline implementation. xOMB achieves this with a *reorder buffer*, data structure, illustrated with an example in Figure 6, that matches server responses with client requests based on their connections. As responses flow in, xOMB pairs the front of the server connection queue with the client reorder buffer, sending the message if they match and shifting the queues, or otherwise buffering the message (the vertical queues in Figure 6). xOMB limits buffered message data by only allowing a fixed number of pipelined requests per connection.

The example in Figure 6 shows six requests from two clients $(c1, c2)$ distributed over three server connections $(s1 - s3)$, and shows both kinds of reordering: 1) requests 1 and 3 from both $c1$ and $c2$ have been interleaved on connection $s1$, and 2) requests from both clients have been spread across different servers. For example, xOMB will buffer responses to $c1$ for messages 2 and 5 until the response for message 1 has been sent.

5. DISTRIBUTED OBJECT STORE

For a detailed example of dynamic request routing, we describe an object store service, xOS, based on Amazon's S3 [1]. To be interface compatible with S3, we used the unmodified Eucalyptus [29] *Walrus* storage components for our back-end application servers. However, as of the latest version (2.0), Eucalyptus does not support more than a single Walrus storage server. By using xOMB middleboxes together with a distributed metadata service, we transparently overcome this limitation while maintaining a unified, scalable storage namespace. We quantify xOS scalability in §6.3.

xOS hosts objects stored in *buckets* named by unique keys. Eucalyptus uses a single *cloud controller* to manage the authentication and metadata for all storage requests. Multiple storage servers will work independently when configured with the same Eucalyptus cloud controller, so we built a xOMB pipeline to consistently forward requests for a given user/bucket to the same storage server. xOS supports any middlebox processing requests for any bucket.

We implement *bucket → server* placement with a distributed metadata service that maps ⟨*userid, bucket*⟩ pairs to storage servers. Metadata servers subscribe to the storage server pool on the xOMB controller. Each metadata server is configured with a portion of a 160-bit key space, which it registers with the xOMB controller. The xOS forwarding module retrieves the key-space to metadata server mapping as part of its configuration state. Additionally, the forwarding module will update its metadata server configuration whenever it receives notification that the set of metadata servers has changed (via addition, removal, or failure).

The xOS request pipeline consists of the standard HTTP parse module followed by our xOS forwarding module. To process a request, the xOS forwarding module takes the SHA1 hash of the ⟨*userid, bucket*⟩ string parsed from the HTTP headers and URL. Using this hash, the module computes the metadata server responsible for that portion of the key-space and makes an RPC to retrieve the storage server. When a metadata server receives a lookup request, it either returns an existing assignment if found or otherwise chooses a new storage server and stores the assignment. The forwarding module caches these assignments to avoid subsequent lookups for the same bucket.

In our design, middleboxes maintain only soft state for their fowarding tables, so no middlebox recovery is necessary. Furthermore, middleboxes can update their bucket-to-server mappings lazily; if they attempt to forward a request to a failed server and receive a socket error, then they can contact the metadata servers to retrieve an updated mapping. When the storage servers are replicated (we did not implement this, as it was not the focus of this example), then storage server failure would be transparent to the client.

Basing xOS on distributed Walrus servers presents an additional challenge for the front-end middleboxes. The S3 interface contains a *ListBuckets* request to list all of a user's buckets. However, no single server contains this information. To support the complete interface, the xOS forwarding module recognizes the *ListBuckets* message and makes an RPC to each metadata server requesting all the user's buckets. Once the forwarding module has received all responses, it generates the HTTP and XML response by combining the separate bucket lists. Although not difficult to implement, we note that such functionality would be impossible in designs limited to event-handler callbacks.

6. EVALUATION

In this section, we evaluate the main xOMB design goals of scalability, performance, and programmability. For these experiments, we use servers with Intel 2.13 GHz Xeon quad-core processors and 4 GB DRAM running Linux 2.6.26. All machines have 1 Gbps NICs connected to the same 1Gbps Ethernet switch.

6.1 Programmability

First, we give a sense of xOMB's programmability. Table 1 shows the number of lines of C/C++ code for modules we implemented, as well as the xOMB core middlebox service framework (excluding the controller) for reference. The majority of the modules are short, although the HTTP parsing module includes 1700 lines of an HTTP parsing library [14], and the NFS parse and Cache modules both use code generated from the XDR protocol file. In addition, all of the pipelines use the default pipeline construction process, meaning they are only a handful of lines each.

Module	Lines of Code
HTTP Parse	111 (+ 1699)
Round-Robin Forward	73
Random Forward	37
URL Pool Forward	31
HTTP Attack Filter	99
HTTP Version Filter	72
xOS Forward	277
NFS Parse	235 (+ 2906)
NFS Cache	413 (+ 602)
xOMB Core	4478

Table 1: Module Code Lengths

6.2 HTTP Performance

We first evaluate xOMB performance with HTTP pipelines. We used Apache 2.2.9 [2] running the MPM worker module, serving files of various sizes. All throughput measurements include the bytes transferred for HTTP headers. In addition, our HTTP client is pipelined, but limits itself to 10 outstanding requests. In these experiments, xOMB uses a basic HTTP pipeline configured to parse requests and forward them, using a "client sticky" round-robin forwarding module, across the available webservers. Our forwarding module ensures that all of a client's requests go to the same webserver, although each client is assigned to a webserver round-robin. In addition, we perform connection collapsing down to at most five connections from each middlebox to each webserver.

We compare our performance against Apache directly, the popular open-source reverse proxy nginx [10], and a programmable hardware switch from F5 Networks [8]. The F5 BIG-IP Local Traffic Manager 1600 (LTM) we used has an Intel 1.8 GHz E2160 dual-core processor and 4 GB DRAM and is running OS Version 9.4.8 and has a single 1 Gb/s NIC. We refer to the LTM simply as "F5" in our experiments. It is difficult to compare the processor employed in the F5 relative to our servers. While our machines are three years old and based on older microarchitecture, they do have four cores and a higher clock speed. One challenge with specialized hardware such as the F5 switch is integrating the latest processors and motherboards into a specialized

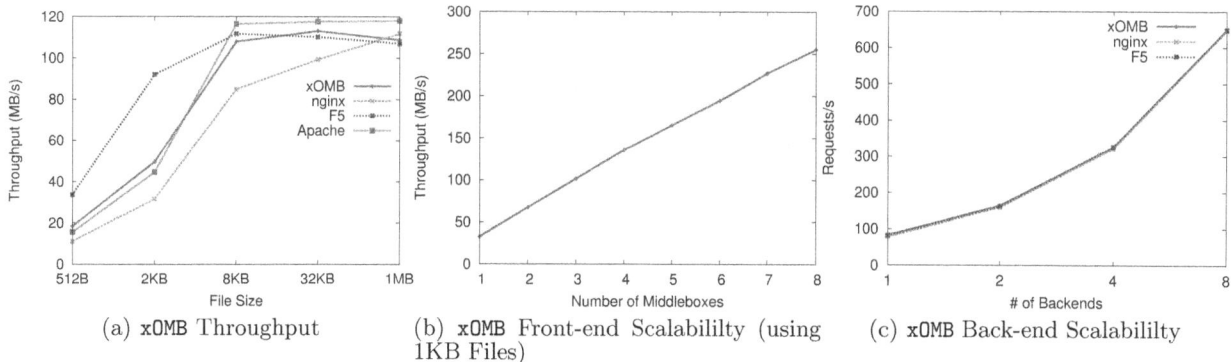

(a) xOMB Throughput   (b) xOMB Front-end Scalabililty (using 1KB Files)   (c) xOMB Back-end Scalabililty

Figure 7: Throughput and Scalability Comparisons

hardware and software environment, a downside endemic to non-commodity solutions.

Figure 7a shows a comparison of client throughput to Apache, through a single xOMB middlebox, a single nginx reverse proxy, and through the F5 switch. For each file size, we used 100 clients, which was enough to maximize throughput. Compared to Apache, xOMB does quite well, only losing out slightly at large file sizes due to the fact that the xOMB middlebox must allocate some of its 1 Gb/s of NIC bandwidth to forwarding the clients' requests to the back-end webserver. Nginx performs noticeably slower in almost all cases, only able to surpass xOMB slightly with 1 MB files. Finally, xOMB performs similarly to the F5 switch for larger files, but the F5 handily outperforms with smaller files.

Intrigued by the F5's impressive performance, we investigated further to determine how the F5 could outperform stand-alone Apache with small files since we had disabled caching. We wrote a very simple TCP proxy that accepts a client connection, parses request streams by looking for a lone CRLF, and simply copies the requests to a fixed destination connection. When running a single HTTP client through this proxy connected to Apache, we saw throughput increase by a factor of two-to-ten compared to the client connecting to Apache directly, depending on the file size. We saw this speedup regardless of whether the client pipelined requests or not. However, we saw zero speedup when we repeated this through a TCP proxy that did not parse requests. While we leave a more in-depth study to future research, our initial conclusion is that Apache appears to be sensitive to the way it receives streams of client requests, and for small files, parsing these requests is the limiting factor for throughput.

Although the presence of a single xOMB middlebox shows reduced performance compared to the F5 switch for small files, one of the key components of our design is the ability to scale well. Figure 7b shows client throughput when requesting 1 KB files with eight back-end webservers behind differing numbers of middleboxes. A single middlebox is not able to handle the extra capacity additional back-ends provide, but we see near-perfect scalability as we add middleboxes.

In our next experiment, clients request a simple CGI application that computes the SHA-1 hash of a 1 MB file on disk. Instead of being limited by middlebox processing capacity, we are now limited by back-end capacity. Figure 7c shows client throughput, now measured in requests per sec-

ond, of 100 clients making requests to the CGI program with varying numbers of webservers behind a single xOMB middlebox. We see that xOMB, nginx, and the F5 switch are all able to achieve near-perfect scalability as back-end resources are added.

One of the most difficult aspects of middlebox design is performance under extremely high numbers of concurrent connections. We ran between 1 and 10,000 clients requesting 1 KB files against xOMB, nginx, the F5, and Apache, with the results shown in Figure 8a. We first note that Apache by itself does not perform well with 1,000 clients, and we could not get it to serve 10,000 concurrent clients at all. Both xOMB and nginx show similar curves, although we outperform nginx for all connection sizes. However, the F5 shows very unusual behavior. We see it is able to perform extremely well with between 100 and 1000 clients. Additionally, although it outperforms xOMB and nginx, we saw around 6,000 of the 10,000 clients' connections closed prematurely by the F5. We have been unable to determine the cause of this performance anomaly and leave further study of the F5 for future work.

Our final benchmark for HTTP traffic is a pipeline for filtering potential malicious requests (§3.1.2). We compare against the F5, which we programmed to perform the same checks. Although F5 offers a firewall module that can filter attack traffic, we chose to implement an attack filter for the F5 in their provided scripting language to both gain experience programming the F5 and to compare the exact same set of rules for both systems.

Our attack filter module loads 285 Snort [15] rules from the controller, each containing a regular expression to search for in the URL of a web request. We check each URL request against all 285 rules. Although we did not introduce any malicious requests in this experiment, we measure the performance hit of checking every request against all rules. Figure 8b compares xOMB's performance against the F5 running this attack filter against a single webserver. Not only does xOMB outperform the F5 across all tested numbers of clients with 1 MB and 4 KB files, but it sustains the throughput achieved when not running the attack filter with 1 MB files and comes within 10% for 4 KB files.

6.3 xOS Performance

Next, we evaluate the performance of xOS with dynamic request routing as described in §5. We ran two experiments,

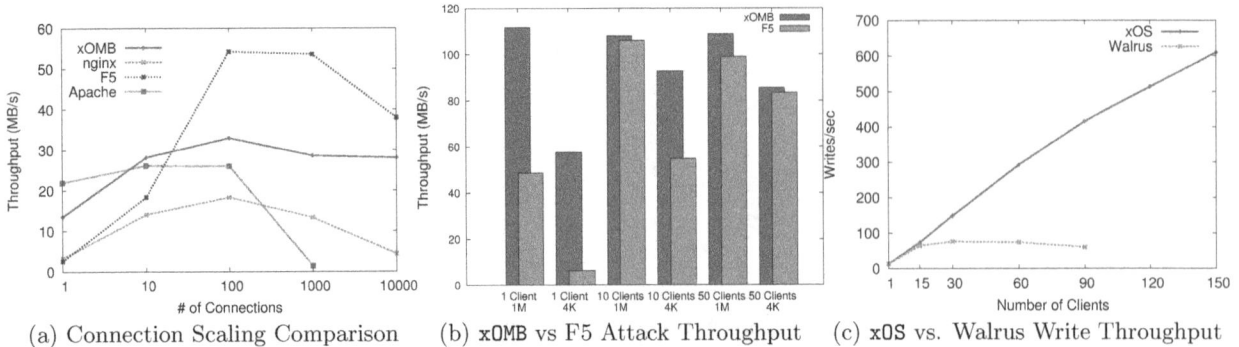

(a) Connection Scaling Comparison (b) xOMB vs F5 Attack Throughput (c) xOS vs. Walrus Write Throughput

Figure 8: Throughput and Scalability Comparisons

one using a single instance of Walrus by itself, and the other using xOS. For both experiments, we used varying numbers of clients to repeatedly write 4 KB blocks to the storage system. We used the Amazon S3 curl client to make the write requests, which do not support pipelining. We had each client create and write to its own unique bucket, a workload that causes the middlebox to do the most dynamic request routing. For the xOS experiment, we used two metadata nodes, eight Walrus nodes, and a single xOMB middlebox.

The results of both experiments are shown in Figure 8c. We see the Walrus write throughput max out at around 74 operations per second. Throughput decreases as the number of clients increases past 30, and tests with more than 90 clients did not finish correctly as we began having connection issues with Walrus. In contrast, xOMB allows xOS clients to write to the eight Walrus backends in parallel, easily scaling past the capacity of a single Walrus machine.

6.4 NFS Acceleration

To evaluate xOMB in a different context than HTTP load balancing, we implemented an NFS protocol accelerator pipeline. Our basic accelerator, designed to speed up wide-area access to an NFS server, caches file attribute structures from lookups and file data from reads or writes. We did not attempt to write a comprehensive accelerator. Rather, our intent is to demonstrate that xOMB can effectively process diverse protocols with varying goals. A limitation of our implementation is the assumption that all client requests pass through the middlebox; it does not attempt to reconcile the cache with the server if some clients connect to the server directly. Some protocol accelerators increase write throughput by responding to the client immediately before forwarding the write to the server; we opted not to implement this.

The file attribute (getattr) and lookup calls form a significant portion of NFS traffic [32]. Typically, clients cache these attributes for a short duration (3 sec). Also, clients need to ensure that file attributes are up to date while serving reads from its local buffer cache. The large number of round trips due to attribute lookups causes significant performance degradation for NFS over a wide-area network. Having NFS clients connect to a xOMB middlebox on the same LAN, which in turn connects to the server, results in better response time for file system operations by caching file attributes and data.

We evaluate performance with the Postmark benchmark [12], modified to bypass the kernel buffer cache. For this experiment, we compare the performance of direct NFS versus xOMB when the clients connect over both a LAN and the an emulated wide-area link. We emulate a 100ms delay using NetEm [25]. To be a worst-case test for NFS, we mount the file system using synchronous writes. Postmark creates 60 files with sizes ranging from 1B to 10KB and performs 300 transactions. Table 2 shows the throughput, average operation latency, and total runtime results for the four runs. xOMB adds a small amount of latency on the LAN, as it must process all the NFS traffic and copy data into the cache. However, the xOMB cache significantly improves the performance of file lookups and read operations with a wide-area delay, completing the workload almost twice as quickly.

Operations	LAN		100ms RTT WAN	
	xOMB	NFS	xOMB	NFS
Read (KB/s)	48	50	3.2	1.6
Write (KB/s)	55	57	3.7	1.9
Create (ms)	1.9	0.3	201	204
Open (ms)	0.6	0.2	0.5	192
Remove (ms)	0.9	0.2	101	101
Read (ms)	0.6	0.2	0.5	100
Write (ms)	7.6	7.8	101	100
Total time (s)	23	22	341	672

Table 2: NFS Latency and Throughput Comparison

7. DISCUSSION

We are encouraged by our experience with xOMB and its evaluation. However, there are a number of important issues that require work for programmable middle boxes to be successful in general. We discuss these in turn below.

Performance. Our focus has been on scalability and extensibility. While our single node is reasonable (and in many cases superior to commercial product offerings), it will fundamentally be limited by our user-level implementation. One could imagine alternate, higher-performance xOMB implementations in kernel or even in programmable line cards. While reasonable for certain scenarios, we believe that such architectures will fundamentally limit the expressibility of the available programming model.

Load Balancing. Devising good load balancing algorithms is a difficult research and engineering challenge by itself. For evaluation, we implemented simple algorithms with no claims of novelty. The goal of this work is not to innovate in developing better algorithms directly, but rather to provide a framework where it becomes easier to innovate in load balancing algorithms.

Debugging. All of our data processing modules have been of moderate complexity thus far, but debugging the behavior of a middle box is a challenge in general. xOMB facilitates pipeline debugging by allowing the programmer to enable progressively more verbose logging. Because xOMB is written in C++, programmers can leverage standard logging techniques, network monitoring (e.g., `tcpdump` and `wireshark`), and tools such as `gdb` to assist with debugging. More specialized middleboxes typically do not have the same rich set of open source tools or perhaps even the ability to log state over the network or to local disk.

Resource Allocation and Isolation. xOMB currently provides no support to isolate individual processing pipelines from one another or to isolate the processing of one flow from another. For example, a rarely-exercised code path could cause the entire pipeline to fail or to slow processing for concurrent flows. Similarly, an administrator may wish to allocate varying amounts of bandwidth or CPU resources to different flows or pipelines. A range of possible techniques are possible for delivering the necessary isolation, from heavyweight solutions employing entire virtual machines on a per-pipeline basis, to individual processes on a per-flow basis, to in-kernel queueing disciplines limiting the bandwidth available to any individual flow. We plan to explore these and other techniques [23] as part of our ongoing work.

8. RELATED WORK

Most closely related to our work in spirit are Click [28], RouteBricks [21], and CoMb [33]. Click provides a modular programming interface for packet processing in extensible routers. The principal difference between Click and xOMB is our focus on extensibility and programmability at the granularity of byte streams rather than individual packets. Our pipelined programming model focuses on efficiently parsing and transforming request/response based communication.

Like xOMB, RouteBricks also focuses on scaling network processing with commodity servers. Their work, like Click, focuses on routing and operates at the granularity of packets. They focus on the more extreme performance requirements of large-scale routers that may require terabits/sec of aggregate communication bandwidth and their in-kernel implementation and VLB-based load balancing delivers significant scalability. Middleboxes typically do not require quite the same level of bandwidth performance and while our architecture similarly scales with additional servers, our user-level implementation trades per-server performance for programmability and overall extensibility.

CoMb [33] shares the goal of using software middleboxes with commodity hardware, but emphasizes consolidation of middlebox hardware and simplifying network deployments, whereas xOMB focuses on programmability and extensibility. CoMb requires modular applications to be written in the Click framework with flow-level processing supported through a session reconstruction module. We see CoMb's goal of consolidation as complementary to ours and expect that xOMB middleboxes could be used with a CoMb controller.

Flowstream [24] also provides an architecture for middleboxes. It employs OpenFlow [11]-controlled hardware switches to separate traffic at flow granularity that is then forwarded to individual servers for further processing. By default these servers would perform network processing at packet granularity. As such, one could view xOMB as the architecture and programming model for extensible transformation of OpenFlow-forwarded byte streams in Flowstream.

There are many commercial hardware/software products for middlebox processing. F5 networks [8] provides popular load balancing switches. Pai et al. [30] performed some of the early academic work in this space. Bivio [4] focuses on deep packet inspection, while Riverbed [13] delivers protocol accelerators among other products. Each product typically focuses on a niche domain and provides a limited extensibility model. In particular, the F5 switch we evaluate uses the Tcl programming language. However, it is not able to support the full generality of our framework, for example with respect to making remote procedure calls or maintaining protocol-specific metadata and state. For instance, the F5 could not be employed to implement functionality for an entirely different protocol such as the S3 service (§5). In contrast, the goal of our work is to provide a unified framework and programming model for a range of traditional middlebox functionality.

Reverse proxies such as [3, 10, 16–18] aim to provide load balancing over a set of web servers. While this is similar in spirit to part of the functionality that xOMB supports, our architecture is much more general and can support arbitrary protocols. Most reverse proxies can only handle a static set of servers, protocols, and have fixed processing options, unlike xOMB which can handle dynamic membership and arbitrary processing. Nginx [10] is fairly extensible, although modules written for it must be compiled into the executable and not loaded dynamically like in xOMB. In addition, nginx does not support general connection collapsing of client requests, which can severely impact performance with large numbers of clients. Finally, RPCs for making dynamic routing decisions cannot be done with their callback model, although they do support passing arbitrary state between callbacks like xOMB.

Allman performed an early performance study of middleboxes [19]. He found that middleboxes are a mixed bag for performance, increasing or reducing performance under different circumstances. The study also found that middleboxes can reduce end-to-end availability, though typically availability remained at an acceptable 99.9%. One goal of xOMB is to develop a framework to increase the performance and availability of middleboxes.

One challenge with middlebox deployment is ensuring that flows are forwarded through an appropriate set of middleboxes based on higher-level policy. Dilip et al. [27] introduced an architecture to ensure such forwarding. OpenFlow provides a general mechanism to intercept flows and forward them through an appropriate set of middleboxes on the way to the destination. Ethane/SANE [20] is one instance of such an approach for enterprise network security and authentication.

DOA [36] is a delegation oriented architecture for more explicitly integrating middleboxes into the Internet architecture. One goal of DOA is to address the transparency issues introduced by non-extensible hardware middleboxes on evolving network flows. xOMB would ideally make it easier for middleboxes to adapt to changing traffic characteristics. Similar to DOA, I3 [34] explicitly introduces indirection in data forwarding, this time at the overlay level using a DHT.

9. CONCLUSIONS

xOMB demonstrates a new design and architecture for building scalable, extensible middleboxes. We show that programmability need not come at the expense of performance; for instance, the xOMB implementation of a load balancing switch achieves performance comparable to a commercial load balancing switch. Beyond load balancing, we have shown how extensible middleboxes can be used to build scalable services by constructing dynamic forwarding tables based on application service state and the effectiveness of a xOMB protocol accelerator for NFS.

10. ACKNOWLEDGMENTS

We thank our reviewers for their feedback on the paper. This work is supported in part by the National Science Foundation (#CSR-1116079).

11. REFERENCES

[1] Amazon S3. http://aws.amazon.com/s3.
[2] Apache HTTP Server. http://httpd.apache.org.
[3] Apache mod_proxy. http://httpd.apache.org/docs/current/mod/mod_proxy.html.
[4] Bivio Networks. http://www.bivio.net.
[5] Boost Asio. http://www.boost.org/doc/libs/1_42_0/doc/html/boost_asio.html.
[6] Cisco Systems. http://www.cisco.com.
[7] Citrix Systems. http://www.citrix.com.
[8] F5 Networks. http://www.f5.com.
[9] F5 OneConnect Guide. http://www.f5.com/pdf/deployment-guides/oneconnect-tuning-dg.pdf.
[10] nginx. http://www.nginx.org.
[11] OpenFlow. http://www.openflowswitch.org.
[12] Postmark Benchmark. http://www.netapp.com/tech_library/postmark.html.
[13] Riverbed. http://www.riverbed.com.
[14] Ry's HTTP Parser. http://github.com/ry/http-parser.
[15] Snort. http://www.snort.org.
[16] Squid. http://www.squid-cache.org.
[17] Traffic Server. http://trafficserver.apache.org.
[18] Varnish Cache. http://www.varnish-cache.org.
[19] ALLMAN, M. On the Performance of Middleboxes. In *IMC* (2003).
[20] CASADO, M., FREEDMAN, M. J., PETTIT, J., LUO, J., MCKEOWN, N., AND SHENKER, S. Ethane: Taking Control of the Enterprise. In *SIGCOMM* (2007).
[21] DOBRESCU, M., EGI, N., ARGYRAKI, K., CHUN, B.-G., FALL, K., IANNACCONE, G., KNIES, A., MANESH, M., AND RATNASAMY, S. RouteBricks: Exploiting Parallelism To Scale Software Routers. In *SOSP* (2009).
[22] GHEMAWAT, S., GOBIOFF, H., AND LEUNG, S.-T. The Google File System. In *SOSP* (2003).
[23] GHODSI, A., SEKAR, V., ZAHARIA, M., AND STOICA, I. Multi-resource Fair Queueing for Packet Processing. In *Proc. SIGCOMM* (2012).
[24] GREENHALGH, A., HANDLEY, M., HOERDT, M., HUICI, F., MATHY, L., AND PAPADIMITRIOU, P. Flow Processing and the Rise of Commodity Network Hardware. In *ACM CCR* (2009).
[25] HEMMINGER, S. Network emulation with NetEm. In *LCA* (2005).
[26] HUNT, P., KONAR, M., JUNQUEIRA, F. P., AND REED, B. Zookeeper: wait-free coordination for internet-scale systems. In *USENIX ATC* (2010).
[27] JOSEPH, D., TAVAKOLI, A., AND STOICA, I. A Policy-aware Switching Layer for Data Centers. In *Proceedings of ACM SIGCOMM* (2008).
[28] MORRIS, R., KOHLER, E., JANNOTTI, J., AND KAASHOEK, M. F. The Click modular router. In *SOSP* (1999).
[29] NURMI, D., WOLSKI, R., GRZEGORCZYK, C., OBERTELLI, G., SOMAN, S., YOUSEFF, L., AND ZAGORODNOV, D. The Eucalyptus Open-Source Cloud-Computing System. In *CCGrid* (2009).
[30] PAI, V. S., ARON, M., BANGA, G., SVENDSEN, M., DRUSCHEL, P., ZWAENEPOEL, W., AND NAHUM, E. Locality-Aware Request Distribution in Cluster-based Network Servers. In *ASPLOS* (1998).
[31] PAXSON, V. Bro: a system for detecting network intruders in real-time. In *USENIX Security* (1998).
[32] SANDBERG, R. Design and implementation of the sun network file system. In *USENIX* (1985), pp. 119–130.
[33] SEKAR, V., EGI, N., RATNASAMY, S., REITER, M. K., , AND SHI, G. Design and Implementation of a Consolidated Middlebox Architecture. In *Proc. NSDI* (2012).
[34] STOICA, I., ADKINS, D., ZHUANG, S., SHENKER, S., AND SURANA, S. Internet Indirection Infrastructure. In *ACM SIGCOMM* (2002).
[35] THALER, D., AND HOPPS, C. Multipath Issues in Unicast and Multicast Next-Hop Selection. RFC 2991, 2000.
[36] WALFISH, M., STRIBLING, J., KROHN, M., BALAKRISHNAN, H., MORRIS, R., AND SHENKER, S. Middleboxes No Longer Considered Harmful. In *SOSP* (2004).
[37] WELSH, M., CULLER, D., AND BREWER, E. Seda: an architecture for well-conditioned, scalable internet services. *SIGOPS Oper. Syst. Rev. 35*, 5 (2001).

NetBump: User-extensible Active Queue Management with Bumps on the Wire

Mohammad Al-Fares*   Rishi Kapoor*   George Porter*   Sambit Das*
Hakim Weatherspoon**   Balaji Prabhakar†   Amin Vahdat*‡

*UC San Diego   **Cornell University   †Stanford University   ‡Google Inc.

Abstract

Engineering large-scale data center applications built from thousands of commodity nodes requires both an underlying network that supports a wide variety of traffic demands, and low latency at microsecond timescales. Many ideas for adding innovative functionality to networks, especially active queue management strategies, require either modifying packets or performing alternative queuing to packets in-flight on the data plane. However, configuring packet queuing, marking, and dropping is challenging, since buffering in commercial switches and routers is not programmable.

In this work, we present *NetBump*, a platform for experimenting with, evaluating, and deploying a wide variety of active queue management strategies to network data planes with minimal intrusiveness and at low latency. NetBump leaves existing switches and endhosts unmodified by acting as a "bump on the wire," examining, marking, and forwarding packets at line rate in tens of microseconds to implement a variety of virtual active queuing disciplines and congestion control mechanisms. We describe the design of NetBump, and use it to implement several network functions and congestion control protocols including DCTCP and 802.1Qau quantized congestion notification.

Categories and Subject Descriptors

C.2.2 [**Network Protocols**]: Protocol Architecture; C.2.3 [**Network Operations**]: Network Management

General Terms

Design, Experimentation, Management, Performance

Keywords

Datapath programming, vAQM, congestion control

1. INTRODUCTION

One of the ultimate goals in data center networking is predictable, congestion-responsive, low-latency communication. This is a challenging problem and one that requires tight cooperation between endhost protocol stacks, network interface cards, and the switching infrastructure. While there have been a range of interesting ideas in this space, their evaluation and deployment have been hamstrung by the need to develop new hardware to support functionality such as Active Queue Management (AQM) [13,21], QoS [45], traffic shaping [17], and congestion control [1,2,18]. While simulation can show the merits of an idea and support publication, convincing hardware manufacturers to actually support new features requires evidence that a particular technique will actually deliver promised benefits for a range of application and communication scenarios.

We consider a model where new AQM disciplines can be deployed and evaluated directly in production data center networks without modifying existing switches or endhosts. Instead of adding programmability to existing switches themselves, we instead deploy "bumps on the wire," called *NetBumps*, to augment the existing switching infrastructure.[1] Each NetBump exports a virtual queue primitive that emulates a range of AQM mechanisms at line rate that would normally have to be implemented in the switches themselves.

NetBump provides an efficient and easy way to deploy and manage active queue management separate from switches and endhosts. NetBumps enable AQM functions to be incrementally deployed and evaluated by their placement at key points in the network. This makes implementing new functions straightforward. In our experience, new queuing disciplines, congestion control strategies, protocol-specific packet headers (e.g. for XCP [18]), and new packets (for a new congestion control protocol we implement) can be easily built and deployed at line rate into existing networks. Developers can experiment with protocol specifics by simply modifying software within the bump.

The NetBump requirements are: rapid prototyping and evaluation, ease of deployment, support for line rate data processing, low latency (i.e. tens of μs), packet marking and transformation for a range of AQM and congestion control policies, and support for distributed deployment to support data center multipath topologies. We greatly reduce the latency imposed by NetBump because our functionality is limited to modifications of packets in flight, with no actual queuing or buffering done within NetBump. We expect these bumps on the wire to be part of the production network that will form a proving ground to inform eventual hardware development (see Fig. 1 for an example deployment scenario).

We based our NetBump implementation on a user-level, zero-copy, kernel-bypass network API, and found that it performed well;

---

[1]The "bump on the wire" term here is unrelated to previous work about IPsec deployment boxes [19].

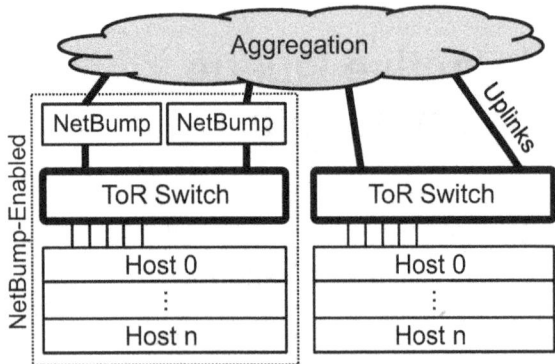

Figure 1: Deployment scenario in the data center. "NetBump-enabled racks" include NetBumps in-line with the Top-of-Rack (ToR) switch's uplinks, and monitor output queues at the host-facing ports.

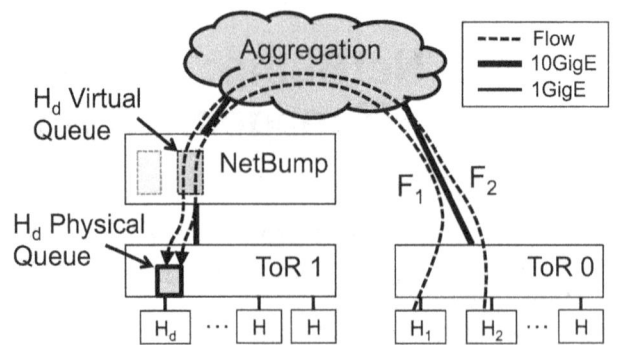

Figure 2: An example of NetBump at ToR switch, monitoring downstream physical queues.

able to support custom active queue management of 64-byte packets at a rate of 14.17Mpps (i.e. 10Gbps line rate) with one CPU core at 20-$30\mu s$. In part this performance is the result of NetBump's simpler packet handling model supporting pass-through functionality on the wire, as compared to general-purpose software routers.

The primary contributions of this paper are: 1) the design of a "bump on the wire" focusing on evaluating and deploying new buffer management packet processing functions, 2) a simple virtual Active Queue Management (vAQM) implementation to indirectly manage the buffers of neighboring, unmodified switches, 3) the evaluation of several new programs implemented on top of NetBump, including an implementation of IEEE 802.1Qau-QCN L2 congestion control, and 4) an extensible and *distributed* traffic update and management platform for remote physical switch queues.

2. MOTIVATION

In this section we first present an example of NetBump functionality in action, and then motivate our requirements for a low-latency implementation.

2.1 NetBump Example

In Fig. 2, we show a simple network where two source hosts H_1 and H_2 each send data to a single destination host H_d (in flows F_1 and F_2, respectively). H_1 and H_2 are connected to Switch0 at 1Gbps. Switch 0 has a 10Gbps uplink to a NetBump (through the aggregation layer), and on the other side of the NetBump is a second 10Gbps link to Switch1. Destination host H_d is attached to Switch1 at 1Gbps. Flows F_1 and F_2 each have a maximum bandwidth of 1Gbps, and since host H_d has only a single 1Gbps link, congestion will occur on H_d's input or output port in Switch1 if $rate(F_1) + rate(F_2) > $ 1Gbps. Without NetBump, assuming Switch1 implements a drop-tail queuing discipline, packets from F_1 and F_2 will be interleaved in H_d's physical queue until it becomes full, at which point Switch1 will drop packets arriving to the full queue. This leads to known problems such as burstiness and lack of fairness.

Instead, as NetBump forwards packets from its input to its output port, it estimates the occupancy of a virtual queue associated with H_d's output port buffer. When a packet arrives, H_d's virtual queue occupancy is increased by the packet's size. Because NetBump has

the topology information and knows the speed of the link between Switch1 and H_d (§ 3.1), it computes the estimated *drain rate*, or the rate that data leaves H_d's queue. By integrating this drain rate over the time between subsequent packets, it calculates the amount of data that has left the queue since the last packet arrival.

Within NetBump, applications previously requiring new hardware development can instead act on the virtual queue. For example, to implement Random Early Detection (RED), the NetBump in Fig. 2 maintains a virtual queue for each physical queue in Switch1. This virtual queue maintains two parameters, `MinThreshold` and `MaxThreshold`, as well as an estimate of the current downstream queue length. According to RED, packets are sent unmodified when the moving average of the queue length is below the `MinThreshold`, packets are marked (or dropped) probabilistically when the average is between the two thresholds, and unconditionally marked (or dropped) when it is above `MaxThreshold`.

Note that in this example, just as in all the network mechanisms presented in this paper, packets are never delayed or queued in the NetBump itself. Instead, NetBump marks, modifies, or drops packets at line rate as if the downstream switch directly supported the functionality in question. Note also that NetBump is not limited to a single queuing discipline or application–it is possible to compose multiple applications (e.g. QCN congestion control with Explicit Congestion Notification (ECN) marking [11]). Furthermore, AQM functionality can act only on particular flows transiting a particular end-to-end path if desired.

2.2 Design Requirements

The primary goal of NetBump is enabling rapid and easy evaluation of new queue management and congestion control mechanisms in deployed networks with minimal intrusiveness. We next describe the requirements NetBump must meet to successfully reach this goal.

Deployment with unmodified switches and endhosts: We seek to enable AQM development and experimentation to take place in the data center itself, rather than separate from the network. This means that NetBump works despite leaving switches and endhosts unmodified. Thus a requirement of NetBump is that it implements a virtual Active Queue Management (vAQM) discipline that tracks the status of neighboring switch buffers. This will differ from previous work that applies this technique within switches [13, 21], as our implementation will be remote to the switch.

62

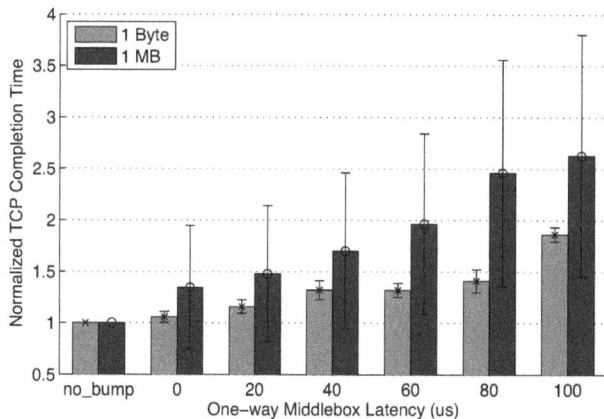

Figure 3: Effect of middlebox latency on completion time of short (1 Byte) and medium-sized (1MB) TCP flows. Baseline (direct-connect) transfer time was $213\mu s$ (1B), 9.0ms (1MB), others are through a NetBump with configurable added delay.

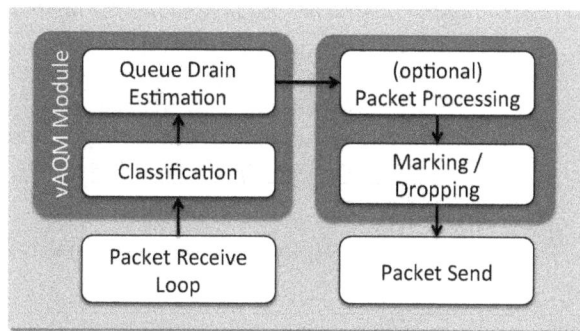

Figure 4: The NetBump pipeline.

Distributed deployment: Modern networks increasingly rely on multipath topologies both for redundancy in the face of link and switch failure, and for improving throughput by utilizing several, parallel links. Left unaddressed, multipath poses a challenge for the NetBump model since a single bump may not be able to monitor all of the flows heading to a given destination. Therefore a requirement for NetBump is that it supports enough throughput to manage a sufficient number of links, and that it supports a distributed deployment model. In a distributed model, multiple bumps deployed throughout the network coordinate with each other to manage flows transiting them. In this way, a set of flows taking separate network paths can still be subjected to a logically centralized, though physically distributed, AQM policy.

Ease of development: Rather than serving as a final deployment strategy, we see NetBump as an experimental platform, albeit one that is deployed directly on the production network. Thus rapid prototyping and reconfiguration are a requirement of its design. Specifically, the platform should export a clear API with which users can quickly develop vAQM applications using C/C++.

Minimizing latency: Many data center and enterprise applications have strict latency deadlines, and any datapath processing elements must likewise have strict performance guarantees, especially given NetBump's target deployment environment of data center networks, whose one-way latency diameters are measured in microseconds. Since the throughput of TCP is in part a function of the network round-trip time [32], any additional latency imposed by NetBump can affect application flows. Fig. 3 shows the completion times of two flows as a function of one-way middlebox latency–one flow transfers a single byte between a sender-receiver pair, the other transfers 1MB. Adding even tens of microseconds of one-way latency has a significant impact on flow completion times when the baseline network RTT is very small.

Since the network layer sits below all data center applications, and since a single application-layer request might cost several round-trips, NetBump's forwarding latency must be very low to minimize the overhead for those applications.

Forwarding at line rate: Although most data center hosts still operate at 1Gbps, 10Gbps has become standard at rack-level aggregation. Deploying a NetBump inline with top-of-rack uplinks and between 10Gbps switches will require an implementation that can support 10Gbps line rates. The challenge then becomes keeping up with packet arrival rates: 10Gbps corresponds to 14.88M 64-byte minimum-sized packets per second, including Ethernet overheads.

3. DESIGN

In this section we describe the design of the NetBump vAQM pipeline, including how it scales to support more links and a distributed deployment for multi-path data centers.

3.1 The NetBump Pipeline

The core NetBump pipeline consists of four algorithms: 1) packet classification, 2) virtual queue (VQ) drain estimation, 3) packet marking/dropping, and optionally 4) extensible packet processing. This pipeline is exported to the user via the NetBump API (Table 1).

Virtual Queue Table Data Structure: Each NetBump maintains a set of virtual queues, which differ from physical queues in that they do not store or buffer packets. Instead, as packets pass through a virtual queue, it maintains state on what its occupancy *would* be if it were actually storing packets. Thus each virtual queue must keep track of 1) the number and sizes of packets transiting it, 2) the packet arrival times, and 3) the virtual rate at which they drain from the queue. Note that packets actually drain at line rate (i.e. 10Gbps), however a virtual queue could be configured with any virtual drain parameter (e.g. 1Gbps, 100Mbps).

The virtual queue table is a simple data structure kept by the Net-Bump that stores these three parameters for each virtual queue at that bump. For the AQM functionality we consider, we only need to know the virtual queue occupancy and drain rate, and so each virtual queue keeps 1) the size in bytes of the queue, 2) the time the last packet arrived to the queue, and 3) the virtual queue drain rate. These values are updated when a packet arrives to the virtual queue.

1. Packet Classification: As packets arrive to the NetBump, they must first be classified to determine into which virtual queue they will be enqueued. This classification API is extensible in NetBump, and can be overridden by a user as needed. A reasonable scheme would be to map packets to virtual queues corresponding to the downstream physical switch output buffer that the packet will reside in when it leaves the bump. In this case the virtual queue is emulating the downstream switch port directly.

Function	Description
void init(vQueue *vq, int drainRate);	Initializes a virtual queue and set the given drain rate.
vQueue * classify(Packet *p) const;	Classifies a packet to a virtual queue.
void vAQM(Packet *p, vQueue *vq);	Updates internal vAQM state during packet reception.
int estimateQlen(vQueue *vq) const;	Returns an estimate of a virtual queue's length.
int process(Packet *p, vQueue *vq);	Defines packet processing. Modify, duplicate, drop, etc.

Table 1: The NetBump API. The user may extend any of the provided functions as needed.

```
Procedure vAQM(Packet *pkt, vQueue *VQ):
1    if (VQ→lastUpdate > 0) {
2        elapsedTime = pkt→timestamp – VQ→lastUpdate
3        drainAmt = elapsedTime * VQ→rate
4        VQ→tokens -= drainAmt
5        VQ→tokens = max(0, VQ→tokens)
6    }
7    VQ→tokens += pkt→len
8    VQ→lastUpdate = pkt→timestamp

Procedure DCTCP(Packet *pkt, vQueue *VQ):
9    if (VQ→tokens > VQ→MaxThresh) {
10       mark(pkt)
11   }
```

Figure 5: The vAQM queue drain estimation and DCTCP. MaxThresh is the ECN marking threshold K.

To make this association, NetBump requires two pieces of information: the mapping of packet destinations to downstream output ports, and the speed of the link attached to that port. The mapping is needed to determine the destination virtual queue for a particular packet, and the link speed is necessary for estimating the virtual queue's drain rate. There are many ways of determining these values: the bump could query neighboring switches (e.g. using SNMP) for their link speeds, or those values could be statically mapped when the bump is configured. For software-defined networks based on OpenFlow [14, 27], the central controller could be queried for host-to-port mappings and link speeds, as well as the network topology. In our evaluation, we statically configure the NetBump with the port-to-host mapping and link speeds.

2. Queue Drain Estimation: The purpose of the queue drain estimation algorithm is to calculate, at the time a packet is received into the bump, the occupancy of the virtual queue associated with the packet (Fig. 5). The virtual queue estimator is a leaky bucket that is filled as packets are assigned to it, and drained according to a fixed drain rate determined by the port speed [43].

Lines 1-6 implement the leaky bucket. First, the elapsed time since the last packet arrived to this virtual queue is calculated. This elapsed time is multiplied by a physical port's rate to calculate how many bytes would have left the downstream queue since receiving the last packet. The physical port's drain rate comes from the link speed of the downstream switch or endhost. This amount is then subtracted from the current estimate (or set to zero, if the result would be negative) of queue occupancy to get an updated occupancy. If this is the first packet to be sent to that port, then the default queue occupancy estimate of 0 is used instead. Lastly, the "last packet arrival" field of the virtual queue is updated accordingly.

A key design decision in NetBump is whether to couple the size of the virtual queue inside the bump with the actual size of the physical buffer in the downstream switch. If we knew the size of the downstream queue, then we could set the maximum allowed occupancy of the virtual queue accordingly. This would be challenging in general, since switches do not typically export the maximum queue size programmatically. Furthermore, for shared buffer switches, this quantity might change based on the instantaneous traffic in the network. In fact, by assuming a small buffer size in the virtual queue within NetBump, we can constrain the flow of packets to reduce actual buffer occupancy throughout the network. Thus, assuming small buffers in our virtual queues has beneficial effects on the network, and simplifies NetBump's design.

3. Packet Marking/Dropping: At line 9 in Fig. 5, NetBump has an estimate for the virtual queue occupancy. Here a variety of actions can be performed, based on the application implemented in the bump. The example code shows the Data Center TCP (DCTCP) application [2]. In this example, there is a "min" limit that results in packet marking, and a "max" limit that results in packet dropping. Packet marking takes the form of setting the ECN bits in the header, and dropping is performed simply in software.

4. Extensible Processing Stage: In addition to the vAQM estimation and packet marking/dropping functionality built into the basic NetBump pipeline, developers can optionally include arbitrary additional packet processing. NetBump developers can include extensions to process packet streams. This API is quite simple, in that the extension is called once per packet, which is represented by a pointer to the packet data and length field. Developers can read, modify, and adjust the packet arbitrarily before re-injecting the packet back into the NetBump pipeline (or dropping it entirely).

Packets destined to particular virtual queues can be forwarded to different extensions, each running in its own thread, and coordinating packet reception from the pipeline through a shared producer-consumer queue. By relying on multi-core processors, each extension can be isolated to run on its own core. This has the advantage that any latency induced by an extension only affects the traffic subject to that extension. Furthermore, correctness or performance bugs in an extension only affects the subset of traffic enqueued in the virtual queues serving that extension. This enables an incremental "opt-in" experimental platform for introducing new NetBump functionality into the production network.

An advantage of the NetBump architecture is that packets travel a single path from the input port to the output port. Thus, unlike multi-port software routers, here packets can remain entirely on a single core, and stay within a single cache hierarchy. The only point

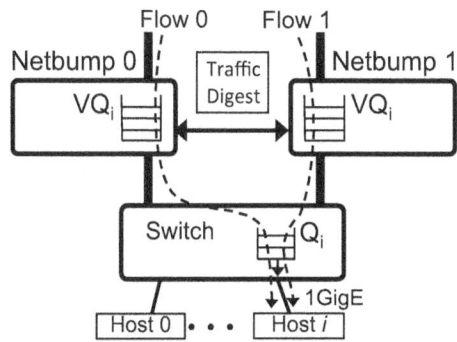

Figure 6: Flow0 and Flow1 both destined to *Host_i*, with two NetBumps monitoring the same Q_i buffer.

of synchronization is the shared vAQM data structure, and we study the overhead of this synchronization and the resulting lock contention in § 6.2.5.

3.2 Scaling NetBump

Managing packet flows in multipath environments requires that NetBump scale with the number of links carrying a particular set of flows. This scaling operates within two distinct regions. First, supporting additional links by adding NICs and CPU cores to a single server, and second, through a distributed deployment model.

3.2.1 Multi-link NetBump

For environments in which packets headed to a single destination might travel over multiple paths, it is possible to scale NetBump by simply adding new NICs and CPU cores. For example, a top-of-rack switch with two 10Gbps uplinks would meet these requirements. Here, a single server is only limited in the number of links that it can support by the amount of PCI bandwidth and the number of CPU cores. Each pair of network interfaces supports a single link (10Gbps in, and 10Gbps out), and PCIe gen 2 supports up to three such bi-directional links. In this case, "Multi-link" NetBump is still conceptually simpler than a software-based router, since packets still follow a single-input, single-output path. Each supported link is handled independently inside the bump, and we can assign to it a dedicated CPU core. The only commonality between these links is the vAQM table, which is shared across the links.

3.2.2 Distributed NetBump

For multi-path environments, where NetBumps must be physically separated, or for those with more links than are supported by a single server, we consider a distributed NetBump implementation. Naturally, if multiple NetBumps contribute packets to a shared downstream buffer, they must exchange updates to maintain accurate VQ estimates. Note that the vAQM table maintains queue estimates for each of neighboring switch's ports (or a monitored subset).

In this case, where we assume the topology (adjacency matrix and link speeds) to be known in advance, NetBumps update their immediate neighbor bumps about the traffic they have processed (Fig. 6). Hence, updates are not the queue estimate itself, but tuples of individual packet lengths and physical downstream switch and port IDs, so that forwarding tables need not be distributed. Each *source* NetBump sends an update to its monitoring neighbors at a given tunable frequency (e.g. per packet, or batched), and each *des-*

tination NetBump calculates a new queue estimate by merging its previous estimate with the traffic update from its neighbor, according to the algorithm in Fig. 5. In this design, updates are tiny; 4B per monitored flow packet (i.e. 2B for packet size and 2B for the port identifier). This translates to about 3MB/s of control traffic per 10Gbps monitored flow. Note also that updates can be transmitted on a dedicated link, or in-band with the monitored traffic. We chose the latter for our Distributed NetBump implementation.

The above technique introduces two possible sources of queue estimation error: 1) batching updates causes estimates to be slightly stale, and since packet sizes are not uniform, the individual packet components of a virtual queue and their respective order would not necessarily be the same, and 2) the propagation delay of the update. Despite this incremental calculation, the estimation naturally synchronizes whenever the buffer occupancy is near its empty/full boundaries.

4. DEPLOYED APPLICATIONS

In this section, we describe the design and implementation of two vAQM applications we developed with NetBump. In addition to Data Center TCP and Quantized Congestion Notification applications described here, the tech report version of this work also includes the implementation of Random Early Detection and rate-limiting applications [34].

4.1 Data Center TCP

We implemented Data Center TCP (DCTCP) [2] on NetBump. The purpose of DCTCP is to improve the behavior of TCP in data center environments, specifically by reducing queue buildup, buffer pressure, and incast. It requires changes to the endhosts as well as network switches. A DCTCP-enabled switch marks the ECN bits of packets when the size of the output buffer in the switch is greater than the marking threshold K. Unlike RED, this marking is based on instantaneous queue size, rather than a smoothed average. The receiver is responsible for signaling back to the sender the particular sequence of marked packets (see [2] for a complete description), and the sender maintains an estimate α of the fraction of marked packets. Unlike a standard sender that cuts the congestion window in half when it receives an ECN-marked acknowledgment, a DCTCP sender reduces its rate according to: $cwnd \leftarrow cwnd * (1 - \alpha/2)$. We support DCTCP in the endhosts by using a modified Linux TCP stack supplied by Kabbani and Alizadeh [15].

Implementing DCTCP in NetBump was straightforward. Here, we mark based on the instantaneous queue size instead of computing a smoothed queue average of the downstream physical queue occupancy. Next, we set both `LowThresh` and `HighThresh` to the supplied K (chosen to be 20 MTU packets, based on the authors' guidelines [2]). We experimented with other values of K, and found that it had little noticeable effect on aggregate throughput or rate convergence time.

4.2 Quantized Congestion Notification

We also implemented the IEEE 802.1Qau-QCN L2 Quantized Congestion Control (QCN) algorithm [1]. QCN-enabled switches monitor their output queue occupancies and when sensing congestion, they send feedback packets to upstream *Reaction Points*. The sender NIC is responsible for adjusting the rate according to a given formula. For every QCN-enabled link, there are two basic algorithms:

Congestion Point (CP): For every output queue, the switch calculates a *feedback measure* (F_b) whenever a new frame is queued. This measure captures the *rate* at which the queue is building up (Q_δ), as well as the *difference* (Q_{off}) between the current occupancy and a desired equilibrium threshold (Q_{eq}, assumed to be 20% of the physical buffer). If Q denotes the current queue occupancy, Q_{old} is the previous iteration, and w is the weight controlling rate build-up, then:

$$Q_{off} = Q - Q_{eq} \qquad Q_\delta = Q - Q_{old}$$

$$F_b = -(Q_{off} + wQ_\delta)$$

Based on F_b, the switch probabilistically generates a congestion notification frame proportional to the severity of the congestion (the probability profile is similar to RED [12], i.e. it starts from 1% and plateaus at 10% when $|F_b| \geq F_{bmax}$). This QCN frame is destined to the upstream reaction point from which the just-added frame was received. If $F_b \geq 0$, then there is no congestion and no notification is generated.

Reaction Point (RP): Since the network generates signals for rate decreases, QCN senders must probe for available bandwidth gradually until another notification is received. The reaction point algorithm has two phases: Fast-Recovery (FR) and Additive-Increase (AI). This is similar to, but independent from, BIC-TCP's dynamic probing.

The RP algorithm keeps track of the sending Target Rate (TR) and Current Rate (CR). When a congestion control frame is received, the RP algorithm immediately enters the Fast Recovery phase; it sets the target rate to the current rate, and reduces the current rate by an amount proportional to the congestion feedback (by at most 1/2). Barring further congestion notifications, it tries to recover the lost bandwidth by setting the current rate to the average of the current and target rates, once every *cycle* (where a cycle is defined in the base byte-counter model as 100 frames). The RP exits the Fast Recovery phase after five cycles, and enters the Additive Increase phase, where the RP continually probes for more bandwidth by adding a constant increase to its target rate (1.5Mbps in our implementation), and again setting the current sending rate to the average of the CR and TR.

5. IMPLEMENTATION

NetBump can be implemented using a wide variety of underlying technologies, either in hardware or in software. We evaluated three such choices: 1) the stock Linux-based forwarding path, 2) the RouteBricks software router, and 3) a user-level application relying on kernel-bypass network APIs to read and write packets directly to the network. We call this last implementation *UNetBump*. We show in Fig. 7 the latency distributions of these systems when forwarding 1500B packets at 10Gbps (except Linux with 9000B). The baseline for comparison being a simple loopback.

All of our implementations are deployed on HP DL380G6 servers with two Intel E5520 four-core CPUs, each operating at 2.26GHz with 8MB of cache. These servers have 24 GB of DRAM separated into two 12GB banks, operating at a speed of 1066MHz. For the Linux and UNetBump implementations, we use an 8-lane Myricom 10G-PCIE2-8B2-2S+E dual-port 10Gbps NIC which has two SFP+ interfaces, plugged into a PCI-Express Gen 2 bus. For RouteBricks, we used an Intel E10G42AFDA dual-port 10Gbps NIC (using an 82598EB controller) with two SFP+ interfaces.

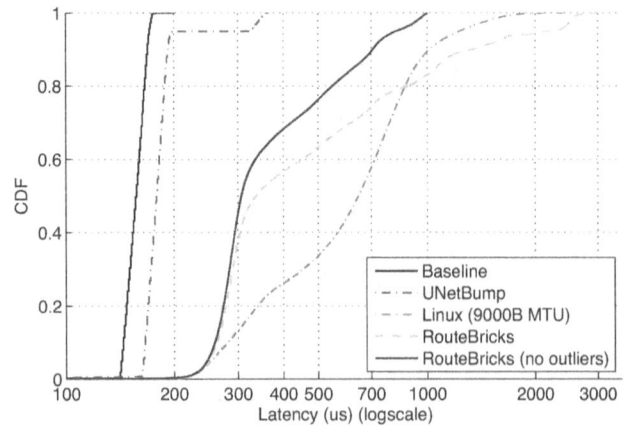

Figure 7: Forwarding latency at line rate of baseline, UNetBump, Linux, RouteBricks (batching factor of 16, and a Click burst factor of 16), with and without an outlier queue.

5.1 Linux

The Linux kernel natively supports a complete IP forwarding path, including a configurable set of queuing disciplines managed through the "traffic control (tc)" extensions [23]. Linux `tc` supports flow and packet shaping, scheduling, policing, and dropping. While `tc` supports a variety of queuing disciplines, it does not support managing the queues of remote switches. This support would have to be added to the kernel. In our evaluation we used Linux kernel version 2.6.32, and found that the latency overheads of the Linux forwarding path were very high, with a mean latency above $500\mu s$, and a 99th percentile above $1500\mu s$. Furthermore, our evaluation found that Linux could not forward non-Jumbo frames at speeds approaching 10Gbps (and certainly not with minimum-sized packets). This is because the kernel implementation incurs high per packet and per byte overhead [35]. Based on these microbenchmarks, we decided not to further consider Linux as an implementation alternative.

5.2 RouteBricks

RouteBricks [8] is a high-throughput software router implementation built using Click's core, extensive element library, and specification language. It increases the scalability of Click in two ways–by improving the forwarding rate within a single server, and by federating a set of servers to support throughputs beyond the capabilities of a single server. To improve the scalability within a single server, RouteBricks relies on a re-architected NIC driver that supports multiple queues per physical interface. This enables multiple cores to read and write packets from the NIC without imposing lock contention, which greatly improves performance [7, 26, 47]. Currently, RouteBricks works only with the `ixgbe` device driver, which delivers packets out of the driver in fixed-size batches of 16 packets each. We built a single-node RouteBricks server using the HP server architecture described above, but with the Intel E10G42AFDA NIC (the only available the RouteBricks driver patch still supported). This server used the Intel `ixgbe` driver (version 1.3.56.5), with a batching factor of 16. The use of this batching driver improves throughput by amortizing the overhead of transferring those packets, at the cost of increased latency on an individual packet basis. Indeed RouteBricks was designed for

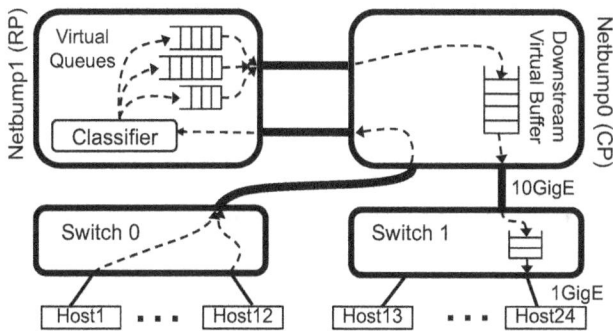

Figure 8: Two-rack 802.1Qau-QCN and DCTCP Testbed

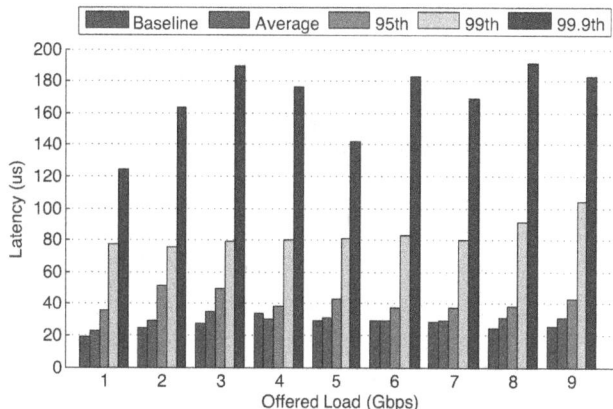

Figure 9: Latency percentiles imposed by UNetBump vs. offered load. Baseline is the loopback measurement overhead.

high throughput, not low-latency. There is nothing in the Click or RouteBricks model that precludes low-latency forwarding, however for this work we chose not to use RouteBricks.

5.3 UNetBump

In user-level networking, instead of having the kernel deliver and demultiplex packets, the NIC instead delivers packets directly to the application. This is typically coupled with kernel-bypass support, which enables the NIC to directly copy packets into the application's memory region.

User-level networking is a well-studied approach that has been implemented in a number of commercially-available products [42]. Myricom offers Ethernet NICs with user-level networking APIs that we use in our evaluation [28]. There have been at least two efforts to create an open and cross-vendor API to user-level, kernel-bypass network APIs [31, 36]. In this paper, we re-evaluate the use of user-level networking to support low-latency applications, especially those requiring low latency variation. Note that it is possible to layer the RouteBricks/Click runtime on top of the user-level, kernel-bypass APIs we use in UNetBump.

6. EVALUATION

Our evaluation seeks to answer the following: 1) How expressive is NetBump? 2) How easy is it to deploy applications? 3) How effective is vAQM estimation in practice? 4) What are the latency overheads and throughput limitations?

To answer these, we built and deployed a set of NetBump prototypes in our experimental testbed. We started by evaluating the baseline latency and latency variation of these prototypes, and based on these measurements, we proceeded with construction of UNetBump, a fully-functional prototype based on user-level networking APIs. We then evaluate a range of AQM functionalities with UNetBump.

6.1 Testbed Environment

Our experimental testbed consists of a set dual-processor Nehalem server described above, using either Myricom NICs, or in the case of RouteBricks, the Intel NIC. The Myricom NICs use the Sniffer10G driver version 1.1.0b3. We use copper direct-attach SFP+ connectors to interconnect the 10Gbps endhosts to our NetBumps. Experiments with 1Gbps endhosts rely on a pair of SMC 8748L2 switches that each have 1.5MB of shared buffering across all ports. Each SMC switch has a 10Gbps uplink that we connect to the appropriate NetBump.

We evaluate NetBump in three different contexts. The first is in microbenchmark, to examine its throughput and latency characteristics. Here we deploy NetBump as a loopback (simply connecting the two ports to the same host) to eliminate the effects of clock skew and synchronization. The second simply puts a NetBump inline between two machines, and tests NetBump's operation at full 10Gbps. Separating the source and destination to different machines enables throughput measurement with real traffic.

The third testbed, Fig. 8, evaluates NetBump in a realistic data center environment in which it might be deployed right above the top-of-rack switch. Here, we have two twelve-node racks of endhosts, each connected to a 1Gbps switch. A 10Gbps uplink connects the two 1Gbps switches and the NetBump is deployed inline with those uplinks. In this case, the NetBump actually has four 10Gbps interfaces–two to the uplinks of each of the two SMC 1Gbps switches, and two that connect to a second NetBump. We use this testbed to evaluate 802.1Qau-QCN, with one NetBump acting as the Congestion Point (CP) and the other as the Reaction Point (RP).

6.2 Microbenchmarks

6.2.1 NetBump Latency

A key metric for evaluation is the latency overhead. To measure this, we use a loopback testbed and had a packet generator on the client host send packets onto the wire, through the NetBump, and back to itself. To calibrate, we also replace the NetBump with a simple loopback wire, which gives us the baseline latency overhead of the measurement host itself. We subtract this latency from the observed latency with the NetBump in place, giving us the latency of just the NetBump. We generated a constant stream of 1500-byte packets sent at configurable rates (Fig. 9).

For UNetBump, the latency is quite low for the majority of forwarded packets. There is a jump in latency at the tail due to NIC packet batching when they arrive above a certain rate. There is no way to disable this batching in software, even though we were only using a single CPU core which could have serviced a higher packet rate without requiring batching. The forwarding performance of UNetBump was sufficient to keep up with line rate using minimum-sized packets and a single CPU core.

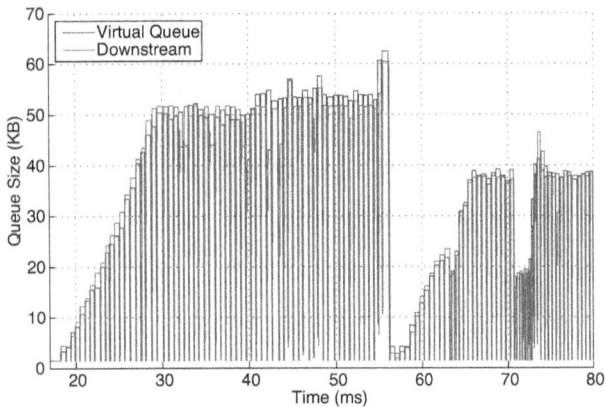

(a) Virtual queue size vs. actual downstream queue. Running an **iperf** TCP session between two 10G hosts, rate-limited to 1Gbps with a 40KB buffer downstream to induce congestion.

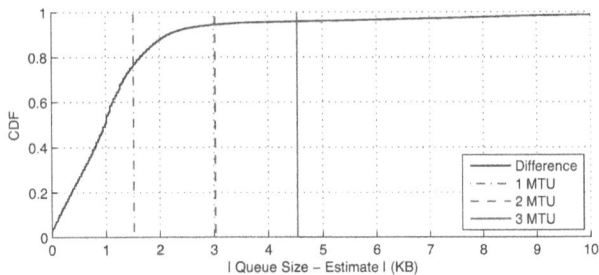

(b) CDF of the queue size difference. The estimate is within two 1500B MTUs 95% of the time.

Figure 10: Downstream vAQM estimation accuracy.

6.2.2 vAQM Estimation Accuracy

To evaluate the accuracy of the vAQM estimation, we ran **iperf** sessions between two hosts, connected in series by a NetBump and another pass-through machine (which records the timestamps of incoming frames). Since we cannot export physical buffer occupancy of commercial switches, we use the frame timestamps and lengths from the downstream pass-through machine to recreate the output buffer size over time, knowing the drain-rate. Fig. 10 shows the NetBump virtual queue size vs. the actual downstream queue. The estimate was within two MTUs 95% of the time.

6.2.3 Distributed NetBump

We also measured the accuracy of queue estimation when multiple NetBumps exchange updates to estimate a common downstream queue. In the first experiment, measure the effect of update latency on queue estimate accuracy. We varied the timestamp interleaving of two TCP **iperf** flows that share a downstream queue in order to simulate receiving delayed updates from a neighboring NetBump. Fig. 11 shows the CDF of the difference between the delayed inter-bump estimation and the in-sync version; Even when update latency was 25μs, the difference was always under 2MTU.

Next, we show the accuracy of NetBump's queue estimating of a downstream queue, based solely on updates from its neighbor. In our implementation, the updates are transmitted in-band with the monitored traffic. Fig. 12 gives the CDF of the difference between the actual queue size and the distributed NetBump estimate. We

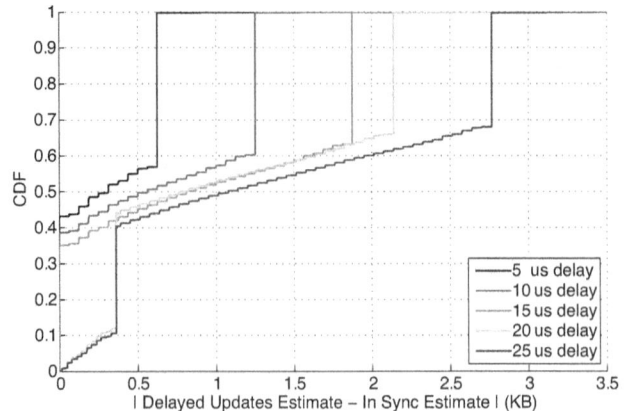

Figure 11: CDF of the absolute difference between the queue estimate with delayed updates and the in-sync version. The combined throughput is rate-limited to 5Gbps, and the downstream buffer is 40KB.

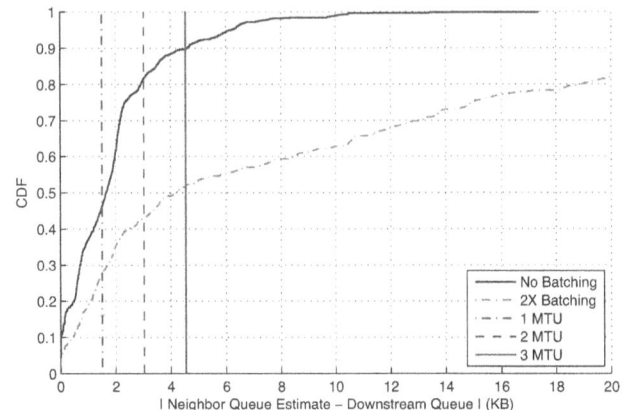

Figure 12: CDF of the difference between actual queue size and the Distributed NetBump estimate using a 1Gbps rate-limited TCP flow and a 40KB buffer. The estimating NetBump does not observe the monitored traffic directly.

observe that the estimate is within 3MTUs 90% of the time. Note, however, the effect of update batching: estimates quickly drift when updates are delayed. Fig. 13 shows a typical relative difference CDF when background elephant flows are present (i.e. some flows are observed directly, and others indirectly through updates).

6.2.4 CPU Affinity Effect

One of the challenges of designing NetBump was not only maintaining a low average latency, but also reducing variance. Modern CPU architectures provide separate cores on the same die and physically separate memory across multiple Non-Uniform Memory Access (NUMA) banks. This means that access time to memory banks changes based on which core issues a given request. To reduce latency outliers, we allocated memory to each UNetBump thread from the same NUMA domain as the CPU core it was scheduled to.

Given the significant additional latency that may be introduced by the unmodified Linux kernel scheduler, we compare latency of NetBump with and without CPU-affinity and scheduler modifications. Our control experiment uses default scheduling. To improve

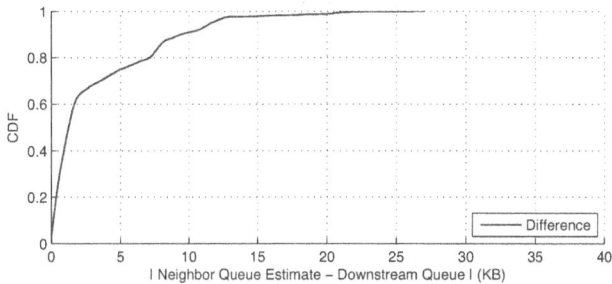

Figure 13: Typical distributed NetBump relative error with background elephant flows.

Latency (μs)	Avg	95th	99th	99.9th	Max
No Affinity	32	39	76	1,322	3,630
With Affinity	30	42	83	169	208

Table 2: UNetBump latency percentiles vs. CPU core affinity.

on this, we exclude all but one of the CPU cores from the default scheduler, and ensure that the UNetBump user-space programs execute on the reserved cores. We then examined the average, 95th, 99th, 99.9th, and maximum latencies through NetBump compared to the baseline (Table 2). CPU-affinity had a minor effect on latency on average, but was most pronounced on outlier packets. The maximum observed latency was 17 times smaller with CPU-affinity at the 99.9th percentile, showing the importance of explicit resource isolation in low-latency deployments.

6.2.5 Multicore Performance

In UNetBump, basic vAQM estimation can be done at 10Gbps using only a single CPU core. However, to support higher link rates, additional cores might be necessary. The NIC itself will partition flows across CPU cores using a hardware hash function. In this scenario, a user-space thread would be responsible for handling each ring pair, and the only time these threads must synchronize would be when updating the vAQM state table. To evaluate the effect of this synchronization on the latency of NetBump in a multi-threaded implementation, we examined the effect of vAQM table lock overhead. As a baseline, a single-threaded forwarding pipeline (FP) has a latency of $29.16\mu s$. Running NetBump with two FPs (two ring pairs in the NIC and each FP running on its own core) increased that latency by 17.9% to $35.5\mu s$. Further running NetBump with four FPs on four cores increased the latency by an additional 1.95% to $36.8\mu s$. Thus we find that the synchronization overhead is minimal to gain back a four-fold increase in computation per packet, or alternatively, a four-fold increase in supported line rate. A key observation is that NetBump avoids some of the required synchronization overheads found in software routers [7, 9, 26] with multiple ports, since in NetBump each input port only forwards to a single output port, preventing packets from spanning cores or causing contention on shared output ports.

6.3 Deployed Applications

One metric highlighting the ease of writing new applications with NetBump is shown in Table 3. Most of our applications took only 10s of lines of code, and QCN, which is much more complex, was

Application	Lines of Code
NetBump core	940
DCTCP	29
QCN	464

Table 3: Coding effort for NetBump and some of its applications.

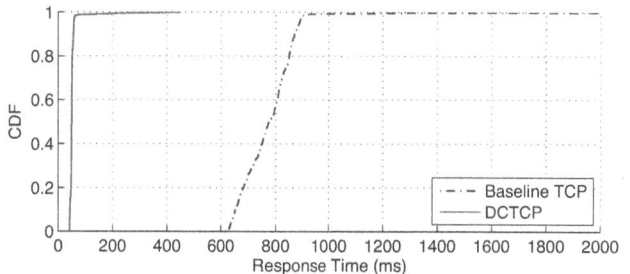

Figure 15: Baseline TCP (CUBIC) and DCTCP response times for short RPC-type flows in the presence of background elephant flows.

written in less than 500 lines of code. The time commitment ranged from hours to a couple of days in the case of QCN. We now examine each application in detail.

6.3.1 Data Center TCP

The next experiment represents a recreation of the DCTCP convergence test presented by Alizadeh et al. [2] performed in our two-rack testbed (Fig. 8). Five source nodes each open a TCP connection to one of five destination nodes in 25 second intervals. In the baseline TCP case (Fig. 14(a)), due to buffer pressure and a drop-tail queuing discipline, the bandwidth is shared unfairly, resulting in a wide oscillation of throughput and unfair sending rate among the flows. Fig. 14(b) shows the throughput of DCTCP-enabled endpoints and a DCTCP vAQM strategy in the NetBump. Like in the original DCTCP work, here the fair sharing of network bandwidth results from the lower queue utilization afforded by senders backing off in response to NetBump-set ECN signals.

Another contribution of reduced queue buildup is better support for mixtures of latency-sensitive and long-lived flows. Fig. 15 shows the CDF of response time for 10,000 RPC-type requests in the presence of two large elephant flows, comparing stock TCP endpoints without NetBump DCTCP support. This figure recreates a key DCTCP result: signaling the long flows to reduce their rates results in smaller queues, lower RTT, and in the end, shorter response times.

6.3.2 Quantized Congestion Notification

Another example of how the NetBump programming model enabled rapid prototyping and evaluation of new protocols was deploying 802.1Qau-QCN. Our implementation of QCN is 464 lines of code, and took around 2-3 days to write and debug. Developing QCN within NetBump enabled us to easily tune parameters and evaluate their effect. This was especially important given QCN's novelty, and the lack of other tools or simulations we could have used to study it. Using the testbed topology of Fig. 8, we use NetBump0 as the CP, and NetBump1 as the RP. In our RP, we chose a virtual queue size of 100KB (and Q_{eq} at 20KB).

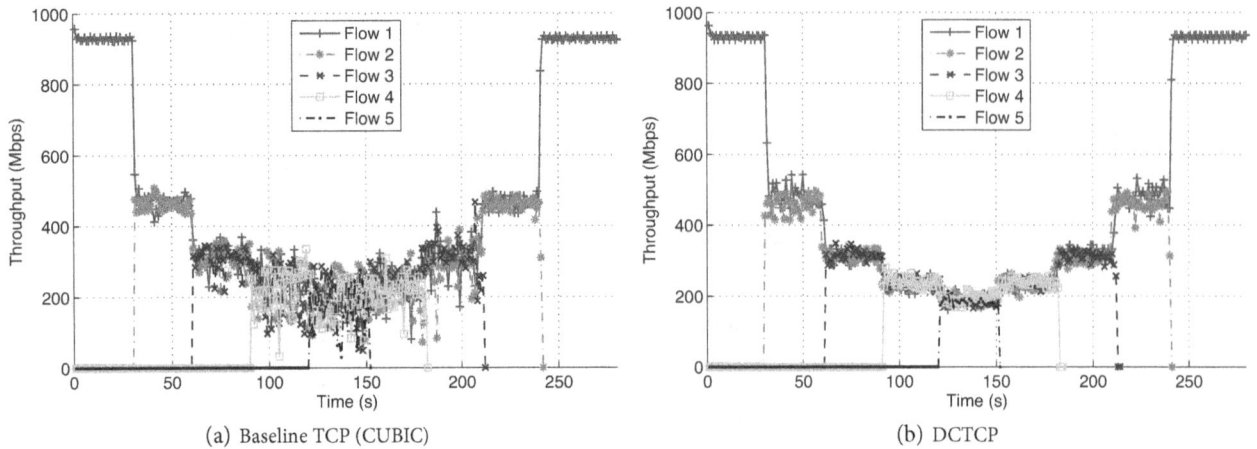

(a) Baseline TCP (CUBIC)   (b) DCTCP

Figure 14: The effect on fairness and convergence of DCTCP on five flows sharing a bottleneck link.

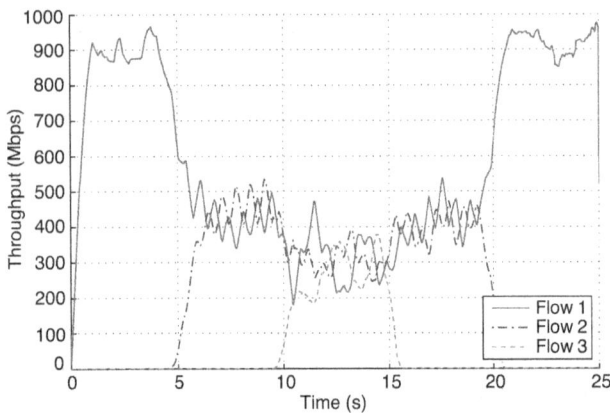

Figure 16: QCN with three 1Gbps UDP flows. With QCN enabled, the RP virtual queue occupancy never exceeded 40%, as opposed to persistent drops downstream without.

We found that the QCN feedback loop tends to be more stable when the frequency of messages is higher and their effect smaller. For this reason, we use $F_{bmax} = 32$, and plateau the probability profile at 20%. Due to the burstiness in packet arrival, we also decreased w to 1 to avoid unnecessary rate drops. Our implementation also considered the relative flow weights in the entire queue when choosing which flow to rate-limit, rather than using just the current packet. We use the byte counter-only model of RP in our implementation. For the Additive Increase phase, we use cycles of 100 packets, and an increase of 1.5Mbps (to show the convergence of the virtual port current rates), and 600 packet cycles for the Fast Recovery phase. We show in Fig. 16 the throughput of three 1Gbps UDP flows sharing the same bottleneck link. Without QCN, the downstream buffer would be persistently overwhelmed by the three UDP flows from 5-20s, but with QCN enabled, congestion is pushed upstream and the virtual queue occupancy never exceeded 40%, thereby preventing drops for potential mice flows.

7. RELATED WORK

Virtual Queuing and AQM: In virtual queuing (VQ), metadata about an incoming packet stream is maintained to simulate the behavior of a hypothetical physical queue. We differ from previous work in that we maintain VQs outside of the switch itself. VQ provides a basis for a variety of active queue management (AQM) techniques. AQM manipulates packets in buffers in the network to enact changes in the control loop of that traffic, typically to reduce packet drops, queue overflows, or buffer sizes. One proposal, Active Virtual Queue [21], reduces queue sizes in traffic with small flows, which typically pose challenges for the TCP control loop. Due to the inefficiency of RED's dropping packets to signal congestion, the Early Congestion Notification (ECN) [22] field was developed to decouple packet drops from congestion indicators. Several proposals for improving on RED have been made [4], including Data Center TCP (DCTCP) [2]. Quantized congestion notification [1] was proposed as a congestion control mechanism for non-TCP traffic, and can respond faster than the round-trip time. Implementations of QCN have been developed on 1Gbps networks [24], as well as emulated within FPGAs at 10Gbps networks [30]. Our deployment is done at 10Gbps and distributed across multiple network hops. Approximate-Fairness QCN (AF-QCN) [16] is an extension that biases input links' feedback by the ratio of their queue occupancy.

Datapath Programming in Software: Software-based packet switches and routers have a long history as a platform for rapidly developing new functionality. Click [20] is a modular software router consisting of a pipeline of simple packet-processing building blocks. Click's library of modules can be extended by writing code in C++ designed to work in the Linux kernel or userspace. RouteBricks [8] focused on scaling out a Click runtime to support forwarding rates in tens of Gbps by distribution of packet processing across cores, and across a small cluster of servers. ServerSwitch [24], allows programming commodity Ethernet switching chips (with matching/modification of standard header fields), but delegates general packet processing to the CPU (e.g. for XCP). Besides avoiding crossing the kernel/user-space boundary, NetBump allows *arbitrary* packet modification at line rate. A key distinction is that these projects are all multi-port software switches focused on packet routing, while NetBump focuses on pass-through virtual queuing within a pre-existing switching layer. SideCar [40], on the other hand, is a recent proposal to delegate a small fraction of traffic requiring special processing from the ToR switch to a companion server. While superficially similar, the redirection and traffic sampling are not applicable for NetBump's vAQM use-case, where low-latency is a key design requirement. For these reasons, we consider these efforts to be orthogonal to this work.

Several efforts have looked at ways of mapping packet handling tasks necessary to support software routers to multi-core, multi-NIC queue commodity servers. Egi et al. [9], and Dobrescu et al. [7] investigate the effects of casting forwarding paths across multiple cores, and find that minimizing core transitions is necessary for high performance. NetBump takes a similar approach to the "split traffic" and "cloning" functionality described, in which an entire forwarding path resides on a single core and cache hierarchy. Manesh et al. [26] study the performance of multi-queue NICs as applied to packet forwarding workloads. They found that increasing the number of NIC queues led to reduced performance, and were not able to forward minimum-sized packets at line rate. We did not find such a limitation with our particular hardware NICs. However, Based on our experiences we fully support their recommendations for new NIC APIs for handling packet forwarding for applications.

Typically, the OS kernel translates streams of raw packets to and from a higher-level interface such as a socket. And While sockets are a useful networking primitive, the required kernel involvement can become a bottleneck, and several alternative user-level networking techniques have been developed [5, 10, 44, 46]. In user-level networking, user-programs are responsible for TCP sequence reassembly, retransmission, etc., and this is typically coupled with zero-copy buffering, where a packet is stored in shared memory with target applications. Kernel-bypass drivers also enable applications to directly access packets from NIC memory, avoiding kernel involvement on the datapath. Commercially-available NICs already support these mechanisms [6, 28, 33, 41, 42]. NetBump is implemented at user-level, and relies on zero-copy, kernel-bypass drivers.

Datapath Programming in Hardware: One drawback of software-based packet forwarding is that historically it has suffered from low performance, and alternative hardware architectures have been proposed. Perhaps the best-known and most widely-used hardware forwarding platform is the NetFPGA [29], a powerful development tool for FPGA devices; however, the complexity of FPGA programming remains a challenge. On top of NetFPGA, the CAFE project [25] enables users to more easily develop forwarding engines based on custom and non-standard packet header formats. RiceNIC [38] is similarly based on an FPGA, but provides additional per-packet computing through two embedded PowerPC processors.

Two recent projects sought to address the programming challenge: Switchblade [3] provides modular building blocks that can support a wide variety of datapaths, and Chimpp [37] converts datapaths specified in the Click language into Verilog code suitable for an FPGA. In addition, network processors (NPs) [39] have been used to prototype and deploy new network functionality. They have the disadvantage of a difficult-to-use programming model and limited production runs. Their primary advantage is their multiple functional units, providing significant parallelism to support faster data rates. Commodity CPUs have since greatly increased their number of cores, and can also provide significant per-packet processing at high line rates.

8. CONCLUSIONS

A major barrier to developing and deploying new network functionality is the difficulty of programming the network datapath. In this work, we presented NetBump, a platform for developing, experimenting with, and deploying alternative packet buffering and queuing disciplines with minimal intrusiveness and at low latency. NetBump leaves existing switches and endhosts unmodified. It acts as a "bump on the wire," examining, optionally modifying, and forwarding packets at line rate in tens of microseconds to implement a variety of virtual active queuing disciplines and congestion control protocols implemented in user-space. We built and deployed several applications with NetBump, including DCTCP and 802.1Qau-QCN. These applications were quickly developed in hours or days, and required only tens or hundreds of lines of code in total. The adoption of multi-core processors, along with kernel-bypass commodity NICs, provides a feasible platform to deploy data modifications written in user-space at line rate. Our experience has shown that NetBump is a useful and practical platform for prototyping and deploying new network functionality in real data center environments.

9. ACKNOWLEDGEMENTS

We would like to thank Brian Dunne for providing the SMC switches, as well as the anonymous reviewers of this work for their valuable insight and advice. This work is supported in part by the National Science Foundation (#CSR-1116079 and #CNS-1053757).

10. REFERENCES

[1] M. Alizadeh, B. Atikoglu, A. Kabbani, A. Lakshmikantha, R. Pan, B. Prabhakar, and M. Seaman. Data Center Transport Mechanisms: Congestion Control Theory and IEEE Standardization. In *Allerton CCC, 2008*.

[2] M. Alizadeh, A. Greenberg, D. A. Maltz, J. Padhye, P. Patel, B. Prabhakar, S. Sengupta, and M. Sridharan. Data center TCP (DCTCP). In *ACM SIGCOMM 2010*.

[3] M. B. Anwer, M. Motiwala, M. b. Tariq, and N. Feamster. SwitchBlade: A Platform for Rapid Deployment of Network Protocols on Programmable Hardware. In *ACM SIGCOMM 2010*.

[4] J. Aweya, M. Ouellette, D. Y. Montuno, and K. Felske. Rate-based Proportional-integral Control Scheme for Active Queue Management. *IJNM*, 16, 2006.

[5] P. Buonadonna, A. Geweke, and D. Culler. An Implementation and Analysis of the Virtual Interface Architecture. In *ACM/IEEE CDROM 1998*.

[6] Chelsio Network Interface. http://www.chelsio.com.

[7] M. Dobrescu, K. Argyraki, G. Iannaccone, M. Manesh, and S. Ratnasamy. Controlling Parallelism in a Multicore Software Router. In *ACM Presto 2010*.

[8] M. Dobrescu, N. Egi, K. Argyraki, B.-G. Chun, K. Fall, G. Iannaccone, A. Knies, M. Manesh, and S. Ratnasamy. RouteBricks: Exploiting Parallelism to Scale Software Routers. In *ACM SOSP 2009*.

[9] N. Egi, A. Greenhalgh, M. Handley, M. Hoerdt, F. Huici, L. Mathy, and P. Papadimitriou. Forwarding Path Architectures for Multicore Software Routers. In *ACM Presto 2010*.

[10] D. Ely, S. Savage, and D. Wetherall. Alpine: A User-level Infrastructure for Network Protocol Development. In *USITS 2001.*

[11] S. Floyd. TCP and Explicit Congestion Notification. *ACM SIGCOMM CCR*, 24(5), 1994.

[12] S. Floyd and V. Jacobson. Random Early Detection Gateways for Congestion Avoidance. *IEEE/ACM TON*, 1(4), 1993.

[13] R. J. Gibbens and F. Kelly. Distributed Connection Acceptance Control for a Connectionless Network. In *Teletraffic Engineering in a Competitive World*. Elsevier, 1999.

[14] N. Gude, T. Koponen, J. Pettit, B. Pfaff, M. Casado, N. McKeown, and S. Shenker. NOX: Towards an Operating System for Networks. *ACM SIGCOMM CCR*, 38(3), 2008.

[15] A. Kabbani and M. Alizadeh. Personal Communication, 2011.

[16] A. Kabbani, M. Alizadeh, M. Yasuda, R. Pan, and B. Prabhakar. AF-QCN: Approximate Fairness with Quantized Congestion Notification for Multi-tenanted Data Centers. In *IEEE Hot Interconnects 2010.*

[17] S. Karandikar, S. Kalyanaraman, P. Bagal, and B. Packer. TCP Rate Control. In *ACM SIGCOMM 2000.*

[18] D. Katabi, M. Handley, and C. Rohrs. Congestion Control for High Bandwidth-delay Product Networks. In *ACM SIGCOMM 2002.*

[19] A. D. Keromytis and J. L. Wright. Transparent Network Security Policy Enforcement. In *USENIX ATC 2000.*

[20] E. Kohler, R. Morris, B. Chen, J. Jannotti, and M. F. Kaashoek. The Click Modular Router. *ACM ToCS*, 18(3), 2000.

[21] S. Kunniyur and R. Srikant. An Adaptive Virtual Queue (AVQ) Algorithm for Active Queue Management. *IEEE/ACM TON*, 12(2), 2004.

[22] A. Kuzmanovic. The Power of Explicit Congestion Notification. In *ACM SIGCOMM 2005.*

[23] Linux Traffic Control howto. `http://tldp.org/HOWTO/Traffic-Control-HOWTO`.

[24] G. Lu, C. Guo, Y. Li, Z. Zhou, T. Yuan, H. Wu, Y. Xiong, R. Gao, and Y. Zhang. ServerSwitch: A Programmable and High Performance Platform for Data Center Networks. In *USENIX NSDI 2011.*

[25] G. Lu, Y. Shi, C. Guo, and Y. Zhang. CAFE: A Configurable Packet Forwarding Engine for Data Center Networks. In *ACM PRESTO 2009.*

[26] M. Manesh, K. Argyraki, M. Dobrescu, N. Egi, K. Fall, G. Iannaccone, E. Kohler, and S. Ratnasamy. Evaluating the Suitability of Server Network Cards for Software Routers. In *ACM PRESTO 2010.*

[27] N. McKeown, T. Anderson, H. Balakrishnan, G. Parulkar, L. Peterson, J. Rexford, S. Shenker, and J. Turner. OpenFlow: Enabling Innovation in Campus Networks. *ACM SIGCOMM CCR*, 38(2), 2008.

[28] Myricom Sniffer10G. `http://www.myricom.com/support/downloads/sniffer.html`.

[29] J. Naous, G. Gibb, S. Bolouki, and N. McKeown. NetFPGA: Reusable Router Architecture for Experimental Research. In *ACM PRESTO 2008.*

[30] NEC/Stanford: 10G QCN Implementation on Hardware. `http://www.ieee802.org/1/files/public/docs2009/au-yasuda-10G-QCN-Implementation-1109.pdf`.

[31] OpenOnload. `http://www.openonload.org`.

[32] J. Padhye, V. Firoiu, D. Towsley, and J. Kurose. Modeling TCP Throughput: A Simple Model and its Empirical Validation. In *ACM SIGCOMM 1998.*

[33] PF_RING Direct NIC Access. `http://www.ntop.org/products/pf_ring/dna/`.

[34] G. Porter, R. Kapoor, S. Das, M. Al-Fares, H. Weatherspoon, B. Prabhakar, and A. Vahdat. NetBump: User-extensible Active Queue Management with Bumps on the Wire. Technical report, CSE, University of California, San Diego, La Jolla, CA, USA, 2012.

[35] L. Rizzo. Netmap: A Novel Framework for Fast Packet I/O. In *USENIX ATC 2012.*

[36] L. Rizzo and M. Landi. Netmap: Memory-mapped Access to Network Devices. In *ACM SIGCOMM 2011.*

[37] E. Rubow, R. McGeer, J. Mogul, and A. Vahdat. Chimpp: A Click-based Programming and Simulation Environment for Reconfigurable Networking Hardware. In *ACM/IEEE ANCS 2010.*

[38] J. Shafer and S. Rixner. RiceNIC: A Reconfigurable Network Interface for Experimental Research and Education. In *ACM ExpCS 2007.*

[39] N. Shah. Understanding Network Processors. Master's thesis, University of California, Berkeley, Calif., 2001.

[40] A. Shieh, S. Kandula, and E. G. Sirer. SideCar: Building Programmable Datacenter Networks without Programmable Switches. In *ACM Hotnets 2010.*

[41] SMC SMC10GPCIe-10BT Network Adapter. `http://www.smc.com/files/AY/DS_SMC10GPCIe-10BT.pdf`.

[42] SolarFlare Solarstorm Network Adapters. `http://www.solarflare.com/Enterprise-10GbE-Adapters`.

[43] J. Turner. New Directions in Communications (or which way to the Information Age?). *IEEE Communications Magazine*, 24(10), 2002.

[44] T. von Eicken, A. Basu, V. Buch, and W. Vogels. U-Net: A User-level Network Interface for Parallel and Distributed Computing. In *ACM SOSP 1995.*

[45] Z. Wang. *Internet QoS: Architectures and Mechanisms for Quality of Service*. Morgan Kaufmann Publishers Inc., San Francisco, CA, USA, 1st edition, 2001.

[46] M. Welsh, A. Basu, and T. von Eicken. ATM and Fast Ethernet Network Interfaces for User-level Communication. In *IEEE HPCA 1997.*

[47] Q. Wu, D. J. Mampilly, and T. Wolf. Distributed Runtime Load-balancing for Software Routers on Homogeneous Many-core Processors. In *ACM Presto 2010.*

Fast Longest Prefix Name Lookup for Content-Centric Network Forwarding

Fu Li[†]   Fuyu Chen[†]   Jianming Wu[†]   Haiyong Xie[†*]

[†]Center for Innovation, Huawei Technologies
[*]Suzhou Institute for Advanced Study, University of Science and Technology of China

ABSTRACT

Longest-prefix name-based forwarding in Content-Centric Networking poses many significant challenges on design of routers and algorithms, because the name space is many orders of magnitude larger and significantly more complex than the IP address space. IP forwarding algorithms are no longer applicable and cannot reach the satisfactory performance. In this paper, we present a framework of fast longest-name-prefix lookup , based on a name space reduction scheme we proposed for content-centric networking. We implement the algorithm using fat tree and extensible hybrid data structures to speed up name lookup. Our evaluations demonstrate that we can achieve name lookup throughput as high as more than 37 Mpps on off-the-shelf general-purpose CPU platforms.

Categories and Subject Descriptors: C.2.0 [**Computer Communication Networks**]: Data communications

Keywords: Content-centric networking, longest-name-prefix lookup

1 Introduction

The emerging Information-Centric Networking (ICN) as a future Internet architecture enables the paradigm shift from the traditional "*host-to-host*" model to the "*object-to-object*" model, where information objects, associated with (potentially human-readable) names, are a first-class abstraction. For instance, Content-Centric Networking (CCN) [2] and Named Data Networking (NDN) [4] are two closely related ICN proposals. In CCN and NDN, routing and forwarding are *name-oriented*. Specifically, in order to retrieve information objects, users send a special type of packets (Interests) bearing the hierarchically structured, URL-like names of those objects. Routers forward these packets based on these names, using a longest-name-prefix-matching algorithm to look up the outbound ports from the name-prefix-based forwarding information base (FIB).

Name-based routing in CCN poses significant new challenges on the design of FIB and lookup algorithms. First, the number of distinct name prefixes is expected to be enormously large. For instance, Google announced in July 2008 that the number of unique URLs it processed had exceeded *1 trillion* [1]. Even after extremely aggressive aggregation, the top-level name prefixes could still be enormously large. Second, statistics have also shown that the number of domain names (*i.e.*, the most aggressively aggregated name prefixes) has *doubled* in the past three years [3] (as of December 2011, the Internet has more than *half a billion* domain names).

As a result, the size of CCN name space is multiple orders of magnitude larger than that of the IP address space (*i.e.*, 2^{32}), and

the complexity of processing structured names with varying lengths is significantly higher than processing 32-bit IP addresses. Therefore, the longest-name-prefix-matching lookup algorithm as well as the data structure for FIB has to be extremely efficient in coping with these challenges. There have been a large number of proposals and designs on IP forwarding algorithms; however, to the best of our knowledge, they either are highly customized for IP's stringent, specific structures and thus are not applicable to CCN, or are tremendously inefficient to deal with CCN name prefixes and thus cannot achieve satisfactory performance.

In this paper, we make a first attempt to address the above challenges. We focus on the general-purpose CPU architectures for the purpose of programmability, flexibility and general availability. We propose a framework of fast longest-prefix name lookup based on a hash-based name space reduction scheme. Inspired by the fat-tree based data structures, we further design an extensible hybrid data structure which combines numerous primitives including hash, fat tree and array. With this hybrid data structure and an optimized lookup algorithm, we can not only substantially increase the number of name prefixes that FIB can store but also significantly improve the lookup throughput. We also leverage the available parallelism in modern CPUs to further improve the lookup performance, achieving name lookup throughput as high as 37 million packets per second on off-the-shelf general-purpose CPU platforms.

2 A Framework for Content-Centric Longest-Name-Prefix Lookup

Our framework for CCN longest-prefix name lookup consists of two key components: name (prefix) transformation and FIB instrumentation, as shown in Figure 1. There are two challenges that we address in this framework: how to do longest-prefix name lookup on the hierarchically structured, length-varying, URL-like name prefixes, and what data structure to efficiently store a large number of name prefixes.

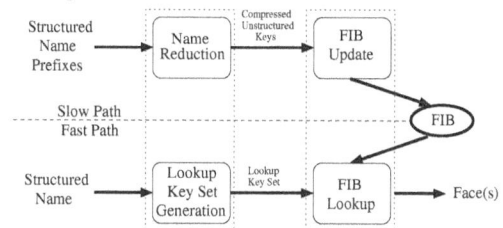

Figure 1: A framework for name prefix lookup.

On the slow path, when a router receives a name prefix announcement, it first applies the *name reduction* algorithm to transform the hierarchically structured name prefix into a compressed key (note that the key space is much smaller than the name prefix space). For the purpose of efficiency and easy manipulation, the transformed

/ccnx.org/demo/video1/v1/s0

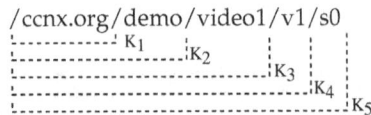

Figure 2: An example of generating the lookup key set.

key may be non-human-readable and unstructured. The FIB update algorithm then inserts into or removes from FIB the transformed key. Many data structures can be adopted to implement FIB and the update algorithm. To address potential hash collisions, we adopt a simple yet efficient dual-hash solution using two different hash functions. Namely, when two name prefixes have collisions using the first hash, we apply the second to differentiate them; by doing so, we can effectively reduce the collision rate to a level comparable to the error rate of the main memory.

On the fast path, upon receiving an Interest packet, a router first applies the *key set generation* algorithm to the name extracted from the packet, producing a set of unstructured keys to facilitate longest-prefix name lookup. Note that a main challenge here is that we *cannot* directly apply name reduction to get a key and then use that key for lookup. The reason is that name reduction loses the internal structure carried in names; without the internal structure, it is completely impossible to look up in the reduced key space the longest name prefix that matches the name.

Figure 2 shows an example of the lookup key generation algorithm that keeps the internal structure in a name. Specifically, always starting from the first segment (segments are separated by the slash /), we iteratively apply the name reduction algorithm (*e.g.*, rolling hashes) to the left most n segments to produce K_n.

Algorithm 1 shows an example of the longest-prefix name lookup algorithm LPM_LOOKUP. It takes the set of lookup keys and iteratively look up for the longest name prefix matching the given name. This algorithm calls a primitive single-key lookup function KEY_LOOKUP to search for a matching key-value pair in FIB.

Algorithm 1 A Longest-Prefix Matching Algorithm

Input: a lookup key set $\{K_1, K_2, \cdots, K_n\}$
Input: FIB storing a set of $\{K, V\}$ pairs ▷ V is the set of outbound face(s) corresponding to K
Output: a set of outbound faces
91: **function** LPM_LOOKUP(FIB, $\{K_1, K_2, \cdots, K_n\}$)
92:　　**for** $i \leftarrow n, 1$ **do**
93:　　　　$V \leftarrow$ KEY_LOOKUP(FIB, K_i)
94:　　　　**if** $V \neq \phi$ **then**
95:　　　　　　**return** V
96:　　　　**end if**
97:　　**end for**
98:　　**return** ϕ
99: **end function**

We adopt a data structure based on the fat tree to store and look up keys efficiently. A k-ary fat tree is a rooted tree where each node can have at most $k - 1$ items and has no more than k children. However, it is highly inefficient to directly adopt a naive implementation of the fat tree data structure. Instead, we significantly optimize its implementation by (1) choosing an optimal k in order to maximize the efficiency of L1 and L2 cache available in the general CPUs, (2) converting the tree representation into an array representation in order for fast traversal, and (3) adopting an extensible hybrid design that combines hashing and fat tree to further improve the efficiency. In the extensible hybrid design, we maintain an array called *directory* to store the addresses of a set of fat trees. For a given key, we first hash it to get an index into the directory. The index indicates the location where the root of a fat tree that might contains the search key is stored. We then traverse the corresponding tree to find a match of the given key.

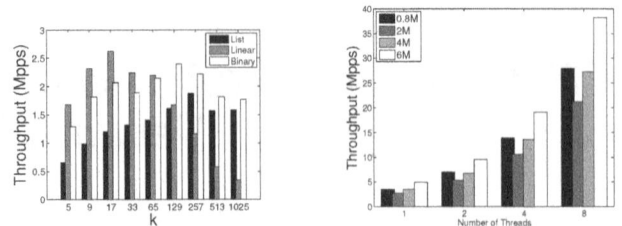

(a) Baseline algorithm.　　　(b) Optimized algorithm.
Figure 3: Throughput of lookup algorithms.

3　Evaluations

We evaluate the proposed framework and fat-tree based longest-prefix name lookup algorithms on Intel Xeon X5675 with 24GB main memory. We use throughput, defined as the number of packet lookups (*i.e.*, longest-prefix matching name lookups) per second, as the metric to measure the performance of our algorithms. We evaluate both the baseline algorithm based on fat trees and the optimized algorithms based on the extensible hybrid data structure. We also exploit the thread-level parallelism available in the general-purpose CPU architectures. We use three data sets of URLs collected from tier-1 ISP networks. These data sets consist of 1.2, 1.0 and 6.0 million URLs, respectively. We report the results on the largest data set. However, results on the remaining data sets are consistent.

We observe in Figure 3 that the baseline algorithms (fat tree nodes implemented using list, linear and binary search, respectively) can achieve approximately 2–2.5 Mpps when fully leveraging L1/L2 caches (*i.e.*, $k = 17$). However, the optimized algorithm can achieve 4–5 Mpps on the same data set (*i.e.*, 100% improvement). With the help of the thread-level parallelism (we use shadow FIBs to allow concurrent FIB lookup and update), we can significantly boost the throughput up to 37 Mpps with only 8 threads on a FIB of 6 million name prefixes.

4　Conclusion

We focused on the challenge of the longest prefix name lookup in content-centric networking, and proposed a framework of fast name lookup based on name space reduction and extensible hybrid data structures. We implemented and evaluated the framework on general-purpose CPU architecture using real-world name prefixes. We demonstrated that the proposed algorithms could achieve 37 Mpps lookup throughput on off-the-shelf general-purpose CPUs, which is equivalent to 40+ Gbps line speed lookup on average if considering the fact that URL lengths typically range between 80 – 200 bytes and assuming the length of URL-like CCN names follows a similar distribution.

5　Acknowledgements

Haiyong Xie is supported in part by the National Natural Science Foundation of China under Grant No. 61073192, by the Grand Fundamental Research Program of China (973 Program) under Grant No. 2011CB302905, by the New Century Excellent Talents Program, and by the Fundamental Research Funds for Central Universities under Grant No. WK0110000014.

6　References

[1] Google. We knew the web was big.
http://googleblog.blogspot.com/2008/07/we-knew-web-was-big.html.
[2] V. Jacobson, D. K. Smetters, J. D. Thornton, and M. F. Plass, et al. Networking named content. In *ACM CoNEXT 2009*, Rome, Italy, Dec 2009.
[3] Netcraft. December 2011 Web Server Survey.
http://news.netcraft.com/archives/2011/12/09/december-2011-web-server-survey.html.
[4] L. Zhang, D. Estrin, J. Burke, and V. Jacobson, et al. Named data networking (NDN) project. Technical Report NDN-0001, Palo Alto Research Center (PARC), Oct 2010.

Hardware Support for Dynamic Protocol Stacks

Ariane Keller   Daniel Borkmann   Stephan Neuhaus

ETH Zurich
Zurich, Switzerland
first.last@tik.ee.ethz.ch

ABSTRACT

Most networking performance enhancements occur through *specific static* solutions, where the structure of the protocol stack remains unchanged. Instead, we focus on a *flexible* software and hardware co-design for the *entire* protocol stack. In this paper, we present EmbedNet, a System-on-Chip implementation of a flexible network architecture for the Future Internet, where parts of the protocol stack can be moved between software and hardware at runtime. This enables the construction of *dynamic* protocol stacks that use available resources optimally.

Categories and Subject Descriptors

C.2 [**Computer-Communication Networks**]: Network Architecture and Design

Keywords

HW/SW Codesign, network node architecture, future Internet, FPGA

1. INTRODUCTION

The tremendous success of the Internet can be attributed to the diversity of supported physical transport media, which allows users to be connected always and everywhere; and its plethora of applications that offers something for anyone. With the addition of more and more features to the original Internet architecture (such as firewalls, VPNs, NATs, P2P, etc.) the question arises whether the Internet architecture ought to be redesigned from scratch. Several research initiatives [1, 2, 4] supported work in the area of clean slate architectures. In the context of such an initiative, we suggested in earlier work [6, 7] to split network functionality into individual *functional blocks* (FBs) that can be assembled into optimized protocol stacks at runtime. However, it was never clear how such protocol stacks could benefit from hardware accelerators, for example for checksum calculation, encryption or intrusion prevention. A state-of-the-art ASIC implementation is unsuitable since the provided functionality is fixed and cannot be optimized at runtime.

In this paper we present EmbedNet, a dynamic protocol stack System-on-Chip (SoC) implementation on a state-of-the-art FPGA (Virtex-6 ML605 evaluation board [5]). This

FPGA hosts a softcore MicroBlaze CPU [3] capable of running Linux, as well as several *hardware threads* that are enabled to do interprocess communication (shared memory, message passing) with the Linux kernel space. These mechanisms are provided by ReconOS, an operating system extension for reconfigurable hardware [8]. The hardware threads can be reconfigured while other parts of the FPGA are processing data.

In EmbedNet, FBs can be executed either in the Linux kernel space or in a hardware thread. Since hardware threads can be reconfigured at runtime, the mapping of functional blocks between hardware and software can be optimized with respect to the current network traffic mix. In this paper we present the architecture of EmbedNet and a performance analysis of the first prototype implementation.

2. EMBEDNET

The overall EmbedNet architecture is depicted in Figure 1. Packets are sent from an application through a BSD socket to the Linux kernel. Here, packets are forwarded between the individual FBs by a *packet processing engine*. Each FB has an associated flag that determines whether the FB is currently executed in hardware or in software. If the FB should be executed in hardware the PPE copies it to the memory region that is shared with the hardware and notifies the hardware of the new packet. The hardware then reads the packet from the shared memory and forwards it to the next FB. From there it can be either sent to another FB in hardware, software or to the Ethernet interface.

In hardware packets are forwarded in a *network on chip* (NoC) consisting of switches arranged in a unidirectional ring. Each switch is also connected to a configurable number of functional blocks. A software controller configures the FBs with the addresses of the other FBs so that packets can be forwarded correctly.

3. EVALUATION

We evaluated the correct functioning of the EmbedNet node with the scenario depicted in Figure 2. In a first step, an application sends packets through the Ethernet FB to another node. In a second step, we add a software FB while the application continues to send data. Eventually, in a third step, we move this FB to the hardware. In order to evaluate the change of the protocol stack, the FBs in software and hardware were implemented slightly different. Thus, on the packet collecting node, we are able to see a change in the payload pattern of the received packets.

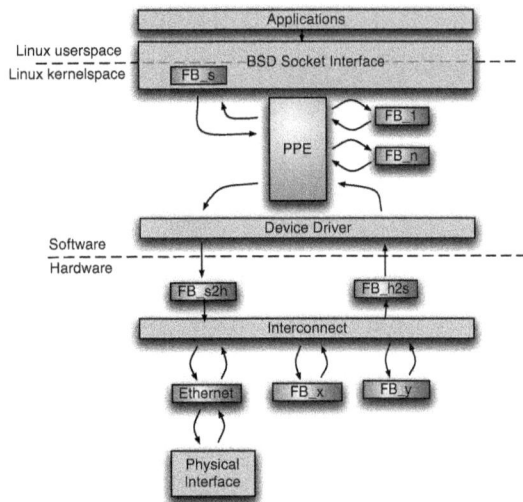

Figure 1: Architecture of EmbedNet.

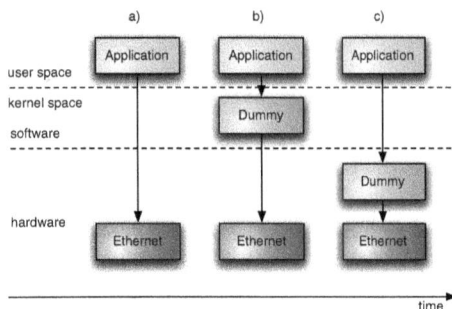

Figure 2: Functional evaluation setup.

We evaluated the maximum sending rate of EmbedNet in different scenarios (see Figure 3). The baseline evaluation reveals a maximum packet sending rate of 4'000 minimum-sized packets per second from normal Linux raw sockets on the ML605 evaluation board. This rather low rate might be due to the low MicroBlaze CPU frequency and a non DMA-capable network IP core. Exchanging the Linux raw sockets with PF_LANA sockets—a new socket class for EmbedNet—shows that PF_LANA (without hardware support) is currently about 1'000 packets per second slower. A second baseline evaluation showed that the sending rate of packets from within the Linux kernel using the EmbedNet hardware is almost independent from the payload size. Sending packets from PF_LANA trough the EmbedNet hardware is again about 1'000 packets per second slower.

This suggests that the limiting factor is the handshake required to transmit data over the software/hardware boundary. Therefore, we plan to implement a ring buffer in which several packets can be transmitted at once. This should increase the performance by a factor corresponding to the number of packets in this buffer.

We also measured the maximum throughput of the hardware by sending packets from an external node to EmbedNet where it was forwarded to a dummy FB (which does nothing) and back to the Ethernet FB. The maximum throughput is currently 0.8 Gbit/s which corresponds to the forwarding rate of the switches (8 bits at 100 MHz). In the future we plan to run the switches at 125 MHz which would allow for line rate forwarding.

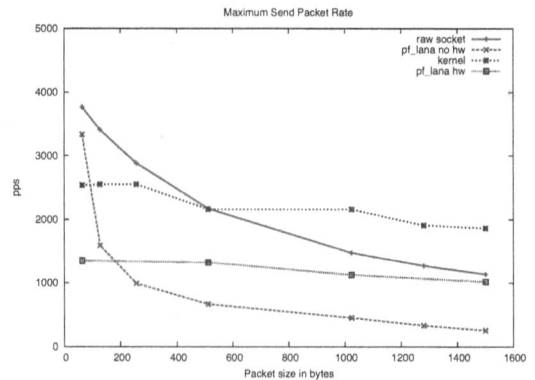

Figure 3: Maximum sending rate comparison.

4. CONCLUSIONS AND FUTURE WORK

We showed that FPGAs can be used to provide hardware support for dynamic protocol stacks, but that the softcore CPUs currently used with FPGAs are rather slow and therefore the overall throughput small. This suggests implementing as much functionality as possible in hardware and to implement a ring buffer between hardware and software to decrease the load on the CPU.

In order to provide an optimal mapping of functional blocks to either hardware and software, we are currently working on a controller that determines the best mapping at runtime based on the packets to process.

5. ACKNOWLEDGMENTS

The research leading to these results has received funding from the European Union Seventh Framework Programme under grant agreement n° 257906.

6. REFERENCES

[1] FIND – Future Internet Design (FIND) - US National Science Foundation. At http://www.nets-find.net/, (May 09).

[2] Future Internet Research and Experimentation (FIRE) initiative. http://cordis.europa.eu/fp7/ict/fire, (May 09).

[3] MicroBlaze Soft Processor Core. http://www.xilinx.com/tools/microblaze.htm.

[4] NWGN – New-Generation Network R&D Project - Japan. At http://nwgn.nict.go.jp/index_e.html, (June 10).

[5] Virtex-6 FPGA ML605 Evaluation Kit. http://www.xilinx.com/products/boards-and-kits/EK-V6-ML605-G.htm.

[6] G. Bouabene, C. Jelger, C. Tschudin, S. Schmid, A. Keller, and M. May. The autonomic network architecture (ANA). *Selected Areas in Communications, IEEE Journal on*, 28(1):4 –14, Jan. 2010.

[7] A. Keller, D. Borkmann, and W. Mühlbauer. Efficient implementation of dynamic protocol stacks. In *ANCS*, page 83â€Ş84, Washington, DC, USA, oct 2011. IEEE Computer Society.

[8] E. Lübbers and M. Platzner. ReconOS: An RTOS supporting hard- and software threads. *IEEE Int. Conf. on Field Programmable Logic and Applications*, 2007.

Low-Latency Modular Packet Header Parser for FPGA

Viktor Puš, Lukáš Kekely
CESNET a. l. e.
Zikova 4, 160 00 Prague, Czech Republic
pus,kekely@cesnet.cz

Jan Kořenek
IT4Innovations Centre of Excellence
Faculty of Information Technology
Brno University of Technology
Božetěchova 2, 612 66 Brno, Czech Republic
korenek@fit.vutbr.cz

ABSTRACT

Packet parsing is the basic operation performed at all points of the network infrastructure. Modern networks impose challenging requirements on the performance and configurability of packet parsing modules, however the high-speed parsers often use very large chip area. We propose novel architecture of pipelined packet parser, which in addition to high throughput (over 100 Gb/s) offers also low latency. Moreover, the latency to throughput ratio can be finely tuned to fit the particular application. The parser is hand-optimized thanks to the direct implementation in VHDL, yet the structure is very uniform and easily extensible for new protocols.

Categories and Subject Descriptors

B.7.1 [**Integrated Circuits**]: Types and Design Styles—*Gate arrays, Algorithms implemented in hardware*

Keywords

Packet Parsing, Latency, FPGA

1. INTRODUCTION

As the computer networks evolve both in terms of speed and diversity of protocols, there is still need for packet parsing modules at all points of the infrastructure. This is true not only in the public Internet, but also in closed, application specific networks. There may be different expectations on packet parsers. Consider for example the multi-million dollar business of fast algorithmic trading, where the parsing latency is often more important than the raw throughput.

Current high-speed FPGA-based parsers can achieve raw throughput over 400 Gb/s at the cost of extreme pipelining, which increases both latency and chip area significantly [1]. Also, the configurability issue is being solved only partially. Configuring the set of supported protocols is often addressed by some higher-level protocol description followed by automatic code generation, but the configuration of implementation details is left unnoticed.

2. MODULAR PARSER DESIGN

Our modular packet parser is designed to meet the demands rising from various applications. Since we realize

that the development of VHDL modules is very low-level and often slow, we start with the design of *Generic Protocol Parser Interface* (GPPI). This interface provides the input information necessary to parse a single protocol header: current data being transferred at the data bus, offset of these data, offset of the protocol header. GPPI also contains output information needed to parse the next protocol header or to read or modify packet header fields. The GPPI output information includes the type and offset of the next protocol header. Fig. 1 shows how modules are connected. By manually adhering to the GPPI, we achieve a hint of object orientation in VHDL – all protocol header parsers are derived from the same "class". This improves the code maintainability and enable easy extensibility of the parser: new protocol header parser is connected just in the same way as the others.

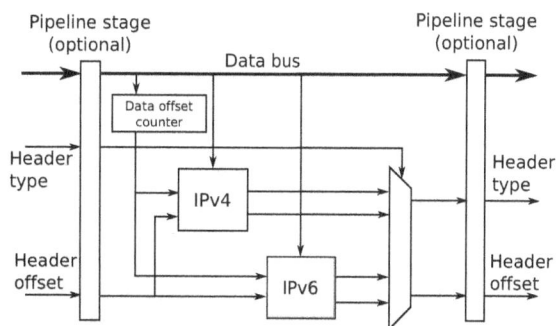

Figure 1: Example of one pipeline stage. Protocol header parsers share inputs, their outputs are selected based on the input type of protocol.

The inner implementation of each protocol header parser is protocol-specific, but the basic parser block is usually the waiting for a specific header field to appear at the data bus, i.e. $po + fo \in \langle do; do + dw \rangle$, where po is protocol offset (module input), fo is field offset (from protocol specification), do is data bus offset (module input), and dw is data bus width. Once the header field is observed at the data bus, it can be used to compute the length of current header, decode the type of the next header, or any other operation.

The output information about types and offsets of protocol headers is more general than having already parsed header field values: The offset is needed for packet editing, and obtaining the header field values can be done simply by multiplexers, with the knowledge of offsets. Our parser offers the option to skip the actual multiplexing of header field values from the data stream. This may save considerable amount of logic resources and is particularly useful for applications where only small number of header fields must be read, or when the packets are modified in a write-only manner.

Similarly to [1], our parser also uses pipelining to achieve high throughput. However, every pipeline step in our design is optional. If many pipelines are enabled, then the frequency (and throughput) rises, but also latency and logic resources increase. By tuning the use of pipelines, designer can find the optimal parameters for the particular use case.

3. RESULTS

We have implemented the parser supporting the following protocol stack: Ethernet, up to two MPLS headers, up to two VLAN headers, IPv4 or IPv6 (with up to two extension headers), TCP or UDP. The header field extraction module (if present) extracts the classical 5-tuple: IP addresses, protocol, TCP or UDP ports. We provide results after synthesis for Xilinx Virtex-7 870HT FPGA, with the different settings of data width, number of pipeline stages and the use of extraction module.

These settings, together with the resulting frequency, latency and resource usage generate a large space of solutions, where the Pareto set can be found and used to pick the best-fitting solution for the application. Fig. 2 shows the throughput and FPGA resources for data widths from 128 to 2048 bits. For each data width, each possible placement of pipelines is shown as a point in the graph and the Pareto set (finding the best throughput and resource utilization, without regard to latency) is highlighted. The lower curve is the Pareto set for the parser without the extraction module. Tab. 1 shows the Pareto set optimized for latency and throughput, without regard to chip area. The last line of Tab. 1 is the estimation of parser from [1] with similar configuration of supported protocols (TcpIP4andIP6).

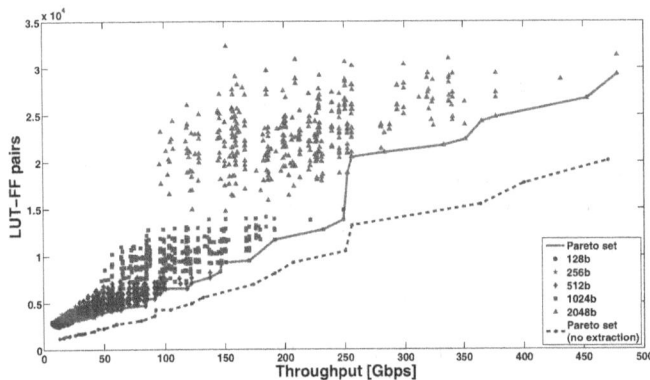

Figure 2: FPGA resource utilization for different parser settings.

Careful design space exploration is very important for our parser. For example, parser optimized for latency uses 17 685 LUT-FlipFlop pairs to achieve near 100 Gb/s throughput with latency only 21.1 ns (see Tab. 1), while parser optimized for chip area uses only 6 536 LUT-FF pairs to achieve throughput just over 100 Gb/s (but with the latency of 35.8 ns).

Data Width	Pipes	Throughput [Gb/s]	Latency [ns]	LUT-FF pairs
256	0	14.5	17.1	3 238
512	0	28.4	18.0	4 053
2 048	0	96.9	21.1	17 685
2 048	1	158.5	25.9	18 547
2 048	2	212.8	28.9	18 317
2 048	4	333.0	30.8	21 775
2 048	5	352.0	34.9	22 373
2 048	7	453.0	36.2	26 728
2 048	8	478.1	38.6	29 301
1 024	?	325	309	67 902

Table 1: Pareto set for best throughput and latency

4. CONCLUSION

Our low-latency modular packet parser for FPGA uses only 1.19 % of the Virtex-7 870HT FPGA to achieve throughput over 100 Gb/s and 4.88 % for throughput over 400 Gb/s with reasonable latency, while most of the FPGA logic is kept for application. These results, together with the latency below 40 ns are better than previous solutions (compare to FPGA utilization over 10 % and latency over 300 ns at 300+ Gb/s throuput in [1]). Our parser uses 68 % less FPGA resources and has 90 % smaller latency than [1] for throughput over 300 Gb/s. Moreover, the parser can be finely tuned trough design space exploration to meet the demands of the particular application.

Acknowledgment

This research has been partially supported by the "CESNET Large Infrastructure" project no. LM2010005 funded by the Ministry of Education, Youth and Sports of the Czech Republic, the research programme MSM 0021630528, the grant BUT FIT-S-11-1 and the IT4Innovations Centre of Excellence CZ.1.05/1.1.00/02.0070.

5. REFERENCES

[1] M. Attig and G. Brebner. 400 gb/s programmable packet parsing on a single fpga. In *Architectures for Networking and Communications Systems (ANCS), 2011 Seventh ACM/IEEE Symposium on*, pages 12 –23, oct. 2011.

M-DFA (Multithreaded DFA): An Algorithm for Reduction of State Transitions and Acceleration of REGEXP Matching

Cheng-Hung Lin
National Taiwan Normal University
Taipei, Taiwan
brucelin@ntnu.edu.tw

Jyh-Charn Liu
Dept. of Computer Science & Engineering
Texas A&M University
liu@cse.tamu.edu

ABSTRACT

This paper proposes a multi-thread based regular expression (regexp) matching algorithm, M-DFA (multithreaded DFA), for parallel computer architectures such as multi-core processors and graphic processing units (GPU). At the thread level, one thread is designated to traverse the DFA of a possible matching path until its termination, and at the task level multiple threads concurrently match each input symbol in parallel. Given a set of regexps, the total number of (DFA) state transitions in M-DFA is significantly smaller than that of its traditional DFA counterpart. The significant saving of state transitions is contributed by elimination of *backtracking transitions*, which commonly occur to mapping of *concurrent active* states in NFA to DFA and other situations. Experimental result shows that the proposed algorithm achieves significant reduction on state and state transition. In addition, the proposed algorithm running on Nvidia® GTX 480 is 35 times faster than the popular regexp library, RE2 performed on Intel Core i7 CPU.

Categories and Subject Descriptors

C.2.0 [**Computer Communication Networks**]: General-Security and protection (e.g., firewalls)

General Terms

Algorithms, Design, Security

Keywords

Regular expression matching, graphics processing units

1. BACKTRACKING STATE TRANSITIONS ELIMINATION

REGEXP matching is typically implemented as a nondeterministic finite automaton (NFA) or its equivalent deterministic finite automata (DFA). NFA has smaller sizes in terms of memory space, but it may take multiple cycles to match an input symbol when multiple states become *active* concurrently. An NFA can be mapped to its DFA equivalence by mapping concurrent active states in NFA to one single active state in the DFA. *Backtracking* state transitions are generated in the mapping process to enumerate mismatched symbols at different stage, in order to direct the DFA to a correct matching outcome. This issue arises during integration of multiple regexps into a DFA, and it is a common cause of state explosion, among other issues.

To eliminate the excessive state transitions in regexp matching, and for acceleration of the matching engines, we note

that a similar problem exists in fixed string based matching, when it becomes necessary to track concurrent active states. In our earlier work, we proposed a Parallel Failureless Aho-Corasick (PFAC) algorithm to perform fixed string matching on GPUs [1]. In PFAC, we first use Aho-Corasick algorithm to generate a DFA, which is then reduced to a loop free DFA (DFA$_{LF}$) by removing all the *failure* transitions. At run time, each input symbol is assigned to a new thread to traverse the DFA$_{LF}$ using subsequent input symbols until it reaches a final state, or aborts the matching. We showed in [1] that the multi-threaded DFA$_{LF}$ based implementation can eliminate all the failure transitions, as well as backtracking transitions of the initial state of an Aho-Corasick DFA.

Next, we discuss how to extend PFAC for regexp matching to achieve similar effects. We note that the traditional DFA model is essentially a single-threaded, monolithic automata generated from a NFA to perform the matching. When taking a closer look at the nature of the DFA, it has forward state transitions for matched symbols, or backward (aka backtracking) transitions when no matching can be made. In M-DFA, we modify the matching process based on a multi-thread execution model, so that (a) each input symbol is treated as the first symbol of a possibly matched substring to the regexp, (b) a DFA free of backtracking transitions is assigned to a thread, which terminates either when it reaches the final state or any mismatch occurs, (c) multiple threads concurrently read in each input symbol for their independent matching.

Conceptually, the M-DFA resembles the behavior of an NFA at the string matching level, yet each thread is a DFA. Given the broad support of multi-thread execution models supported by contemporary processor architectures, e.g., multicore processors, GPU, etc., M-DFA can fully explore the parallel processing power of advanced hardware for acceleration, and it also reaps the benefit reduced memory space requirements for DFA based regexp matching models. The proposed algorithm is well poised to handle the demanding performance required of high throughput applications, such as network intrusion detection systems and spam filters.

2. M-DFA DESIGN

We use matching of two regexps "[abc]x" and "c[xy]", where the bracket pair "[]" contains a character set, to illustrate relationship between concurrent states, backtracking transitions, and the M-DFA algorithm. The first regexp matches "ax", "bx", and "cx", while the second one "cx" and "cy". Figure 1 shows an NFA that can match the two regexps, where the ε transition causes the initial state to be always activated. For an input substring "cx" which matches both regexps, states 4 and 5 will be activated after "cx" is consumed. One will need to track both active states and their subsequent possibilities when the NFA is to be mapped to its equivalent DFA, as the states and their transitions shown in Figure 2. State 6 in the DFA represents the final state of matching the substring "cx", which is equivalent to reaching states 4 and 5 of the NFA simultaneously.

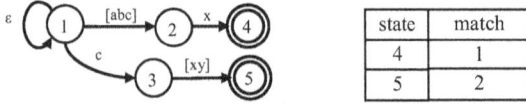

Figure 1. NFA of "[abc]x" and "c[xy]"

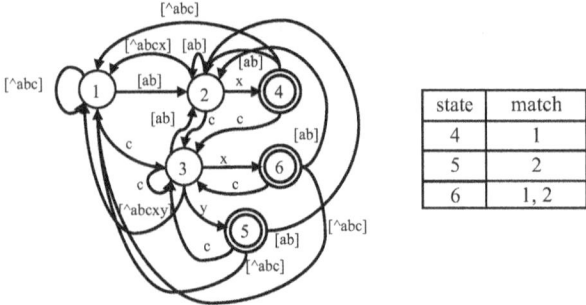

Figure 2. DFA of "[abc]x" and "c[xy]"

Consider to match the two regular expressions "[abc]x" and "c[xy]". The DFA in Figure 2 can be reduced to the DFA$_{LF}$ as shown in Figure 3 (a).

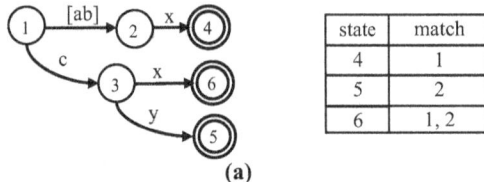

(a)

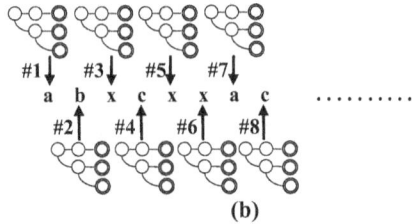

(b)

Figure 3. The multi-threaded M-DFA model. (a) The DFA$_{LF}$ for the NFA. (b) Multi-threaded execution of multiple DFA$_{LF}$.

Figure 3(b) shows the parallel architecture where each input symbol is assigned a thread to perform matching by traversing the same state machine. In Figure 3(b), the second thread pointing to 'b' matches the expression "[abc]x" while the fourth thread pointing to 'c' matches both "[abc]x" and "c[xy]."

For the 6 *matched* transitions generated in this simple example, a large number of backtracking transitions can be generated in the DFA. The exact number of backtracking transitions is dependent upon the binary patterns of the matching/rejection symbols. In the worst scenario, where a byte

is needed for a symbol, and each rejected symbol requires a backtracking transition, there will be a total of 1,530 backtracking transitions. Although one could reduce the number of backtracking transitions by certain compaction techniques, it will remain difficult to guarantee the reduction. That being said, the multi-threaded DFA$_{LF}$ algorithm is guaranteed to eliminate backtracking transitions, which are essentially the house keeping overheads for single-threaded regexp matching DFAs.

Next, we discuss how to construct DFA$_{LF}$ from a DFA. Constructing DFA$_{LF}$ has three steps: we first apply Thompson's algorithm [2] to convert regexps into an NFA and then convert the NFA into an equivalent DFA using Rabin-Scott powerset constructionalgorithm [3]. Finally, the DFA$_{LF}$ is achieved by removing all backtracking transitions of the DFA.

3. EXPERIMENTAL RESULTS

The proposed M-DFA is implemented on Nvidia® GeForce® GTX480 GPU. We extract 16 regular expression patterns and 2 exact string patterns from a public benchmark [4]. The results are compared with a popular regular expression library, RE2 [5] performed on Intel Core™ i7-950 CPU. Table 1 shows the memory and performance comparisons of RE2 with M-DFA$_{GPU}$ which denotes the proposed M-DFA running on GPUs. The speedup is defined as the ratio of RE2 elapsed time over the total elapsed time of M-DFA$_{GPU}$ which including data transfer time via PCIe and GPU elapsed time. On average, M-DFA$_{GPU}$ achieves 35 times faster than the RE2 library performed on CPU. In terms of memory, the M-DFA$_{GPU}$ achieves 44% of state reduction and 99.8% of state transition reduction compared to RE2.

4. CONCLUSIONS

In this paper, we have proposed a multi-thread based regexp matching algorithm which significantly reduces the memory space by removing backtracking transitions. The experimental results show promising results on memory reduction and acceleration of regexp matching on GPUs.

5. REFERENCES

[1] C.-H. Lin, *et al.,* "Accelerating String Matching Using Multi-Threaded Algorithm on GPU," in *GLOBECOM 2010, 2010 IEEE Global Telecommunications Conference*, 2010, pp. 1-5.

[2] K. Thompson, "Regular expression search algorithm," *Communications of the ACM 11(6)*, 1968, pp. 419–422.

[3] M. O. Rabin and D. Scott, "Finite automata and their decision problems," *IBM Journal of Research and Development*, 3(2):114–125, 1959

[4] Regexdna benchmark, http://shootout.alioth.debian.org/u32q/performance.php?test=regexdna

[5] RE2, http://code.google.com/p/re2/

TABLE 1: MEMORY AND PERFORMANCE COMPARISONS OF RE2 WITH THE PROPOSED APPROACH M-DFA$_{GPU}$

Input size (bytes)	RE2 library			M-DFA$_{GPU}$						
	# of states	# of state transition	elapsed time(ms)	# of states	state reduction	# of state transition	Transition reduction	GPU elapsed time (ms)	Total elapsed time (ms)	speed up
100K	162	41,472	14.08	91	44%	90	99.8%	0.05	0.35	40.23
32M			2556.62					5.59	69.24	36.92
64M			4980.15					11.16	139.32	35.74

A New Embedded Platform for Rapid Development of Network Applications

Jan Korenek, Pavol Korcek, Vlastimil Kosar, Martin Zadnik, Jan Viktorin

Brno University of Technology
Faculty of Information Technology
IT4Innovations Centre of Excellence
Bozetechova 1/2,612 66 Brno, Czech Republic
{korenek, ikorcek, ikosar, izadnik}@fit.vutbr.cz, xvikto03@stud.fit.vutbr.cz

Categories and Subject Descriptors

C.3 [Special-purpose and application-based systems]:
Real-time and embedded systems

General Terms

Design, Embedded, Networking

Keywords

FPGA, Zynq, Embedded, Networking

1. INTRODUCTION

NetFPGA-1G [2] has shown its potential in enabling fast traffic processing while introducing no packet loss and minimal delay. Now it is time to scale down in order to enable a massive deployment of the FPGA solutions in networking. We propose and build a low-cost and low-power platform which is be capable of hosting embedded applications with FPGA support. Such a platform might enable faster deployment of new ideas in networking and might prove useful for large-scale experiments (stacks of platforms) as its size, power consumption and cost are expected to be ten times lower of the NetFPGA-cube. It is recognized that the FPGA coupled with the host processor comprises a powerful platform for network traffic processing. The logic of such a solution is clear. Computational intensive tasks are handled by the FPGA logic whereas more complex tasks by the processor. The proposed platform aims at such a design in which the FPGA and the processor are even more tightly coupled together on a single die SoC solution.

As a proof of this concept, we build a platform called uG4-150. Its FPGA hosts a synthesized softcore processor (e.g. Xilinx MicroBlaze). But our final goal is to utilize the hardcore processor (e.g. ARM-based) integrated with the FPGA such as Xilinx Zynq [1].

2. SYSTEM CONCEPT

The concept of the whole system is based on a modular and layered design (see Fig. 1) to minimize the efforts when porting it to a new platform or when new components are introduced. The concept comprises hardware platform,

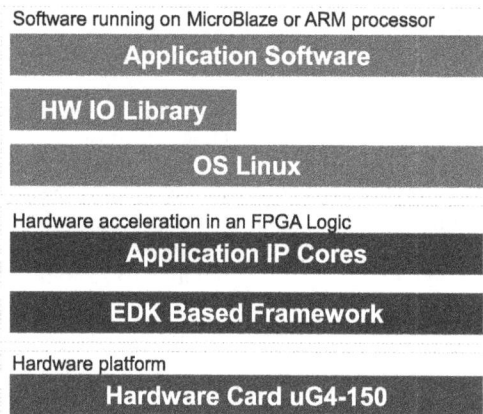

Figure 1: Concept of the system.

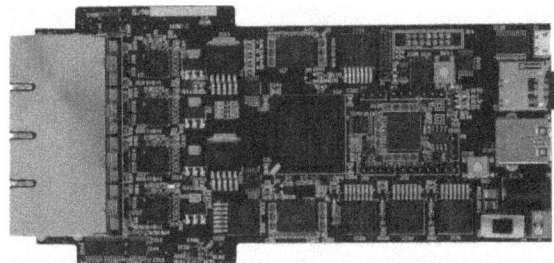

Figure 2: uG4-150 hardware platform.

EDK-based framework, firmware, application IP cores, operating system and application software libraries.

The first layer represents the hardware platform itself. In our case it is uG4-150 platform (depicted in Fig. 2) with its main processing element Xilinx Spartan-6 FPGA but other similar platforms may fit as well. Based on the analysis of FPGA components (i.e. Ethernet cores, DMA cores, MicroBlaze soft processor core, AXI interconnection system and packet processing system) we have equipped uG4-150 with the high-end Spartan-6 (XC6SLX150-3FGG484C).

uG4-150 also hosts additional components such as:

- 2x 256MB of DDR3 for the processor and the FPGA processing system,

- 4x 1 Gbps Ethernet over metallic RJ45 connectors,

- USB 3.0 (type A) connector to a fast data storage,

- 16 MB 4-bit serial flash memory (FPGA boot),

- 512 kB I^{2}C PROM (USB3.0 controller boot),

- slot for microSD memory cards, etc.

The EDK-based framework abstracts application IP cores from specifics of the given platform by providing controllers to the various peripheral components. The most important modules are: AXI Ethernet (transforming RGMII interface to AXI-Stream), SDRAM controller (AXI S6 DDRX controls DDR3 memories), SD card controller and others.

The application IP cores implement various tasks of basic packet processing. These tasks ranges from protocol analysis, packet classification, to filtration and queuing. In case of protocol analysis, the structure and encapsulation of protocols is described in XML and synthesized into VHDL. Despite that the extracted fields may be changed on-the-fly. Filter and classification components identify packets based on the match of its IP addresses and/or ports and protocol number against the given set of rules. The matching of prefix rules is implemented using Tree bitmap algorithm [6] whereas the matching of exact rules is implemented as Cuckoo hashing [5]. Further the packets may be queued in a fairly large buffer in DDR3 memories. The size of these memories provide capacity large enough to store more than a second of traffic at the wire-speed. Finally, there is a module implementing fast and secure packet encoding and secure delivery. All the cores may be assembled in a modular way to perform more complex network operations. The modularity has been achieved through utilization of standard AXI-Stream bus to pass data among modules and AXI-Lite for management purposes.

The Linux OS [3] provides standard ways to access the whole system. It allows to run a wide set of UNIX-based software such as ssh server, iptables or the Click Modular Router [4]. The current version of the operating system can be used on the Zynq platform without any significant modifications.

The access to the non-standard IP cores, which come without any standard kernel-space driver, is done over an abstract layer called HWIO. HWIO is a tiny library that utilizes *device-tree* provided by the Open Firmware kernel [1] to access memory mapped registers of application-specific components.

The application software initializes the HWIO and looks for a compatible core which performs required job. If such is found it is initialized and configured according to the needs of user.

3. USE CASES

The uG4-150 has been tested in application scenario aiming at legal network traffic interception. The system of the probe is built using the proposed system. It utilizes the IP cores to delay incoming packets, parse and extract their headers, filter incoming packets and send them to the collector. Tab. 1 shows synthesis results in ISE 14.1 xst tool for 32-bit wide processing pipeline and 128 filtering rules.

The platform may serve other applications as well. The striking one are specific routers or switches (OpenFlow), firewalls, various proxies, gateways and monitoring probes.

---

[1] www.openfirmware.org

Type	BRAMs	LUTs	FFs	Freq.
uG4-150	6(2%)	8887(9%)	2482(1%)	111 MHz
Zynq-7030	3(1%)	8046(7%)	2419(1%)	218 MHz

Table 1: Real and expected FPGA resource consumption on uG4-150 and Zynq platforms respectively.

4. CONCLUSIONS

This paper proposed a platform for rapid prototyping of high-speed and low-power embedded applications in networking. The concept utilizes the FPGA with the embedded processor to benefit from software flexibility and high performance of hardware processing. In comparison with the NetFPGA-cube, the proposed uG4-150 platform has significantly lower power consumption, cost and size.

As the processor is running OS Linux, the applications can use standard Linux tools and libraries to shorten software development. Moreover, hardware development is simplified by the prepared set of IP cores. These cores accelerate basic time-critical operations, provide AXI-Stream interface, and can be simply connected with Xilinx EDK into the processing pipeline. All IP cores are configurable from user space by the HWIO library.

Currently, the uG4-150 platform utilizes Spartan-6 FPGA with the MicroBlaze processor. But the aim is to utilize Xilinx Zynq with hardcore ARM processor. The uG4-150 is meant to be a functional testbed for upcoming development of platform with Zynq.

5. ACKNOWLEDGMENTS

This work has been partially supported by the Research Plan MSM 0021630528, IT4Innovations Centre of Excellence project CZ.1.05/1.1.00/02.0070, the grant BUT FIT-S-12-1 and Sec6net project VG20102015022.

6. REFERENCES

[1] Xilinx Inc., *Zynq-7000 EPP Overview*, Advance Product Specification, DS190 (v1.1.1) June 11, 2012 http://www.xilinx.com/support/documentation/ data_sheets/ds190-Zynq-7000-Overview.pdf.

[2] NetFPGA Team, *NetFPGA-1G*, http://www.netfpga.org/php/specs.php.

[3] J. Viktorin, *MicroBlaze Simple Linux*, https://github.com/jviki/mbsl

[4] Click Modular Router documentation - http://pdos.csail.mit.edu/click/doc/

[5] Rasmus Pagh and Flemming Friche Rodler. Cuckoo hashing. *J. Algorithms*, 51(2):122-144, 2004.

[6] W. Eatherton and G. Varghese and Z. Dittia. Tree bitmap: hardware/software ip lookups with incremental updates. SIGCOMM Comput. Commun. Rev., 34:97-122, April 2004.

Structural Compression of Packet Classification Trees

Xiang Wang[1,2], Zhi Liu[1,2], Yaxuan Qi[2] and Jun Li[2]
[1]Department of Automation, Tsinghua University, China
[2]Research Institute of Information Technology, Tsinghua University, China
{xiang-wang11, zhi-liu12}@mails.tsinghua.edu.cn, {yaxuan, junl}@tsinghua.edu.cn

ABSTRACT

Most of state-of-the-art packet classification algorithms employ heuristics to trade off between classification speed and memory usage. However, intelligent heuristics often result in complex data structures in algorithm implementation. This brings difficulties to the deployment and optimization of packet classification algorithms. In this poster, a structural compression approach is presented for decision tree based packet classification algorithms. This approach exploits the similarity in real-life filter sets to achieve high compression ratio without loss of tree semantics.

Categories and Subject Descriptors

C.2.3 [**Computer-Communication Networks**]: Network Operations – *Network management, Network monitoring*

General Terms

Algorithms, Design, Performance

Keywords

Packet Classification, Algorithms, Data Structure

1. INTRODUCTION

Packet classification has been studied for a long period, and many decision tree based packet classification algorithms have been proposed. Most of them trade worst-case search time for memory space. They implement various strategies of search space decomposition, because consistent cuttings on the other hand usually introduce excessive overhead of memory usage. For example, several algorithms set *binth* [1, 2] to control the load of linear search at leaf nodes, and take variable-stride cuttings at internal nodes. Those techniques not only hamper the guarantee of worst-case classification time among all types of filter sets, but also limit the optimization of search data structure for hardware acceleration.

In this poster, we argue that the consistent cutting strategy is the key to the improvement of both processing speed and memory usage. The memory overhead of consistent strategy is categorized into two types of redundancy: the *local redundancy* usually exists in pointer arrays in each internal node, and the *global redundancy* lurks in the space decomposition model among all internal nodes.

This poster investigates those redundancies in fixed-stride cutting trees, and illustrates the method to remove them individually.

ANCS'12, October 29-30, 2012, Austin, Texas, USA.
ACM 978-1-4503-1685-9/12/10.

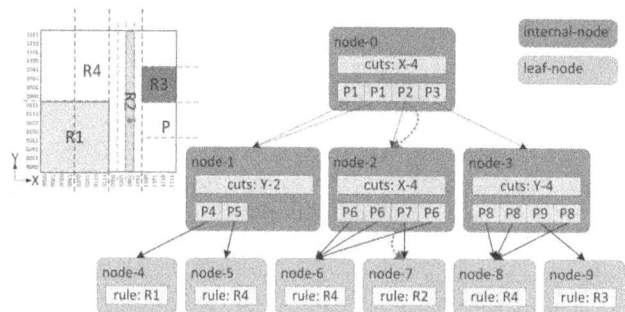

Figure 1. 4-rule Classifier and HiCuts Tree

2. ALGORITHM

Structural redundancy can be illustrated based on a typical HiCuts [1] tree. Figure 1 shows a 4-rule classifier and its corresponding HiCuts Tree. Most of implementations of HiCuts and similar algorithms use pointer array to address child nodes and store the mapping between fixed-stride cut unit-spaces and aggregated sub-spaces. It is obvious that direct pointer addressing scheme brings too much overhead of memory usage, as each pointer takes one word length memory, which is regarded as local redundancy. Furthermore, due to the similarity of the rule distribution in the whole search space, several nodes share the same space decomposition performed in their own sub-spaces. For example, node-2 and node-3 have the same numbers of both unit-space and sub-space. Besides, they both aggregate the 1st, 2nd and 4th unit-space to the 1st sub-space, and map the 3rd unit-space to the 2nd sub-space. Each internal node needs to store the mappings, which is regarded as global redundancy. We propose a 3-step structural compression algorithm to eliminate the redundancy. The first step removes the pointer array. The second step compresses the local redundancy by using a bitmap technique. And the third step extracts the space decomposition from all nodes, which is further aggregated into two shared memory.

STEP-1: Elimination of pointer array

We argue that it is the pointer array that hampers the reduction of both local and global redundancy. To reduce the overhead introduced by the pointer array, the tree building procedure bounds each tree node into fixed size and stores the nodes in consecutive memory. All nodes sharing the same parent are viewed as siblings. The parent node only needs to store its first child node address, and uses offset to address other child nodes. As a consequence, the original pointer array can be expressed using one base pointer and one offset array. If the procedure takes 256-stride cutting strategy, each array element can be stored within single byte. The significance of this step is not only reducing the overhead of storing pointers, but also providing the possibility of compressing the global redundancy.

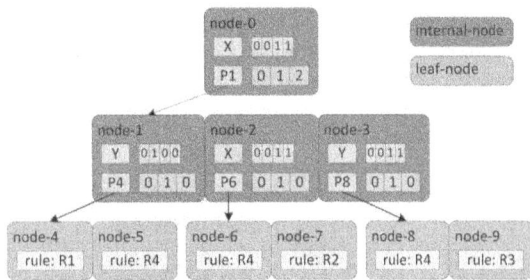

Figure 2. Classification Tree Ready for Elimination of Global Redundancy

STEP-2: Elimination of local redundancy

In the previous step, each node uses a base pointer with an offset array to address its child nodes, and the offset array takes the majority of memory space in each node. In practice, massive internal nodes have only fewer child nodes, which mean many consecutive unit-spaces will be aggregated into the same sub-space [3]. As a consequence, the bitmap technique is employed to compress the offset array, which generates a bitmap and a compressed offset list. Figure 2 shows the classification tree which is ready for the third compression step.

STEP-3: Elimination of global redundancy

After the formal two steps, it is observed that a great deal of internal nodes has same bitmaps and offset lists, and the number of unique bitmap and offset list are both relatively small in practical classifier. Based on this observation, unique bitmaps and offset lists are extracted and stored in two consecutive shared memory blocks respectively, leaving two indices in each internal node.

After the 3-step structural compression, the original classification tree is reshaped into three lookup tables, and a compact packet classification tree is obtained. Figure 3 shows the compressed search data structure of classifier in Figure 1.

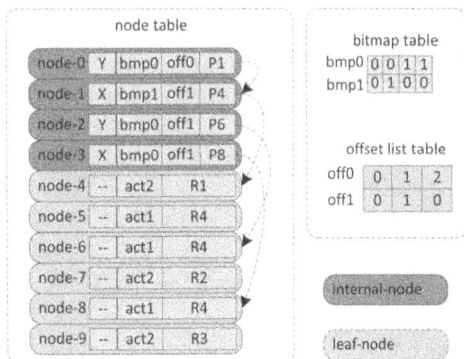

Figure 3. Classification Tree for Structural Compression

3. EVALUATION

We carry out the preliminary evaluation test on FW classifier generated by classbench [4]. Table 1 compares the numbers of bitmap and offset before and after elimination of global redundancy. Figure 4 shows the memory size comparison between original HiCuts tree, AggreCuts [3], HyperSplit [2] and structural compressed tree. When bounding to the same memory access time, HiCuts tree needs about 2 orders memory usage than structural compressed tree. Besides, the memory of structural compressed tree is comparable to the one of HyperSplit, where the memory access time of the latter is two times larger than the former one.

Table 1. Global Redundancy of Bitmap and Offset List

rule	*FW_100*	*FW_1K*	*FW_5K*	*FW_10K*
total	6012	304185	1585.3K	4818.4K
uni bmp	116	703	2612	3713
uni off	23	109	293	499

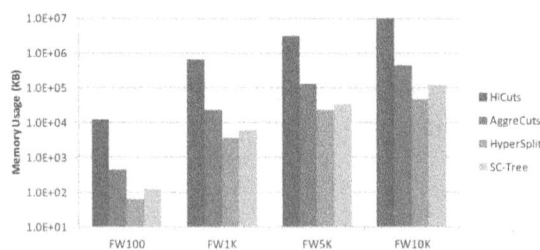

Figure 4. Memory size of HiCuts, AggreCuts, HyperSplit and structural compressed tree

4. CONCLUSION

This poster presents a structural compression approach to eliminate redundancy in packet classification trees. It exploits the similarity residing in real-life filter sets, and achieves high compression ratio in preliminary evaluation. Our future work will apply structural compression to other equal-sized space decomposition packet classification algorithms with grouping of rules [5, 6].

5. REFERENCES

[1] P. Gupta and N. McKeown. "Classifying Packets with Hierarchical Intelligent Cuttings," in IEEE Micro, 2000.

[2] Y. Qi, L. Xu, B. Yang, Y. Xue and J. Li. "Packet Classification Algorithms: From Theory to Practice," in Proc. of INFOCOM, 2009.

[3] Y. Qi, B. Xu, F. He, B. Yang, J. Yu and J. Li. "Towards High-performance Flow-level Packet Processing on Multicore Network Processors," in Proc. of ANCS, 2007.

[4] ClassBench: A Packet Classification Benchmark. http://www.arl.wustl.edu/classbench/

[5] B. Vamanan, G. Voskuilen and T. Vijaykumar. "EffiCuts: Optimizing Packet Classification for Memory and Throughput," in Proc. of SIGCOMM, 2010.

[6] J. Fong, Y. Qi, J. Li and W. Jiang. "ParaSplit: A Scalable Architecture on FPGA for Terabit Packet Classification," in Proc. of HOTI, 2012.

Toward Fast NDN Software Forwarding Lookup Engine Based on Hash Tables

Won So, Ashok Narayanan, Dave Oran and Yaogong Wang
Cisco Systems, Inc. Boxborough, MA, USA
{woso, ashokn, oran}@cisco.com, ywang15@ncsu.edu

ABSTRACT

In Named Data Networking (NDN), forwarding lookup is based on tokenized variable-length names instead of fixed-length host addresses, and therefore it requires a new approach for designing a fast packet forwarding lookup engine. In this paper, we propose a design of an NDN software forwarding lookup engine based on hash tables and evaluate its performance with different design options. With a good hash function and table design combined with Bloom filters and data prefetching, we demonstrate that our design reaches about 1.5MPPS with a single thread on an Intel 2.0GHz Xeon processor.

Categories and Subject Descriptors

C.2.1 [**Network Architecture and Design**]: Store and forward networks; C.2.6 [**Internetworking**]: Routers

Keywords

Named data networking, packet forwarding engine, hash table

1. INTRODUCTION

Named Data Networking (NDN) is a networking paradigm that tries to address issues with the current Internet by using named data instead of named hosts for the communication model. NDN communication is driven by the consumer of data; a consumer asks for a content via an *Interest* packet with a unique name, then any node hearing the interest and having data that satisfies it can respond with a *Data* packet [5]. Among various fundamental issues in NDN, this paper addresses the NDN forwarding problem focusing on forwarding lookups.

NDN forwarding lookups are different from that of IP mainly for the following 2 reasons. First, NDN forwarding lookup is based on unbounded tokenized names rather fixed length addresses. NDN uses variable-length hierarchical names that consist of a series of delimited components. Interest packets are forwarded based on the longest prefix match at any component. For example, suppose a name '/c1/c2/c3' delimited by '/', a lookup begins with the longest prefix '/c1/c2/c3' then continues to shorter prefixes such as '/c1/c2' and '/c1'.

Second, NDN uses 3 forwarding tables: CS, PIT and FIB. *PIT (Pending Interest Table)* is a table that stores unsatisfied Interest; an entry is added when a new Interest packet arrives and removed when it is satisfied by the corresponding Data packet. *Content Store (CS)* is a buffer (or cache) memory (or storage) to store previously processed Data packets that can be re-requested. If we simplify the forwarding algorithm in terms of lookups, Interest packet forwarding consists of 1) CS lookup 2) PIT lookup & add 3) repeated FIB lookups from the longest prefix until it finds a match. Data packet forwarding is composed of 1) CS lookup 2) PIT lookup & delete 3) CS delete & add [5].

2. BASIC DESIGN

Since NDN lookup is based on variable-length names, it can be seen as a string match problem. After investigating 2 approaches – DFA (Deterministic finite automata) vs. hash table (HT), we choose HT because it is simpler and fits better to the nature of the NDN lookup problem. The difficulty of a DFA based approach is that each prefix in NDN names is associated with an unbounded string, and hence it requires a special encoding scheme. (Refer [10] for an example that uses a Trie for NDN lookup.)

For HT design, we use a chained hash table with linked lists (a.k.a closed addressing). Our key design goal is to avoid unnecessary string comparison which negatively affects the performance. For this, the 64-bit hash value generated from a string is stored at every hash entry. To search a matching entry from a chain, it is enough to compare two hash values when there is no hash collision. It turns out that string comparison is rarely necessary because the hash collision probability is extremely low with a good hash function as we see in Section 4. Additionally, linked lists are sorted by hash values to speed up the search. The HT entry size is fixed to 64 bytes, the same as the cache line size of our experimental platform.

3. EXPERIMENTAL PLATFORM

We built an NDN forwarding simulator, written in C language. It simulates NDN Interest and Data forwarding lookups by taking packets from an input file and reports its performance. The simulator runs on a Linux machine with an Intel Xeon processor. We defined a fixed-length packet format which consists of a 64-byte header, and a 64-byte name field. (We avoided the CCNx format due to XML decoding overhead. For CCNx forwarding, refer [11].)

For input packets, synthetic names are generated by a Python script. The number of components in names uni-

Table 1: Performance and quality of various hash functions

Hash function	CCNx[1]	CRC32[7]	CityHash64[3]	Spooky[6]	Jenkins[6]	MD5[9]	LANE[2]	ECHO[2]	Grøstl[4]
Cycles/byte hash	4.93	1.08	0.69	0.77	3.29	7.67	5.41	3.18	23.67
Cycles/HT lookup*	436.73	323.23	296.66	380.12	389.28	906.53	1316.51	2237.78	2407.47
Avg. length of chain	1.60	1.58	1.57	1.58	1.58	1.58	1.57	1.58	1.58
Max. length of chain	9	7	7	8	8	8	7	8	9
Empty bucket %	38.18	37.17	36.98	37.25	37.16	37.34	37.03	37.25	37.22
Hash collision	164	1	0	0	0	0	0	0	0

* HT lookup tests are done with 64K hash buckets and 64K pre-populated entries (i.e. load factor = 1).

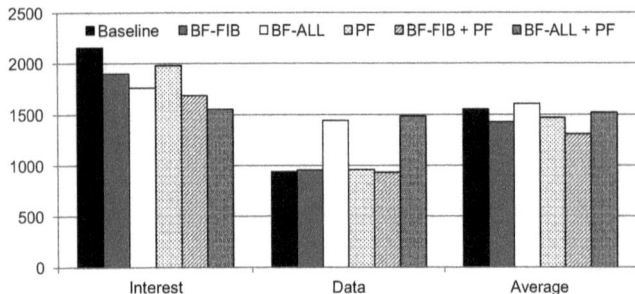

Figure 1: CPU cycles of Interest/Data forwarding

formly ranges between 2 and 12, and each component consists of random 2 to 6 alphabets; the average name length is 33.9 bytes. Matching FIB entry names are constructed by randomly choosing the subset components from full names. The CPU cycle counts are measured with 5 runs of 64K packets using on-chip HW PMU counters via *perfmon* library [8].

4. PERFORMANCE & IMPROVEMENT

We compared the performance and quality of various hash functions as summarized in Table 1. The performance of cryptographic hash functions (4 right columns) is noticeably worse despite their uses of AES-NI. The quality metrics (lower 4 rows) show similarity while CCNx and CRC32 suffer hash collisions that force string comparison. Among the rest, CityHash64 is the best in terms of both byte hash and HT lookup performance.

With hash tables using CityHash64, we simulated NDN forwarding lookups. All HTs are set to 1M buckets and half-populated before simulation. Since the average number of FIB name components is 4 while that of input packet names is 7, Interest forwarding involves average 4 FIB lookups. Therefore, our simulation workload represents 'CS lookup + PIT lookup & add + average 4 FIB lookups' for Interest and 'CS lookup + PIT lookup & delete + CS delete & add' for Data. The leftmost solid bars in Figure 1 present CPU cycles of Interest, Data forwarding and the average of our baseline design. The Interest forwarding takes longer time because it involves more lookups. Simple breakdown analysis discovers that each HT lookup approximately takes 200+ cycles while HT lookup & add (or delete) consumes 300+ cycles.

To improve the baseline design, we evaluated 2 different methods. First, a basic Bloom filter (BF) is stored at every HT bucket and maintained to detect lookup miss earlier without chain walk. For Interest, BF is quite effective because it saves most chain walks; it gives 13.50% speedup for [BF-FIB: BF in FIB HT only] and 22.35% speedup for [BF-ALL: BF in all HT] as shown in Figure 1. However, BF negatively affects Data forwarding due to the PIT, CS

deletion overhead that requires the whole chain walk as [BF-ALL] lowers the average performance.

Second, we applied data prefetching. Due to the nature of sequential lookups on HT (CS, PIT, FIB×n) for Interest, the HT bucket for the next lookup can be prefetched during the previous lookup. For Interest, prefetching [PF] gives 8.83% speedup over the baseline design while it results in more significant gains – 27.83% for [BF-FIB + PF] and 39.10% for [BF-ALL + PF] – when combined with BF as shown in Figure 1. It is because BF is located at the same cache line as the list head pointer, and hence any extra data cache misses can be avoided in case of consecutive lookup misses.

Overall, [BF-FIB + PF] results in the highest average forwarding rate of 1.53MPPS with a single thread on a 2.0GHz processor; 1.19MPPS Interest and 2.14MPPS Data forwarding rates respectively.

5. CONCLUSION & FUTURE WORK

In this paper, we proposed the design of an NDN software forwarding lookup engine based on hash tables and evaluated its performance with simulation. With a good hash function and table design combined with Bloom filters and data prefetching, we demonstrated that our design reaches about 1.5MPPS with a single thread on an Intel 2.0GHz Xeon processor. Our next step is to extend this design and build an engine running on multiple parallel threads with real packets.

6. ACKNOWLEDGMENTS

We would like to thank Dave Barach at Cisco Systems for insightful discussions on this work.

7. REFERENCES

[1] CCNx 0.6.0. http://www.ccnx.org.
[2] ECHO Design Team. http://crypto.rd.francetelecom.com/ECHO/sha3/AES.
[3] Google Inc. http://code.google.com/p/cityhash.
[4] Grøstl Team. http://www.groestl.info.
[5] V. Jacobson, D. K. Smetters, J. D. Thornton, M. F. Plass, N. H. Briggs, and R. L. Braynard. Networking named content. In *Proceedings of the 5th International Conference on Emerging Networking Experiments and Technologies*, CoNEXT '09, 2009.
[6] B. Jenkins. http://www.burtleburtle.net/bob.
[7] B. Kittridge. http://byteworm.com/2010/10/13/crc32.
[8] Perfmon2. http://perfmon2.sourceforge.net.
[9] The OpenSSL Software Foundation. http://www.openssl.org/docs/crypto/md5.html.
[10] Y. Wang, K. He, H. Dai, W. Meng, J. Jiang, B. Liu, and Y. Chen. Scalable name lookup in NDN using effective name component encoding. In *Proceedings of the 32nd International Conference on Distributed Computing Systems*, ICDCS '11, 2012.
[11] H. Yuan, T. Song, and P. Crowley. Scalable NDN forwarding: Concepts, issues and principles. In *Proceedings of the 21st International Conference on Computer Communications and Networks*, ICCN '12, 2012.

Host-based Multi-tenant Technology for Scalable Data Center Networks

Koichi Onoue
Fujitsu Laboratories Ltd.
koichi@labs.fujitsu.com

Naoki Matsuoka
Fujitsu Laboratories Ltd.
nmatsu@labs.fujitsu.com

Jun Tanaka
Fujitsu Laboratories Ltd.
ucx14@labs.fujitsu.com

ABSTRACT

These days, various academic and industrial institutions are sharing the computing resources of cloud data centers. For the sake of security, data center networks need to be separated by institution or department. One conventional approach is using tag-based VLAN standardized in IEEE 802.1Q. However, this approach cannot accommodate scalable networks because of a limitation on the number of VLAN IDs. To address this problem, we propose *HostVLAN*, a novel multi-tenant technique for scalable networks. To provide logical isolated networks for individual tenants (e.g., academic and industrial institutions), end host servers deployed in a data center filter receiving network data and forward them to designated virtual machines on the basis of isolation information. To reduce the broadcast traffic for the protocols, ARP and DHCP, end host servers convert the broadcast data into unicast data. Unlike conventional approaches that work in cooperation with switches, HostVLAN provides multi-tenant environments at the end-host-server side. To build a HostVLAN-based network architecture, we extended three virtual network switches supported by KVM and Xen VMMs. The results of performance evaluation demonstrate that HostVLAN can be scaled up to large numbers of multi-tenant networks with little overhead.

Categories and Subject Descriptors

C.2.1 [**Computer Communication Networks**]: Network Architecture and Design

General Terms

Design, Management, Performance, Security

Keywords

Cloud Computing, Network Virtualization, Multi Tenancy, Layer 2 Network, Host-based Approach

1. INTRODUCTION

"Infrastructure as a service" (IaaS), such as Amazon EC2 [1], is one category of cloud-computing service. Data centers that support IaaS can elastically provide computing resources such as server machines, storage, and networks in accordance with user demand. IaaS enables users, e.g., administrators of academic and industrial organizations, to build and manage their system infrastructures cost-effectively and cost-efficiently.

Owing to research and development of server virtualization technologies over the past few years [3,13,16,23], many current data centers can provide virtualized computing resources (more concretely, logically segment hardware resources such as processors, memory, and networks) with acceptable performance. In data centers, virtualization technologies are beneficial since they allow for scaling, namely, deploying huge numbers of servers (e.g., hundreds of thousands of virtual machines (VMs)), and they take advantage of dynamic server allocation/de-allocation and migration.

For the purpose of security enhancement and performance isolation, data centers are required to support *tenant isolation*, which is a mechanism for isolating computing resources shared among different users, called *tenants*. For example, a data center needs to prevent one tenant from leaking and/or tampering with confidential data on the servers of other tenants. In addition, a data center might be required to control the available network bandwidth so that a tenant does not disturb the service level agreement of other tenants by exceeding the network capacity that has been allocated to it. tenant.

As shown in Figure 1 (i), one of the approaches to achieving tenant isolation is using VLAN tagging (VLAN) defined by IEEE in the 802.1Q standard [10]. To distinguish between different tenants, a VLAN ID field in a network-layer-2 (L2) frame header is used as a tenant ID. An L2 network switch (L2 switch) only transmits a network frame to a destination VM that has the same VLAN ID as the source one. Since most commodity L2 switches accommodate VLAN, this approach enables data center networks to maintain a multi-tenant function at low cost. Consequently, several existing virtual network switch functions [17,21,24,25] provide multi-tenancy based on VLAN tagging.

However, VLAN-based tenant isolation poses a scalability problem [20]: a limitation on the number of tenants. The proliferation of today's cloud-computing services implies that data centers need to scale to a large number of not only VMs but also networks and tenants, e.g., a thousand or more tenants. The problem is that the size of the

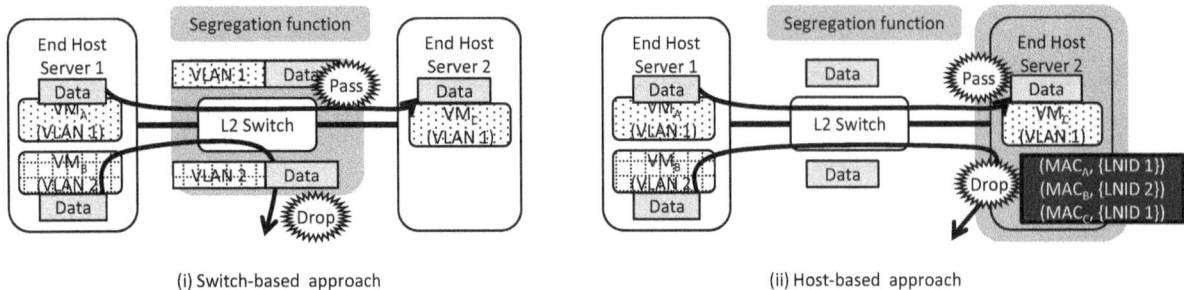

(i) Switch-based approach   (ii) Host-based approach

Figure 1: Comparison of our approach (ii) with a conventional one (i)

VLAN ID space is only 4,094 because the VLAN ID field is 12 bits and reserved IDs 0 and 4095 are excluded. Data centers using VLAN, therefore, cannot scale to large multi-tenant networks.

It is interesting to note that data centers can support a large L2 flat network, i.e., a single large L2 domain, for the following reason. A single large L2 network can more easily and flexibly deploy and migrate resources than a network composed of multiple L2 domains can, although a single large L2 network can excessively consume network resources owing to the amount of broadcast traffic. Several existing data center networks [8,12,19] offer L2 network semantics. Dunbar et al. [5] asserted that huge L2 data center networks can facilitate VM migration, network services such as firewalls and load balancers, and a fail-over mechanism based on active/standby VMs. Moreover, data center networks that provide IaaS generally allocate reserved computing resources for each L2 domain in order to scale the amount of available computing resources up or down on demand. This condition results in inefficient resource allocation because it increases the amount of unused reserved computing resources.

Given the above circumstances, we have developed a host-based multi-tenant technique, *HostVLAN*, that overcomes the scalability limitation of VLAN. As shown in Figure 1 (ii), HostVLAN segregates data center networks according to logical network IDs (LNIDs) associated with MAC addresses. A destination end host server then forwards a receiving network frame to only the destination VM belonging to the same LNID as the source VM. To determine the effectiveness of HostVLAN, we designed, implemented, and evaluated a HostVLAN-based network architecture. The three existing types of virtual switches, *bridge*, *Open vSwitch*, and *macvtap*, were extended and applied to the virtual machine monitors (VMMs), KVM [13] and Xen [3].

The HostVLAN-based approach differentiates between different tenants on the end host server, while the conventional approach based on VLAN differentiates between different tenants according to the switches that they use. HostVLAN does not need to encapsulate network data or rewrite network data headers. Furthermore, a host-based network architecture incurs no cost for configuring switches, including managing L2 domains by using VLAN and IP subnets.

The remainder of this paper is organized as follows. Section 2 introduces HostVLAN-based architecture, and Section 3 presents a HostVLAN in detail. The feasibility of HostVLAN is examined in Section 4. Section 5 discusses related work. Finally, Section 6 concludes the paper.

2. HostVLAN-BASED ARCHITECTURE

2.1 Design Goals

The goal of this study is to build a multi-tenant network environment for data center networks that simultaneously fulfills the following requirements.

Provision for a large number of tenants at low cost: A network environment with a hundred or so tenants, at least more than the maximum number supported by VLAN, 4,094, should be supported. One of the approaches to achieving this goal is to use the networking communication standard "Provider Backbone Bridge "(PBB), defined by the IEEE 802.1ah [11]. In the case of PBB, the sender switch adds a PBB-specific L2 frame header to a network frame transmitted by a VM. This scheme is called MAC-in-MAC encapsulation. If a PBB-based approach is adopted, that is, if an I-SID field (24 bits) in a PBB-specific header is used as a tenant ID, it would be possible to provide a sufficient number of tenants (namely, more than approximately 16 million). This approach, however, requires expensive dedicated switches. Accordingly, it is more realistic and less expensive to create scalable multi-tenant networks by using commodity switches.

Transparent data transmission: There are two possible schemes for achieving logical segregation: encapsulation of or modification of network data. In an encapsulation scheme such as MAC-in-IP and MAC-in-UDP used by an end host server, logical networks are segregated according to tenant IDs contained in the outer header. However, the encapsulation scheme has two problems. First, it incurs an overhead as a result of adding a specific outer header. Second, end host servers are unable to work with hardware offload engines such as checksum offloading.

In the modification scheme, the multicast addresses (including broadcast ones) in the destination L2 frame header are changed into specific addresses for each tenant. This scheme, however, incurs the following problem. With one method, if a multicast frame were converted into unicast frames and forwarded only to VMs belonging to the same tenant as the source address, the network load would increase in proportion to the number of destination VMs. With an alternative method, if a destination address were overwritten with a specific multicast address that is mapped to each tenant in advance, the servers would be unable to define the original multicast address since it can overlap with the multicast addresses mapped to the tenants. A multi-tenant network with neither encapsulation nor modification should therefore be provided.

High compatibility with existing network infrastructure: To transport entire network environments to cloud data centers easily and quickly, some tenant administrators might configure the same network addresses (such as IP addresses and IP subnets) before and after the relocation. In this case, multi-tenant networks would be under the requirement that IP addresses are uniquely assigned, such as with Diverter [6], since the addresses of different tenants might overlap. To deal with this possibility, the proposed multi-tenant architecture should enable individual administrators to setup their own network addresses without having to consider the network-address configurations of other tenants.

High resistance against compromised VMs: VMs can be hijacked by exploiting vulnerabilities in OSes and the applications running on them. Malicious users with access to compromised VMs might attempt to subvert a multi-tenant isolation mechanism in order to leak and tamper with data related to other tenants. Hence, it is necessary to devise a multi-tenant mechanism that is difficult to disable or abuse even if a VM is taken over.

Broadcast-traffic restraining: In communications protocols such as "address resolution protocol" (ARP) and "dynamic host configuration protocol" (DHCP), broadcast data are used to acquire information on unknown target machines. Broadcast traffic normally has a much greater impact on network domains in proportion to the number of VMs or virtual network interface cards (NICs) in the domain [5, 7]. The developed network architecture is particularly affected by broadcast traffic since it assumes a single large L2 network domain. Moreover, it is possible for malicious users to send massive broadcast data to intentionally congest the network. To address these problems, we need to reduce the amount of broadcast traffic in data center networks.

Easy management of multi-tenant networks: In the HostVLAN-based architecture, each end host server has information tables on multi-tenant networks, which will be described in detail in the next sections. When multi-tenant networks are changed, all of the information tables managed by the end host servers need to be synchronously updated. The management of the information tables should be consolidated via a central management server, like OpenFlow [15], instead of configuring information for each individual end host server, which would be somewhat cumbersome. To minimize the need for manual administrator configuration, which is more likely to be error-prone, the update mechanism should be automated.

2.2 Structure

An example of a multi-tenant network environment to which HostVLAN is applied is shown in Figure 2. This environment runs two types of physical servers: a *management server* and *work servers*. Each server has two types of multi-tenant information tables: an *LNID table*, which is used in the network isolation process, and *BtoU tables*, which are used in the process of restraining broadcast traffic.

Management server: A *management agent* residing at this server manages the LNID tables in cooperation with *work agents* residing at the work servers. Command-line utilities are provided at the management server so that the system administrator of a multi-tenant environment can consolidate the networks. Multiple management servers can also be deployed for load balancing and fault tolerance.

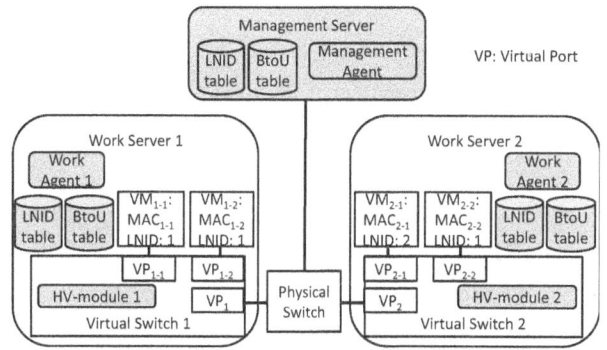

Figure 2: HostVLAN-based architecture. Colored components are the proposed extensions.

Work server: This server isolates data center networks for each VM, or to be more precise, a virtual NIC. *HV-module* is an original extension component of an existing virtual switch. The virtual switch resides outside or within a VMM depending on the type of implementation. The HV-module uses an LNID table to forward received network frames to only VMs that have the same LNID. It also reduces the broadcast traffic of ARP and DHCP by using each BtoU table. The *work agent* updates the LNID table and BtoU table in cooperation with the management agent.

The HV-module is isolated from the target VMs and runs at a higher privilege level than VMs. The proposed multi-tenant mechanism, therefore, cannot be subverted by direct attacks from compromised VMs.

3. PROPOSED TECHNOLOGY: HostVLAN

3.1 Logical Network ID (LNID)

An LNID is assigned to each an isolated logical network. Each tenant can have multiple LNIDs. To enable flexible configuration of logical networks, we opted for many-to-many relationships in the associations between the LNIDs and MAC addresses. One LNID can associate with multiple MAC addresses, i.e., virtual NICs, while one MAC address can associate with multiple LNIDs. This means that one NIC can belong to multiple logical networks. This capability is particularly useful in a situation in which there is a need to transiently share networks for a project between multiple enterprises. Furthermore, a *global LNID* is defined as an LNID that belongs to all logical networks. Virtual NICs with the global LNID are allowed to forward all received network frames.

LNID table entries are *pairs* composed of the MAC addresses of virtual NICs and a tuple of LNIDs. For example, (00:11:22:33:44:55, {1,2}) indicates that a VM with the MAC address 00:11:22:33:44:55 belongs to LNIDs 1 and 2. The LNID table owned by the management server needs to include all the pairs related to the MAC addresses of virtual NICs used in data center networks. Meanwhile, the LNID tables used by the work servers only need to include the pairs involved with the MAC addresses of the virtual NICs within the data center with the same LNIDs as the VMs running on the work servers.

In the following explanation, we shall assume that the

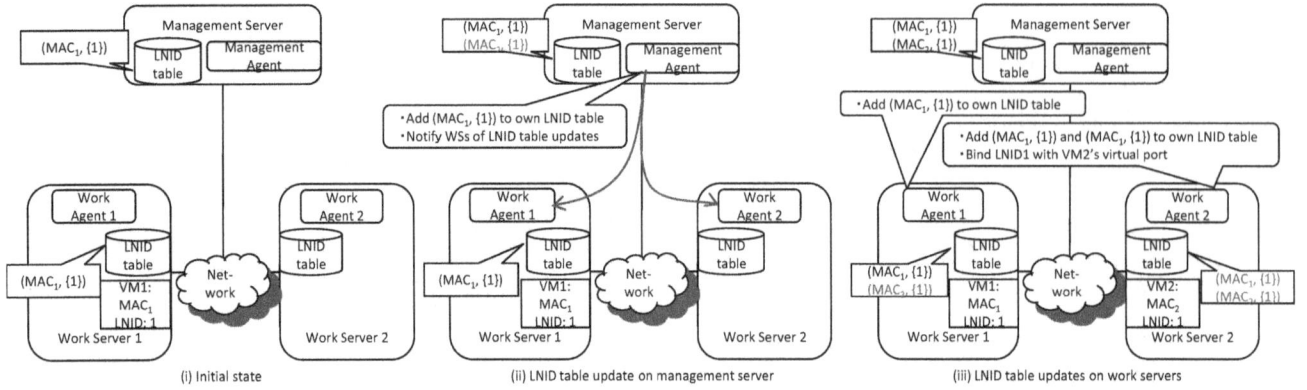

Figure 3: State transition for LNID update at the start of VM2

data center comprised of 1,000 VMs, each of which has one virtual NIC, and 10 logical networks, each of which includes 100 VMs. Given that each VM deployed on a work server belongs to one of two logical networks, the number of entries of LNID table owned by the work server is 200, while the number of entries of the LNID table owned by the management server is 1,000. In present-day data centers, there are some cases in which VMs belonging to only some of the tenants are deployed on one end host server. In such a case, the above approach to LNID table management is effective because it reduces the size of the LNID table managed by the work servers.

Figure 3 illustrates how LNID tables on the management server and work servers 1 and 2 are updated when VM2 starts. Figure 3 (i) assumes that VM1 with MAC address, MAC_1, and LNID1 is running on work server1. When VM2 start running on work server 2, the management agent adds a new pair, $(MAC_2, 1)$ to its own LNID table (Figure 3 (ii)). Next, the management agent tells work agent 2 to store two pairs, $(MAC_1, 1)$ and $(MAC_2, 1)$. In addition, the management agent orders work agent 1 to add the new pair, $(MAC_2, 1)$. After receiving the notification, work agents 1 and 2 update their own LNID tables. At this point, all the LNIDs have been synchronized (Figure 3 (iii)).

3.2 Control Flow for Network Isolation

3.2.1 Flow Mechanism

The mechanism by which a receiving HV-module forwards received network data to the appropriate virtual switch ports is shown Figure 4. The forwarding flows depend on the results of pattern matching using the LNID table. When a virtual switch receives an L2 network frame, forwarding virtual ports are determined on the basis of the destination MAC address of the received L2 network frame. An original virtual switch provided by server virtualization has this forwarding function. The HV-module then identifies the LNID corresponding to the source VM by using the LNID table. If the source LNID cannot be identified, the HV-module judges that the forwarding virtual port does not belong to the source logical network and discards the L2 network frame.

To identify an LNID associated with the destination VM, the HV-module uses the MAC address bound with the desti-

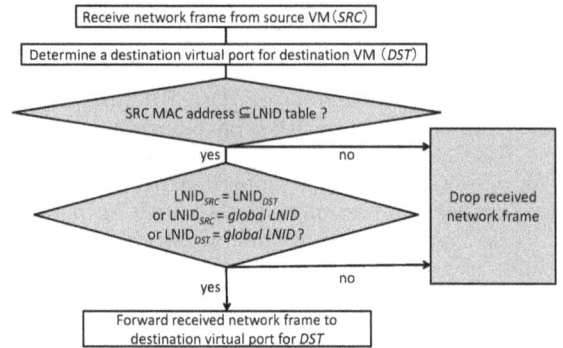

Figure 4: Procedure of network isolation in a virtual switch. Colored parts are the process of the HV-module.

nation virtual port instead of the destination MAC address in the received frame header. If the destination MAC address in the header were used, the HV-module would not be able to identify the LNID corresponding to the broadcast address. By using the MAC address of the destination port, the HV-module can prevent the received broadcast frame from being forwarded to VMs belonging to different logical networks. To reduce the time needed for identifying the LNID associated with a destination, the HV-module uses not the LNID extracted from the LNID table but rather one coupled with the destination virtual port. The binding between an LNID and the virtual port is configured via a command provided by the management server.

Finally, the HV-module permits the received frame to be forwarded to the destination port when the LNID associated with the source VM, $LNID_{SRC}$, is the same as the LNID associated with the destination VM, $LNID_{DST}$. The HV-module also forwards the received frame when either $LNID_{SRC}$ or $LNID_{DST}$ is global LNID. Otherwise, the received frame is discarded.

3.2.2 Examination of Source Address

As described in Section 3.2.1, the proposed network isolation scheme is based on a source MAC address in a L2 network frame header. If a VM is compromised, malicious

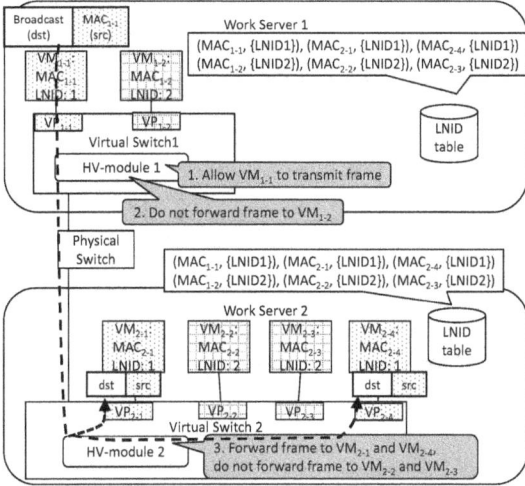

Figure 5: Network isolation flow for broadcast frame

users can transmit a spoofed L2 network frame with the aim of mounting a DoS attack and breaking into the system.

To prevent attacks using such source-MAC-address spoofing, a source MAC address in a L2 network frame header is checked to see whether it was spoofed at the time of the transmission. If the source MAC address does not match one bound with the source virtual port, the HV-module judges the frame as spoofed and prohibits it from being forwarded.

3.2.3 Example of Network Isolation Flow

For ease of comprehension of the flow mechanism described in Section 3.2.1, the manner in which a broadcast frame transmitted from VM_{1-1} is controlled is shown in Figure 5. First, to check for source-MAC-address spoofing HV-module 1 on work server 1 compares the source MAC address in the header with one bound with the source virtual port VP_{1-1}. In this case, HV-module 1 authorizes the frame to be forwarded because the source MAC address in the header is the same as that of VP_{1-1}. In addition, HV-module 1 compares the LNID of the VM_{1-1} with that of VM_{1-2}. In this case, HV-module 1 forbids the frame from being forwarded to VM_{1-2} because LNID 2 is different from LNID 1 of VM_{1-1}.

Next, HV-module 2 on work server 2 compares LNID 1 of the source VM_{1-1} with the LNIDs of the destination VMs. The HV-module forwards the broadcast frame to the VMs with LNID 1, i.e., VM_{2-1} and VM_{2-4}, but does not forward it to VMs with a different LNID 2 from LNID 1, i.e., VM_{2-2} and VM_{2-3}.

3.3 Broadcast Suppression

3.3.1 Background

Broadcast traffic has a potentially significant effect on large-scale networks and can be used by malicious users to flood networks. Limiting broadcast traffic is more important for a HostVLAN-based architecture (in which networks are segregated at the receiving side of an end host server) than for architectures based on VLAN (in which networks are isolated at the transmitting side of a switch). In HostVLAN-based networks, broadcast frames are forwarded to all the work servers in the same L2 domain. Here, we will focus

on ARP and DHCP, because these protocols generally account for a large fraction of broadcast traffic - according to a report by by Elmeleegy et al. [7].

ARP: ARP is used to derive MAC addresses from IP addresses. Broadcast frames are used since the target location is unknown when the ARP requests are transmitted. If a MAC address corresponding to an IP address is found in an ARP cache table, the end host immediately finishes the ARP procedure without exchanging any ARP frames. If no such address is found, the end host scatters the broadcast ARP requests within the L2 domain.

Using ARP cache tables helps to suppress broadcast ARP request frames. In conventional system architectures, however, ARP cache entries are deleted from the table unless they are used within a specified period (e.g., with Linux, the default timeout value is 60 seconds). Broadcast ARP requests are, therefore, transmitted again after cache entries have expired.

DHCP: DHCP is used to dynamically set up and update the network parameters of a DHCP client (such as an IP address and a subnet mask) according to information acquired from DHCP servers. A subnet might be managed by multiple servers. Broadcast frames are used since a client cannot identify where the servers are running. Furthermore, to maintain the consistency of the network information on clients shared among multiple servers, DHCP messages can be transmitted as broadcast frames.

When a new IP address is assigned, the interaction between the servers and clients proceeds as follows. First, a client transmits a broadcast DHCPDISCOVER message in order to request the DHCP servers to assign an IP address. Next, each server responds with a DHCPOFFER message that includes an available IP address. The client then chooses a new IP address from the DHCPOFFER messages and transmits a broadcast DHCPREQUEST message stating the choice to the servers. Finally, the server responds with a DHCPACK message in order to commit this assignment. DHCPOFFER and DHCPACK are broadcast or unicast messages based on the IP headers of messages received from clients.

3.3.2 Suppression Mechanism

We took two approaches to reducing broadcast traffic: (i) conversion of broadcast frames into unicast ones and (ii) rate limiting for each virtual port. The former approach depends on network protocols while the latter does not. In the former approach, a HV-module transforms broadcast frames received from *local* VMs, i.e., VMs running on top of the HV-module, into unicast ones in accordance with the *BtoU* tables. As with the LNID table, the BtoU tables only store entries associated with the LNIDs of VMs deployed on a work server. Unnecessary network traffic is eliminated since the converted broadcast frame is forwarded only to the target VM.

In addition, to mitigate the impact of deliberate broadcast flooding by malicious users, the proposed technique includes rate limiting on virtual ports, similar to what conventional switches do. Concretely, broadcast traffic is kept below a given threshold for each destination address.

3.3.3 Suppression of ARP Requests

Before forwarding an ARP request frame, the HV-module converts the broadcast frame into a unicast one if the target

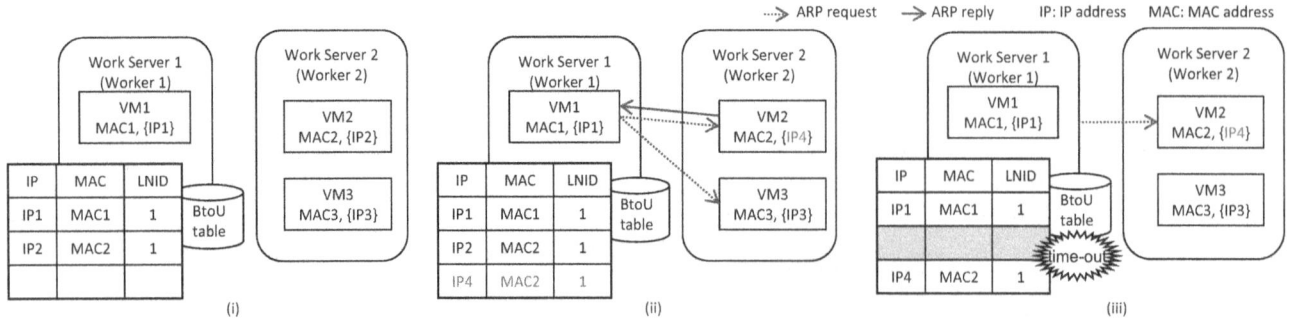

Figure 7: State transition of ARP BtoU table update for changing the IP address

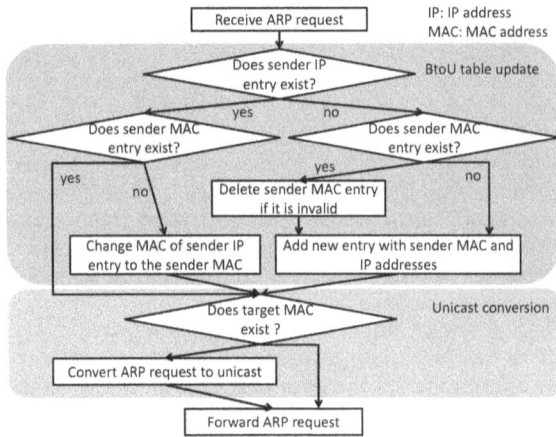

Figure 6: Procedure for ARP request suppression in the HV-module

IP address within the ARP message is contained in the BtoU table. The entries in the BtoU table for ARP messages are tuples consisting of a MAC address, an IP address, and an LNID. The LNID is required to distinguish the same IP addresses of different LNIDs. The BtoU table entries related to sender addresses in the ARP messages are updated upon receipt of ARP requests while the ones related to target addresses are updated upon receipt of ARP replies.

The procedure for reducing ARP requests is shown in Figure 6. A broadcast ARP request is converted into a unicast frame after the BtoU table is updated. The procedure for updating the entries in the BtoU table for an ARP reply is the same as the one for updating an ARP request when the "sender"s in Figure 6 are replaced with the "target"s.

First, the HV-module checks whether a tuple of a sender IP address in an ARP message and an LNID associated with the source MAC address of the L2 frame header exists when the ARP request frames are received (*C1*). If such a tuple exists and the MAC address of the tuple is different from the sender MAC address, the MAC address is rewritten with the sender MAC address.

If *C1* is false and if there is a tuple with the sender MAC address of the ARP request, the HV-module creates a unicast ARP request frame for each IP address of the tuple and transmits it. The unicast frame is used to confirm that the tuple related to the sender MAC address is available. In the unicast frame, the destination MAC address of the L2 frame header, the target MAC address, and the IP address in the ARP message are, respectively, the MAC address, the MAC address and the IP address of the tuple. The unicast ARP requests created for the confirmation are called *probe ARP requests*. Furthermore, the HV-module registers a time-out handler for each probe ARP request in order to delete the entry in the BtoU table unless the corresponding ARP reply is received within a given period (the default value is three seconds.). The time-out handler is necessary for erasing invalid entries, e.g., entries related to an IP address before modification.

If *C1* is false and there is no tuple with the sender MAC address of the ARP request, the HV-module stores a tuple consisting of the sender MAC address and the IP address of the ARP message and the LNID associated with the source MAC address of the L2 frame header in the BtoU table.

The received ARP requests are converted into unicast ones if there are entries related to the target MAC addresses of the ARP requests. The converted ARP requests are only forwarded to the target VM.

Although the BtoU table is similar to the ARP cache table, it differs in the way that entries are deleted. In the ARP cache table, entries are deleted if they have not been used within a specified period, while in the BtoU table, entries are deleted if ARP replies are not returned within a specified period.

To make the procedure for suppressing ARP requests easier to understand, Figure 7 shows a case in which VM2's IP address changes from IP2 to IP4. In this example, VM1 running on work server 1 (worker 1) and VM2 and VM3 running on worker 2 belong to the logical network with LNID1. We assume that the BtoU table contains entries for both IP1 and IP2 (Figure 7 (i)). After the ARP reply for IP2 has been received, a new association between IP4 and MAC2 is stored (Figure 7 (ii)). Since an entry associated with MAC2 - namely, a tuple {IP2, MAC2} - has already been stored, worker 1 then sends the probe ARP request whose target IP address is IP2 to VM2 and sets the time-out handler for the probe frame. Since the IP address changes to IP4, VM2 does not respond to the probe ARP request. Consequently, {IP2, MAC2} is deleted from the BtoU table by the time-out handler (Figure 7 (iii)).

3.3.4 Suppression of DHCP Messages

Figure 8 is an overview of the procedure for suppressing DHCP messages. Upon receiving broadcast DHCP mes-

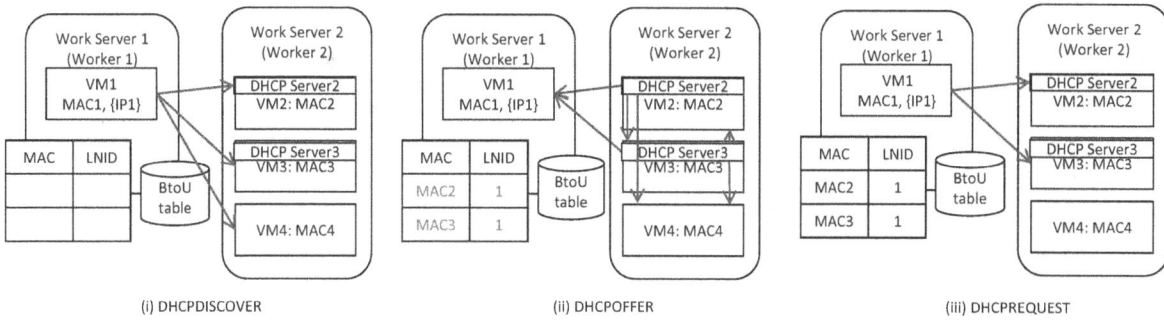

Figure 9: Procedure for suppressing DHCP messages when a new IP address is assigned

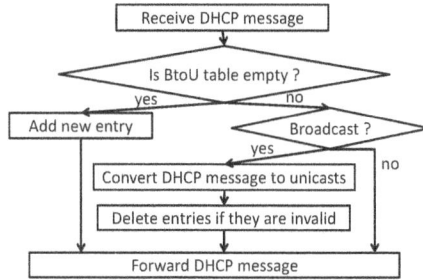

Figure 8: Procedure for DHCP suppression

sages, such as *DHCPDISCOVER, DHCPOFFER, DHCPRE-QUEST*, and *DHCPACK*, from local VMs, the HV-module converts them into unicast messages unless the BtoU table is empty. For messages sent from DHCP clients to DHCP servers (e.g., DHCPDISCOVER and DHCPREQUEST), the unicast messages are forwarded to DHCP servers. For messages sent from DHCP servers to DHCP clients (e.g., DHCP-OFFER and DHCPACK), the unicast messages are forwarded to DHCP clients and DHCP servers. After converting the broadcast messages into unicast ones, the time-out handler for each DHCP server is registered in order to remove the entry for the DHCP server if it is not available.

If the BtoU table for DHCP is empty and the HV-module has received DHCPOFFER, the HV-module adds a new entry associated with the DHCP server in DHCPOFFER to the BtoU table. To synchronize all BtoU tables, the new entry is transferred through the management agent to the other work servers.

The BtoU table entries for DHCP are pairs composed of the MAC address and LNID of a DHCP server. An LNID is necessary for identifying DHCP servers that have the same LNID as VMs sending DHCP messages. The HV-module updates the entry, and the work agent notifies the management agent of the update when it receives the DHCPOF-FER.

Figure 9 shows the BtoU table update and unicast conversion procedures for when a new IP address is assigned. We assume that within the logical network with LNID1, VM1 runs on work server1 (worker 1) and VM2, VM3, and VM4 run on work server2 (worker 2). We also assume that the DHCP servers are running on VM2 and VM3. First,

VM1 locates the DHCP servers and asks them to allocate a new address by transmitting the broadcast DHCPDIS-COVER to VM2, VM3, and VM4 (Figure 9(i)). After that, worker 1 stores the MAC addresses (MAC2 and MAC3) and LNID1 in the BtoU table (Figure 9(ii)) when it receives the DHCPOFFER from the DHCP servers residing in VM2 and VM3. The work agent then notifies the management agent of the new table entries.

After receiving the available IP addresses (Figure 9(iii)), VM1 selects one of them and transmits the broadcast DHCP-REQUEST as a notification of its choice of DHCP server. At this time, worker 1 duplicates the broadcast DHCPRE-QUEST and converts it into unicast ones bound for two DHCP servers that have the same LNID as the source MAC address, MAC1, and it transmits the duplicated unicast frames to the two DHCP servers since the BtoU table contains two entries (pairs related to MAC2 and MAC3).

There is one difficulty with this approach for DHCP suppression: the HV-module cannot detect a DHCP server that starts running under the condition that the BtoU table is not empty. Accordingly, the HV-module converts the broadcast messages into unicast ones bound for only the DHCP servers stored in the BtoU table as long as the table contains some entries. Consequently, DHCPDISCOVER cannot be transmitted to a DHCP server that starts running when the BtoU table is not empty.

To overcome this issue, each work server (worker) periodically sends a *probe DHCPDISCOVER* to all of the local VMs. When a worker receives a DHCPOFFER from DHCP servers with MAC addresses that are not contained in the BtoU table, it conveys the MAC addresses and LNIDs of these DHCP servers to the management server. The management server then forwards the received information to the other workers. The other workers store this information in their own BtoU tables.

3.4 Limitations

The HostVLAN-based architecture assumes a large-scale and single L2 domain. Therefore, it cannot segregate networks across multiple L2 domains. Since the source MAC addresses of network frames forwarded via routers contain the address of an edge router, the HV-module in the destination work server cannot discriminate between the LNIDs of the source VMs running on the other L2 domains. However, provided that the routers are equipped with a "virtual routing and forwarding" (VRF) functionality, the HostVLAN-based architecture can cooperate with them in order to segregate networks across L2 domains. Accordingly, the destination HV-module can determine the LNIDs of the source VMs on the basis of virtual paths segregated for each LNID by the routers equipped with VRF.

Since the HostVLAN architecture provides logical networks based on MAC addresses, it cannot be applied to data center networks in which usage of the same MAC address is allowable. This limitation can be alleviated by assuming that the data center administrators assign unique MAC addresses to VMs. We consider that this assumption is not stringent because the VM configurations are usually managed by the administrators with little impact on VM users.

HostVLAN neither encapsulates nor rewrites network frames for network segregation. Consequently, when switches attempt to store a large number of VM MAC addresses in the MAC address tables, there is the potential for flooding due to forwarding-data-base overflow. In the HostVLAN-based architecture, however, L2 switches can simply control received data at a fine level granularity, i.e., in accordance with the MAC addresses of the VMs (e.g., control depending on ACLs).

4. EVALUATION

To demonstrate that HostVLAN can be applied to a wide range of virtual switches, we added it to three existing virtual switches; *bridge*, *Open vSwitch*, and *macvtap*. The HostVLAN-based architecture was implemented on commodity VMMs, KVM, and Xen, which support these network functions. The virtual switches for KVM were implemented in the hypervisor, i.e., the Linux kernel, while those for Xen were implemented in the OS kernel of the privileged VM, termed *dom0*. KVM supported all three functions, while Xen supported bridge and Open vSwitch.

We appraised the feasibility of HostVLAN from two viewpoints: (i) its impact on performance and (ii) the effectiveness of its broadcast-traffic suppression mechanism. We used KVM 0.14.1 and Xen 4.1.1 with full- and para-virtualization as the VMM, Linux OS kernel 3.0 as the hypervisor OS kernel of KVM and dom0 OS kernel of Xen, Linux OS kernel 2.6.34 as the guest OS kernels, and *Open vSwitch* 1.2.0.

4.1 Performance Impact

We gauged the impact on scalability of HostVLAN from three aspects: amount of memory consumption, processor overhead, and processor utilization.

4.1.1 Memory Consumption

HostVLAN's scalability is estimated in terms of the number of VMs per work server ($\#(VM_{ws})$), VMs per tenant (LNID) ($\#(VM_{LNID})$), and work servers ($\#(ws)$). The number of LNIDs in a data center is represented as $\#(LNID) = (\#(VM_{ws}) * \#(ws)) / \#(VM_{LNID})$, and the number of VMs in the data center is $\#(VM) = \#(VM_{ws}) * \#(ws)$. The LNID table and the BtoU table on each work server include only the entries associated with the LNIDs to which the VMs running on the work server belong. Here, we consider the worst case: the maximum number of table entries. That is, each VM running on a work server has a different LNID. The amount of memory consumed by a work server depends on the number of LNID table entries ($\#(entry_{LNID}) = \#(VM_{ws}) * \#(VM_{LNID})$) and BtoU table entries for ARP ($\#(entry_{ARP}) = \#(VM_{ws}) * \#(VM_{LNID})$)

Number of work servers	8,192
Number of VMs per work server	128
Number of VMs per tenant	64
Number of tenants in data center	16,384
Number of VMs in data center	1,048,576

Table 1: Data center networks used in the evaluation

and DHCP ($\#(entry_{DHCP}) = \#(VM_{ws}) * \#(DHCP_{LNID})$, where $DHCP_{LNID}$ stands for DHCP servers per tenant). We assume that $\#(VM_{LNID})$ and $\#(DHCP_{LNID})$ are both constant, so the amount of memory is determined by $\#(VM_{ws})$. This means that the amount increases as a result of scaling the data center not horizontally (a scale-out scheme) but vertically (a scale-up scheme).

Next, we assume the data center that could provide logical networks in excess of 4,094 (the maximum number of logical networks provided by using VLAN), as shown in Table 1. $\#(VM_{ws})$, $\#(VM_{LNID})$, and $\#(ws)$ are 128, 64, and 8,192, respectively. Accordingly, $\#(LNID)$ and $\#(VM)$ are 16,384 (2^{14}) and 1,048,576 (2^{20}), respectively.

We implemented a chaining hash table whose key was the lower 16 bits of a MAC address as the data structure of LNID table. For the BtoU table for ARP, we used two chaining hash tables whose keys were the lower 8 bits of the MAC address (ARP-MAC) and the IP address (ARP-IP). For the BtoU table for DHCP, we used a chaining hash table whose key was the lower 8 bits of the MAC address. The memory consumed by each hash table was calculated as (bucket size of hash table) * 8 (pointer size) + (entry size of hash table) * (numbers of entries in hash table).

Assuming that the association between the MAC address and LNID is a one-to-one mapping, the size of an entry in the hash table for LNID would be 96 bytes, while those of the hash tables for ARP-MAC, ARP-IP, and DHCP would be 112, 152, and 136 bytes, respectively. For the data center network indicated in Table 1 and under the above association and $\#(DHCP_{LNID})$ of 1, $\#(entry_{LNID})$, $\#(entry_{ARP})$, and $\#(entry_{DHCP})$ would be, respectively, 8,192, 8,192, and 128. Consequently, the memory required for the prototype would approximately 3 MB (the size of the LNID table and the BtoU tables for ARP and DHCP would about 852 KB, 2,166 KB, and 19 KB, respectively). This shows that the amount of memory consumed by HostVLAN would be acceptable to users.

4.1.2 Processor Overhead and Utilization

The network-isolation mechanism used for the received frames mostly affected the performance of the HostVLAN-based architecture. To clarify the overhead of HostVLAN under the conditions in Table 1, we ran microbenchmark and application benchmark programs. The benchmark server programs ran on the VM. The benchmark client programs were executed on a different physical machine, and two physical machines were deployed within the same LAN.

Setup: The physical machines had two Quad-Core Intel Xeon 2.93 GHz processors with 48 GB of RAM, and a 1-Gbps NIC. For KVM, the hypervisor OS kernel was configured with 1 CPU and 48 GB of memory and the VM was configured with 1 CPU and 4 GB of memory, and a *virtio-net* NIC. For Xen, the dom0 OS kernel was configured with 1

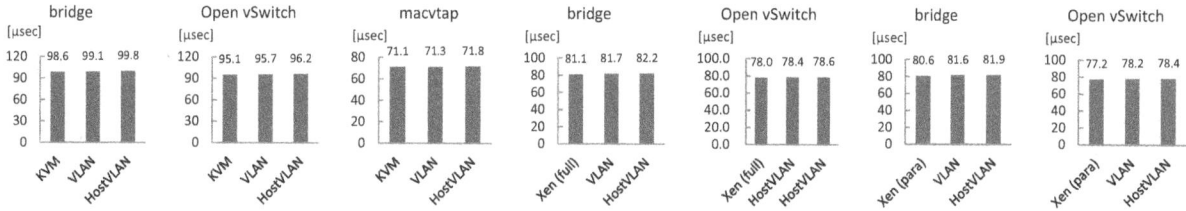

Figure 10: Microbenchmark results for network latency

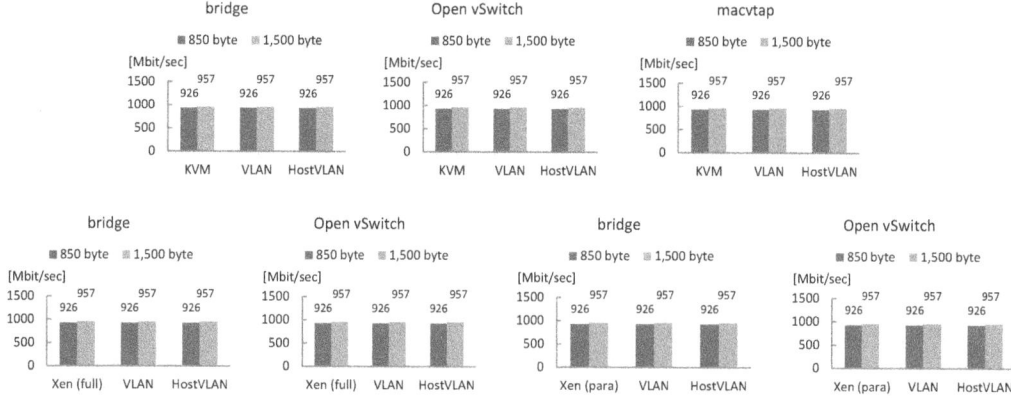

Figure 11: Microbenchmark results for network throughput

CPU, 4 GB of memory and the VM (domU) was configured with 1 CPU and 4 GB of memory running a para-virtualized Linux OS kernel. DomU was bound with a different physical CPU from dom0.

We measured the execution times of the benchmarks for KVM with *bridge*, *Open vSwitch*, and *macvtap* and for the full- and para-virtualization versions of Xen with *bridge* and *Open vSwitch*. In regard to *macvtap*, a *vhost-net* function was used to improve the I/O network performance of KVM. We also compared the microbenchmark results of the HostV-LAN with those of unmodified VMMs (KVM, Xen (full), and Xen (para)) and VMMs using VLAN. Although the network isolation of VLAN could not support 8,192 logical networks, we quantified its execution time, which is independent of the number of logical networks, for comparison.

Microbenchmarks: We examined the network latency and throughput for the above settings. First, we measured the latencies of UDP communications by using *lat_udp* of *lm-bench*. The transmitted data size was 46 bytes, the minimum size of an L2 frame payload. Next, *Netperf* was used to measure the throughput of UDP communications. We chose two sizes of transmitted L2 frame payload; 850 bytes and 1,500 bytes. The former size was the minimum size at which the processor did not saturate for all experimental settings when the benchmark program ran. The latter is the maximum size of an L2 frame payload without IP fragmentation.

The experimental results on network latency and throughput are shown in Figures 10 and 11. The network latencies of our implementations were 2% or less than those of unmodified VMMs. The network throughput did not decrease. *Application Benchmarks*: The throughput for Web ser-

vices was measured using the *ApacheBench* program. *ApacheBench* sent requests for static content (1-KB file) to an *Apache* Web server running on a different physical machine. The number of requests was 10,240. As shown in Figure 12, HostVLAN's overhead for Web services was 1% or less.

We also examined processor utilization on the KVM hypervisor and dom0 of Xen when we measured the above network throughput microbenchmarks. For KVM and Xen, *mpstat* and and *xenmon* commands were respectively used for assessing processor utilization. As shown in Figure 13, the HostVLAN-based architecture had little impact on the processor load, which increased by at most 3.6% and 2.3% for KVM and Xen relative to the unmodified VMMs.

These experimental results indicate that HostVLAN is a useful technique for supporting multi-tenant networks consisting of thousands of tenants, and its use only entails a minor performance degradation.

4.2 Network Load

In the case of unicast frames, the number of frames transferred in the HostVLAN-based network infrastructure is the same as in the VLAN-based one. In the case of broadcast frames, HostVLAN restrains the amount of broadcast traffic of ARP and DHCP by converting broadcast frames into unicast ones and by limiting the traffic rate. The conversion is applicable when there are BtoU table entries for the received broadcast frames. The number of destination VMs of a broadcast frame is one in HostVLAN, whereas in VLAN it is however many VMs that have the same VLAN ID, $\#(VM_{LNID})$.

In a similar manner to the performance evaluation described above, we estimate the decrease in network traffic

95

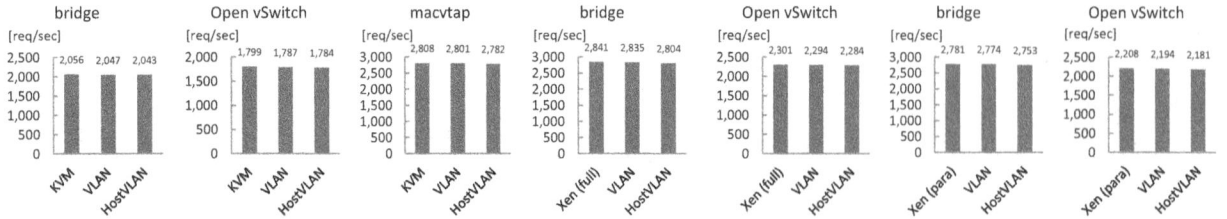

Figure 12: Experimental results for Web service throughput

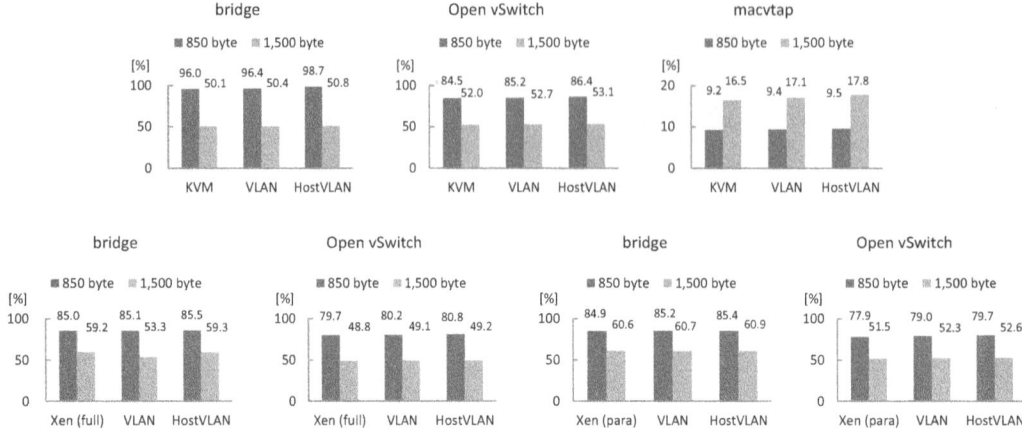

Figure 13: Experimental results for processor utilization

in the data center networks listed in Table 1. At the peak load of broadcast traffic, to acquire the MAC address of the Web server and dynamic IP addresses, 63 VMs intensively transmit ARP and DHCP broadcast frames for one VM on which a Web server and DHCP servers are running in a tenant, *tenantA*. In this case, broadcast frames for all VMs, about one million (1,048,576) in the data center, are converted into unicast ones for each VM in *tenantA*. Consequently, the amount of network frames in the entire data center decreases from about 63 million (= 63 * 1,048,576) to 63 (63 * 1). In contrast, the amount of network frames in the entire data center decreases to 3,969 (= 63*63) as a result of using the VLAN-like approach.

It is noteworthy that the first broadcast frames transmitted from the VMs are not always broadcasted. We assume that 64 VMs in a tenant are running on different work servers and that one VM, VM1, transmits an ARP request to another VM, VM2. 63 work servers (excluding the work server on which VM1 runs) store the entry related to VM1 in their BtoU tables upon receipt of the broadcast ARP request. Thereafter, if 63 VMs other than VM1 transmit the first ARP request for VM1, the requests are converted into unicast ones because there is a BtoU table entry related to VM1.

We close this section with a description of the DHCP server detection. Each work server can transmit additional network frames related to *pseudo* frames. Regarding ARP, *pseudo* frames are used when VMs add or change MAC and IP addresses. Regarding DHCP, when a work server detects local DHCP servers, the work server notifies the others via the management server. However, we consider that

these extra data transmissions have an insignificant impact on network performance because their number is generally small.

5. RELATED WORK

There have been several studies on leveraging an overlay network scheme to scale a large number of logical networks [2,4,9,14,18,22]. VXLAN [14] and DOVE [2] provide scalable multi-tenant networks with MAC-in-UDP encapsulation. The header within the UDP payload includes the tenant ID with space for about 16 million IDs (24 bits). NVGRE [22] utilizes MAC-in-GRE encapsulation to render large multi-tenant environments. The tenant ID in NVGRE is contained in the GRE header; its ID field is 24 bits. STT [4] adds a TCP-like header to an L2 frame transmitted from VMs (MAC-in-TCP encapsulation); the field of the tenant ID is 64 bits. In VXLAN, DOVE, NVGRE, and STT, a network endpoint residing in the source VMM creates encapsulated network frames with the extra header and forwards them to the network endpoints residing in the destination VMMs. The destination network endpoint decapsulates the frame and forwards it only to the VMs that belong to a tenant ID included in the extra header. To limit broadcast traffic, these systems assign one broadcast domain to each tenant.

VXLAN, DOVE, NVGRE, and STT place three restrictions on encapsulation. Firstly, the size of the transferred network frames is increased by the extra encapsulation header. Secondly, hardware offload engines such as TCP checksum offloading cannot be applied to the inner headers of encapsu-

96

lated frames in the currently available NICs. Thirdly, if the four isolation schemes coexist, the virtual switch cannot segregate logical networks based on the other schemes because the tenant isolation scheme depends on the encapsulation scheme. For example, a virtual switch for VXLAN cannot interpret network data encapsulated by DOVE, NVGRE, or STT. Moreover, to reduce broadcast traffic, these systems map multicast addresses, including a broadcast address, to the specified IP multicast ones. This scheme imposes two restrictions. One is that the tenant ID spaces are limited by the number of physical IP multicast addresses, and the other is that the segregation of broadcast domains cannot limit broadcast traffic within the same tenant.

NetLord [18] is a scalable multi-tenant network architecture using MAC-in-IP encapsulation. In this architecture, a tenant ID is contained in the outer IP destination address. The default ID field is 24 bits. The sender network agent running in the VMM transmits encapsulated frames in which the outer MAC address is a MAC address of an egress edge switch to which the destination VMM connects. When receiving the encapsulated frames, the egress edge switch removes the original L2 header and adds a new L2 header whose destination MAC address is the MAC address of the destination VMM. It then forwards the new encapsulated frames to the destination agent. As with VXLAN, DOVE, NVGRE, and STT, NetLord has two restrictions due to the encapsulation scheme. Moreover, NetLord requires an IP forwarding function because an edge switch forwards network frames to destination agents according to an IP forwarding table.

Hao et al. [9] proposed SEC2, a multi-tenant network infrastructure combined with multiple edge networks, which separates different tenants on the basis of VLAN. The VLAN IDs are unique for each edge network but not for each data center. To communicate across an edge network, a forwarding element (a gateway) of the source edge network forwards network frames to the forwarding element of the destination edge network with MAC-in-MAC encapsulation in conjunction with the controller that manages the configurations of a data center. The problem here is that although SEC2 enables a data center as a whole to exceed the limit on the maximum number of VLAN IDs, the edge networks still have this limitation. The transferred frame size is increased due to MAC-in-MAC tunneling. In addition, forwarding elements need to have flow forwarding functions and tunneling capabilities.

SEATTLE [12] and EtherProxy [7] provide a technique for limiting broadcast traffic. SEATTLE is a network architecture that provides scalable self-configuring routing. It uses a directory service based on a network-layer distributed hash table to proxy ARP requests and convert them into unicast ARP messages. EtherProxy [7] is a network device that responds to ARP requests with its own ARP cache entries and converts broadcast DHCP messages into unicast ones. However, it increases network traffic due to the extra unicast ARP requests used to update the expired ARP cache entries and the DHCPDISCOVER messages required to update the DHCP server list. In addition, it transfers broadcast DHCPDISCOVER messages throughout networks to detect DHCP server updates in the networks. In contrast, in HostVLAN, the work agent only transfers the broadcast DHCPDISCOVERs to local VMs, i.e., without transmitting any data to physical networks.

6. SUMMARY

We proposed HostVLAN, a multi-tenant technique for isolating a huge number of logical networks without encapsulating or rewriting network frames. Unlike conventional approaches which work on the switch side, this multi-tenant technique works on the end-host-server side. Scalability is becoming a concern as data centers using server virtualization and supporting multi-tenant networks continue to grow in size. HostVLAN is applicable to any physical network topology. Its architecture enables administrators to easily mange and control the overall multi-tenant environment via a central management server.

In HostVLAN, an HV-module residing inside VMMs stops data from being transferred among different tenants through its use of an isolation information table, LNID table. HostVLAN also includes functions for preventing feigned frame transmission by compromised VMs and for suppressing ARP and DHCP broadcast traffic. We designed and implemented an HostVLAN-based architecture by extending three virtual network-switch functions. In a simulation featuring approximately 16,000 tenants and 1,000,000 VMs, the overhead due to HostVLAN for Web services was less than 1%. We believe that HostVLAN will be a useful technique for supporting large-scale multi-tenant networks.

7. REFERENCES

[1] amazon.com. *Amazon Elastic Compute Cloud.* http://aws.amazon.com/ec2.

[2] K. Barabash, R. Cohen, D. Hadas, V. Jain, R. Recio, and B. Rochwerger. A Case for Overlays in DCN Virtualization. In *Proceedings of the 3rd Workshop on Data Center - Converged and Virtual Ethernet Switching*, San Francisco, September 2011.

[3] P. Barham, B. Dragovic, K. Fraser, S. Hand, T. Harris, A. Ho, R. Neugebauer, I. Pratt, and A. Warfield. Xen and the Art of Virtualization. In *Proceedings of the 19th ACM Symposium on Operating Systems Principles*, New York, October 2003.

[4] B. Davie and E. Gross. A Stateless Transport Tunneling Protocol for Network Virtualization (STT) (draft-davie-stt-02), August 2012.

[5] L. Dunbar, S. Hares, M. Sridharan, N. Venkataramaiah, and B. Schliesser. Address Resolution for Large Data Center Problem Statement (draft-dunbar-armd-problem-statement-01.txt), March 2011.

[6] A. Edwards, A. Fischer, and A. Lain. Diverter: A New Approach to Networking Within Virtualized Infrastructures. In *Proceedings of the 1st Workshop on Research on Enterprise Network*, Barcelona, August 2009.

[7] K. Elmeleegy and A. Cox. EtherProxy: Scaling Ethernet By Suppressing Broadcast Traffic. In *Proceedings of the 28th IEEE International Conference on Computer Communications*, Rio de Janeiro, April 2009.

[8] A. Greenberg, J. Hamilton, N. Jain, S. Kandula, C. Kim, P. Lahiri, D. Maltz, P. Patel, and S. Sengupta. VL2: A Scalable and Flexible Data Center Network. In *Proceedings of the ACM SIGCOMM 2009*, Barcelona, August 2009.

[9] F. Hao, T. Lakshman, S. Mukherjee, and H. Song.

Secure Cloud Computing with a Virtualized Network Infrastructure. In *Proceedings of the 2nd Workshop on Hot Topics in Cloud Computing*, Boston, June 2010.

[10] IEEE Computer Society. IEEE Standard for Local and metropolitan area networks - Virtual Bridged Local Area Networks, May 2006.

[11] IEEE Computer Society. IEEE Standard for Local and metropolitan area networks - Virtual Bridged Local Area Networks Amendment 7: Provider Backbone Bridges, June 2008.

[12] C. Kim, M. Caesar, and J. Rexford. Floodless in SEATTLE: A Scalable Ethernet Architecture for Large Enterprises. In *Proceedings of the ACM SIGCOMM 2008*, Seattle, August 2008.

[13] KVM. *Kernel Based Virtual Machine.* http://www.linux-kvm.org/page/Main_Page.

[14] M. Mahalingam, D. Dutt, K. Duda, P. Agarwal, L. Kreeger, T. Sridhar, M. Bursell, and C. Wright. VXLAN: A Framework for Overlaying Virtualized Layer 2 Networks over Layer 3 Networks (draft-mahalingam-dutt-dcops-vxlan-02.txt), August 2012.

[15] N. McKeown, T. Anderson, H. Balakrishnan, G. Parulkar, L. Peterson, J. Rexford, S. Shenker, and J. Turner. OpenFlow: Enabling Innovation in Campus Networks. *ACM SIGCOMM Computer Communication Review*, 38(2):69–74, April 2008.

[16] Microsoft. *Hyper-V server.* http://www.microsoft.com/en-us/server-cloud/hyper-v-server/.

[17] Microsoft. *Understanding Networking with Hyper-V.* http://www.microsoft.com/downloads/en/details.aspx?displaylang=en&familyID=3fac6d40-d6b5-4658-bc54-62b925ed7eea.

[18] J. Mudigonda, P. Yalagandula, J. Mogul, B. Stiekes, and Y. Pouffary. NetLord: A Scalable Multi-Tenant Network Architecture for Virtualized Datacenters. In *Proceedings of the ACM SIGCOMM 2011*, Toronto, August 2011.

[19] R. Mysore, A. Pamboris, N. Farrington, N. Huang, P. Miri, S. Radhakrishnan, V. Subramanya, and A. Vahdat. PortLand: A Scalable Fault-Tolerant Layer 2 Data Center Network Fabric. In *Proceedings of the ACM SIGCOMM 2009*, Barcelona, August 2009.

[20] T. Narten, M. Sridharan, D. Dutt, D. Black, and L. Kreeger. Problem Statement: Overlays for Network Virtualization (draft-narten-nvo3-overlay-problem-statement-01), October 2011.

[21] Open vSwitch. *VLANs.* http://openvswitch.org/?page_id=146.

[22] M. Sridharan, A. Greenberg, N. Venkataramiah, Y. Wang, K. Duda, I. Ganga, G. Lin, M. Pearson, P. Thaler, and C. Tumuluri. NVGRE: Network Virtualization using Generic Routing Encapsulation (draft-stridharan-virtualization-nvgre-01.txt), July 2012.

[23] VMware, Inc. *VMware vShpere Hypervisor (ESXi).* http://www.vmware.com/products/vsphere-hypervisor/.

[24] VMware, Inc. *VMware ESX Server 3: 802.1Q VLAN Solutions*, June 2006. http://www.vmware.com/resources/techresources/412.

[25] T. Wood, A. Gerber, K. Ramakrishnan, P. Shenoy, and J. Merwe. The Case for Enterprise-Ready Virtual Private Clouds. In *Proceedings of the 1st Workshop on Hot Topics in Cloud Computing*, San Diego, June 2009.

Distributed Adaptive Routing for Big-Data Applications Running on Data Center Networks

Eitan Zahavi
Mellanox Technologies & Technion
eitan@mellanox.com

Isaac Keslassy
Technion
isaac@ee.technion.ac.il

Avinoam Kolodny
Technion
kolodny@ee.technion.ac.il

ABSTRACT

With the growing popularity of big-data applications, Data Center Networks increasingly carry larger and longer traffic flows. As a result of this increased flow granularity, static routing cannot efficiently load-balance traffic, resulting in an increased network contention and a reduced throughput. Unfortunately, while adaptive routing can solve this load-balancing problem, network designers refrain from using it, because it also creates out-of-order packet delivery that can significantly degrade the reliable transport performance of the longer flows.

In this paper, we show that by throttling each flow bandwidth to half of the network link capacity, a distributed-adaptive-routing algorithm is able to converge to a non-blocking routing assignment within a few iterations, causing minimal out-of-order packet delivery. We present a Markov chain model for distributed-adaptive-routing in the context of Clos networks that provides an approximation for the expected convergence time. This model predicts that for full-link-bandwidth traffic, the convergence time is exponential with the network size, so out-of-order packet delivery is unavoidable for long messages. However, with half-rate traffic, the algorithm converges within a few iterations and exhibits weak dependency on the network size. Therefore, we show that distributed-adaptive-routing may be used to provide a scalable and non-blocking routing even for long flows on a rearrangeably-non-blocking Clos network under half-rate conditions. The proposed model is evaluated and approximately fits the abstract system simulation model. Hardware implementation guidelines are provided and evaluated using a detailed flit-level InfiniBand simulation model. These results directly apply to adaptive-routing systems designed and deployed in various fields.

Categories and Subject Descriptors

C2.1 [**Computer Communication Networks**]: Network Architecture and Design – *Data Center Networks' Adaptive Routing*. **Keywords**
Data Center Networks, Big-Data, Adaptive Routing.

1. INTRODUCTION

1.1 Background

Nearly all currently-deployed state-of-the-art Data Center Networks (DCNs) rely on layer-3 Equal Cost Multipath (ECMP) routing to evenly distribute traffic and utilize the aggregated bandwidth provided by the multi-tier network [2]. ECMP routing is *deterministic and static*, because it is based on constant hash functions of the flow's identifier. The obtained bandwidth from these techniques is close to the network cross-bisectional-bandwidth as long as flow granularity is small, i.e. the routing algorithm spreads many flows that are short and/or long but low-bandwidth.

In recent years a new challenge has emerged for DCNs: *support "big-data" applications like MapReduce* [18][30]. In measurements conducted on the Shuffle and Data-Spreading stages of MapReduce applications, it was shown that up to 50% of the run time may be consumed by these stages [7]. In fact, these stages transmit the intermediate computation results with sizes up to 10's of gigabytes between each pair of servers participating in the computation. Therefore, the long and high-bandwidth flows characterizing these phases break the nice traffic spreading provided by the ECMP hash functions. For static routing there are always adversary patterns exhibiting high link over-subscription [28]. The probability of over-subscription follows the balls-and-bins max-load distribution. The contending flows result in a low effective bandwidth [6].

Adaptive routing can provide a solution to this contention problem [25][26]. In fact, adaptive routing can reach efficient traffic spreading in the DCN, even when flow granularity is high. Furthermore, when using adaptive routing, switches need to know little about the global state of the network or about the states of other switches. Unfortunately, adaptive routing can also cause high out-of-order packet delivery in long flows, which greatly degrades window-based transport protocols like TCP, and can result in a significant degradation of throughput and latency [23][31]. Due to this limitation, adaptive routing is often considered irrelevant for DCNs running big-data applications.[1]

In this paper, our goal is to determine conditions under which a distributed adaptive routing DCN algorithm can cause minimal out-of-order packet delivery in big-data applications, while achieving high throughput.

---

[1] Although new reliable transports may be designed to tolerate some out-of-order delivery, they are limited by a basic tradeoff between the allowed out-of-order window size and the resources required to maintain it.

1.2 Related Work

The network contention caused by a relatively small number of high-bandwidth flows is also a long outstanding problem of static routing in High Performance Computing (HPC) clusters. The scientific applications run on these clusters resemble big-data applications, because most parallel scientific applications are coded according to the BSP model [11], where computation and communication are separated into non-overlapping phases [16]. Under such a model, network contention directly impacts the overall program runtime since the slowest flow dictates the length of the entire communication phase [17][32]. For these reasons, efforts have been made to provide adaptive-routing, together with heuristics and mechanisms to improve both throughput and latency [20][22]. Most commercial interconnection networks like Cray BlackWidow[27], IBM BlueGene [1] and the InfiniBand-based Mellanox InfiniScale switch devices [5] provide adaptive routing. The most scalable systems are designed such that each switch knows little or nothing about the traffic or queues of the other elements in the network. Such systems are thus denoted oblivious-adaptive-routing. Mechanisms to enhance the adaptive-routing hardware in switches by relying on a complete or partial view of the entire network state were also proposed [15][26]. Nevertheless, to maintain scalability, even when complex feedback mechanisms are suggested, the self-routing principle, where each switch makes its own independent decisions, is maintained in most published works.

One of the key drawbacks of adaptive routing is the resulting out-of-order packet delivery produced by the modification of the forwarding path of different packets of the same message [12]. To overcome the out-of-order delivery, which greatly degrades TCP (or any other packet window-based transport protocol), some studies propose the use of re-order input-buffers and limit the number of in-flight packets to control their cost [25][23]. However, limiting the number of in-flight packets degrades bandwidth, increases latency, and requires additional hardware buffers.

Another challenge for adaptive routing is that previous network states cannot be used for deciding about the new routing when the entire traffic pattern changes synchronously. Unfortunately, this is the exact behavior of BSP model programs as well as for the shuffle stages of a MapReduce.

Adaptive-routing stability was studied in the context of the Internet [3][10]. In these studies, a centralized adaptive-routing algorithm is employed to optimize the network performance for some figure of merit. The computed routing slowly changes when compared to the traffic message times. The stability of such systems is then defined as the ability to avoid fluctuations in routing assignments when a computed routing is applied. Thus, adaptive-routing stability is different from our definition of adaptive-routing convergence, i.e. the ability of the system

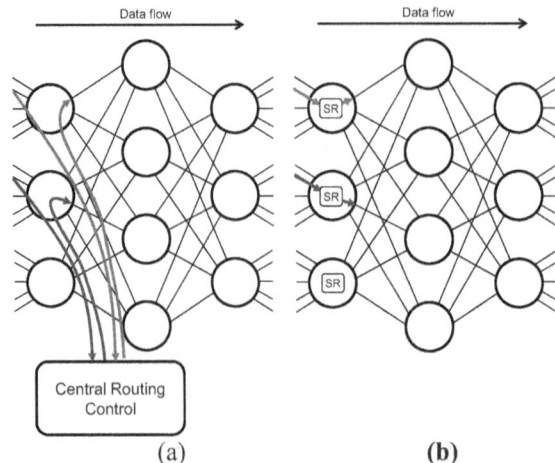

Figure 1: Centralized Routing versus Self Routing: (a) In a centrally controlled Clos, input switches request an output-port assignment for each arriving flow from the central controller. (b) In a distributed adaptive routing system, a Self Routing unit within each input switch provides that decision in an autonomic manner.

to reach a non-blocking routing assignment for any given traffic permutation.

We limit our discussion to Clos and folded-Clos networks (also known as fat trees), which are the most commonly used topologies for DCNs. Routing in Clos networks was mostly studied in the context of systems where a *centralized* control unit allocates virtual circuits to injected flows [19] (see Figure 1(a)). Consequently, the network's topological properties that support strict sense non-blocking (SNB), rearrangeable non-blocking (RNB) and wide-sense non-blocking (WSNB) [4] were defined. When packet switching was proposed as an alternative to virtual circuits, Clos research focused on the properties obtained for multi-rate traffic injected into a network of multiple-capacity links [9][24]. In Clos networks, the term *n-rate* represents the number of different ratios of the traffic bandwidth to the system link capacity [24].

The centralized controller approach used for Clos routing does not scale with the cluster size, and therefore can hardly apply to our DCNs: To estimate the time available for the central controller for handling a single flow, consider a DCN of 10K nodes each running 10 VMs. An optimistically long flow length of 64KB on 40Gbps link provides 12.8usec flow lifetime. Further, if we assume communication is only 10% of application runtime; the flows arrival rate is 1/128usec on average. Under the above conditions the central routing unit has to handle a request rate of ~1G[req/sec] which allows roughly 2 operations per request on 2GHz CPU. Even a parallel routing algorithm will have to use more than a single OP for handling a request.

For this reason, a distributed adaptive routing approach was also proposed in Clos networks, and denoted as "self-routing" [8]. In this approach, each switch can make its own routing decisions such that no central control unit is required (see Figure 1(b)). The self-routing study [8] was mainly focused on reducing the non-scalable overhead of the central routing controller. A probability analysis conducted by [33] on some specific self-routing Clos systems also provides an upper-bound on the number of contending flows (with high probability), and thus provides an upper-bound for the expected network queue length and service time. Our work is different as it shows that under some conditions, adaptive-routing can actually converge into a non-blocking routing assignment where no networking queuing is formed.

1.3 Contributions

To the best of our knowledge, *no work in the literature examines the conditions under which adaptive routing converges to a non-blocking forwarding assignment;* and in case of convergence, *we also know of no result on its convergence speed.*

By analyzing convergence time, we are able to show that under some traffic conditions the adaptive-routing can converge to a non-blocking routing assignment within a very short time. After this convergence time, there is no out-of-order delivery and no network contention. Indeed, there will be some performance degradation due to re-transmission of the first few packets of the message. However, even for 256KB messages, re-transmission would introduce very small bandwidth degradation for the entire flow.

To reach the above conclusion we have developed approximated Markov process models for Clos self-adaptive-routing system. The importance of these approximated models is the insight they provide about the speed of convergence. These models predict the extreme difference between convergence time of oblivious-self-routing Clos with flows of full link capacity, and flows of half the link capacity. These models are then compared to simulation. We define a set of features that are practical to implement and provide converging oblivious-adaptive-routing system. The proposed hardware is then evaluated by simulation. We claim the following contributions:

- We present an approximate Markov chain model for a three-level Clos network to evaluate the convergence rate of the adaptive-routing process.

- Based on this model, we provide a lower bound on the convergence time for the case where the bandwidth of each flow equals the link capacity. The convergence time under such conditions for rearrangeably-non-blocking Clos networks is more than exponential with the number of input switches, so it typically does not converge within any practical message size.

- Conversely, for the case where each flow bandwidth equals half the link capacity, the model shows fast convergence with weak dependency on the network size. Under these conditions, adaptive routing causes very little out-of-order packet delivery.

- We propose a set of system features that provide an oblivious-adaptive-routing. A detailed simulation model of InfiniBand hardware, enhanced with these mechanisms, confirms the above results.

The rest of the paper is organized as follows: Section 2 provides a description of an oblivious-self-routing system Section 3 analyzes that system using a Markov chain model for predicting the convergence time. Section 4 discusses implementation guidelines for adaptive routing system and Section 5 provides an evaluation of both the model and the proposed implementation. A discussion and conclusions are presented in Section 6.

2. A DISTRIBUTED ADAPTIVE-ROUTING SYSTEM MODEL

In this section, we introduce and define our architecture model and adaptive routing algorithm, before analyzing their performance in the next section.

As illustrated in Figure 2, consider a *1-rate* and uniform symmetrical Clos network. Assume that it has r input (and output) switches, of n x m ports each, and denote it as $CLOS(n, r, m)$. Further assume that all links have an equal capacity, and that all flows have an equal bandwidth demand, such that this bandwidth equals $1/p$ of the link capacity. For instance, $p = 2$ means that each flow bandwidth requires half of the link capacity.

Assume that the network carries a full-permutation traffic pattern, i.e. each source sends a continuous flow of data to a single destination, and each destination receives data from a single source. When more than p flows are routed through a link, we declare these flows as *bad flows*, and that link as a *bad link*. We declare the routing as a *good routing* if there are no bad links in the system.

We now want to define the adaptive-routing algorithm. There are many different adaptive-routing systems defined in studies and implemented by hardware, as described in the related work section of the introduction. Most of these systems are *hard to model* mathematically. Some use complex criteria for selecting output ports, some use state history, and some even rely on the distribution of the global network state. Since we seek to learn about the conditions under which convergence is fast enough to support big-data applications, we want to define an adaptive-routing system that is simple enough to be modeled.

Assume that the adaptive-routing system behaves as follows: At $t = 0$, a new full-permutation traffic pattern is applied at the input switches. Each input switch assigns an output port to each of its flows (on Clos and folded Clos topologies this output port defines the complete route to the

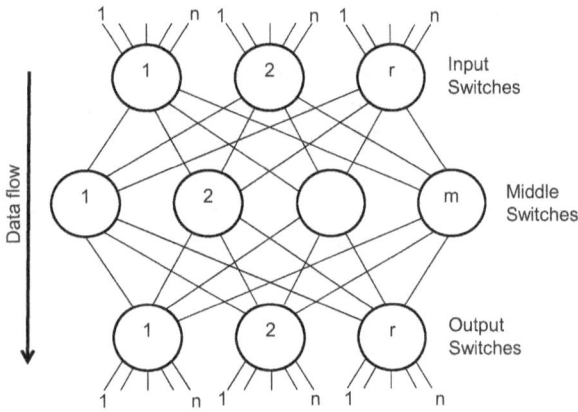

Figure 2: A Symmetrical Clos network CLOS(*n*, *r*, *m*): **The top row consists of *r* crossbar switches of *n* x *m* ports, denoted as input switches. The middle row includes *m* crossbar switches of *r* x *r* ports. The bottom row consists of *r* crossbar switches of *m* x *n* ports, named output switches.**

destination). The output port assignment performed by the input switches is semi-random as a reasonable approach for spreading their traffic with no global knowledge about the flows in other switches. The assignment is termed semi-random since, as input switches do know their own flows, they never assign more than p flows to any of their outputs. This means that bad links are only possible between the middle and output switches where flows from multiple input switches may congest.

Once the initial routing is defined, the system iterates synchronously through the following phases. Each iteration takes exactly one time unit. In the middle of the jth iteration at $t = j + 0.5$, each output switch selects a random bad flow that belongs to its input link with the largest number of flows. It then sends to the input switch at the origin of this bad flow a request to change its routing. The notification process and the change of routing happen before the end of the iteration period $t = j + 1$, when the operation repeats itself. The system keeps adapting routes until no more bad links exist.

When an input switch receives a bad-flow notification, it moves that flow to a new randomly-selected output port. If that new port is already full, the input switch swaps the moved flow with another flow on that output port to avoid congestion. As a result, the swapped flow may cause a new oversubscription on some middle-switch-to-output-switch link.

In the above model the middle switches do not perform any adaptive routing. All input switches are active at the beginning of each iteration period, and all output switches at the middle of each iteration period. Packets continuously flow through the network during the routing adaptation in order to provide the switches with the information about the flows routed through their links.

3. ANALYSIS

This section presents Markov chain models for the convergence time of the system presented in Section 2. Even for that simple system, an accurate model is hard to provide since the system state should represent all the flows on every link. Since the size of the Markov model grows exponentially with the number of flows and the number of links, we must provide an approximation instead.

The first model below takes the unrealistic but simplifying assumption that each output switch may be treated as an *independent system*. Due to the interdependency of the output-switch convergence times, as imposed by the input switches, this model is only useful to describe the convergence process of a single output switch.

Then, to better predict the convergence time, we present *two other models*, for full- and half-bandwidth flows. These models track the dependency between the output switches, and focus on the last stages of convergence when that inter-dependency has its greatest impact. Finally, in the evaluation section, we will use a simulation program that mimics the analyzed system behavior to evaluate these approximations.

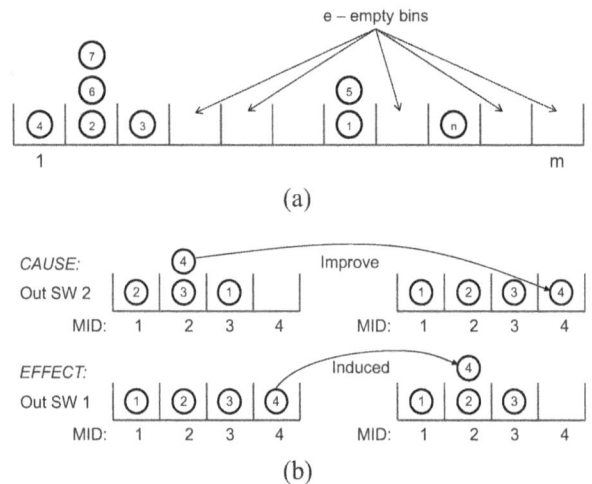

Figure 3: (a) A balls-and-bins representation of the *m* output-switch input links as bins and the *n* flows as balls. The state variable *e* represents the total number of empty bins. (b) When a contending ball is requested to be moved in order to *improve* one output-switch state it may contend with another flow on the new input-switch output. Resolution of that contention may cause some *induced* move on another output-switch. For example: Balls are numbered by their source input switch. Output switch 2 requests to move ball 4 since it is a bad ball. Input switch 4 moves that ball to middle switch 4. This move *improves* switch 2 situation. But since ball 4 of output switch 1 is previously routed from input switch 4 to middle switch 4, these two balls are contending and are swapped. This causes an *induced* move of ball 4 in output switch 1.

As shown in Figure 3(a), we propose a model to represent all the links that feed into the same output switch as a *balls-and-bins* problem: each input link is considered as a bin, and each flow as a ball. We start with a random spreading of the n balls into the m bins, and want to obtain the expected time t at which there are at most p balls in each bin.

Inspecting the changes to flows routed through the links feeding into a specific output switch, there are two processes that happen concurrently: an *improvement process* and an *induced-move process*. The improvement process results from the request of that output switch to move one of its worst-link bad flows, when that bad flow appears on a new link. In the balls-and-bins representation this would cause the movement of one of the worst bad-bin balls into some new bin (not necessarily an empty one). The induced-move process results from one of that output-switch flows being involved in a flow-swap on some input-switch. This change is denoted *induced* since the originator for the swap may be some other output switch. Figure 3(b) provides an example for how improvement in one output switch causes an induced move on the other. The distribution of induced moves on the different output switches resembles the random throw of k balls into r bins, where k is the number of output switches that have not reached their steady state.

3.1 A Single Output-Switch Markov Chain Model for Full-Link Bandwidth per Flow

This sub-section presents an approximate Markov chain model for a single output switch for the case of full-link bandwidth per flow (i.e., $p = 1$). Based on the system symmetry, this model considers each output switch independently. To model the interactions between output switches, the model assumes that each time an output switch kicks some bad ball, an induced move will happen with probability of n/m.

As depicted in Figure 4 we define a state variable e that represents the number of empty bins, and another state variable g that counts the number of good bins. So the Markov state for the single output switch can be represented using the pair (e, g). To simplify the analysis, this model makes the following approximations, as further explained below: 1) Induced moves are evenly distributed over all output switches, such that all the r output-switch systems are identical and can be treated as uncorrelated. 2) When the number of bad bins is small compared to the number of bad balls, such that there are in average over 2.5 bad balls per bad bin, the induced-move impact is modeled as if all bad bins have at least 3 balls. The concurrent processes that affect the state of the balls are:

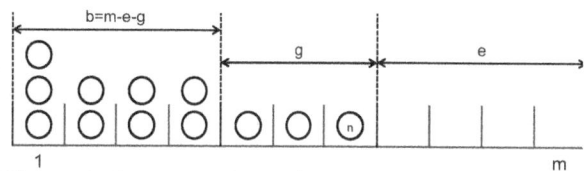

Figure 4: An approximated model assuming an even distribution of bad balls in bad bins: In the shown case, the *Improvement* process must take one of the balls in the left-most bin, as it is the worst bin. The *Induced-Move* process may move any ball. In the shown case, if a bad ball is selected by the *Induced-Move* process from a bad bin with 2 balls (i.e. bin 2, 3 or 4) and then falls on an empty bin, 2 good bins are created and g increases by 2. Else if it is selected from the first bin and then falls on an empty bin, the first bin does not become good and g only increases by 1.

The *improvement* process: Each output switch selects one of the balls in one of the bins with the highest number of balls, and randomly places it into a bin. As long as there are any bins with at least 3 balls, it is guaranteed that the number of bad bins cannot decrease by this process (since the selected bin will only lose one ball). The probability for the selected ball to fall on an empty, good or bad bin depends on the number of such bins.

The *induced-move* process: This process takes a random ball and randomly places it in some bin. The probability for such moves to occur depends on the number of empty links on the input switches. Since there are only n flows spread on m links, and the selection of the flow to be moved is random, the distribution of the moved flows to input switches should follow the distribution of throwing n balls into m bins. In order to avoid the complexity of using this distribution, the model takes the approximation of the average probability of an induced move, which is n/m.

To predict if the induced move changes the number of good bins by +1, -1 or -2, we should have tracked the exact number of bins with 0, 1 and 2 balls. As shown in the example of Figure 4, if the moved ball is from bins with just 2 balls and it falls in an empty bin, g is increased by 2. But if the bad ball is from a bin with more than 2 balls, the number of good bins is only increased by 1. However, to compute using a Markov chain the number of bins with 2 balls means we need to track the number of bins with 3 balls, and so forth. If we define these n state variables, *the state space explodes*. As the distribution of the number of balls per bin is a sharp function, such that the probability drops significantly with the number of balls, we claim that a reasonable approximation is to assume that all the bad bins have the same number of balls. The output switch policy of re-routing a ball from the bin with the highest number of balls strengthens this assumption.

Table 1 Possible state changes with $p=1$ and their probability

Process	b-balls/ b-bins	Who moves	Move Where	New State	Probability
Induced Move	Any case	good ball	empty bin	e, g	$g/n*e/m*n/m=ge/m^2$
		good ball	other good bin	e+1, g-2	$g/n*(g-1)/m*n/m = g(g-1)/m^2$
		good ball	same good bin	e, g	$g/n*1/m*n/m=g/m^2$
		good ball	bad bin	e+1, g-1	$g/n*(m-g-e)/m*n/m = g(m-g-e)/m^2$
	> 2.5	bad ball	empty bin	e-1, g+1	$(n-g)/n*e/m*n/m=(n-g)e/m^2$
		bad ball	good bin	e, g-1	$(n-g)/n*g/m*n/m=(n-g)g/m^2$
		bad ball	bad bin	e, g	$(n-g)/n*(m-g-e)/m*n/m=(n-g)(m-g-e)/m^2$
	≤ 2.5	bad ball	empty bin	e-1, g+2	$(n-g)/n*e/m*n/m = (n-g)e/m^2$
		bad ball	good bin	e, g	$(n-g)/n*g/m*n/m= (n-g)g/m^2$
		bad ball	other bad bin	e, g+1	$(n-g)/n*(m-e-g-1)/m*n/m=(n-g)(m-e-g-1)/m^2$
		bad ball	same bad bin	e, g	$(n-g)/n*1/m*n/m= (n-g)/m^2$
Improvement	> 2	bad ball	empty bin	e-1, g+1	e/m
		bad ball	good bin	e, g-1	g/m
		bad ball	bad bin	e, g	$(m-e-g)/m$
	≤ 2	bad ball	empty bin	e-1, g+2	e/m
		bad ball	good bin	e, g	g/m
		bad ball	same bad bin	e, g	$1/m$
		bad ball	other bad bin	e, g+1	$(m-e-g-1)/m$

Let us consider each state transition induced by the request of some other output switch to improve its state. The ball selected may be a good or bad ball, and we assume by symmetry that it may be moved into any bin with the same probability. The probability for a good ball to be selected is g/n. The probability for a bad ball is the complementary $(n - g)/n$. The possible moves and their respective probabilities are described in Table 1.

The combined impact of the two processes is obtained by considering each possible pair of the *Improve* and *Induced-move* state transitions, and adding their probability product to the Markov state transition matrix.

3.2 Last-Step Model for Flows of Full Link Bandwidth

A full model of the entire Markov matrix of all states of all output switches is infeasible due to its size. In order to obtain an upper bound approximation on the convergence time, we suggest inspecting the r output switches *just before they reach convergence*. We call this the *Last-Step* model.

As we suspect that the long convergence times are a result of the induced moves forced by one output switch on another, we focus the model on the last steps of convergence. Only when all the output switches reach together their good state, they stop forcing each other back into bad states. The model only uses a single bad state that is closest to the good state. In that sense, it is an optimistic model, as a sequence of bad induced moves is not modeled and the output switch stays close to its good state. Yet, we will later show that it correctly models the *exponential* convergence time of our system.

Figure 5(a) shows the Markov states of a single output switch. There are only two states for an output switch: 0 (good) and 1 (bad). Only a single bad bin is possible one step away from the good state as shown on the balls-and-bins systems drawn below the state graph. The probabilities for transition represent the improvement and induced move processes, but their values depend on the other output switches states. The Markov system contains r approximated output switch sub-systems each with a single state variable which is either 1 (bad) or 0 (good) as shown in Figure 5(b). The system state is coded as a binary variable of r bits. Bit S_i represents the state of the i output switch. The resulting state space has 2^r system states. The entire system has a single absorbing state which is when all the bits of the binary representation are 0. We can also assume all output switches start with some bad bins so the initial state value is $2^r - 1$. The total number of induced moves denoted by U is the number of '1's in the state binary value. Before the observing state is reached, the probability for each output switch to leave a good state B equals the probability for one of the induced moves to throw a ball in that output switch. We assume the induced moves are equally spread, and thus

$$B = U/r$$
$$A = 1 - B = 1 - U/r$$

The probability to improve a bad state is denoted C and is built from the impact of the two processes C_{imp} and C_{ind}. The improvement process always selects a bad ball and thus improvement depends on the number of empty bins.

$$C_{imp} = (m - n + 1)/m$$

(a)

Output switch 1

Output switch 2

Output switch r

(b)

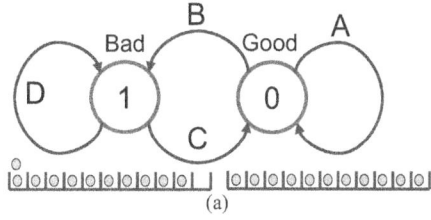

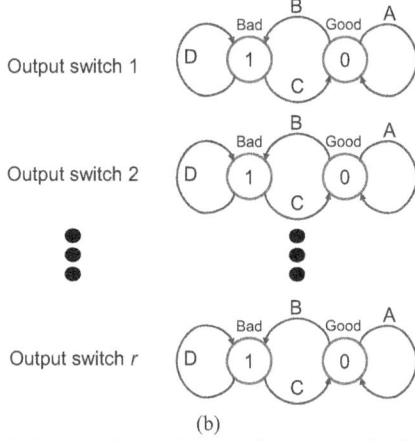

Figure 5: Approximated Last Step model of a single output-switch: (a) A single bad state implies a lower bound on convergence time as it does not let the bad state degrade further than the last step. (b) The set of r state pairs describes the entire system.

For the induced move to improve we need to multiply the chance for induced move by the probability a bad ball will be selected and the probability the move will be into an empty bin:

$$C_{ind} = \frac{u}{r} \frac{2}{n} \frac{m-n+1}{m}$$

$$C = C_{imp} + C_{ind} - C_{imp} \cdot C_{ind}$$

$$D = 1 - C$$

To build the Markov state transitions matrix the present state is represented as a binary variable: $S = S_{r-1} S_{r-2} \dots S_1 S_0$ and the next state is represented as $Q = Q_{r-1} Q_{r-2} \dots Q_1 Q_0$

Define E as the number of digits j where $0 = S_j = Q_j$, F the number of digits j where $0 = S_j \neq Q_j$, G the number of digits j where $1 = S_j = Q_j$ and H the number of digits j where $1 = S_j \neq Q_j$. The probability to move from S to Q is given by:

$$P_{ij} = A^E B^F C^H D^G$$

3.3 Last-Step Model for Flows of Half Link Bandwidth

Unlike the Last-Step model for $p = 1$ that has just one absorbing state for the output switches, the case of two flows per link has several good states. As illustrated in Figure 6, to distinguish these states we introduce the following variables describing the ball distributions in each

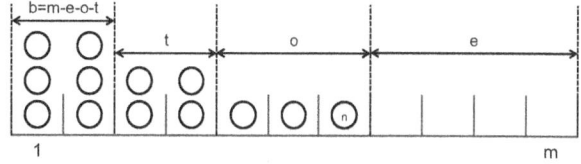

Figure 6: For $p=2$ we introduce new state variables: e=empty, o=one, t=two.

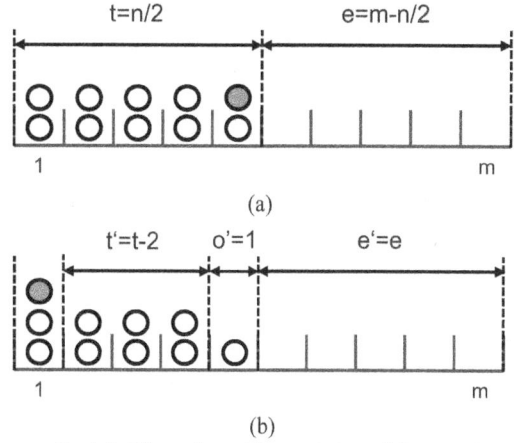

(a)

(b)

Figure 7: (a) The absorbing state with maximal B and (b) its neighbor "bad" state. The moved ball is shadowed.

output switch: e, the number of empty bins; o, the number of bins with one ball; and t, the number of bins with two balls.

It can be shown that for a state to be good the following should be met:

$$n = 2t + o$$

$$m = e + t + o$$

Or:

$$t = e - (m - n)$$
$$o = m - 2e + (m - n)$$

As we are interested in an upper bound for the number of iterations to convergence (for the $p = 2$ case only) we pick the worst absorbing state that is the state with highest B probability shown in Figure 7(a). For that state:

$$B = \frac{o}{n} \frac{t}{m} + \frac{2t}{n} \frac{(t-1)}{m} = \frac{t}{nm} (o + 2t - 2)$$
$$= \frac{t}{nm} (n - 2)$$

So the maximal B is obtained for the maximal $t' = n/2$:

$$B_{max} = \frac{(n/2 - 1)}{m} = \frac{n}{m} \left(\frac{1}{2} - \frac{1}{m} \right)$$

Consider the improvement probability for the bad state closest to the worst B good state as shown in Figure 7(b). The probability C is then a combination of the probability

On Flits Queue or De-Queue (numFlits, TP, RP, to DEST)
  Update the number of flits queued for the Transmit Port (TP)
  If enough time from port change for this flow and TP is congested
    If switch is a middle switch
      Send EAR though the Receive Port (RP) flits were received on
    Else
      If TP is an up-going-port adapt the output port for DEST
        Possibly swap another DEST if the new output-port is busy

On Receiving an EAR (RP, DEST)
  If RP is an up-going-port adapt the output port for DEST
    Possibly swap another DEST if the new output-port is busy

Figure 8: Packet Forwarding Module Algorithms on the queueing of new flits on a Transmit Queue and on receiving an Explicit Adaptive Routing Request

of the improvement process and for the induced move to actually improve the state. We can see that:

$$C_{imp} = \frac{m - t'}{m} = 1 - \frac{n/2 - 2}{m}$$

$$C_{ind} = U \frac{3}{n} \frac{m - t'}{m}$$

The probability for induced moves with $p = 2$ has to take into account the probability for a ball to move without requiring a swap. To that end we could use the balls-and-bins distribution to predict the probability for a bin to have more than two balls.

The construction of the Markov state transition matrix follows the same procedure as in Section 3.2.

4. IMPLEMENTATION GUIDELINES
This section discusses the feature set required for the implementation of an oblivious-adaptive-routing system.

To meet the required behavior described in Section 2, the switches need to extend the deterministic routing and provide random assignment of output ports for flows. Reassignment of a flow output port is only allowed on input switches and is triggered either when receiving an Explicit Adaptation Request (EAR) from an output switch, or when congestion is observed on the previously assigned output port. A timer is used to throttle the number of reassignments in the latter case to avoid multiple reassignments before the congestion is relieved.

To enable an efficient hardware implementation, the proposed mechanisms differs from the model described in Section 2 in several aspects. The first difference is that a real system does not likely count "flows" assuming they are all of the same bandwidth. Instead, it makes more sense to evaluate a transmit-port congestion, in mechanisms similar to those proposed by the IEEE 802.1Qau known as *QCN* [36].

The second aspect is about *concurrency* of bad flow re-routes: congestion-based bad link detection means that the knowledge about bad links is not available at the receiving

Randomize a random or worst permutation as dst[src]
MiddleSwitch[src, dst] = mod(src,n) for each (src, dst) pair
Iterations = 0
While any bad link (depends on P)
  Iterations++
  Randomly select one (src, dst) from the worst link for each out-switch
    For each bad (src, dst) selected, in random order
      Randomly select the new middle switch
      Move the (src, dst) to the new middle switch
      Optionally swap with other (src, dst) going through the same link
Report Iterations

Figure 9: Distributed Adaptive Routing Simulation Main Loop

switch (output switch in our model) but in the middle switch. In the model used in previous sections the output switch requires this information in order to choose a single worst bad flow to be re-routed. Selecting the worst bad link implies the existence of a protocol for each middle switch to notify the output switches to which it connects about bad links and their severity. To avoid the latency and complexity of such a protocol, the proposed implementation does not enforce a single bad flow transition per output switch per iteration. Instead, the responsibility for requesting re-routing of bad flows is given to the middle switches that use the same QCN-like monitoring to detect congestion. When congestion is detected, the middle switches send EAR requests to the relevant input switches. These notifications do require a special signaling protocol to be delivered.

The algorithms for EAR generation and forwarding, as well as for determining when to adapt to output congestion, are depicted in Figure 8.

5. EVALUATION
Evaluation of the worst convergence time using simulations relies on the ability to check many permutations. As the number of possible permutations is extremely large, it is important to focus on the permutations that are presumed to have the worst convergence time, or at least a large one. These so-called *worst permutations* are derived by contrasting them to the set of permutations that are fastest to converge. The fastest converging permutations have all flows originating on the same input switch destined to the same output switch. For such permutations it is enough that each input switch spreads its own outputs to avoid bad links and provide good routing. Such assignment is possible if $mp \geq n$ (this is also the rearrangeable non-blocking condition for *1-rate* Clos with p flows per link). Intuitively, a permutation where each output switch is fed by flows from different input switches will be the hardest to converge as it will require the most synchronization between the input switches that actually do not talk to each other.

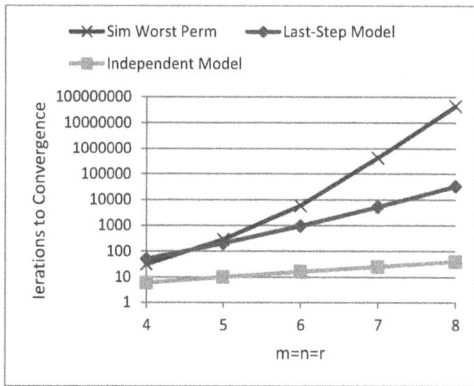

(a)

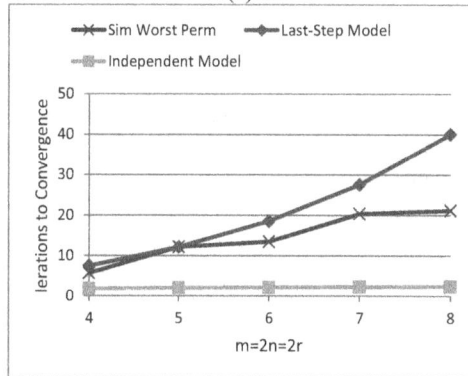

(b)

Figure 10: Expected Convergence Time predicted by the single-output-switch, the p=1 last-step model compared to simulated results of worst permutations as a function of (a) the m=n=r of RNB and (b) m=2n=2r SNB topologies.

5.1 Analyzed System Simulation Model

A dedicated simulation program was written to model the system described in Section 2. The data structure used is a simple matrix $M[s,d]$, where $s, d \in \{0 \cdots nr\}$ represents the source and destination for each flow and $M[s,d]$ denotes the middle switch assigned to that flow. The simulation algorithm is depicted in Figure 9. The program optionally starts from a fully randomized permutation or one that meets the condition of a worst permutation. Each point of simulation result is obtained by simulating a batch of 1000 permutations and then continues to simulate new batches until the average number of iterations required to converge changes by less than 1%.

The models of Sections 3.1, 3.2 and 3.3 were coded in matlab[2], to form an observable Markov chain matrix following the theory presented in [21].

---

[2] The presented Matlab-based evaluation is limited by the exponential state space nature of Markov representations and the capacity of our version of Matlab

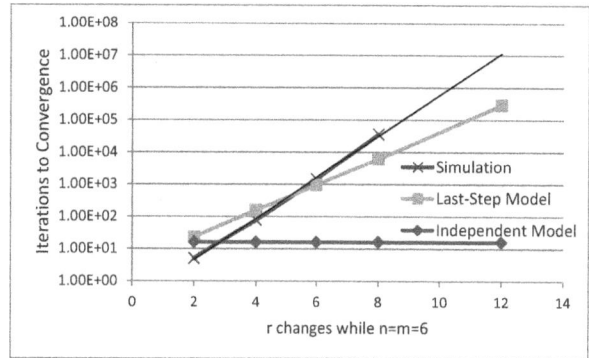

Figure 11: The dependency of the number of iterations to convergence on r for a constant m=n=6.

The first comparison made is for the RNB topologies of the p=1 case. The number of iterations to reach convergence is provided in Figure 10(a). As expected, the Independent output switch model of Section 3.1 is greatly optimistic, while the Last-Step model of Section 3.2 is closer to the simulated results. The Last-Step model is pessimistic for very small n as it assumes all output switches start in a bad state, which is not the case for very small n.

For the SNB case in Figure 10(b), it can be seen that the simulation predicts convergence times many orders of magnitude smaller than the RNB case but still mostly above 10. The Independent output-switch model is very optimistic, and the Last-Step model is pessimistic, probably due to its assumption that all output switches start at a bad state. For SNB, due to the over provisioning of the network, there are many chances that some output switches will start in a good state. Also note that the change of slopes on the simulation curve may be attributed to the change between even and odd port numbers, and how routing in Clos best fits even number of ports.

To strengthen the point that the systems are dependent we show the dependency of the convergence time in the number of parallel output switches in Figure 11. The convergence is plotted for different values of r and a fixed $n = m = 6$. It can be observed that *the dependency on r is exponential*. The Last-Step model shows a lower slope, which we suspect is a result of the approximation of using a single bad state. Note that the Independent model does not depend on r at all, which exhibits yet another limitation of this model.

The Last-Step Markov model of Section 3.3 for the case of half-link capacity flows was simulated on RNB topology and the number of iterations to convergence is provided in Figure 12. The number of iterations required to reach convergence *is shown to be very small even for large values of n*. As can also be seen the Markov model is optimistic for larger networks. We attribute this behavior to the approximation used by this model which defines a single bad state for each output-switch sub-system.

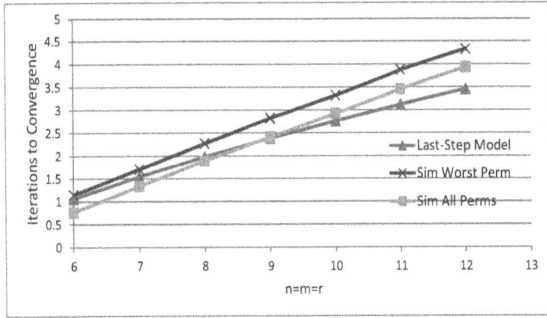

Figure 12: Comparison of time to convergence for Clos(n=m=r) with half rate flows (p=2) cases using Last Step Model and a dedicated simulation model.

5.2 Implementable System Simulation Model

A large compute cluster would have been the obvious choice for the evaluation of an implemented adaptive routing system. However, hardware that implements our proposed Explicit Adaptation Request messaging described above was not available to us.

Instead, we have used a well-known flit level simulator for InfiniBand that accurately models flow dynamics, network queuing and arbitration. These simulators are described in the following sections.

The OMNet++ [29] based InfiniBand flit-level simulation model [35] is commonly used for predicting bandwidth and latency for InfiniBand networks [34][13][14]. In order to evaluate the proposed implementation guidelines presented in Section 4, a new packet-forwarding module was added to the switches. This module implements all the algorithms to detect link capacity overflow, provide flow re-route, swap output ports and introduce a signaling protocol to carry the Explicit Adaptation Requests (EAR) between switches.

The overhead and timing of the EAR protocol is accurately modeled by encapsulating EARs as 8-byte messages similarly to the flow-control packets of the InfiniBand, and sending them through the regular packet send queues.

The simulations performed are of two topologies containing 1152 hosts: The rearrangeably non-blocking (RNB) topology is a folded CLOS(24, 48, 24) equivalent to XGFT(2; 24,48; 1,24) fat-tree topology. The strictly non-blocking topology (SNB) has double the number of middle switches, i.e. is a folded CLOS(24, 48, 48) equivalent to XGFT(2; 24,48; 1,48) fat-tree. The SNB topology is simulated to provide a fair comparison to the half-flows case as it provides double the links. The model assumes a link capacity of 40Gbps and hosts may send data at that speed or be throttled to 20Gbps +/- 0.8Gbps.

The traffic pattern applied to the system is a sequence of random permutations. In each permutation each host sends data to a single random destination and receives data from a single random source. The hosts progress through their sequence of destinations in an asynchronous fashion sending 256KB to each destination.

The simulation tracks the number of routing changes performed by each switch in periods of 10μsec as well as the final throughput at each of the network egress ports. The ratio of packets delivered out-of-order to those provided in-order is also measured. Another measured variable is the out-of-order packet window size, defined as the gap in the number of packets, as observed by the receiving host. This variable is a clear indication for the feasibility of implementing a re-order buffer. The bandwidth, latency and out-of-order percent and window size results are represented in Table 2. The values are taken as average, max or min value over all receivers of the average or max value measured on each receiving host. For example, the *min of average throughput* means that each egress port throughput is averaged over time, and the reported number is the minimal values over all the egress ports. To establish a fair comparison we focus on the results of the two cases when only half of the network resources are used: SNB 40Gbps (first data column) and RNB 19.2Gbps (third data column).

It is shown that although the bandwidth provided by the SNB $p = 1$ case is higher, any transport that would require retransmission due to out-of-order delivery would actually fail to work on the SNB $p = 1$ case since *only one of three packets is provided in order* (ratio of ~2). The latency of the network is also impacted by not reaching a steady state, thus showing a much longer latency.

The routing convergence provided by the $p = 2$ case is most visible when inspecting the number of routing changes per 10μsec. Figure 13 shows on each line the number of adaptations conducted by a specific switch in each 10μsec period for (a) SNB 40Gbps, (b) RNB 20.8Gbps during a single permutation. It can be observed that *routing is constantly changing for the SNB 40Gbps and the RNB 20.8Gbps cases.* A long sequence of 256KB message permutations on RNB 19.2Gbps is shown in Figure 13(c). It can be observed that *adaptive-routing reaches a non-blocking assignment for all permutations in less than 80μsec.*

Table 2 Simulation results for a 1152-hosts cluster

Parameter	SNB Full BW: 40Gbps	RNB Full BW: 40Gbps	RNB Half BW: 19.2Gbps
Avg-of-Avg Throughput	37.7Gbps	23.0Gbps	18.9Gbps
Min-of-Avg Throughput	36.5Gbps	20.7Gbps	18.9Gbps
Avg-of-Avg Network Latency	17.0μsec	32μsec	5.8μsec
Max-of-Max Network Latency	877μsec	656μsec	11.7μsec
Avg-of-Avg Out-of-order / In-order Ratio	**1.87**	**5.2**	**0.0023**
Max-of-Avg Out-of-order / In-order Ratio	3.25	7.4	0.0072
Avg-of-Avg Out-of-order Window	6.9pkts	5.1pkts	2.3pkts
Max-of-Avg Out-of-order Window	8.9pkts	5.7pkts	5.0pkts

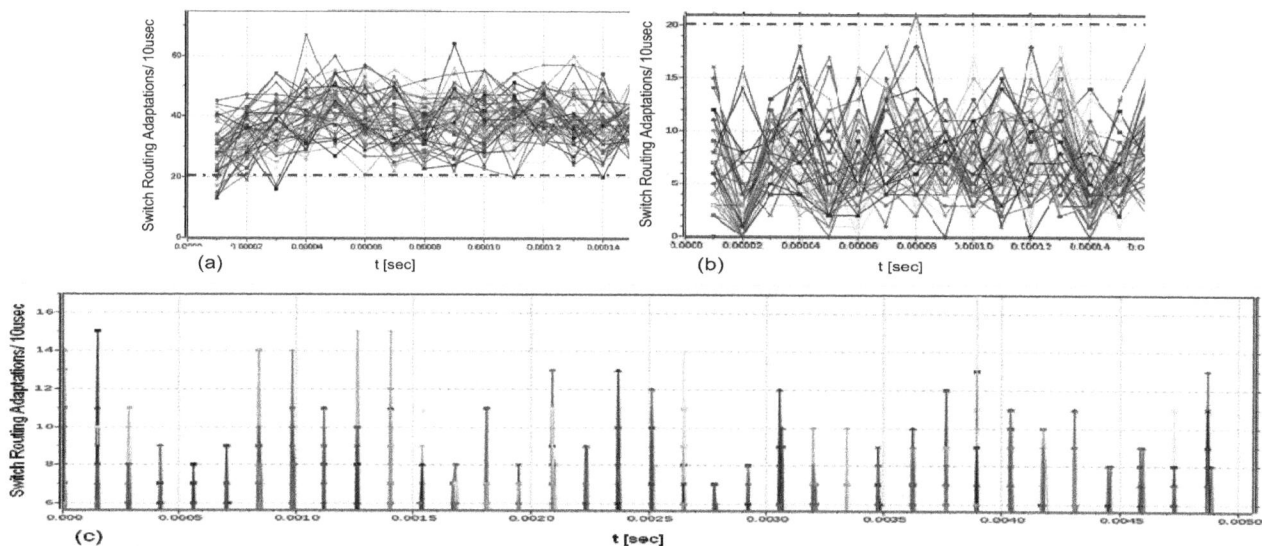

Figure 13: Simulated 1152 nodes cluster, number of re-routing events on each input-switch averaged over 10μsec time periods for 256KB messages for (a) SNB 40Gbps (b) RNB 20.8Gbps and (c) RNB 19.2Gbps. The plots in (a) and (b) focus on a single 150μsec permutation period and show that convergence is not met. For the RNB 19.2Gbps case (c), convergence is reached within a few tens of microseconds from the start of each of the applied random permutations.

6. DISCUSSION AND CONCLUSIONS

In this paper we find sufficient conditions allowing distributed oblivious-adaptive-routing to converge to a non-blocking routing assignment within a very short time, thus making it a viable solution for adaptive routing for medium-to-long messages on fat-trees.

Convergence is shown to require flows that do not exceed half of the link capacity, which raises the question of whether it is worth to pay that high price. Note that actually in our proposed Adaptive Routing only the edge links of the network are operated at half the core network link bandwidth. Many of the network links do route more than one flow and thus utilize the full link capacity. Alternative approaches to deal with the contention caused by high-volume correlated flows may seem cheaper but they are not scalable. They propose the introduction of a centralized traffic engineering engine that can throttle traffic as necessary or perform re-routes. However, as the number of flows correlates to the number of cluster nodes, a central unit is likely to become a bottleneck. Other attempts [26] to provide adaptive routing based on protocols that convey the system state to each switch are also not scalable due to the state size on every switch, the number of messages to provide state updates and the synchronous change of traffic which makes previous states irrelevant.

The developed approximate model provides the insight that the origin of the long convergence time is the *interdependency of re-route events on the different output switches*, as imposed by the topology. It was shown that the time it takes to converge to a non-blocking routing is *exponential with the number of input or output switches*. For that reason, the probability of creating bad links on a single output switch has a major impact on the convergence time. For rearangeable-non-blocking CLOS, limiting the

traffic flows to half or less of the link bandwidth reduces this probability for creating bad links to less than 0.5, and therefore provides fast convergence.

Finally, we propose a simple system architecture for the signaling needed for adaptation, and simulate it to show how it converges within 20-80μsec on a 1152-host network. The insights provided by this research should help in providing self-routing solution to long messages in data-center applications of various fields.

7. ACKNOWLEDGMENTS

We would like to thank Marina Lipshteyn of Mellanox and Israel Cidon, Yossi Kanizo and Erez Kantor from the Technion for their support and insight. This work was partly supported by the Intel ICRI-CI Center and by European Research Council Starting Grant No. 210389.

8. REFERENCES

[1] Adiga, N.R. et al. 2005. Blue Gene/L torus interconnection network. *IBM Journal of Research and Development*. 49, 2.3 (Mar. 2005), 265–276.

[2] Al-Fares, M. et al. 2010. Hedera: dynamic flow scheduling for data center networks. *Proceedings of the 7th USENIX conference on Networked systems design and implementation* (Berkeley, CA, USA, 2010), 19–19.

[3] Anderson, E.J. and Anderson, T.E. 2003. On the stability of adaptive routing in the presence of congestion control. *INFOCOM 2003. Twenty-Second Annual Joint Conference of the IEEE Computer and Communications. IEEE Societies* (Apr. 2003), 948–958 vol.2.

[4] Benes, V.E. 1965. *Mathematical theory of connecting networks and telephone traffic.* Academic press New York.

[5] BLOCH, G. et al. High-Performance Adaptive Routing. Publication number: US 2011/0096668 A1 U.S. Classification: 370/237.

[6] Chen, Y. et al. 2009. Understanding TCP incast throughput collapse in datacenter networks. *Proceedings of the 1st ACM workshop on Research on enterprise networking* (New York, NY, USA, 2009), 73–82.

[7] Chowdhury, M. et al. 2011. Managing data transfers in computer clusters with orchestra. *Proceedings of the ACM SIGCOMM 2011 conference* (New York, NY, USA, 2011), 98–109.

[8] Douglass, B.G. and Oruc, A.Y. 1993. On self-routing in Clos connection networks. *Communications, IEEE Transactions on*. 41, 1 (1993), 121–124.

[9] Du, D.Z. et al. 1998. On multirate rearrangeable Clos networks. *SIAM J. Comput*. 28, 2 (1998), 463–470.

[10] Gamarnik, D. 1999. Stability of adaptive and non-adaptive packet routing policies in adversarial queueing networks. *Proceedings of the thirty-first annual ACM symposium on Theory of computing* (New York, NY, USA, 1999), 206–214.

[11] Gerbessiotis, A.V. and Valiant, L.G. 1994. Direct bulk-synchronous parallel algorithms. *J. Parallel Distrib. Comput*. 22, 2 (Aug. 1994), 251–267.

[12] Gomez, C. et al. 2007. Deterministic versus Adaptive Routing in Fat-Trees. *2007 IEEE International Parallel and Distributed Processing Symposium* (Long Beach, CA, USA, Mar. 2007), 1–8.

[13] Gran, E.G. et al. 2010. First experiences with congestion control in InfiniBand hardware. *2010 IEEE International Symposium on Parallel & Distributed Processing (IPDPS)* (Apr. 2010), 1–12.

[14] Gran, E.G. et al. 2011. On the Relation between Congestion Control, Switch Arbitration and Fairness. (May. 2011), 342–351.

[15] Gusat, M. et al. 2010. R3C2: Reactive Route and Rate Control for CEE. *High-Performance Interconnects, Symposium on* (Los Alamitos, CA, USA, 2010), 50–57.

[16] Hoefler, T. et al. 2007. A Case for Standard Non-blocking Collective Operations. *Recent Advances in Parallel Virtual Machine and Message Passing Interface*. F. Cappello et al., eds. Springer Berlin / Heidelberg. 125–134.

[17] Hoefler, T. et al. 2008. Multistage switches are not crossbars: Effects of static routing in high-performance networks. *2008 IEEE International Conference on Cluster Computing* (Oct. 2008), 116–125.

[18] Isard, M. et al. 2007. Dryad: distributed data-parallel programs from sequential building blocks. *Proceedings of the 2nd ACM SIGOPS/EuroSys European Conference on Computer Systems 2007* (New York, NY, USA, 2007), 59–72.

[19] Jajszczyk, A. 2003. Nonblocking, repackable, and rearrangeable Clos networks: fifty years of the theory evolution. *Communications Magazine, IEEE*. 41, 10 (2003), 28–33.

[20] Jiang, N. et al. 2009. Indirect adaptive routing on large scale interconnection networks. *ACM SIGARCH Computer Architecture News*. 37, (Jun. 2009), 220–231.

[21] Karlin, S. and Taylor, H.M. 1998. *An Introduction to Stochastic Modeling, Third Edition*. Academic Press.

[22] Kim, J. et al. 2006. Adaptive routing in high-radix clos network. (2006), 92.

[23] Koibuchi, M. et al. 2005. Enforcing in-order packet delivery in system area networks with adaptive routing. *Journal of Parallel and Distributed Computing*. 65, 10 (Oct. 2005), 1223–1236.

[24] Liew, S.C. et al. 1998. Blocking and nonblocking multirate Clos switching networks. *Networking, IEEE/ACM Transactions on*. 6, 3 (1998), 307–318.

[25] Martínez, J.C. et al. 2003. Supporting Fully Adaptive Routing in InfiniBand Networks. *Proceedings of the 17th International Symposium on Parallel and Distributed Processing* (Washington, DC, USA, 2003), 44.1–.

[26] Minkenberg, C. et al. 2009. Adaptive Routing in Data Center Bridges. *Proceedings of the 2009 17th IEEE Symposium on High Performance Interconnects* (Washington, DC, USA, 2009), 33–41.

[27] Scott, S. et al. 2006. The BlackWidow High-Radix Clos Network. *Proceedings of the 33rd annual international symposium on Computer Architecture* (Washington, DC, USA, 2006), 16–28.

[28] Towles, B. 2001. *Finding Worst-case Permutations for Oblivious Routing Algorithms*.

[29] Varga, A. *OMNET++. http://www.omnetpp.org*.

[30] White, T. 2010. *Hadoop: The Definitive Guide*. O'Reilly Media, Inc.

[31] Wu, W. et al. 2009. Sorting Reordered Packets with Interrupt Coalescing. *Computer Networks*. 53, 15 (Oct. 2009), 2646–2662.

[32] Xin Yuan 2011. On Nonblocking Folded-Clos Networks in Computer Communication Environments. *Parallel & Distributed Processing Symposium (IPDPS), 2011 IEEE International* (May. 2011), 188–196.

[33] Youssef, A. 1993. Randomized self-routing algorithms for Clos networks. *Computers & Electrical Engineering*. 19, 6 (Nov. 1993), 419–429.

[34] Zahavi, E. Fat-Trees Routing and Node Ordering Providing Contention Free Traffic for MPI Global Collectives. *Journal of Parallel and Distributed Computing*. Communication Arch for Scalable Systems.

[35] Zahavi, E. *InfiniBand(TM) Macro Simulation Model*. http://www.omnetpp.org/omnetpp/doc_details/2070-infiniband.

[36] IEEE Std 802.1Qau-2010 (Amendment to IEEE Std 802.1Q-2005) DOI: 10.1109/IEEESTD.2010.5454063, 2010 , Page(s): c1- 119.

Efficient Buffering and Scheduling for a Single-Chip Crosspoint-Queued Switch

Zizhong Cao
Polytechnic Institute of NYU
5 MetroTech Center, Brooklyn, NY 11201
zcao02@students.poly.edu

Shivendra S. Panwar
Polytechnic Institute of NYU
5 MetroTech Center, Brooklyn, NY 11201
panwar@catt.poly.edu

ABSTRACT

The single-chip crosspoint-queued (CQ) switch is a self-sufficient switching architecture enabled by state-of-art ASIC technology. Unlike the legacy input-queued or output-queued switches, this kind of switch has all its buffers placed at the crosspoints of input and output lines. Scheduling is also performed inside the switching core, and does not rely on instantaneous communications with input or output line-cards. Compared with other legacy switching architectures, the CQ switch has the advantages of high throughput, minimal delay, low scheduling complexity, and no speedup requirement. However, since the crosspoint buffers are small and segregated, packets may be dropped as soon as one of them becomes full. Thus how to efficiently use the crosspoint buffers and decrease the packet drop rate remains a major problem that needs to be addressed. In this paper, we propose a novel chained structure for the CQ switch, which supports load balancing and deflection routing. We also design scheduling algorithms to maintain the correct packet order caused by multi-path switching. All these techniques require modest hardware modifications and memory speedup in the switching core, but can greatly boost the overall buffer utilization and reduce the packet drop rate, especially for large switches with small crosspoint buffers under bursty and non-uniform traffic.

Categories and Subject Descriptors

C.2.1 [**Computer Communication Networks**]: Network Architecture and Design—*Packet-switching networks*; C.2.6 [**Computer Communication Networks**]: Internetworking—*Routers*

General Terms

Algorithms, Design

Keywords

Single-Chip, Crossbar, Load Balancing, Deflection Routing

1. INTRODUCTION

In the past decade, modern Internet-based services such as social networking, video streaming and cloud computing have brought about a continuous, exponential growth in Internet traffic. The recent boom in smartphones, tablets and other portable electronic devices has made all these remote services more accessible to people, while imposing even larger traffic burdens on the backbone networks. To accomodate the increasing demands, the capability of Internet core switches must grow commensurately. Consequently, there has been interest in designing high-performance switching architectures and scheduling algorithms.

Many types of switching architectures have been proposed. The first kind is the output-queued (OQ) switch [24], in which an arriving packet is always directly sent to its destination output, and then buffered there if necessary. The OQ switch may achieve 100% throughput with infinite buffers, but requires an impractically high speedup. Specifically, the switching fabric of an $N \times N$ OQ switch may need to run N times as fast as the single line rate in the worst case, when all inputs target the same output.

Another popular kind of architecture is the input-queued (IQ) switch. In an IQ switch, packets are buffered at the input and served in a first-in-first-out (FIFO) manner if the target output is idle. IQ switches require no speedup, but suffer from the head-of-line (HOL) blocking problem, which limits the throughput to 58.6% [24]. This problem was later solved by implementing virtual output queues (VOQ) at each input. Various scheduling algorithms such as iSLIP [30], DRRM [28], and maximum weight matching (MWM) [36] have been proposed to achieve high throughput. However, many of these algorithms are complex, or require nearly instantaneous communications among input and output schedulers that are usually placed far apart on different line-cards. This might become a bottleneck for high-speed switches, in which the round-trip latency between different line-cards may span several time slots and thus is no longer negligible. For instance, the round-trip latency can be as high as about $100ns$ assuming $10m$ inter-rack cables, while each time slot lasts at most about $50ns$, assuming OC-192 or higher line speeds and $64byte$ fragmentation. A combination of IQ and OQ switches, the combined-input-and-output-queued (CIOQ) switch, has been proposed to achieve high throughput with minimal delay [13], but suffers from similar problems as an IQ switch.

In recent years, a new kind of structure called the buffered crossbar has been widely studied. Typically, one or a few buffers are placed at each crosspoint, while others are still placed at the inputs of a switch, which effectively becomes a combined-input-and-crosspoint-queued (CICQ) switch [31]. With the help of crosspoint buffers, scheduling becomes much easier for CICQ switches since input scheduling and output scheduling can now be performed separately. Many of the scheduling algorithms for IQ switches can be directly applied to CICQ switches at a lower complexity, e.g., the distributed MWM algorithm DISQUO [39] and the push-in-first-out (PIFO) policy [14]. On the other hand, a CICQ switch suffers from the same problem as an IQ switch due to the need for nearly instantaneous control communications between the input line cards and the switching core. Kanizo et al. [23] argue that the power-consuming input buffers are usually placed far away from the switching core, which makes it impractical for an input scheduler to keep track of the real-time buffer occupancies at its associated crosspoints.

To avoid such implementation difficulties, Kanizo et al. [23] consider a self-sufficient single-chip crosspoint-queued (CQ) switch whose buffering and scheduling are performed solely inside the switching core, and argue for its feasibility given state-of-art ASIC technologies [25, 20, 7]. Unlike an IQ or OQ switch which may spread its buffer space on multiple input/output line-cards, the total buffer space of a single-chip CQ switch is limited by the chip size.

This may seem like a severe deficiency at first glance, since it has long been believed that Internet routers should provide one round-trip-time's equivalent of buffering to prevent link starvation. However, recent studies on high-speed Internet routers by Wischik et al. [37, 38] and McKeown et al. [1, 3] challenge this commonly used approach, and suggest that the optimal buffer size can be much smaller than that was previously believed. The reason lies in the fact that the Internet backbone links are usually driven by a large number of different flows, and multiplexing gains can be obtained under the congestion and flow control mechanisms. They also argue that short-term Internet traffic approximates the Poisson process, while long-range dependence holds in large time-scales. As a result, a much smaller amount of buffering is required as long as the traffic load is moderate, and thus can readily be accomodated on a single chip.

The single-chip CQ switch has many distinct features. On the one hand, using small segregated on-chip buffers instead of large aggregated off-chip memory allows much faster memory access on ASICs, which could have been a bottleneck for high speed switches. It also divides the scheduling and buffering tasks into small chunks, which are then fulfilled by a large number of crosspoints with low hardware requirement at each node. On the other hand, because its buffers are small and segregated, a basic CQ switch with simple scheduling algorithms, such as round-robin (RR), oldest-cell-first (OCF) and longest-queue-first (LQF), may experience far more packet drops than an IQ or OQ switch with the same total amount of buffering. Previous analyses and simulations done by Kanizo et al. [23] and Radonjic et al. [32, 34] have shown that LQF provides the highest throughput for a CQ switch in many cases, but its performance is still worse than an OQ switch with the same total buffer space. This problem is more severe when there are more ports and thus the buffer size at each crosspoint is more restricted.

A key observation here is that when a certain crosspoint experiences packet overflow, other crosspoint buffers can still be quite empty, i.e., the buffer utilizations are unbalanced. The unbalanced-utilization problem becomes worse when the incoming traffic is bursty or non-uniform. As reported in [23, 33], even LQF scheduling works poorly under these conditions. Unfortunately, analyses of real Internet traffic traces often reveal such burstiness and non-uniformity. As a result, how to efficiently use the crosspoint buffers so as to reduce packet drops remains a major issue before single-chip CQ switches can be widely accepted.

One possible solution to lessen the problem is to add an extra *load-balancing* stage in front of the original switching fabric [10, 11]. As incoming traffic passes through the first load-balancing stage, its burstiness and non-uniformity can be greatly reduced.

However, the extra load-balancing stage can also introduce a mis-sequencing phenomena, i.e., packets of the same flow may not leave in the same order as they arrive. Mis-sequencing may cause unwanted performance degradation in many Internet services and applications, e.g., TCP-based data transmission. TCP remains the most dominant transport layer protocol used in the public Internet [8], but it performs poorly if the correct packet order is not maintained end-to-end, because such out-of-order packets might be treated as lost and trigger unnecessary retransmissions and congestion control [35, 4]. As a result, many network operators insist that packet ordering must be preserved in designing packet-switched Internet routers. Previous approaches to restore packet ordering include extra re-sequencing buffers [11, 26] and frame-based scheduling [14, 22], but at the cost of higher delay and buffer requirement.

Another candidate is *deflection routing*. This concept was proposed in the networking area as early as in the 1980s. The general idea is to reroute a packet to another node or path when there is no buffer available on its regular (shortest) path. Topologies proposed for deflection routing include Manhattan Street Network [29], Shuffle-Exchange Network [27], etc. All these designs effectively share distributed buffers at different nodes and lower the packet drop rate, but they also alter the packet order due to multi-path routing. Although work has been done to bound the maximum delay with deflection routing [6, 17], they have not solved the mis-sequencing problem completely.

Considering these points, we propose a chained crosspoint-queued (CCQ) switching architecture, apply load balancing and deflection routing techniques, and jointly design buffer sharing and in-order scheduling to meet the goals of low packet drop rate and correct packet order. In order to resolve the major constraints of buffering, some modifications to the basic CQ switching architecure are made, but to a modest and feasible extent. Some fast message passing and cell deflection need to be supported between adjacent crosspoints, but they can be implemented easily since the crosspoints are linked on a single chip and thus such communication is purely internal to the switch core.

We mainly consider four different configurations in this paper:

- *CQ-LQF:* This is the basic single-stage CQ switch (Section 2.1) with LQF scheduling at each output [23]. The crosspoint buffers are segregated, and no speedup is required. This serves as a benchmark (worst-case performance) to be compared with other schemes.

- *CCQ-OCF:* This refers to a two-stage CCQ switch (Section 2.2) with OCF scheduling at each output (Section 3). A load-balancing stage is added to the front, and the segregated crosspoints associated with common outputs are connected into daisy chains to support deflection routing, which requires an internal memory speedup of 2.

- *CCQ-RR:* We also propose a less-demanding RR algorithm (Section 4) to take the place of OCF scheduling in the CCQ switch. In this scheme, a *wait-counter* is tagged to each cell, aligned with other cells of the same flow through just-in-time notifications, and preserved during deflection. Upon departure, the wait-counter of a HOL cell is compared with another *RR-counter* maintained by each output scheduler to determine its eligibility of departure. In this way, the correct packet order is ensured.

- *OQ:* Finally, a typical OQ switch with the same total buffer space as the CQ switches above is considered. It can also be viewed as a CQ switch in which crosspoints associated with the same output use a shared memory. A speedup of N is required in the worst case. This also serves as a benchmark (best-case performance) to be compared with our proposed schemes.

The rest of this paper is organized as follows. In Section 2, the basic single-chip CQ switch is briefly reviewed, and an augmented CCQ switching architecture is introduced. In Sections 3 and 4, two kinds of buffering and scheduling schemes suitable for load balancing and deflection routing are proposed and analyzed. Then we run some numerical simulations with different traffic patterns and system configurations in Section 5, verifying the effectiveness of the proposed techniques. Finally, we conclude our work in Section 6.

2. SYSTEM ARCHITECTURE

2.1 Basic Crosspoint-Queued Switch

The single-chip CQ switch [23] is a self-sufficient architecture which has all its buffers placed at the crosspoints of input and output lines. There is no buffering at input or output line-cards, as shown in Fig. 1.

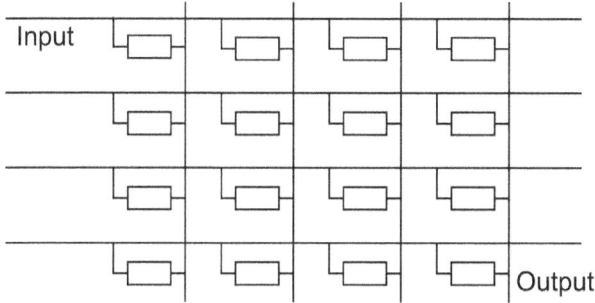

Figure 1: The single-stage basic CQ switch.

Assume an $N \times N$ CQ switch with crosspoint buffers of size B each. Then $0 \leq B(i, j) \leq B$ denotes the buffer occupancy at crosspoint (i, j), $i, j = 1, 2, ..., N$. We assume that the CQ switch works in a slotted manner, i.e., packets are fragmented into fixed-length cells before entering the switch core. The buffer occupancies and sizes are also measured in units of such cells. Usually a header is appended to each cell before entering the switching fabric. Such headers may contain a cell ID, source/destination ports, a time-stamp, etc.

The basic *CQ-LQF* scheduling scheme can be described as two phases in each time slot:

- *Arrival Phase:* For each input i, if there is a newly arriving cell destined to output j, it is directly sent to crosspoint (i, j). If buffer (i, j) is not full, i.e. $B(i, j) < B$, the new cell is accepted and buffered at the tail of line (TOL). Otherwise, this cell is dropped.

- *Departure Phase:* For each output j, if not all crosspoints $(*, j)$ are empty, the output scheduler picks the one with the longest queue, and serves its HOL cell.

The point of LQF rule is that it always serves the fullest buffer which is most likely to overflow. Since each output must determine the longest queue among all N crosspoints in each time slot, its

worst-case time complexity is at least $O(\log N)$, assuming parallel comparator networks.

In this paper, we define that a cell belongs to flow (i, j) if it travels from input i to output j. Thus for *CQ-LQF*, cells that belong to the same flows are always served in the same order as they arrive.

2.2 Chained Crosspoint-Queued Switch

The basic CQ switch is simple and elegant. However, its buffers are small and segregated, which results in a high packet drop rate and a low buffer utilization. We therefore develop an augmented architecture, the CCQ switch, which is suitable for load balancing and deflection routing when combined with the scheduling schemes to be proposed in subsequent sections.

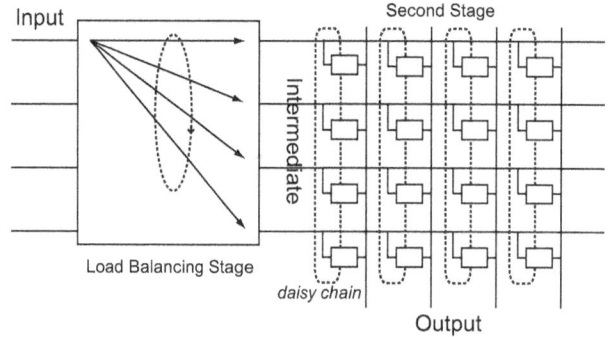

Figure 2: The two-stage CCQ switch.

In the CCQ switch, crosspoints associated with a common output port are single-connected into a daisy chain (in the order of their associated input port indices), as shown in Fig. 2. Specifically, crosspoint (i, j) is connected with its predecessor $(i - 1, j)$ and successor $(i + 1, j)$. Note that since the input and output port indices should always be within 1 through N, thus $(i - 1, j)$ should actually be $(mod(i - 2, N) + 1, j)$, while $(i + 1, j)$ should actually be $(mod(i, N) + 1, j)$. For ease of presentation, we shall use the $(i - 1, j)$ and $(i + 1, j)$ notation in the rest of this paper.

With this modification, message passing and cell deflection can be easily supported between adjacent crosspoints along the daisy chains. In terms of the hardware requirement, by adding an extra layer of connections, we introduce an extra memory read/write speedup for each crosspoint buffer. The extra memory speedup and inter-crosspoint connections are purely internal to the switch core, implemented on a single chip, and thus do not impose extra burdens on the links between the input/output line-cards and the switching core (card-edge and chip-pin limitations [7]).

To further reduce the probability of buffer overflow, we also place an extra load-balancing switch (first stage) in front of the CQ switching fabric (second stage), as shown in Fig. 2. The load-balancing stage walks through a fixed sequence of configurations: at time t, it connects each input i to intermediate port $i + t$, which is also input $i + t$ of the second-stage. Note that $i + t$ is an abbreviation of $mod(i + t - 1, N) + 1$ for ease of presentation.

3. OLDEST-CELL-FIRST SCHEDULING FOR THE CCQ SWITCH

In [23, 32], it has been recognized that LQF provides a lower packet drop rate for the basic CQ switch than any other simple scheduling algorithms like random, RR and OCF. However, its performance can still be far worse than an OQ switch with the same total buffer space, if the incoming traffic is bursty or non-uniform.

In this section, we propose a scheme that allows different cross-points in the same daisy chain to share packets evenly, and use the OCF scheduling algorithm to ensure correct packet ordering.

3.1 CCQ-OCF Scheduling Design

OCF is a popular scheduling algorithm which always picks the oldest cell to serve. Compared with LQF, OCF usually incurs a larger packet drop rate since it does not always serve the buffer that is most likely to overflow. Compared with RR, OCF is much more complicated since it requires repeated comparisons of time-stamps at each time slot. Despite these disadvantages, OCF is still attractive since it can easily maintain the packet order across all flows. This advantage makes OCF a good candidate to solve the mis-sequencing problem caused by load balancing and deflection routing. The performance loss due to using OCF rather than LQF can be negligible since load balancing and deflection routing already do a good job in equalizing the buffer utilizations.

In this scheme, we use the two-stage CCQ switch. Every incoming cell is assigned a time-stamp to record its arrival time. Each crosspoint needs to maintain the buffered cells in the order of non-decreasing time-stamps (i.e., first-come-first-serve). Then the output schedulers will only need to compare the time-stamps of HOL cells to determine the oldest one in each time slot.

The detailed scheme for *CCQ-OCF* is described below:

- *Arrival Phase:*

  At time t, for each input i, if there is a newly arriving cell destined to output j, then after passing the load-balancing stage that connects input port i to intermediate port $i + t$, it is directly sent to crosspoint $(i + t, j)$ of the second stage. If the buffer is not full, i.e., $B(i + t, j) < B$, the new cell is accepted and buffered at TOL with time-stamp t. Otherwise, this overflowing cell is dropped.

- *Departure Phase:*

  For each output j, if there is at least one non-empty cross-point buffer $(*, j)$, the output scheduler picks the one with the oldest HOL cell, and serves this cell.

- *Deflection Phase:*

  Each crosspoint (i, j) does the following step by step:

  1. Report buffer occupancy $B(i, j)$ to its predecessor cross-point $(i - 1, j)$;
  2. Receive a buffer occupancy report $B(i + 1, j)$ from its successor crosspoint $(i + 1, j)$;
  3. If $B(i, j) > B(i + 1, j)$, deflect the TOL cell to its successor crosspoint $(i + 1, j)$;
  4. Receive a deflected cell from its predecessor crosspoint $(i - 1, j)$. If there is one, insert the deflected cell into the ordered queue according to its time-stamp.

An example is illustrated in Fig. 3. Different flows are marked with different colors and alphabets, e.g., $yellow - a, red - b$, and $green - c$. The time-stamps are indicated by integer subscripts, e.g., $1, 2, 3$. During the departure phase, crosspoint $(2, j)$ with the oldest cell b_1 is served by output j. Then in the deflection phase, crosspoint $(4, j)$ deflects its TOL cell c_2 to its less occupied successor $(1, j)$, whereas crosspoint $(2, j)$ and $(3, j)$ do not deflect because $B(2, j) = 1 = B(3, j) < B(4, j) = 2$ at the beginning of this deflection phase. The load balancing and deflection routing mechanisms in *CCQ-OCF* aim to equalize the buffer occupancies of all crosspoints throughout the daisy chain, and thus

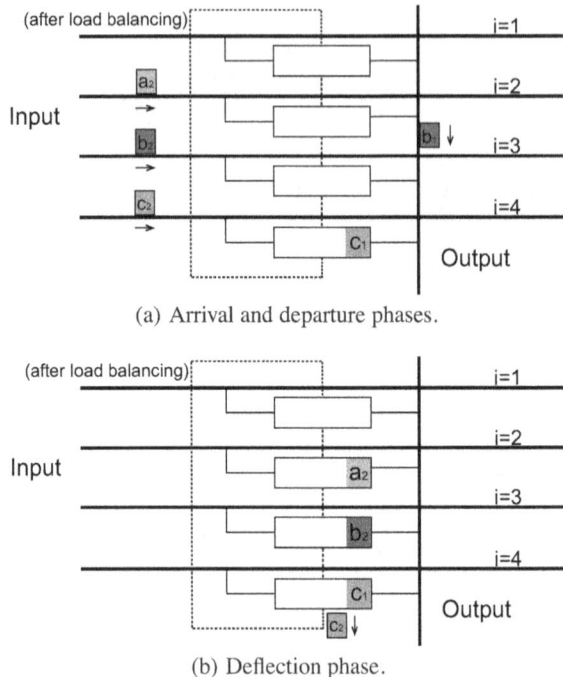

(a) Arrival and departure phases.

(b) Deflection phase.

Figure 3: An example of *CCQ-OCF*.

fully utilize the limited buffer space. All cells leave the switch in non-decreasing order of time-stamps, irrespective of where they are actually buffered.

3.2 Features

We list some important features of *CCQ-OCF* as follows.

Property 1: There is no cell drop during deflection.

Deflections only take place when $B(i, j) > B(i + 1, j)$, and the direction of deflections can only be from (i, j) to $(i + 1, j)$. So $B(i + 1, j)$ increases at most by 1 during the deflection phase. Since $B(i + 1, j) < B(i, j) \le B$ before deflection, it will never cause an overflow. This applies to all $i, j = 1, 2, ..., N$.

Property 2: Cells of the same flow always leave the switch in the same order as they arrive.

According to *CCQ-OCF*, incoming cells are always directly sent to the corresponding crosspoints. Cells in these crosspoints are always kept in non-decreasing order of their arrival time, even when deflections take place. Then each output scheduler picks the HOL cells according to the OCF scheduling algorithm, which strictly maintains the order of cells, even across different flows.

Insight 1: *CCQ-OCF* gains the most advantage over *CQ-LQF* in large switches with moderate crosspoint buffer size under bursty or non-uniform traffic.

First, we assume uniform Bernoulli i.i.d. traffic. Then according to the law of large numbers, the long-term buffer occupancy of each crosspoint tends to be equalized asymptotically if the buffer size is large enough, even without load balancing or deflection routing. In this case, the proposed scheme is of limited efficacy. By contrast, *CCQ-OCF* can be much more effective under bursty or non-uniform traffic, since the arrival rates at each crosspoint can be very different, especially in short time scales.

On the other hand, a large switch size boosts the multiplexing gain since the bursty traffic can be evenly distributed to more crosspoints. In terms of the crosspoint buffer size, neither should it be so large that the short-term bursty traffic is smoothed through long-term averaging, nor should it be too small to sustain regional traffic fluctuations before deflection routing takes place to resolve them.

Insight 2: The worst-case time complexity at each crosspoint is $O(\log B)$ in each time slot, and each output scheduler needs $O(\log N)$ time to determine the oldest cell.

As mentioned before, the crosspoints need to maintain the cells in non-decreasing order of time-stamps. Observing that such ordering may only be disturbed upon cell arrival, departure and deflection, we consider the following cases:

- Newly arriving cells should always be placed at TOL since they have the largest time-stamps so far. Thus cell arrivals do not break the ordering.

- Only HOL cells can leave the switch. These cells always have the smallest time-stamps. Thus cell departures do not break the ordering either.

- Only the newest cells (TOL cells), can be deflected from highly-utilized crosspoints (sender) to their successor crosspoints (receiver). They always have the largest time-stamps at the senders. Thus cell deflections do not break the ordering at these senders.

- In each time slot, each crosspoint may receive at most one deflected cell from its predecessor. This deflected cell should then be searched and inserted into the ordered queue at the receiver according to its time-stamp.

Summing up all three phases, each crosspoint needs to perform at most $O(1)$ search, $O(1)$ insertion, and $O(1)$ deletion operations in each time slot. All these can be done in $O(\log B)$ time with a self-balancing binary search tree.

In terms of the output schedulers, they need to determine the oldest HOL cell in each time slot. One way to accomplish this is to implement a hardware-based comparator network which works like a single-elimination tournament. In each round we eliminate half of the HOL cells. Then the total amount of comparisons is $O(N)$ and the number of rounds is $O(\log N)$. An alternative way is to maintain a heap of HOL cells according to their time-stamps. In each time slot we extract the oldest HOL cell and insert some new HOL cells into the heap. A drawback of this approach is that the number of new HOL cells to be inserted can be $O(N)$ in the worst case if all crosspoints were empty in the previous time slot.

4. ROUND-ROBIN SCHEDULING FOR THE CCQ SWITCH

In the previous section, the OCF scheduling algorithm has been used to maintain the correct packet order. This method is straightforward and promising, but requires considerable computation due to repeated sorting in each time slot. On the other hand, the global packet ordering guaranteed by OCF is too strict, since we only need per-flow packet ordering. In this section, we propose a new scheme that relies on a less-demanding RR polling algorithm and an explicit notification mechanism between adjacent crosspoints to preserve per-flow packet ordering. The underlying idea is partly inspired by the Mailbox Switch [9] and Padded Frame [22], but it is implemented in a very different way here that avoids extra delays.

4.1 CCQ-RR Scheduling Design

4.1.1 Wait-Counter and RR-Counter

In this scheme, every crosspoint should maintain a *wait-counter* for each of its buffered cells, denoted by $W(i,j,k)$, $1 \leq k \leq$

$B(i,j)$. Another anticipatory wait-counter for the next incoming cell, denoted by $W(i,j,B(i,j)+1)$, is also maintained by crosspoint (i,j). When a new cell arrives at (i,j), it is assigned $W(i,j,B(i,j)+1)$ upon acceptance. Then $B(i,j)$ gets incremented, and a new anticipatory wait-counter should be generated as $W(i,j,B(i,j)+1) = W(i,j,B(i,j))+1$.

As a counterpart of the wait-counters, we also let each output j maintain a *RR-counter* $R(j)$, in addition to its arbiter position $1 \leq A(j) \leq N$ which always points to the last crosspoint it has polled. $R(j)$ is incremented during each RR polling cycle when $A(j) = 1$.

Both the wait-counters $W(i,j,k)$ and the RR-counters $R(j)$ are ever-increasing (the grow-to-infinity problem can be resolved by dropping the carry when these counters exceed a sufficiently large value), but they should be maintained in non-decreasing order, so that $R(j) \leq W(i,j,k) \leq W(i,j,k+1)$ for any $1 \leq i,j \leq N$ and $1 \leq k \leq B(i,j)$ at any time.

An arbitrary cell k stored at a non-empty crosspoint (i,j) is eligible to leave the switch, if and only if, $W(i,j,k) = R(j)$; thus crosspoint (i,j) should refrain from being served by output j until its HOL cell becomes eligible. In terms of any empty crosspoint (i',j), it should update $W(i',j,1) = R(j)+1$ every time when output j polls it and proceeds to subsequent crosspoints.

4.1.2 Counter-Alignment Notification

We also design an explicit counter-alignment notification mechanism, which coordinates the correct packet ordering under load balancing. Such a notification is initiated by any crosspoint (i,j) upon acceptance of a newly arriving cell. It is then passed down to $(i+1,j)$ and subsequent crosspoints along the daisy chain. Upon reception, the receiver crosspoint should examine the contents, make necessary updates to its own anticipatory wait-counter, and determine whether to drop the notification message or to relay it to subsequent crosspoints.

Information contained in a notification message consists of two parts: a counter-alignment field $CA(i,j)$, which indicates the minimum wait-counter for the next incoming cell to crosspoint $(i+1,j)$, and a source-of-notification field $SN(i,j)$, which denotes the crosspoint that has initiated the message.

Specifically, when crosspoint (i,j) accepts a new cell, it immediately initiates a counter-alignment notification with $CA(i,j) = W(i,j,B(i,j))$ (increment if $i = N$) and $SN(i,j) = i$, and sends it to the successor crosspoint $(i+1,j)$ in the same daisy chain.

Then for crosspoint $(i+1,j)$, if $CA(i,j) \geq W(i+1,j,B(i+1,j)+1)$ and not $SN(i,j) = i+1$ (discard the message if it has traversed the daisy chain and come back to its origination), it updates $W(i+1,j,B(i+1,j)+1) = CA(i,j)$, and decides to relay the notification message with $CA(i+1,j) = CA(i,j)$ (increment if $i+1 = N$) and $SN(i+1,j) = SN(i,j)$ to its own successor $(i+2,j)$ in the next time slot, if by that time it has not accepted a new cell and generated a new notification message.

In this way, the mis-sequencing problem caused by load balancing can be solved. Cells of the the same flow are always assigned with non-decreasing wait-counters through just-in-time notifications between any two consecutive arrivals.

4.1.3 Deflection Routing with Counter Preserved

Deflection routing may also introduce mis-sequencing. With wait-counters, it is straightforward to resolve the issue.

Similar to *CCQ-OCF*, each crosspoint (i,j) is allowed to deflect one TOL cell to its successor $(i+1,j)$ in each time slot if $B(i,j) > B(i+1,j)$. The deflected cell should carry its own wait-counter $DW(i,j) = W(i,j,B(i,j))$ (increment if $i = N$)

with it. When crosspoint $(i+1, j)$ receives the deflected cell, it compares $DW(i, j)$ with its own cells, and inserts the deflected cell to the appropriate position to maintain non-decreasing order of wait-counters. If it has one or more cells with wait-counters equal to $DW(i, j)$, the deflected cell should be inserted in front of all of them to preserve their relative order of departure. In case $DW(i, j) \geq W(i+1, j, B(i+1, j)+1)$, update $W(i+1, j, B(i+1, j)+1) = DW(i, j) + 1$.

Now that there may be multiple cells with the same wait-counters at each crosspoint (i, j), output j should adopt an exhaustive RR algorithm, serving all cells k at crosspoint (i, j) with $W(i, j, k) = R(j)$ before proceeding to the next eligible crosspoint. In this way, deflection routing will not alter the order of cells to be served.

4.1.4 CCQ-RR Scheme

- **Arrival Phase:**

  Same as in *CCQ-OCF* except that the wait-counters should be assigned and updated according to Section 4.1.1 instead of the time-stamps.

- **Notification Phase:**

  Each crosspoint (i, j) sends and receives a counter-alignment notification message according to Section 4.1.2.

- **Departure Phase:**

  Each output j polls its associated crosspoints $(*, j)$ in an exhaustive RR fashion, starting from its final position $A(j)$ in the previous time slot. The polling process continues until output j serves an eligible crosspoint with $W(i, j, 1) = R(j)$, or it finds all buffers empty.

- **Deflection Phase:**

  Same as in *CCQ-OCF* except that wait-counters take the place of time-stamps according to Section 4.1.3.

An example is illustrated in Fig. 4. Different flows are marked with different colors and alphabets, e.g., $yellow - a$. The time-stamps (for illustration, not required in implementation) are indicated by integer subscripts, e.g., 1,2,3. Wait-counters are represented by their positions on the time-line, while vacancies (cross-marked squares) in the time-lines do not occupy real buffer positions. During the arrival phase at time $t = 1$, the new cell b_1 is tagged with wait-counter $W(3, j, 1) = 0$, and $W(3, j, 2) = W(3, j, 1) + 1 = 1$ is generated. Similar tagging and updates occur for cells a_1 and c_1. Next, during the notification phase, crosspoint $(3, j)$ initiates a counter-alignment notification with $CA(3, j) = W(3, j, 1) = 0$ for the newly accepted cell b_3, and sends it to its successor $(4, j)$, but this message is discarded because $CA(3, j) = 0 < W(4, j, 2) = 1$. On the other hand, crosspoint $(4, j)$ also initiates a counter-alignment notification $CA(4, j) = W(4, j, 1) + 1 = 1$ (note that $i = 4 = N$ here) for c_1. Crosspoint $(1, j)$ accepts the notification, updates $W(1, j, 2) = CA(4, j) = 1$ so that the next incoming cell c_2 will be served later than c_1, and decides to relay this message to subsequent crosspoints in future time slots. Then during the departure phase, the first eligible cell a_1 with $W(1, j, 1) = 0 = R(j)$ is served by the output, leaving a vacancy in the time-line. Finally, during the deflection phase, crosspoint $(4, j)$ finds its successor $(1, j)$ less occupied, so it deflects the TOL cell c_1 with $DW(4, j) = W(4, j, 1) + 1 = 1$ (again, $i = 4 = N$). As a result, the new cell c_2 to arrive at time $t = 2$ will be pushed back to the $3rd$ time-line position, although it is stored in the $2nd$ buffer position. The cells shall leave the switch in the order of a_1, b_1, c_1, c_2, etc.

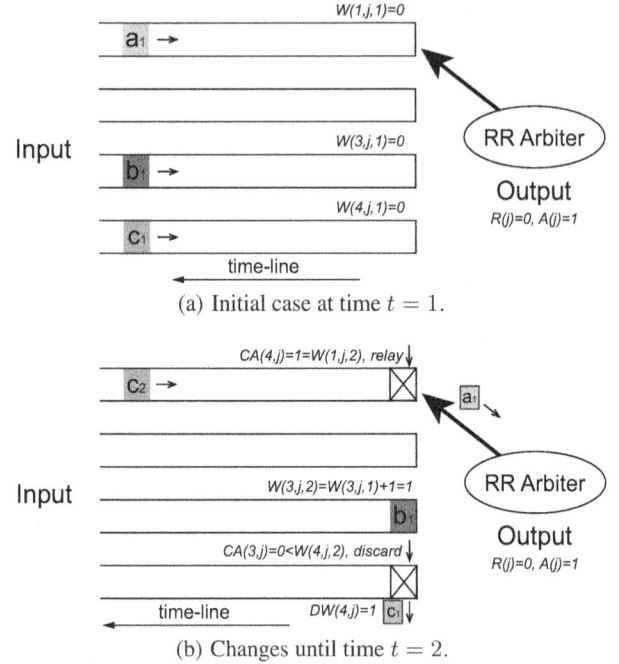

(a) Initial case at time $t = 1$.

(b) Changes until time $t = 2$.

Figure 4: An example of *CCQ-RR*.

4.2 Features

Some key features of *CCQ-RR* are listed below.

Property 1: The proposed *CCQ-RR* scheme is work-conserving, if the maximum number of deflections is restricted to K, and each output can perform $N + K + 1$ polls in each time slot.

First, consider the situation without deflection routing. Pick any arbitrary cell $\mathcal{X}_1$ that arrives at crosspoint (i, j) and gets wait-counter $W(i, j, k)$.

- If $W(i, j, k)$ was updated upon acceptance of a newly arriving or deflected cell $\mathcal{Y}$, then $\mathcal{Y}$ must have been exactly $N + 1$ polls away at that time, and may have become even closer after deflections. If $\mathcal{Y}$ is served immediately before $\mathcal{X}$, then the output arbiter needs at most $N + 1$ polls to reach $\mathcal{X}$. Otherwise, if any other cell $\mathcal{Z}$ is the immediate predecessor (in the order of departure), $\mathcal{Z}$ must be at most $N + 1$ polls away.

- If $W(i, j, k)$ was updated through counter-alignment initiated for cell $\mathcal{Y}$, then $\mathcal{Y}$ must have been at most N polls away at that time, otherwise the counter-alignment notification should have already been discarded after traversing the daisy chain. If $\mathcal{Y}$ is served immediately before $\mathcal{X}$, then the output arbiter needs at most N polls to reach $\mathcal{X}$. Otherwise, if any other cell $\mathcal{Z}$ is the immediate predecessor (in the order of departure), $\mathcal{Z}$ must be at most N polls away from $\mathcal{X}$.

- Otherwise, $W(i, j, k)$ must have been updated when crosspoint (i, j) was empty through $W(i, j, 1) = R(j) + 1$, then $k = 1$ and it is at most N polls away from the output arbiter.

Summing up all three conditions, the output arbiter needs at most $N + 1$ polls (starting from its last polled crosspoint) in each time slot to ensure it is work-conserving.

We next take deflection routing into account. If the number of deflections is limited to K, then the gap between any two consecutive cells (in the order of departure) is enlarged by at most K polls. As a result, each output arbiter needs at most $N + 1 + K$ polls in each time slot to be work-conserving.

Property 2: Cells of the same flow always leave the switch in the same order as they arrive.

For load balancing, cell order is preserved through just-in-time counter-alignment notifications between any two consecutive arrivals of the same flow. In terms of deflection routing, it will not alter the order of departure if the wait-counters are preserved and adjusted when necessary. These are elaborated in Sections 4.1.2 and 4.1.3, and some boundary conditions should be taken care of. Specifically, the last crosspoint (N, j) in each daisy chain j must always increment the counter-alignment field $CA(N, j)$, as well as the wait-counter of its deflected cell $DW(N, j)$, so as to match with the starting point of a new RR polling cycle.

Insight 1: *CCQ-RR* gains the most advantage over *CQ-LQF* in large switches under bursty or non-uniform traffic.

The reasons are similar as for *CCQ-OCF* and thus omitted.

Insight 2: The worst-case time complexity at each crosspoint is $O(\log B)$ in each time slot, and each output scheduler can find the next eligible HOL cell in $O(\log N)$ time.

The operations at each crosspoint in *CCQ-RR* are exactly the same as those in *CCQ-OCF*, except that the time-stamps are replaced with the wait-counters, and that $O(1)$ additional updates to the anticipatory wait-counters need to performed. All these can still be accomplished in $O(\log B)$ time using a self-balancing binary search tree.

In terms of the output scheduler, each RR arbiter may find the next eligible crosspoint within $O(\log N)$ time using a hardware-based priority encoder [18] (typically a few nanoseconds). Although the magnitude of time complexity for RR looks the same as that for OCF, the constant factor can be much smaller, and it has been widely recognized that RR is much easier to implement than OCF. On the other hand, in order to utilize the priority encoder, each output arbiter j may need to broadcast its RR-counter $R(j)$ and arbiter position $A(j)$, so that each crosspoint may determine its own eligibility in a distributed manner.

5. NUMERICAL SIMULATIONS

In this section, we perform numerical simulations with MAT-LAB to show the performance improvements through load balancing and deflection routing. Specifically, we compare the cell drop rates and critical buffer utilizations of the CCQ switches against a basic LQF-based CQ switch and an OQ switch with the same total buffer space. The latter two systems are used as benchmarks in our comparison.

The *cell drop rate* is the average probability a random cell is dropped by the switch. We shall focus on the drop rate of fixed-length cells after fragmentation. Cell drop rate should be as low as possible, but we set up a reasonable target at 10^{-5} for the following reasons:

- The state-of-art Internet end-to-end loss rate for IP packets is in the order of 10^{-3} to 10^{-2} [5, 21];

- Empirical results reveal that TCP/IP protocols may tolerate an end-to-end loss rate of 10^{-3} and still yield satisfactory performances [19];

- Measurements on the Internet show that the average end-to-end hop-count is of the order of tens [2, 12];

- Assume that the variable-length IP packets are fragmented into $64byte$ cells, then the average number of segments for each IP packet is of the order of tens [8].

The *critical buffer utilization* is defined as the average utilization of all buffers $(*, j)$ when a cell destined to output j is dropped, i.e., $\eta_{cq} = E(\frac{\sum_{i=1}^{N} B(i,j)}{N \times B} |$ cell drop at daisy chain j) for a CQ switch, and $\eta_{oq} \equiv 100\%$ for an OQ switch. We shall see later that the critical buffer utilization is negatively correlated with the cell drop rate.

In the rest of this paper, we shall investigate the impact of traffic load, non-uniformity, burstiness and switch size on *CQ-LQF*, *CCQ-OCF*, *CCQ-RR* and *OQ* schemes using various synthesized traffic patterns and real Internet traces.

5.1 Impact of Traffic Load

First, we evaluate the effectiveness of the proposed schemes under uniform bursty traffic. The destinations of incoming cells are evenly distributed among all N output ports, i.e., $\lambda_{ij} = \frac{\lambda}{N}$, $i, j = 1, 2, ..., N$, where $0 \leq \lambda \leq 1$ is the normalized traffic load.

Since real Internet traffic is usually bursty and long-range dependent (LRD), we shall focus on this kind of traffic. Specifically, we use the Markov Chain model in [16] to generate LRD traffic with Hurst parameter $H = 0.75$ and maximum length $L = 1000$, i.e., each single burst of cells belonging to the same flow may last for at most 1000 time slots. Subsequently, we shall use this traffic-generating model, and adjust H, L, λ_{ij} to control the traffic pattern.

We consider 32×32 switches with crosspoint buffer size $B = 40$ cells. The simulation lasts $T = 10^7$ time slots.

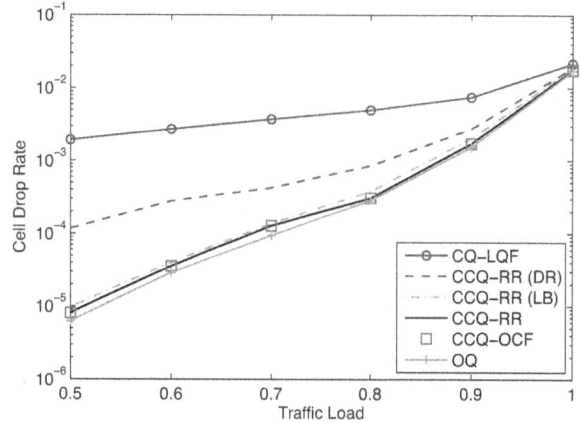

Figure 5: Cell drop rate of 32×32 switches with $B = 40$ under uniform bursty traffic with $H = 0.75$, $L = 1000$ and $0.5 \leq \lambda \leq 1.0$.

Fig. 5 compares the cell drop rates of various schemes. The abbreviation "*CCQ-RR (LB)*" stands for "*CCQ-RR* with load balancing only", and "*CCQ-RR (DR)*" means "*CCQ-RR* with deflection routing only". These two degenerate versions of *CCQ-RR* are compared here so as to demonstrate the respective effectivenesses of load balancing and deflection routing. They also preserve the correct packet order.

Simulation results show that *CCQ-OCF* and *CCQ-RR* have the lowest cell drop rates, which are much better than that of *CQ-LQF* and very close to that of *OQ*, going down to about 10^{-5} when the traffic load is $\lambda = 0.5$. Similar performances can also be achieved under higher traffic loads if larger buffers are implemented.

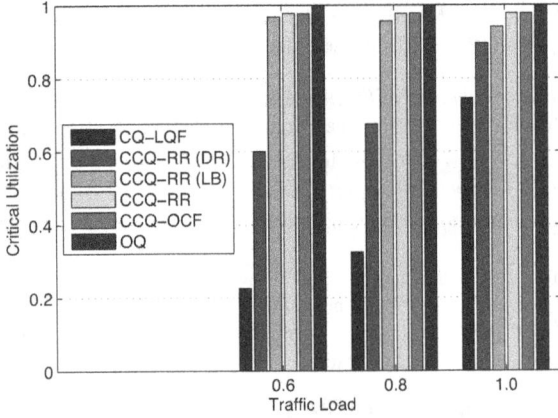

Figure 6: **Critical buffer utilization of** 32×32 **switches with** $B = 40$ **under uniform bursty traffic with** $H = 0.75$, $L = 1000$ **and** $0.5 \leq \lambda \leq 1.0$.

Comparing *CCQ-RR (LB)* with *CCQ-RR (DR)*, we can find that deflection routing does not contribute as much as load balancing in this case. However, one cannot conclude that deflection routing is ineffective if load balancing is employed. In fact, the superiority of load balancing could largely be attributed to how we model the LRD traffic. As mentioned before, our model generates separate bursts of cells that belong to different flows, which makes load balancing especially effective. On the other hand, in real Internet traffic, such bursts should be interleaved, showing Poisson characteristics in short time scales, and leaving more time for deflection routing to propagate. Besides, load balancing is a passive mechanism, while deflection routing is a reactive strategy and its advantage can be very significant under adversarial traffic patterns.

We also compare the buffer utilizations of different schemes in Fig. 6. Here we can see that the critical utilization of *CQ-LQF* is fair when the traffic load is high (about 70% when $\lambda = 1.0$), but drops quickly as the traffic load becomes lower (only 20% when $\lambda = 0.6$). To understand this, we must realize that a lower traffic load does not necessarily lead to less burstiness according to our model, since the Hurst parameter does not change at all. Ironically, when the traffic load is lower, the incoming traffic at different cross-points can be even more unbalanced in a short time-scale. This kind of low buffer utilization leads to a larger performance degradation when the traffic load is low (as compared with *OQ*). By contrast, *CCQ-OCF* and *CCQ-RR* are not affected by the change of traffic load, showing robustness against various traffic loads.

Comparing Fig. 5 and Fig. 6, we can see a clear trend that the cell drop rate is negatively correlated with the critical buffer utilization given the same incoming traffic. The critical buffer utilizations of *CCQ-OCF* and *CCQ-RR* are close to 100%, which is only achievable by the OQ switch. Thus the significant performance improvements of the proposed schemes can be attributed to their efficient buffer sharing mechanisms, i.e., load balancing and deflection routing.

5.2 Impact of Non-uniformity

In addition to the uniform bursty traffic, we also test the proposed buffering and scheduling techniques under non-uniform traffic. In this case, the destinations of incoming cells are not evenly distributed among all N outputs. Instead, we adopt a hot-spot traf-

fic model as follows:

$$\lambda_{ij} = \begin{cases} \rho\lambda & \text{if } i = j \\ \frac{(1-\rho)\lambda}{N-1} & \text{otherwise} \end{cases}$$

We still focus on 32×32 CQ switches with buffer size $B = 40$ cells. The incoming traffic is LRD with $H = 0.75$ and $L = 1000$, and the hot-spot parameter is set to $\rho = 0.5$.

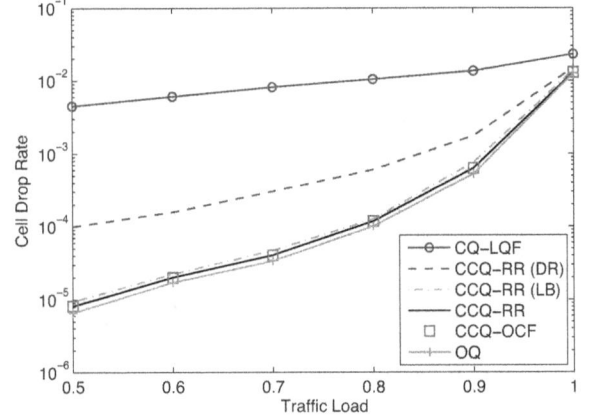

Figure 7: **Cell drop rate of** 32×32 **switches with** $B = 40$ **under non-uniform bursty traffic with** $H = 0.75$, $L = 1000$ **and** $0.5 \leq \lambda \leq 1.0$.

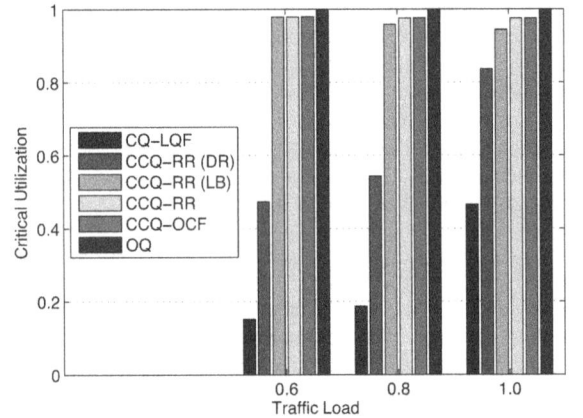

Figure 8: **Critical buffer utilization of** 32×32 **switches with** $B = 40$ **under non-uniform bursty traffic with** $H = 0.75$, $L = 1000$ **and** $0.5 \leq \lambda \leq 1.0$.

The cell drop rates and critical buffer utilizations of the proposed schemes under hot-spot LRD traffic are illustrated in Fig. 7 and Fig. 8 respectively. Comparing these two figures with their counterparts under uniform bursty traffic in Section 5.1, we find that *CQ-LQF* performs worse under non-uniform traffic, as indicated by a higher cell drop rate and a lower critical buffer utilization. By contrast, *CCQ-OCF* and *CCQ-RR* have slightly lower cell drop rates and higher critical buffer utilizations under non-uniform bursty traffic, demonstrating the same trend as *OQ*. These results show that the proposed schemes are relatively better under non-uniform traffic. We also notice that deflection routing suffers more from the non-uniformity of the incoming traffic, e.g., the critical buffer utilization drops from 60% under uniform bursty traffic with $\lambda = 0.6$ to below 50% in this case.

5.3 Impact of Burstiness

The impact of burstiness on the performance of different schemes is also investigated. Here we set the crosspoint buffer size to $B = 40$, fix the maximum length to $L = 1000$, then change the Hurst parameter $0.6 \leq H \leq 0.9$.

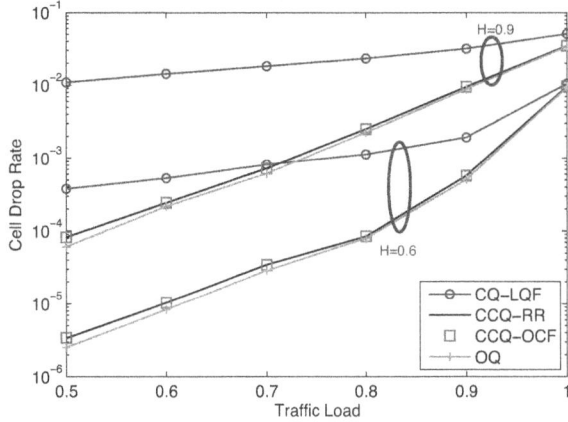

Figure 9: Cell drop rate of 32×32 **switches with** $B = 40$ **under uniform bursty traffic with** $0.6 \leq H \leq 0.9$, $L = 1000$ **and** $0.5 \leq \lambda \leq 1.0$.

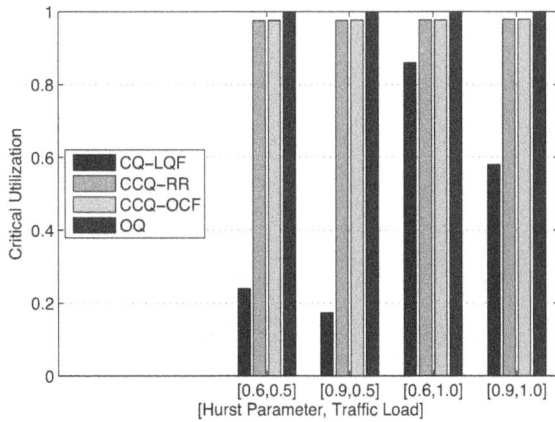

Figure 10: Critical buffer utilization of 32×32 **switches with** $B = 40$ **under uniform bursty traffic with** $0.6 \leq H \leq 0.9$, $L = 1000$ **and** $0.5 \leq \lambda \leq 1.0$.

Simulation results in Fig. 9 and Fig. 10 show that *CQ-LQF* performs worse when the traffic is more bursty but lower loaded. On the other hand, the proposed *CCQ-OCF* and *CCQ-RR* schemes are not affected much (as compared with *OQ*), demonstrating their robustness against different burstiness levels. The underlying reason is that the small crosspoint buffers become less capable to sustain the traffic fluctuations as the incoming cells become more bursty and intermittent, and depend more on load balancing and deflection routing to smooth the traffic. As a conclusion, the proposed schemes gain relatively larger advantages under highly bursty and intermittent traffic.

5.4 Impact of Large Switch Size

Till now, we have examined the performances of 32×32 switches under various traffic patterns. What if the switch becomes larger, i.e., with more input and output ports? Here we consider a large 128×128 CQ switch, and investigate the impact of large N on different switch configurations.

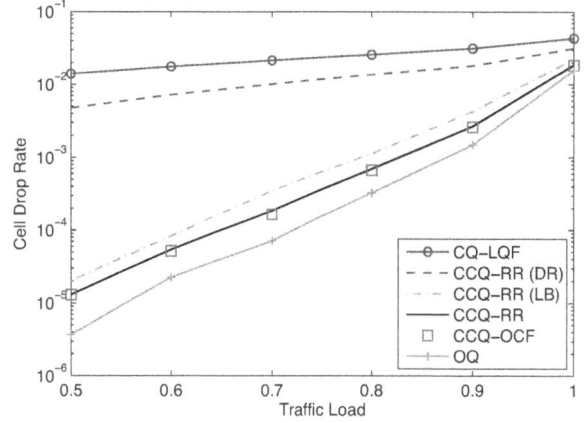

Figure 11: Cell drop rate of 128×128 **switches with** $B = 10$ **under uniform bursty traffic with** $H = 0.75$, $L = 1000$ **and** $0.5 \leq \lambda \leq 1.0$.

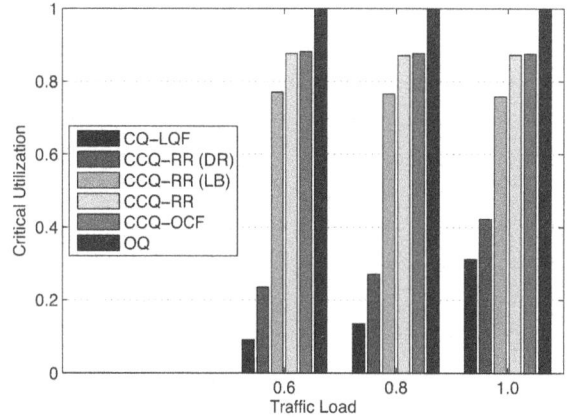

Figure 12: Critical buffer utilization of 128×128 **switches with** $B = 10$ **under uniform bursty traffic with** $H = 0.75$, $L = 1000$ **and** $0.5 \leq \lambda \leq 1.0$.

From Fig. 11 and Fig. 12, we can see that the legacy *CQ-LQF* method suffers from a higher cell drop rate due to a smaller crosspoint buffer size. *CCQ-OCF* and *CCQ-RR* gain a larger advantage over *CQ-LQF* in this case, but are inferior to *OQ* due to increased difficulties in buffer-sharing along longer daisy chains of smaller crosspoint buffers. Notwithstanding this issue, we may still infer that the proposed schemes are more suitable for large switches with small crosspoint buffers. We also notice that deflection routing becomes much less effective when N grows larger, because its buffer-sharing effect is local and requires more time to propagate (due to the constraint of $B(i, j) \geq B(i + 1, j) + 1$) than load balancing.

A larger switch size of $N = 128$ needs additional buffer space to achieve the same satisfactory cell drop rates as before. For

CQ-LQF, the total buffer space required to achieve similar performances may scale as $\Theta(N^2)$, since each crosspoint buffer should at least tolerate a single burst, whose length does not shrink much as N increases. By contrast, for *CCQ-OCF*, *CCQ-RR* and *OQ*, the total buffer space required to achieve similar performances does not scale so poorly. Compared with the case in Section 5.1, even though the switch size is 4 times larger than before, the aggregated buffer size for each output does not change at all, i.e., $N \times B = 128 \times 10 = 32 \times 40 = 1280 cells$, and the total buffer space of all outputs scales as $\Theta(N)$.

For an OQ switch, this is easy to understand, since the traffic load at each output always equals to $0.5 \leq \lambda \leq 1$, and does not change with different switch sizes. If we assume Poisson arrival processes at each input, the output queue length distributions are always the same, irrespective of N. The LRD arrival process is certainly different, but as long as the burst length is not too large compared with the output buffer size, the performance of *OQ* stays approximately the same. *CCQ-OCF* and *CCQ-RR* may also share the segregated crosspoint buffers efficiently. That is why the total amount of buffers in each daisy chain stays almost the same for a given traffic level and loss performance.

5.5 Real Internet Traces

Finally, we test the proposed schemes using real Internet traces. In the simulation, a different CAIDA OC-192 (10Gbps) trace [15] is fed into each input port of the CQ switch. The incoming packets are hashed according to a fixed look-up table, so that the outputs work at approximately the same load. Variable-length IP packets are fragmented into fixed-length cells of 64byte each, which is a common value used in Internet core switches.

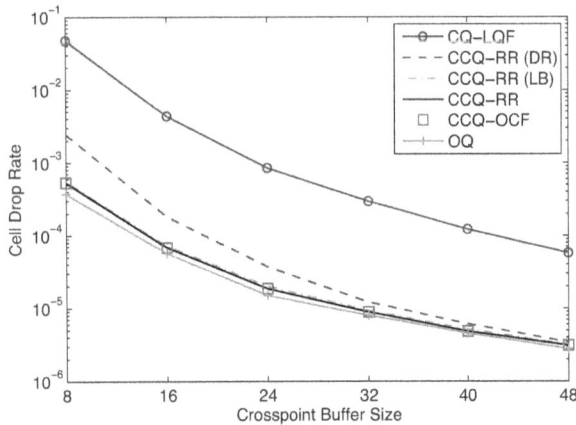

Figure 13: Cell drop rate of 32×32 switches with $8 \leq B \leq 48$ under real Internet traces with $\lambda \approx 0.45$ and $H \approx 0.75$.

First we consider a 32×32 CQ switch, and use the original traces from CAIDA with an average traffic load of $\lambda \approx 0.45$ and a measured Hurst parameter of $H \approx 0.75$. The simulation period is $T = 10^7$ time slots. Examination of the packet headers reveals that over $50,000$ flows with different source/destination IP addresses are multiplexed into each link during the simulation period. As displayed in Fig. 13, *CCQ-OCF* and *CCQ-RR* ensure very low cell drop rates, about 10 to 100 times lower than the basic LQF-based CQ switch, and close to the OQ switch with the same total buffer space. To support an average cell drop rate of 10^{-5}, only about $32 \times 32 \times 40 \times 64byte = 2.5Mbyte$ total buffer space is needed, thus even larger Internet core switches can be accommodated onto

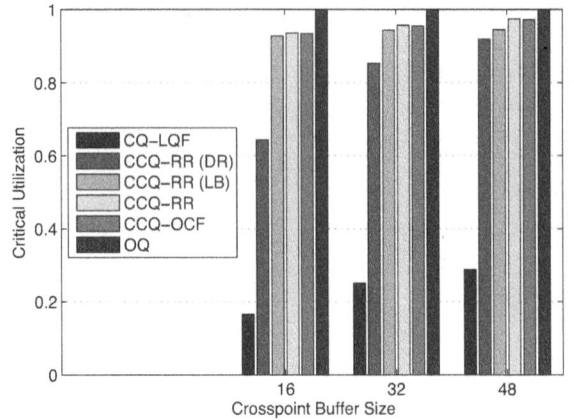

Figure 14: Critical buffer utilization of 32×32 switches with $8 \leq B \leq 48$ under real Internet traces with $\lambda \approx 0.45$ and $H \approx 0.75$.

a single chip. Also note that deflection routing contributes more as the crosspoint buffer size grows larger, and becomes almost as effective as load balancing when $B = 48$.

Comparison of the critical buffer utilizations show that all schemes achieve higher buffer utilizations with larger crosspoint buffers. However, *CCQ-OCF* and *CCQ-RR* can achieve about 90% buffer utilization when the buffer size is as small as $B = 8$, whereas *CQ-LQF* requires a much larger buffer size to smooth the traffic. Deflection routing also requires a modest buffer size to achieve high utilization, as predicted in Section 3.2, with a boost in the buffer utilizations after load balancing is applied as well. The advantages of the proposed schemes are clearly demonstrated.

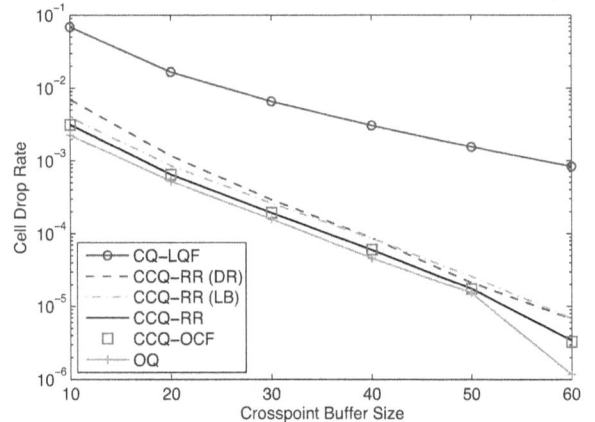

Figure 15: Cell drop rate of 128×128 switches with $10 \leq B \leq 60$ under real Internet traces with $\lambda = 0.7$.

We then consider a larger 128×128 CQ switch. We use the same Internet traces (with a different look-up table), but reduce the core switching speed and place throttles right before the input ports so that the system effectively works at a higher traffic load of $\lambda = 0.7$. The cell drop rates and buffer utilizations are shown in Fig. 15 and Fig. 16 respectively. In this case, a much larger buffer space, $128 \times 128 \times 60 \times 64byte = 60Mbyte$, is required to achieve the same cell drop rate of under 10^{-5}, but it is still feasible using state-of-art ASIC technologies [25, 20, 7]. The relative

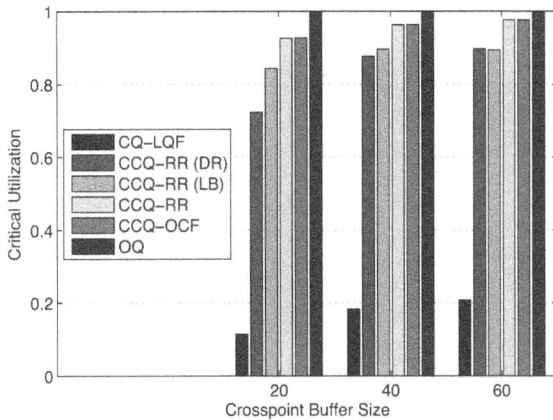

Figure 16: Critical buffer utilization of 128×128 **switches with** $10 \leq B \leq 60$ **under real Internet traces with** $\lambda = 0.7$.

performance gains of the proposed schemes over *CQ-LQF* are even higher in this case. Also note that the deflection routing mechanism in *CCQ-RR* works better than load balancing in this case, showing the robustness of such a reactive strategy against varying burstiness and non-uniformity.

6. CONCLUSION

In this paper, we address the crucial buffering constraints in a single-chip CQ switch. At the cost of some modest hardware modifications and memory speedup, we make it possible for the segregated buffers at different crosspoints to be dynamically shared along daisy chains, effectively mimicking an OQ switch. At the same time, the proposed scheduling schemes can also maintain the correct packet ordering with low complexity, which is also important in designing packet-switched networks. Exploiting the benefits of load balancing and deflection routing, we significantly improve the buffer utilization and reduce the packet drop rate, especially for large switches with small crosspoint buffers under bursty and non-uniform traffic. Extensive simulations have been performed to demonstrate that the memory sizes available using current ASIC technology is sufficient to deliver a satisfactory packet loss performance with a single-chip CQ architecture.

As part of the future work, we may improve the scheduling algorithms, and also mathematically evaluate or bound their performances. In addition, we will explore other efficient buffering techniques, and push the limits of buffer sharing across different outputs to achieve higher multiplexing gains. Other extensions like the support of QoS for packets with different priorities and multicasting are also worthy of investigation.

7. ACKNOWLEDGEMENT

This work is supported by the New York State Center for Advanced Technology in Telecommunications (CATT) and the Wireless Internet Center for Advanced Technology (WICAT) at Polytechnic Institute of New York University, Brooklyn, NY, USA.

8. REFERENCES

[1] G. Appenzeller, I. Keslassy, and N. McKeown. Sizing router buffers. In *Proc. ACM SIGCOMM*, pages 281–292, 2004.

[2] F. Begtasevic and P. V. Mieghen. Measurements of the hopcount in Internet. In *Passive and Active Measurements*, pages 183–190, 2001.

[3] N. Beheshti, Y. Ganjali, R. Rajaduray, D. Blumenthal, and N. McKeown. Buffer sizing in all-optical packet switches. In *Proc. Optical Fiber Communication Conference (OFC)*, Mar. 2006.

[4] E. Blanton and M. Allman. On making TCP more robust to packet reordering. *ACM SIGCOMM Comput. Commun. Rev.*, 32:20–30, 2002.

[5] M. Borella, D. Swider, S. Uludag, and G. Brewster. Internet packet loss: measurement and implications for end-to-end Qos. In *Proc. ICPP*, pages 3–12, 1998.

[6] J. Brassil and R. Cruz. Bounds on maximum delay in networks with deflection routing. *IEEE Trans. Parallel Distrib. Syst.*, 6(7):724–732, 1995.

[7] C. Minkenberg and R.P. Luijten and F. Abel and W. Denzel and M. Gusat. Current issues in packet switch design. *ACM SIGCOMM Comput. Commun. Rev.*, 33(1):119–124, 2003.

[8] CAIDA. CAIDA Internet Data - Realtime Monitors. Online: http://www.caida.org/data/realtime.

[9] C. Chang, D. Lee, and Y. J. Shih. Mailbox switch: a scalable two-stage switch architecture for conflict resolution of ordered packets. In *Proc. IEEE INFOCOM*, volume 3, pages 1995–2006, 2004.

[10] C.-S. Chang, D.-S. Lee, and Y.-S. Jou. Load balanced Birkhoff-von Neumann switches, Part I: one-stage buffering. *Comput. Commun.*, 25:611–622, 2002.

[11] C.-S. Chang, D.-S. Lee, and C.-M. Lien. Load balanced Birkhoff-von Neumann switches, Part II: multi-stage buffering. *Comput. Commun.*, 25:623–634, 2002.

[12] X. Chen, L. Xing, and Q. Ma. A distributed measurement method and analysis on Internet hop counts. In *Proc. ICCSNT*, volume 3, pages 1732–1735, 2011.

[13] S.-T. Chuang, A. Goel, N. McKeown, and B. Prabhakar. Matching output queueing with a combined input/output-queued switch. *IEEE J. Sel. Areas Commun.*, 17(6):1030–1039, 1999.

[14] S.-T. Chuang, S. Iyer, and N. McKeown. Practical algorithms for performance guarantees in buffered crossbars. In *Proc. IEEE INFOCOM*, volume 2, pages 981–991, 2005.

[15] K. Claffy, D. Anderson, and P. Hick. The CAIDA Anonymized 2011 IPv6 Day Internet Traces. Online: http://www.caida.org/data/passive/passive_2011_ipv6day_dataset.xml.

[16] R. G. Clegg and M. Dodson. Markov chain-based method for generating long-range dependence. *Phys. Rev. E*, 72(2), Aug. 2005.

[17] W. Dobosiewicz and P. Gburzynski. A bounded-hop-count deflection scheme for Manhattan-street networks. In *Proc. IEEE INFOCOM*, volume 1, pages 172–179, 1996.

[18] P. Gupta and N. McKeown. Designing and implementing a fast crossbar scheduler. *IEEE Micro*, 19(1):20–28, 1999.

[19] D. Hayes and G. Armitage. Improved coexistence and loss tolerance for delay based TCP congestion control. In *Proc. IEEE LCN*, pages 24–31, 2010.

[20] International Technology Roadmap for Semiconductors (ITRS). Executive summary. 2011.

[21] S. Jaiswal, G. Iannaccone, C. Diot, J. Kurose, and D. Towsley. Measurement and classification of out-of-sequence packets in a tier-1 IP backbone. *IEEE/ACM Trans. Networking*, 15(1):54–66, 2007.

[22] J. Jaramillo, F. Milan, and R. Srikant. Padded frames: a novel algorithm for stable scheduling in load-balanced switches. *IEEE/ACM Trans. Networking*, 16(5):1212–1225, 2008.

[23] Y. Kanizo, D. Hay, and I. Keslassy. The crosspoint-queued switch. In *Proc. IEEE INFOCOM*, pages 729–737, 2009.

[24] M. Karol, M. Hluchyj, and S. Morgan. Input versus output queueing on a space-division packet switch. *IEEE Trans. Commun.*, 35(12):1347–1356, 1987.

[25] M. Katevenis, G. Passas, D. Simos, I. Papaefstathiou, and N. Chrysos. Variable packet size buffered crossbar (CICQ) switches. In *Proc. IEEE ICC*, volume 2, pages 1090–1096, 2004.

[26] I. Keslassy and N. McKeown. Maintaining packet order in two-stage switches. In *Proc. IEEE INFOCOM*, volume 2, pages 1032–1041, 2002.

[27] D. Lawrie and D. Padua. Analysis of message switching with shuffle/exchange in multiprocessors. In *Proc. Workshop on Interconnection Networks*, pages 116–123, 1980.

[28] Y. Li, S. Panwar, and H. J. Chao. On the performance of a dual round-robin switch. In *Proc. IEEE INFOCOM*, volume 3, pages 1688–1697, 2001.

[29] N. Maxemchuk. Routing in the Manhattan street network. *IEEE Trans. Commun.*, 35(5):503–512, 1987.

[30] N. Mckeown. The iSLIP scheduling algorithm for input-queued switches. *IEEE/ACM Trans. Networking*, 7:188–201, 1999.

[31] M. Nabeshima. Performance evaluation of a combined input-and-crosspoint-queued switch. *IEICE Trans. Commun.*, E83-B(3):737–741, 2000.

[32] M. Radonjic and I. Radusinovic. Average latency and loss probability analysis of crosspoint queued crossbar switches. In *Proc. ELMAR*, pages 203–206, 2010.

[33] M. Radonjic and I. Radusinovic. Buffer length impact to crosspoint queued crossbar switch performance. In *Proc. IEEE MELECON*, pages 119–124, 2010.

[34] M. Radonjic and I. Radusinovic. Impact of scheduling algorithms on performance of crosspoint-queued switch. *Ann. Telecommun.*, 66(5-6):363–376, 2011.

[35] S. Rewaskar. *Real world evaluation of techniques for mitigating the impact of packet losses on TCP performance.* PhD thesis, Univ. North Carolina, Chapel Hill, 2008.

[36] L. Tassiullas and A. Ephremides. Stability properties of constrained queueing systems and scheduling policies for maximum throughput in multi-hop radio networks. *IEEE Trans. Autom. Control*, 37(12):1936–1949, 1992.

[37] D. Wischik. Buffer requirements for high-speed routers. In *Proc. ECOC*, volume 5, pages 23–26, 2005.

[38] D. Wischik and N. McKeown. Part I: buffer sizes for core routers. *ACM SIGCOMM Comput. Commun. Rev.*, 35(3):75–78, 2005.

[39] S. Ye, Y. Shen, and S. Panwar. DISQUO: A distributed 100% throughput algorithm for a buffered crossbar switch. In *Proc. IEEE Workshop on HPSR*, 2010.

Efficient Traffic Aware Power Management for Multicore Communications Processors

Muhammad Faisal Iqbal
University of Texas at Austin
faisaliqbal@utexas.edu

Lizy K. John
University of Texas at Austin
ljohn@ece.utexas.edu

ABSTRACT

Multicore communications processors have become the main computing element in Internet routers and mobile base stations due to their flexibility and high processing capability. These processors are designed and equipped with enough resources to handle peak traffic loads. But network traffic varies significantly over time and peak traffic is observed very rarely. This variation in amount of traffic gives us an opportunity to save power during the low traffic times. Existing power management schemes are either too conservative or are unaware of traffic demands. We present a predictive power management scheme for communications or network processors. We use a traffic and load predictor to pro-actively change the number of active cores. Predictive power management provides more power efficiency than reactive schemes because it reduces the lag between load changes and changes in power adaptations since adaptations can be applied before the load changes. The proposed scheme also uses Dynamic Voltage and Frequency Scaling (DVFS) to change the frequency of the active cores to adapt to variation in traffic during the prediction interval. We perform experiments on real network traces and show that the proposed traffic aware scheme can save up to 40% more power in communications processors as compared to traditional power management schemes.

Categories and Subject Descriptors

C.2.6 [**Internetworking**]: Routers; C.1.4 [**Processor Architectures**]: Parallel Architectures

General Terms

Performance, Design

Keywords

Power Management, Network Processors, P-States, C-States

1. INTRODUCTION

The Internet infrastructure contributes to about 2% of world's energy consumption [13, 49, 32]. This contribution is likely to

increase in future with exponential growth in number of users and high bandwidth services. According to different studies, routers are major contributors of power in Internet infrastructure [16, 32]. The energy consumption in routers is reaching the limits of air cooling [16, 14]. For example, a fully configured Cisco CRS-1 router can consume up to one megawatt of power [14]. A typical router has a set of line cards and each line card has one or more network processors [8, 2, 42]. Multicore network processors have become the major computing element in routers due to their flexibility and high processing capability. FreeScale's P4080 [5], Intel IXP [12] and Tilera processors [10] are some examples of multicore processors being used in networking applications. Power consumption of a single line card can reach up to 500 Watts [2, 6]. Modern routers can have hundreds of line cards. For example, a CISCO CRS-1 router can house up to 1152 line cards in different chassis. These line cards are densely packed in routers. High power consumption can result in high temperature of parts and failure due to thermal stress. Such failures affect the reliability and availability of networks. This results in lower quality of service and increased expenditures in replacement parts. High power consumption of equipment leads to higher cooling costs and results in increased operational expenditure of the network. According to Erricson's vice president, "The cost of electricity over lifetime of network equipment is more than the cost of network equipment itself" [23]. With increasing traffic rate demands and computational complexity, the number and complexity of cores in network processors are on the rise resulting in more and more power consumption. Tight power budgets and dense integration requirements call for design of power efficient network processors.

The multicore packet processing systems are usually designed and provisioned with enough resources to satisfy peak traffic load. But network traffic varies with time and reaches the peak value for only a small portion of time. Figure 1 shows traffic observed over two days by CAIDA monitor [20] at Internet backbone in Chicago. There is a huge variation in packet rates and thus different processing requirements at different times of the day. Most of the time the traffic rate is below the maximum traffic and we do not need to run the processors at full capabilities. The low activity periods can be exploited to save power in network processors by running them in low power modes and/or by turning off some processing cores.

We propose a predictive power management scheme whereas previous schemes proposed for Network Processors [42, 41] are reactive in nature. Predictive power management provides more power efficiency than reactive schemes because it reduces the lag between load changes and changes in power adaptations since adaptations can be applied before the load changes. Power management policies used in general purpose processors are unaware of traffic

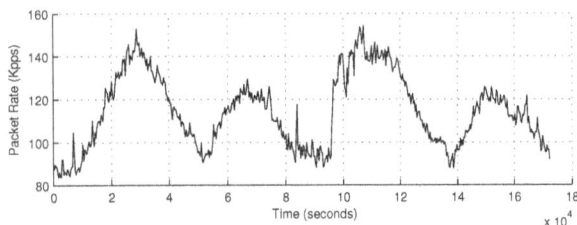

Figure 1: Variation in traffic arrival rates(Kilo Packets Per Sec) over 2 days at equinix-chicago Internet backbone

demands and provide sub-optimal results for network processors. In this paper we make following contributions:

- We propose a predictive power management scheme for communications processors which uses a low cost traffic and load predictor.

- The proposed scheme aims at reducing both active and idle power by utilizing P and C-states of the processor.

- We propose a new parameter called *traffic_factor* which combines traffic prediction and application processing requirements into a single parameter for efficiently predicting required number of active cores.

- We perform experiments on real network traces and show that the proposed scheme can save up to 40% more power as compared to traditional schemes (Section 5).

2. BACKGROUND

Modern processors are equipped with capabilities to save power during active periods (P-states) and during idle periods (C-states). P and C-states are part of an open industrial standard called Advanced Configuration and Power Interface (ACPI) [33]. ACPI was proposed by Intel, Microsoft and Toshiba to facilitate the development of Operating System based power management. Policies to manage P and C-states are usually implemented as Operating System modules. Almost all modern operating systems have such modules or governors. In this section we give short background of P and C-states. We also provide an overview of existing policies for C and P-states and explain why these policies may result in un-optimal power management in case of network processors.

2.1 Active Power Management Using P-states

P-states refer to the different performance states of the processor and provide choices for different power/performance points to adapt to dynamic processing requirements. P-states are an implementation of Dynamic Voltage and Frequency Scaling (DVFS) and are aimed at reducing dynamic power. Recall that dynamic power is given as

$$P_{dynamic} = k \times v^2 \times f \qquad (1)$$

where k is a workload and processor dependent parameter determined by switching capacitance and activity of the processor. Dynamic power can be saved if we lower frequency and voltage. P-states define frequency and voltage levels of the processor so that during times of low processing requirements, frequency and voltage are lowered to save power and energy. P-states are named numerically form P_0 to P_N. P_0 is the highest performance state. Performance and power consumption reduces with increasing P-state numbers. Table 1 shows example P-states for a typical processor

[18]. These P-states are similar to the P-states of AMD Opteron Processor [1].

P-state	Frequency	Voltage
P0	$F0$	$V0$
P1	$F0 \times 0.85$	V0 x 0.96
P2	$F0 \times 0.75$	V0 x 0.90
P3	$F0 \times 0.65$	V0 x 0.85
P4	$F0 \times 0.50$	V0 x 0.80

Table 1: Example P-states of a typical processor

2.2 Idle Power Management Using C-states

Processor's C-states represent the capability of processor to save power during idle periods. States are named numerically starting from C_0 to C_N, where C_0 represents the active state. As the C-state number increases, the power consumption of the processor decreases and wakeup latency increases. Designers employ different techniques to implement C-states. Low latency techniques include clock and fetch gating whereas high latency techniques include voltage scaling and power gating. Table 2 shows an example of C-states. The table shows only three C-states. Modern processors have a large number of C-states. For example, Intel Core 2 Duo has five C-states [50] and some processors even have up to eight C-states [3]. Wakeup latency of processors increases as we

C-state	Response Latency	Relative Power
C0	0	100%
C1	10 uS	40%
C2	100 uS	5%

Table 2: Example C-states

move to deeper sleep states. It only makes sense to enter a C-state if inactive time is equal or greater than break even time T_{BE}[17]. The break even time is composed of two terms: the total transition time (i.e., $T_{tr} = T_{enter} + T_{exit}$) and minimum time that has to be spent in that state to compensate for the additional power during transition. If power consumption during transition is less than or equal to on-state power (this is what we assume in this study) then $T_{BE} = T_{tr}$. This break even time is usually used as a threshold for transitioning into deeper states. For Table 2, the break even time will be 20 μs for C1 and 200 μs for C2. Also note that modern operating systems support a tick-less kernel i.e., idle CPUs do not have to respond to periodic ticks. These CPUs are allowed to remain idle and are woken up by interrupts when a new job arrives for them. We assume such a tick-less kernel in this study.

2.3 Policies for C-state Management

2.3.1 Using Idle Time

Most implementations of C-state management use *Fixed Timeout* policy. For example, Ladder governor in Linux is used to implement C-state management [11, 9]. This governor uses elapsed idle time to predict the total duration of the current idle period. When a processor becomes idle a counter starts. This idle time counter is then compared with pre-defined thresholds. When the counter reaches $C1_{th}$ value, the system is forced into a sleep state C1. The counter continues counting until the processor is woken up by an external event. If the counter reaches $C2_{th}$, the system transitions to C2 and so on. The processor keeps transitioning to deeper C-states until it reaches the lowest power state or it is woken up by an external event. The CPU starts from C1 again when it becomes idle the next time.

As described in Section 2.2, the threshold values are decided based on break even times which are of the order of hundreds of microseconds. This scheme works well to exploit idle time in general purpose applications e.g., time waiting for user input or response from I/O subsystem where the waiting times are very high. But in case of NPs, the inter-packet arrival times viewed by multiple cores are usually smaller than these thresholds even if the number of active cores is more than the required to sustain a certain amount of traffic. In other words, this scheme might be too conservative and wastes a lot of power saving opportunity which could be exploited if we directly have information about traffic and processing demands. Furthermore, this is a reactive scheme and some power saving opportunity is lost in waiting from threshold times to elapse. In contrast, we present a predictive scheme which uses direct information about traffic to more efficiently manage power.

2.3.2 Using Number of Idle Threads

The scheme proposed by Luo et al. [42, 43] targets multicore NPs and is the closest related work to our proposal. This scheme monitors the number of idle threads in the thread queue during an interval. If the number of idle threads is more than the required number of cores for majority of the interval, it shuts down the additional threads and cores. Using the number of idle threads works fine if we assume that each processor runs at maximum frequency. But if each core is allowed to change its frequency, the number of idle threads does not effectively represent the load. For example, consider a situation where two processors are active and running at half of the maximum frequency. These processors will be utilized 100% of the time to handle a traffic which a single processor could handle at full speed. But the threads running on slow cores will never enter the thread queue and hence we will never be able to turn off any cores. Hence the number of idle threads does not represent processing requirements in this situation.

Furthermore, this scheme is also a reactive scheme and uses only clock gating. The transition overheads for clock gating are small so the reactive scheme works well. Since this scheme is targeted for 280 nm, leakage power is not a big issue and clock gating works fine. But in modern technologies, the power consumption is usually dominated by leakage power and hence it is important to utilize deep sleep states which have additional power savings even beyond DVFS and clock gating. These deep sleep states have high overheads in terms of transition delays and reactive scheme may result in opportunity loss for power saving. A predictive scheme is needed so that the lag between power adaptation and load changes is minimized.

2.4 Policies for P-State Management

2.4.1 Using Processor Utilization

P-state management policies are generally aimed at saving power during performance-insensitive phases of programs. For example, power can be saved during memory-bound phase of a program by reducing clock frequency. Many implementations of these policies use processor utilization to drive P-states. These policies try to maintain processor utilization within a certain range [18, 15, 30, 9]. Processor utilization or activity level represents the ratio of code execution time (active time) to wall clock time (active + idle time) i.e.,

$$utilization = \frac{Time_{active}}{(Time_{active} + Time_{idle})} \quad (2)$$

Listing 1 shows an implementation of Linux "Ondemand" governor [11]. The algorithm monitors processor utilization for an interval

and then makes a decision whether to increase or decrease the frequency.

```
#define up_threshold 0.90

for (each sampling interval){

  if (utilization > up_threshold)
     freq = max_freq;

  else
     freq = next_lower_freq;
}
```

Listing 1: Linux ondemand frequency Governor

Ondemand governor is the most aggressive governor in Linux implementations because it tries to settle to lowest frequency in case of zero load and will settle to the highest frequency at peak load. Another relevant governor is a conservative governor [11] which is similar to power management module found in Vista [15]. This governor tries to maintain the processor utilization within a range say 0.3 to 0.6.

The policy which uses CPU utilization does not factor in traffic demands and suffers from several pitfalls. First, CPU utilization is a function of mixture of events (e.g., performance of memory, I/O devices etc.) and does not directly indicate load requirements. Second, if this policy is too conservative, it will lose a lot of opportunities to save power and if it is too aggressive then it may result in performance degradation in terms of traffic loss.

2.4.2 Minimizing Energy Per Instruction

Herbert et al. proposed a greedy search method to minimize Energy Per Instruction (EPI) [30, 31] for CMPs. This method is an extension of the technique proposed by Magklis et al. [44]. The P-state controller attempts to operate at frequency level which minimizes EPI assuming EPI is a bath-tub shaped function of voltage and frequency levels. This algorithm assumes the availability of current sensors which help to approximate EPI. After each interval, the controller compares current EPI with the EPI of previous interval. If EPI is improved, the controller makes a move in the same direction as last one. If the EPI has increased, it is assumed that controller has overshot the optimal frequency level. It makes a transition in opposite direction as the last one and stay there for N = 5 intervals. After the holding period the controller continues exploration in a direction opposite to one which preceded the hold. This type of scheme is un-aware of traffic demands and may result in running at lower frequency than needed and may result in extra packet loss.

3. PREDICTIVE TRAFFIC AWARE POWER MANAGEMENT (PTM)

An efficient power management scheme for NP has to make the following decisions:

1. Predicting load for the next interval

2. Deciding required number of active cores N_{opt} for the predicted load

3. Deciding frequency f_i for each active core i.

In this section we provide details of our proposed PTM scheme and explain how PTM makes the three above mentioned decisions.

3.1 Prediction of Load

The computational requirements for NPs depend on both traffic arrival rate and the computation complexity of the applications. We propose a new parameter called *Traffic_factor* which combines both traffic rate and computational complexity to give a true estimation of load to be handled by the NP.

3.1.1 Traffic Prediction

PTM uses Double Exponential Smoothing Predictor (DES) for traffic prediction. A more complex predictor may result in greater accuracy in some situations but a low overhead predictor is desirable for energy efficiency. Recently, Iqbal et al. [34] studied many real network traces and have shown that Double Exponential Smoothing (DES) predictor is a low overhead predictor with accuracy comparable to complex predictors like Artificial Neural Network (ANN) or Wavelet transform based predictors. This makes DES very suitable for our application. This study uses DES predictor but designers can use the proposed scheme with any other predictor based on their requirements. We will give a brief introduction to DES predictor in this section and a short comparison of different predictors is presented in Section 5.2.

3.1.2 Double Exponential Smoothing (DES) Predictor

Exponential Smoothing assigns exponentially lower weights to older observations. Single exponential smoothing does not work well when there is a trend in data [4]. Trend means that the average value of the time series increases or decreases with time. However, Double Exponential Smoothing adds trend component for estimation and is considered more appropriate for data with trends. The equation for DES based prediction for a time series $X(t)$ is given as

$$X_{t+1} = S_t + b_t \qquad (3)$$

where

$$S_t = \alpha X_t + (1 - \alpha)(S_{t-1} + b_{t-1}) \qquad (4)$$

and

$$b_t = \gamma(S_t - S_{t-1}) + (1 - \gamma)b_{t-1} \qquad (5)$$

S_t and b_t are smoothed value of stationary process and trend value respectively. S_t and b_t are added together to get the prediction for next interval. α defines the speed at which older values of S_t are damped. When α is close to 1, dampening is quick and when α is close to 0, dampening is slow. γ is similar smoothing constant for b_t. The values of α and γ are obtained using non-linear optimization techniques and are learned during the training phase of the predictor. Note that this is a very low cost predictor. It requires only four registers for storing α, γ, S_{t-1} and b_{t-1}. To make a prediction it requires six multiplications and four addition operations. This low overhead and reasonable accuracy makes it an appropriate predictor for the purpose of power management.

3.1.3 Traffic Factor

We propose a new parameter called *Traffic_Factor* which combines traffic rate and application's processing capability to give a true estimation of processing requirement. Traffic rate is the rate at which packets arrive at the input and is represented as Packets Per Second (PPS). We are naturally tempted to use this parameter directly to exploit traffic variability. But different applications have different processing requirements and hence can support different packet rates i.e., a complex application will require more resources to sustain a particular traffic rate when compared to a simple application. This means packet rate directly cannot be used for power

management purposes. But if we can incorporate applications processing requirements with input packet rate, we can use this information to drive the power management scheme. If we know Cycles Per Instruction (CPI) and Instructions Per Packet (IPP), we can directly find Cycles Per Packet (CPP) i.e., $CPP = IPP \times CPI$. We define *Traffic_Factor* as

$$Traffic\_Factor = \frac{PPS_{predicted} \times CPP}{(max\_cpu\_MHZ \times num\_cores)} \qquad (6)$$

where $PPS_{predicted}$ is the traffic predicted using DES predictor explained above. *Traffic_Factor* incorporates both application performance requirements and traffic rate into a single parameter and is an excellent parameter for use in power management schemes. Note that CPP is independent of the frequency level i.e., cycles per packet will remain constant with changing frequency and hence this parameter does not suffer from the same limitations as the processor utilization used by previous schemes. Processor utilization is a direct function of frequency whereas CPP does not depend upon frequency if we assume that there are limited off-chip memory accesses. This assumption is valid in network processors since the packet processing applications are small and fit into caches and the packet queues are also implemented on on-chip memories [12]. Furthermore, we do not need any additional resources to measure this parameter. Many network processors like P4080 provide performance counters to measure PPS, IPP and CPI directly [7]. Table 3 lists the performance counters which can be used for measuring these parameters.

Metric	Performance Counters	Formula
Core Cycles	CE:Ref:1	CE:Ref:1
Ins. Count	CE:Ref:2	CE:Ref:2
Pkt. Count	SE:Ref:36	SE:Ref:36
CPI	CE:Ref:1 CE:Ref:2	CE:Ref:2/CE:Ref:1
IPP	SE:Ref:36 CE:Ref:2	CE:Ref:2/SE:Ref:36
PPS	SE:Ref:36,CE:Ref:1 CE:Ref:1	SE:Ref:36/A A=(CE:Ref:1/Freq)

Table 3: P4080 Counters to measure CPI, IPP and PPS

3.2 Deciding Number of Active Cores

Required number of active cores P can be directly calculated from the Traffic Factor i.e., $P = Traffic\_Factor \times total\_cores$ and number of active cores are adjusted as shown in Listing 2. The sampling interval used is 500 μS.

```
for (every sampling interval){
    p = traffic_factor * total_cores;

    if (p > active_cores )
        wakeupCores(p - active_cores);

    else if (p < active_cores)
        killCores(active_cores - p);
}
```

Listing 2: Algorithm to decide number of active cores

If P is less than the current number of active cores, we can shutdown all the extra cores. When a core goes into a sleep state, it first goes into C1, where it stays for 2 sampling intervals and then it goes to state C2. In our scheme, we have used C2 as the deepest sleep state. Note that the difference between ladder governor and our proposed scheme is different input parameters. Instead of using idle time, we are using *traffic_factor* to drive C-state management.

In order to wakeup cores, we look at the input queue length in addition to *traffic_factor* (See Listing 3). And if the size of the queue length reaches a certain threshold, we wakeup one of the sleeping cores.

3.3 Finding P-states of Active Cores

If the load predictor over-predicts, or there is variation in traffic during prediction interval, we can make use of DVFS to further save power at smaller timescales. At regular intervals (50 μS), we check the size of the input queue and based on size of the queue we decide whether to increase or decrease the power levels. When a packet arrives at the input interface, the interface control logic stores the packet in the input queue until it is picked and serviced by an available processor. Figure 3 shows a simplified model of a network processor. If the input queue is nearly empty most of the time, we have enough resources to handle the traffic load and if it is near full, it means we need more processing capability. Length of the input queue gives a direct indication of whether we need more resources or not. The algorithm for finding the appropriate P-state is shown in Listing 3. If the queue is nearly empty, we assume that we have excess processing capability and we go to a lower frequency level. When the queue starts to grow, it means that the current processing capability is lower than what is needed and we increase the frequency level.

```
int pstate[numcores];
for (every sampling interval){

  if (avg < low_th){
    core = findMin(pstate);
    pstate[core] = pstate[core]+1;
  }

  else if (avg < high_th){}

  else{
    core = findMax(pstate);
    if(core == -1)
      WakeUpCore(1);
    else
      pstate[core] = pstate[core]-1;
  }
}
```

Listing 3: Algorithm to find P-state values

Note that Listing 3 uses a global governor which makes decisions based on queue size instead of having a separate governor for each core. The array $pstate[numcores]$ holds the p-states of each core. Each core can have five possible P-states similar to Table 1. The function $findMin()$ and $findMax()$ return indices of fastest and slowest cores respectively. In case the queue size is less than the threshold value low_{th}, we lower the frequency of the fastest core and in case it is higher than the $high_{th}$ we increase the frequency of the slowest core. Also note that if frequency cannot be increased further, we increase the number of cores. This allows us to adjust to dynamic variation in traffic during the interval or under prediction and helps us avoid dropping any packets.

3.3.1 Measuring Queue Length

We use the average queue length during the sampling interval to make decision about choosing the appropriate P-states. We use a low pass filter to calculate the average queue size as proposed in the RED algorithm [24]. Thus short term increase in traffic which results from bursty traffic or transient network congestion does not affect the average queue length. The filter used is exponential smoothing filter and is given as

$$avg = \alpha \times qlength + (1 - \alpha) \times avg \qquad (7)$$

where qlength is the instantaneous size of the queue and α defines the speed at which older values are dampened. We use an α of 0.025 in this study. Many congestion control algorithms like RED [24], rely on occupancy of the input queue and modern network processors provide hardware support for these congestion control algorithms. P4080 processor provides a dedicated hardware (QMan) for queue management. QMan implements a variant of RED algorithm and keeps track of the input queue length. Thus queue length can be used for power management purposes and it does not require additional resources since it is already monitored for congestion control purposes. Even if it is not readily available in any processor, it is easy to add functionality in the interface logic to keep track of the queue length or simple dedicated hardware to do the purpose.

3.3.2 Deciding Threshold Values

The algorithms proposed in Listing 2 and 3 monitor length of the input queue and compare it with the predefined threshold values to decide the state of the processor. We allow the thresholds to change dynamically to adjust to changing traffic load and thus do not need to find a common value for all traffic rates and applications. The optimum value of $high_{th}$ depends on the wake up time for a core in deep sleep state. When qlength reaches this threshold value and all the active processors are running at highest frequency, we need to turn on additional cores in order to avoid dropping packets. Assuming the wakeup time of 200 μS, we need to find extra buffer space which is enough such that no additional packets are dropped before an additional core is on line. In our experiments we found the maximum packet rate to be 200 KPPS. Assuming this is the worst case increase in the packet rate, we need an additional buffer space for 40 packets before the core is on line. If maximum queue size is Q_{max}, we use $Q_{max} - 40$ as the value of $high_{th}$. We have used a Q_{max} value of 80 in our simulations. The value of low_{th} can be chosen from a wide range. Essentially, it should not be too close to $high_{th}$ to avoid $avg\_qlength$ to oscillate between low and high thresholds. We chose low_{th} to be such that $high_{th} = 4 \times low_{th}$. Essentially, P-state manager tries to keep input queue to be less than 50% full all the time. But if the traffic rate is higher than what P-states can manage, the queue length will increase more than $high_{th}$. When queue length exceeds this value we know that current number of processor are not enough to manage the traffic and an additional processor is woken up. If value of queue length reaches 95%, we know that we turned on a new processor too late and C_{th}, low_{th} and $high_{th}$ value are decreased by 10%. If qlength never reaches C_{th} value for 10 consecutive intervals, we increase all the thresholds by 10%.

4. EVALUATION METHODOLOGY

4.1 Simulation Infrastructure

Our simulation infrastructure is based on SCE simulator from UC Irvine [22] and follows host compiled simulation approach proposed in [27, 28]. This approach pairs a high level functional model with back annotation of statitically determined timing and power estimates in order to achieve fast and accurate simulation. At the highest level, the user application consists of a set of back annotated tasks that are controlled and interact with underlying OS model. Figure 2 shows the back annotation flow which is similar to the flow proposed in [28]. The packet processing applications are compiled

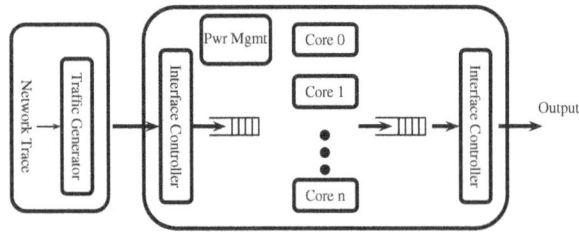

Figure 3: Simulation Model of Network Processor

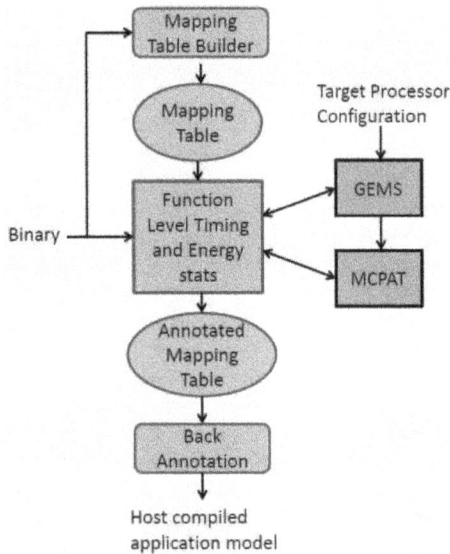

Figure 2: Back Annotation Flow[28]

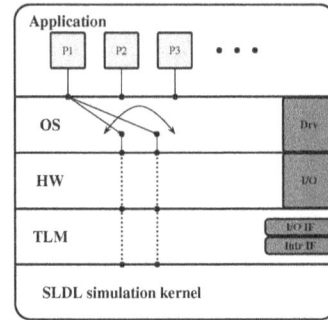

Figure 4: Host compiled multicore platform model [27]

[26] which is similar in design and philosophy to SystemC [29]. Each core in GEMS simulator is modeled after the processing core in P4080 network processor and we model a network processor with 32 processing cores in this study. The detailed configuration of the processing cores is listed in Table 4. For power modeling in MCPAT we use 45nm technology.

L1 Cache	L2 Cache	DRAM	Pipeline	Branch Prediction
16KB	64KB	4 GB	10 stages	YAGS/
2way	4 way		2-wide	64entry BTB

Table 4: Configuration of the processing cores

Figure 3 shows the block diagram of our SpecC based simulation model. The traffic generator module replays the actual trace with accurate timing. Each core is an instance of back annotated application. Power Management module monitors different statistics and control the C and P-states of the cores. The power and delay numbers are looked up from the mapping table of each core based on its P and C-states.

4.2 Network Traces and Benchmarks

We chose seven packet processing applications from NPBench [39] and PacketBench [48]. Table 6 lists the application used for evaluation in this study. We use real network traces from CAIDA's *equinix-chicago* and *equinix-sanjose* monitors [20] as inputs to these benchmarks. We performed experiments with large set of traces but the results of two representative traces (one from each) are presented in this study. All other traces show similar behavior. CAIDA monitors capture the traffic traces on OC192 links. The traces in this data set are of one hour long duration and are captured in year 2011. The network traces are replayed with actual arrival times and packet sizes in the SpecC to model the real traffic scenario. We also present some results with synthetic traces to show the effectiveness of scheme at different traffic rates.

for target platform and are simulated in GEMS full system simulators [45] to gather detailed performance statistics. GEMS simulator provides hooks to find statistics for particular piece of code using special unused opcodes (called magic instructions in GEMS terminology) from the target ISA. We add these magic instructions at the function boundaries in the source code of the application to get per function statistics. Performance statistics from GEMS are fed into MCPAT [40] Power simulator to get power statistics. Power is multiplied by time to get energy for each iteration of the function execution. We use average delay and energy consumption of all iterations of the function for back annotation. MCPAT also allows to get power numbers for different P-states. The final back annotated application has power and delay numbers for each function for each possible P-states. This back annotated application is then simulated on top of an abstract OS model. Figure 4 shows the high level picture of simulation model. The OS model is responsible for controlling the power states of the processor which we study in this work. A Hardware Abstraction Layer (HAL) consists of necessary I/O drivers and implements an interrupt handling mechanism. A HAL combined with an abstract Transaction Level Modeling (TLM) layer provides a high level processor model that interfaces with the TLM backplane. The complete simulation model is based on System Level Design Language (SLDL) simulation kernel. This kernel provides basic concurrency and event handling on the host machine. The SLDL kernel used in this study is based on SpecC

Network Trace	Description
equinix-chicago	One hour long traffic trace at Internet backbone in Chicago
equinix-sanjose	One hour long traffic trace at Internet backbone in Sanjose
Synthetic	One minute long traffic synthetically generated traffic traces at different rates

Table 5: Network Traces used in this study

Application	Description
FRAG	Packet fragmentation Algorithm
IPV4T	IPV4 routing based on trie
IPV4R	IPV4 routing based on radix tree structure
IPSEC	IP Security protocol
MPLS	Multi Protocol Layer Switching is forwarding technology based on short labels
SSLD	Secure Socket Layer dispatcher is an example of content based switching
WFQ	Queue scheduling algorithm to serve packets in order of their finish time

Table 6: Benchmark Applications used in this study

5. RESULTS

We implemented different power management policies for comparison purposes. Table 7 lists the policies under consideration. The C-state policy in Base scheme is similar to Linux ladder governor. The thresholds used are based on Table 2. The P-state management policy is based on Linux ondemand governor explained in Listing 1. Greedy scheme uses similar ladder governor for C-state management but the P-state manager is a greedy algorithm of Section 2.4.2 which tries to minimize EPI. The IdleT policy uses a scheme similar to one proposed by Luo et al. [42, 43] for managing number of active cores based on number of idle threads in the given interval. For fair comparison with PTM, we modified this scheme to go into deeper sleep state if it remain in the current state for two consecutive intervals. Note that the original scheme just made use of clock gating and did not utilize deeper sleep states. We further augmented this scheme to use ondemand governor for frequency management of individual cores. PTM is the proposed scheme based on traffic prediction.

Policy	C-state Policy	P-state Policy
Base	ladder	ondemand
Greedy	ladder	greedy EPI
IdleT	idle threads	ondemand
PTM	Traffic Factor based	Queue Length based

Table 7: Power Management Policies Implemented for Comparison

Figure 5(a) and Figure 5(b) show power savings for the equinix-sanjose and equinix-chicago traces respectively. We use Base policy as the baseline and results are presented relative to Base. Table 8 shows the absolute numbers for power consumption during the trace. We can see that the proposed PTM consistently beats the other policies on all applications. Major portion of the energy saving comes from having right number of active cores. Table 9 shows average number of active cores throughout the trace for different strategies. PTM scheme has the lowest average number of active cores on all benchmarks. *Traffic Factor* allows PTM to estimate minimum number of active cores as compared to other schemes which base their decisions on idle time. PTM changes number of

cores pro actively and thus reduces the lag between load changes and power adaptation. Other schemes, being reactive in nature lose some power saving opportunities since the cores are turned off after some idle time has been elapsed. Base and Greedy schemes result in highest number of active cores. C-state management in Base and Greedy is too conservative in the sense that it waits for break even time to elapse before going to deep sleep state. Also note that number of active cores depends on the p-state management as well. For example, if the active cores are running at lower speed than needed, they will result in activating more cores than required. PTM calculates required number of active cores directly using traffic factor and results in minimum number of cores being active. For IdleT also, the number of active cores is a function of frequency of individual cores and results in higher number of cores than needed. In some situations, Greedy results in more number of cores than any other policy. The reason is that P-state policy in this scheme is unaware of traffic at all. It tries to minimize energy per instruction irrespective of the traffic rate. Thus it results in more cores since individual cores are running at lower frequency. Consider the situation presented in Figure 6. The input packet rate is such that it is required that a single core operates at power level P1 to sustain that traffic.

	equinix-sanjose				sanjose-chicago			
	Base	Greedy	IdleT	PTM	Base	Greedy	IdleT	PTM
FRAG	7.82	7.29	5.93	3.53	7.4	7.32	6.6	3.8
IPV4T	6.73	7.79	6.11	4.10	6.73	6.91	6.23	4.5
IPV4R	29.80	27.80	28.1	25.23	31.5	29.1	28.3	26.1
IPSEC	63.20	60.69	59.80	56.59	71.5	69.3	68.7	62.4
MPLS	45.03	46.6	45.2	40.8	55.5	52.2	51.1	46.3
SSLD	101.6	95.4	96.1	90.1	99.4	94.1	96.1	90.1
WFQ	15.94	16.73	16.02	12.1	15.8	16.3	14.7	11.1

Table 8: Power Consumption in Watts for different schemes

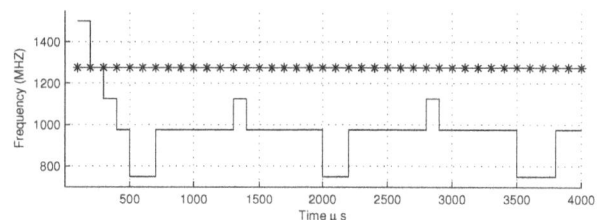

Figure 6: Greedy Algorithm to minimize EPI. Asterisks show optimum power level

Figure 6 shows the response of greedy algorithm described in Section 2.4.2 to this input traffic. Since this greedy algorithm is unaware of the traffic requirement, it continues to move in the direction of lowering EPI and overshoots the required power level and operates at frequency lower than the required frequency for rest of the trace. This scheme will results in running 2 cores instead of 1 for the above situation and will result in more power consumption.

Application	Base	Greedy	IdleT	PTM
Frag	3.10	2.81	2.21	1.30
IPV4T	3.34	3.69	3.10	1.42
IPV4R	7.70	7.90	7.62	5.50
IPSEC	29.10	30.10	29.00	24.30
MPLS	13.81	15.72	13.81	10.31
SSLD	30.53	31.21	31.11	26.11

Table 9: Average Number of active cores

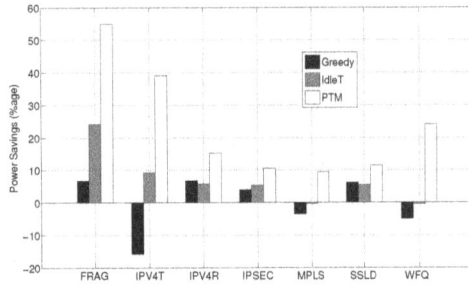

(a) equinix-sanjose   (b) equinix-chicago

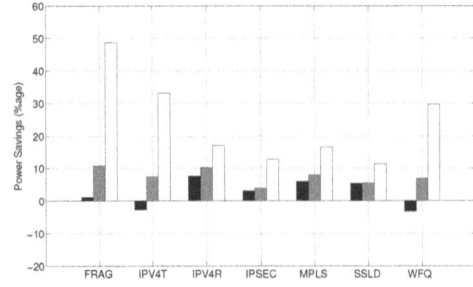

Figure 5: Effectiveness of proposed methodology on two real traces.

5.1 Effectiveness of DVFS

Although most of the benefit comes from having right number of active cores, the ability of individual cores to change frequency also provide some power benefits. Figure 7 presents a comparison of proposed PTM scheme with and without the capability of DVFS. In the scheme without DVFS, the number of cores are controlled by the traffic aware scheme and there is no DVFS i.e., all the active cores run at full speed. Although most of the benefit comes from having right number of active cores but DVFS still has significant impact on performance. From the figure we see that in most cases DVFS improves the power consumption by 15% and as much as by 39% in case of FRAG. This benefit comes from the fact that if we over predicted the workload or there is variation in traffic during the prediction interval, then DVFS helps us at reducing power by lowering the frequencies of the cores. Figure 8 shows poten-

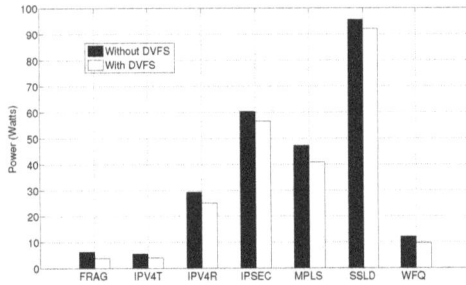

Figure 7: Comparison of proposed scheme with and without DVFS

tial benefit of having the ability to change frequency and voltage of the cores in addition to adjusting the number of active cores. The plot is for a 16 core system running IP4R benchmark when traffic is varied from 0 to 100% which the given configuration can handle. The power consumption is plotted relative to the system which has ability to change number of active cores but does not have a per core DVFS. We see that there is a lot of potential power saving at low to medium traffic. At high traffic, obviously there is less room for power savings. The dotted line shows potential power savings if we have global DVFS, i.e., all cores change their frequency in unison and at any moment all cores are at the same frequency. Global DVFS provides a good tradeoff between design complexity and power savings since it decreases the design complexity and is able to exploit most of the power saving opportunities. Our P-state algorithm becomes even simpler if global DVFS is used instead

of per core DVFS i.e., instead of using findMin() and findMax() function we can increase or decrease the frequencies of all cores together based on queue length.

5.2 Effectiveness of DES Prediction

A distinguishing feature of PTM is that the traffic_factor is based on DES predictor for traffic (Section 3.1). Figure 9 compares the accuracy of different predictors. The predictors under comparison are listed in Table 10.

Predictor	Description
LV	Last observed value is used as prediction for next interval
MA	Moving average of last 8 observations
AR	Auto Regression based prediction
ARMA	AutoRegressive Moving Average of order
ANN	Artificial Neural Network. (3 layers)
DES	Double Exponential Smoothing

Table 10: Predictors used for comparison

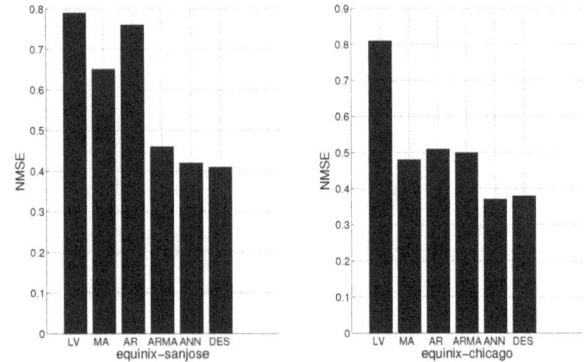

Figure 9: Accuracy Comparison of Predictors. Graphs show NMSE values (lower bar is better).

We use Normalized Mean Square Error ($NMSE$) to compare the performance of predictors. This metric is widely used for evaluating prediction performance. It is the ratio of mean square error to the variance of the series.

$$NMSE = \frac{1}{\sigma^2} \frac{1}{M} \sum_{t=1}^{M} (X_t - \hat{X}_t)^2 \qquad (8)$$

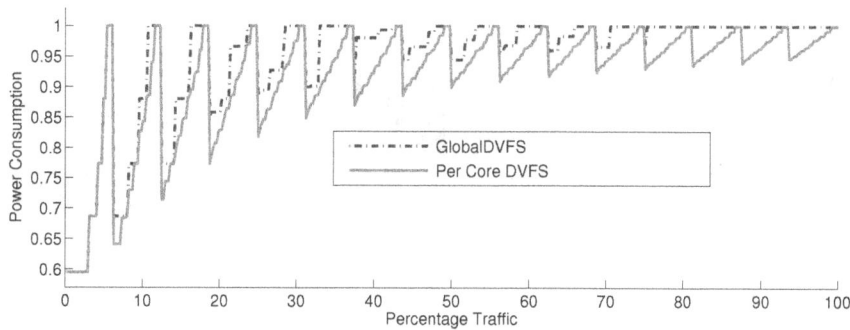

Figure 8: Benefit of having DVFS

where X_t is the actual value of traffic during interval t, $\hat{X}_t$ is the predicted value of X_t and M is the total number of predictions. σ^2 is the variance of time series during prediction. This metric compares the performance of the predictor with a trivial predictor (one which always predicts mean of the time series). In case of this trivial predictor (mean predictor) $NMSE = 1$. If $NMSE > 1$, this means that the predictor is worse than the trivial. $NMSE = 0$ in case of a perfect predictor.

From Figure 9, we can see that DES performs comparably well as compared to more complex ANN based predictor and it outperforms other predictors in the study by big margin. More details about different traffic prediction techniques can be found in [34]. Figure 10 shows power savings with DES predictor over LV pre-

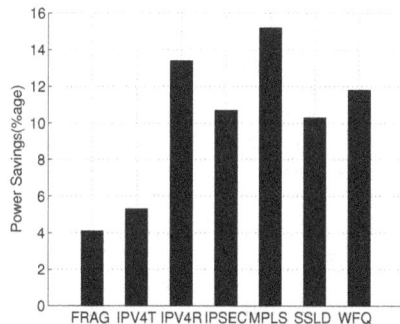

Figure 10: Power Saving when using DES Predictor compared to LV predictor. The trace used is equinix-snjose.

dictor. The figure shows that a good predictor can result in more efficient management of power.

5.3 Packet Queue Behavior

Figure 11 shows the queue length values during a portion of the equinix-sanjose trace. We see that the filtered queue length effectively neglects the short term variation in traffic and is effective in preventing the system from oscillating between states. Also, the scheme is effective to adapt with increasing traffic i.e., if the queue length increases above the threshold values, the system is able to adapt its resources and brings back the queue length within desired limits.

5.4 Power Saving at Different Traffic Rates

Figure 12 shows comparison of different power management schemes at different traffic rates with synthetic traces. The proposed scheme

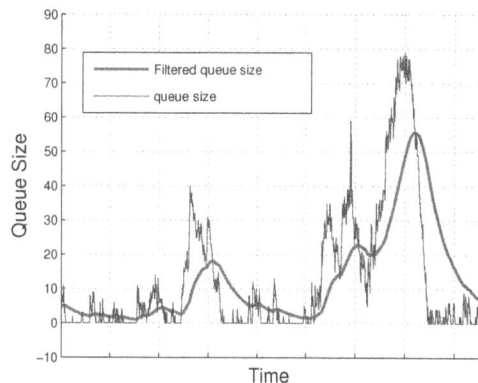

Figure 11: Queue Size and Filtered Queue Size

adapts well at different traffic rates and is the best performer for all applications at all rates. It is important to note that PTM does not gain much benefit from prediction in these traces. These traces are of constant data rate and reactive schemes, which behave similar to LV predictor, are also able to accurately predict traffic. For expensive applications like SSLD and IPSEC, there is not much room for power saving since load is already high but PTM is able to get benefit around 6-10% even for those applications. In general greedy scheme runs individual cores at lower power levels but results in activating more number of cores. For IPSEC and MPLS Greedy scheme seems to perform similar to PTM at some data rates. Although power numbers are similar but Greedy scheme results in 11% packet loss while PTM scheme does not drop any packets.

6. RELATED WORK

Most prior work on power management [18, 31, 44, 35] are not in communication processors domain, but we discuss the network related work in this section.

Need for power efficient Internet infrastructure has fueled studies on design of power efficient network processors. A modeling framework for network processors was presented by Crowley et al. [21]. Franklin et al. also developed an analytical model to explore design space for power efficient network processors [25]. Memick et al. proposed techniques of data filtering to reduce bus accesses [46]. Reduction in bus activity can help save significant power. Zane et al. [52] and Kaxiras et al. [36] propose power efficient TCAM structures to be used in packet processors. Wu et al. [51] investigate a runtime management system for NPs to exploit

131

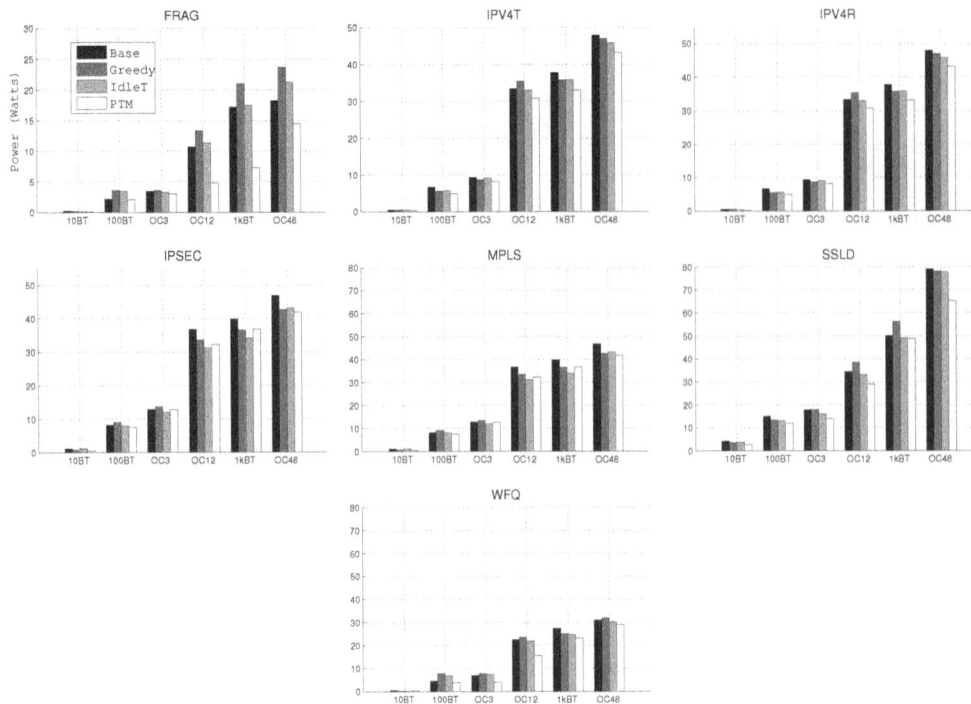

Figure 12: Power consumption comparison of different power management techniques at different Traffic Rates

variability in network traffic for efficient allocation of processing tasks to processing cores. The system is able to adapt to varying demands of traffic and balance utilization of all resources to maximize throughput. Kuang et al. [38] use DVFS for scheduling of pipelined networking applications. This scheme allocates frequencies to the pipeline stages statically and is not aimed at exploiting dynamic traffic variations. Kokku et al. [37] show that variability in traffic can be used for run time scheduling of tasks for conserving energy. Our proposed scheme is not related to scheduling and can complement the power efficient scheduling schemes. Another set of studies which can be used to complement our proposed scheme targets various parts of Internet infrastructure for power efficiency. Chiaraviglio et al. [19] present a scheme to turn off links and nodes during periods of low activity while still guaranteeing full connectivity. Nedevschi et al. [47] show that in addition to putting elements to sleep, link rate adaptation with varying traffic can help further to save power. Luo et al. [41] used processor idle time for DVFS in network processors. All cores are active all the time in this scheme. In another study, Luo et al. [42, 43] applied clock gating to change number of active cores in network processors. This scheme is explained in Section 2.3.2. In contrast to both the above scheme, we propose a predictive scheme to use traffic information directly to manage power in Network Processors and our proposed methodology makes use of both changing number of active cores and DVFS to save power and is simple and general enough to apply to any network processor.

7. CONCLUSIONS

We have proposed and evaluated a predictive power management scheme for Network Processors to exploit variation in traffic. This scheme is able to recognize optimum number of active cores and frequency level of each active core to sustain the input traffic. Traditional power management schemes either result in high power consumption or result in packet loss. Our proposed methodology does not have this weakness as it uses traffic information directly for power management purposes. We have shown that predictive power management schemes work well for packet processing. A simple predictor like DES is able to capture the trends in traffic behavior. Our scheme adapts to variation within the prediction interval using DVFS by monitoring the queue length. DVFS provides us the capability to adaptation at a finer time scale. Furthermore by using filtered queue length we minimize the impact of short term variation in traffic. Experiments on real network traces show that the proposed scheme can save up to 40% more power than traditional schemes.

8. ACKNOWLEDGEMENTS

We would like to thank Heather Hanson (IBM) for her many helpful comments. We would also like to thank anonymous reviewers for their useful suggestions to improve the text. This work was sponsored in part by Semiconductor Research Consortium (SRC) under Task ID 2155.001. The opinions and views expressed in this paper are those of the authors and not those of SRC.

9. REFERENCES

[1] AMD Opteron Processor Power and Thermal Datasheet. http://support.amd.com/us/Processor_TechDocs/30417.pdf.

[2] Cisco 7609-s router. http://www.cisco.com.

[3] Energy Efficient Platforms - Considerations for Applications Software and Services. http://download.intel.com/technology/pdf/Green_Hill_Software.pdf.

[4] NIST/SEMATECH e-handbook of statistical methods.

[5] The P4080 processor. http://www.frescale.com.

[6] Power consumption for MX960. http://www.juniper.net.

[7] PowerQuiccIII Monitors. http://www.freescale.com.

[8] Quidway NetEngine 5000e. http://www.huawei.com.

[9] Red hat enterprise linux 6 power management guide. http://docs.redhat.com/.

[10] The Tilera processor. http://www.tilera.com.

[11] Using the linux cpufreq subsystem for energy management. http://publib.boulder.ibm.com/infocenter/lnxinfo /v3r0m0/topic/liaai/cpufreq/liaai-cpufreq_pdf.pdf.

[12] Intel IXP hardware reference manual, January 2003.

[13] Global action plan report. http://www.globalactionplan.org.uk, 2006.

[14] System power challenges. http://www.slidefinder.net/c/ cisco_routing_research/seminar_august_29/1562106, 2006.

[15] Processor power management in windows vista and windows server 2008. http://www.microsoft.com, November 2007.

[16] J. Baliga, K. Hinton, and R. Tucker. Energy consumption of the internet. In *Optical Internet, 2007 and the 2007 32nd Australian Conference on Optical Fibre Technology. COIN-ACOFT 2007. Joint International Conference on*, pages 1 –3, june 2007.

[17] L. Benini, A. Bogliolo, and G. De Micheli. A survey of design techniques for system-level dynamic power management. *Very Large Scale Integration (VLSI) Systems, IEEE Transactions on*, 8(3):299 –316, june 2000.

[18] W. L. Bircher and L. K. John. Analysis of dynamic power management on multi-core processors. ICS '08, NY, USA.

[19] L. Chiaraviglio, M. Mellia, and F. Neri. Reducing power consumption in backbone networks. In *In IEEE International Conference on Communications (ICC 09)*, 2009.

[20] K. Claffy, D. Andersen, and P. Hick. The caida anonymized 2011 internet traces.

[21] P. Crowley and J. L. Baer. A modeling framework for network processor systems. In *Network Processor Workshop*, 2002.

[22] R. Dömer, A. Gerstlauer, J. Peng, D. Shin, L. Cai, H. Yu, S. Abdi, and D. D. Gajski. System-on-chip environment: a specc-based framework for heterogeneous mpsoc design. *EURASIP J. Embedded Syst.*, 2008, January 2008.

[23] U. Ewaldsson. Cut your network's electricity bill and carbon footprint. http://wwww.globaltelecomsbusiness.com /Article/2436697/Cut-your-networks-electricity-bill-and-carbon-footprint.html, February 2010.

[24] S. Floyd and V. Jacobson. Random early detection gateways for congestion avoidance. *Networking, IEEE/ACM Transactions on*, 1(4):397 –413, aug 1993.

[25] M. A. Franklin and T. Wolf. Power considerations in network processor design. In *In Network Processor Workshop in conjunction with Ninth International Symposium on High Performance Computer Architecture (HPCA-9)*, 2003.

[26] D. D. Gajski, J. Zhu, R. Domer, A. Gerstlauer, and S. Zhao. *SpecC: Specification Language and Methodology*. Kluwer Academic Publishers Boston, MA, 2000.

[27] A. Gerstlauer. Host-compiled simulation of multi-core platforms. In *Rapid System Prototyping (RSP)*, 2010.

[28] Gerstlauer, A. et al. Abstract system-level models for early performance and power exploration. In *ASPDAC*, 2012.

[29] T. Grotker. *System Design with SystemC*. Kluwer Academic Publishers, Norwell, MA, USA, 2002.

[30] S. Herbert and D. Marculescu. Analysis of dynamic voltage/frequency scaling in chip-multiprocessors. ISLPED '07, pages 38–43, New York, NY, USA, 2007. ACM.

[31] S. Herbert and D. Marculescu. Variation-aware dynamic voltage/frequency scaling. In *HPCA*, Feb. 2009.

[32] K. Hinton, J. Baliga, M. Feng, R. Ayre, and R. Tucker. Power consumption and energy efficiency in the internet. *Network, IEEE*, 25(2):6 –12, march-april 2011.

[33] Intel, Microsoft, and Toshiba. Advanced configuration and power interface specifications. http://www.intel.com/iam/powermgm/specs.html, 1996.

[34] M. F. Iqbal and L. K. John. Power and performance analysis of network traffic prediction techniques. In *ISPASS*, 2012.

[35] C. Isci, A. Buyuktosunoglu, C.-Y. Cher, P. Bose, and M. Martonosi. An analysis of efficient multi-core global power management policies: Maximizing performance for a given power budget. In *MICRO-39*, dec. 2006.

[36] S. Kaxiras. Ipstash: A set-associative memory approach for efficient ip-lookup. *IEEE Infocom*, pages 992–1001, 2005.

[37] R. Kokku, T. L. Riché, A. Kunze, J. Mudigonda, J. Jason, and H. M. Vin. A case for run-time adaptation in packet processing systems. *SIGCOMM Comput. Commun. Rev.*, 34:107–112, January 2004.

[38] J. Kuang and L. Bhuyan. Optimizing throughput and latency under given power budget for network packet processing. In *Proceedings of the 29th conference on Information communications*, INFOCOM'10, pages 2901–2909, Piscataway, NJ, USA, 2010. IEEE Press.

[39] B. Lee and L. John. Npbench: a benchmark suite for control plane and data plane applications for network processors. In *Computer Design*, oct. 2003.

[40] S. Li, J. H. Ahn, R. D. Strong, J. B. Brockman, D. M. Tullsen, and N. P. Jouppi. Mcpat: an integrated power, area, and timing modeling framework for multicore and manycore architectures. MICRO 42, NY, USA, 2009.

[41] Y. Luo, J. Yang, L. Bhuyan, and L. Zhao. Nepsim: a network processor simulator with a power evaluation framework. 24:4 – 44, 2004.

[42] Y. Luo, J. Yu, J. Yang, and L. Bhuyan. Low power network processor design using clock gating. In *DAC*, pages 712–715, 2005.

[43] Y. Luo, J. Yu, J. Yang, and L. N. Bhuyan. Conserving network processor power consumption by exploiting traffic variability. *ACM Trans. Archit. Code Optim.*, 4(1), Mar. 2007.

[44] G. Magklis, P. Chaparro, J. Gonzalez, and A. Gonzalez. Independent front-end and back-end dynamic voltage scaling for a gals microarchitecture. In *ISLPED*, oct. 2006.

[45] M. M. K. Martin, D. J. Sorin, B. M. Beckmann, M. R. Marty, M. Xu, A. R. Alameldeen, K. E. Moore, M. D. Hill, and D. A. Wood. Multifacet's general execution-driven multiprocessor simulator (gems) toolset. CAN, 2005.

[46] G. Memik and W. H. Mangione-smith. Increasing power efficiency of multi-core network processors through data filtering. In *In Intl. Conf. on Compilers, Architecture, and Synthesis for Embedded Systems*, pages 108–116. ACM Press, 2002.

[47] S. Nedevschi, L. Popa, G. Iannaccone, S. Ratnasamy, and D. Wetherall. Reducing network energy consumption via sleeping and rate-adaptation. In *Proceedings of the 5th USENIX Symposium on Networked Systems Design and Implementation*, NSDI'08, pages 323–336, Berkeley, CA, USA, 2008. USENIX Association.

[48] R. Ramaswamy and T. Wolf. PacketBench: A tool for workload characterization of network processing. In *Proc. of IEEE 6th WWC*, Austin, TX, Oct. 2003.

[49] B. Sanso and H. Mellah. On reliability, performance and internet power consumption. In *Design of Reliable Communication Networks, 2009. DRCN 2009. 7th International Workshop on*, pages 259 –264, oct. 2009.

[50] A. V. D. Ven. Absolute Power. http://software.intel.com/sites/oss/pdfs/absolutepower.pdf.

[51] Q. Wu and T. Wolf. On runtime management in multi-core packet processing systems. In *Proceedings of the 4th ACM/IEEE Symposium on Architectures for Networking and Communications Systems*, ANCS '08, pages 69–78, New York, NY, USA, 2008. ACM.

[52] F. Zane, G. Narlikar, and A. Basu. Coolcams: Power-efficient tcams for forwarding engines. In *IN IEEE INFOCOM*, pages 42–52, 2003.

Software is the Future of Networking

Teemu Koponen

VMware, Inc.
Palo Alto, CA

ABSTRACT

In this talk, I'll revisit the role of Software in Software-Defined Networking and argue how not only control plane but also forwarding is becoming increasingly only a matter of software development. In short, I'll discuss how x86 is already on its way to transform the networking as we know it.

Categories and Subject Descriptors

D.3.3 [**Computer Systems Organization**]: Computer Systems Implementation – *General*

General Terms

Design

Keywords

Networking, SDN, Virtalization

Attendre: Mitigating Ill Effects of Race Conditions in OpenFlow via Queueing Mechanism

Xiaoye Sun   Apoorv Agarwal   T. S. Eugene Ng

Rice University
Houston, Texas, USA

Categories and Subject Descriptors

C.2.2 [**Computer-Communication Networks**]: Network Protocols; C.2.1 [**Computer-Communication Networks**]: Network Architecture and Design

Keywords

OpenFlow, protocol design, verification

1. INTRODUCTION

According to the specification of the OpenFlow protocol, whenever a flow table entry is absent for any arriving data packet, a packet-in message is sent to the controller. This behaviour results in various types of race conditions in accord with various packet and message orderings. Increased forwarding delay of data packets, increased complexity in performing software verification, and increased load on switch and controller processors are the ill effects of these races.

We identify three types of races inherent in the OpenFlow protocol. The **Type 1** race is between data packets sent by a host and the command messages sent by the controller. Consider the scenario in which a host has sent two data packets sequentially. The switch, on receiving the first data packet, finds out that there is no entry matching the packet header, so according to the default action, a packet-in message is sent to the controller. The controller processes this packet-in message and sends out a flow-mod message combined with a packet-out message (denoted by flow-mod/packet-out) to the switch.

The state diagram of a system experiencing the Type 1 race is shown in Figure 1. The q_0 state is when a switch has sent a packet-in message to the controller on arrival of the first data packet. Now, the next event at the switch is either processing the flow-mod/packet-out message (or command for short) from the controller, or processing the second data packet from the same flow. Consider case 1 when the switch processes the command before the data packet. In this case, the switch inserts an entry in its flow table and processes the next data packet from this flow according to the installed entry. This scenario causes the system state to change along $q_0 \rightarrow q_1 \rightarrow q_2$. Consider case 2 when the second data packet is addressed by the switch before the command message. In this case, the switch will sent another packet-in message to the controller. The switch on receiving flow-mod/packet-out

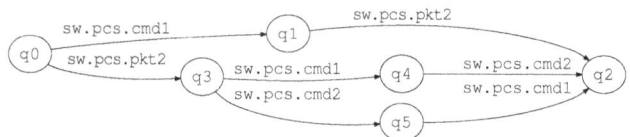

Figure 1: Race Types 1 & 2 state diagram.

from the controller corresponding to this second packet-in will process the second packet. This makes the system state to change from $q_0 \rightarrow q_3 \rightarrow ... \rightarrow q_2$.

Figure 1 also illustrates the **Type 2** race, which is a race between different command messages issued by the controller for a particular switch. This is possible when the controller is *multi-threaded*. A multi-threaded controller can have multiple threads processing several requests which may be contending for resources. This causes threads to finish their tasks in an arbitrary order. Branching at state q_3 in Figure 1 happens because of this reason. The branch $q_3 \rightarrow q_4 \rightarrow q_2$ corresponds to the state when the controller is able to process the packet-in for the first packet before processing the packet-in for the second packet. The other branch $q_3 \rightarrow q_5 \rightarrow q_2$ happens when the controller processes them in reverse.

The **Type 3** race (not depicted due to space limit) is experienced by the network when different command messages are sent by the controller to different switches, such as when a "route flow" controller configures a network path $S_1, S_2, ...$. Consider a case when a data packet arrives at switch S_{n+1} earlier than the command message meant for it. This is possible when the command message arrives at switch S_n before the arrival of the command message for switch S_{n+1}. In this scenario, a packet-in message will be sent by switch S_{n+1}, leading to a series of extra state transitions compared with a scenario in which the command message arrives at switch S_{n+1} before the data packet.

Problems due to the Race Conditions

(a) **Increased complexity in software verification:** The identified race conditions lead to a serious state space explosion problem in controller software verification because the verifier needs to exercise all the possible state transitions. Figure 2 shows the numbers of state transitions (hollow markers) generated during the verification of an OpenFlow "route flow" application for various number of switches and packets. The data are collected from NICE – an OpenFlow controller software verifi-

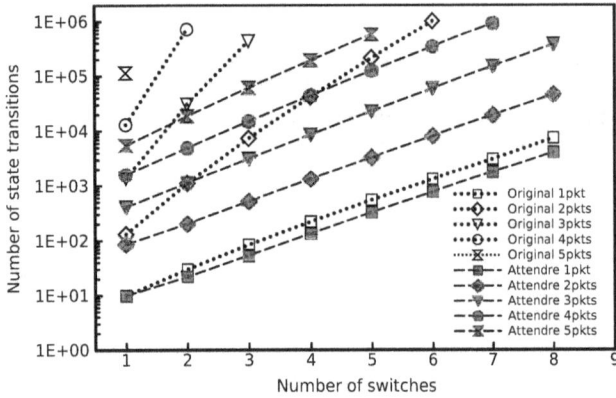

Figure 2: Number of state transitions incurred by NICE verification under various scenarios for original OpenFlow and Attendre.

cation tool[1]. It is important to note that NICE assumes a single-threaded controller, therefore the figures are highly conservative. If a multi-threaded controller is assumed, the occurrences of Type 2 races will cause further explosion in the state space.

(b) **Increased forwarding delay of packet:** The races increase the forwarding delay of a packet as whenever a switch receives a data packet without a matching entry in the flow table, a packet-in is sent to the controller. In the worst case, this could happen at every hop in the path. This increases the forwarding delay of the data packet significantly since in some cases the command message may arrive at the switch just after a packet-in is dispatched for the controller.

(c) **Increased processing load on switches and controller:** The extra packet-in, flow-mod/packet-out, and flow-mod messages caused by the races increase the processing load on switches and the controller. When a packet-in is sent, it is possible that the controller has already processed another packet-in for the same flow and the command message is on its way to the switch. Processing related to subsequent packet-in messages is thus redundant.

2. ATTENDRE MECHANISM

Attendre is used to allay the ill effects of the race conditions in an OpenFlow network by avoiding sending redundant packet-in messages. The core idea of Attendre is that if the switch is expecting a command message, the incoming packets that can be handled according to this command message should be queued at the switch buffer until the command comes.

To achieve this, each flow table entry should be tagged with a **version number** which is issued by the controller platform and sent to the switch together with the flow-mod message. The version number is a unique identifier used to differentiate old and new entries. The command message to a switch S_n should carry an additional match field and the version number of the flow table entry for the next hop switch S_{n+1}. This occurs when the application on the controller intends to install a forwarding path for the packet.

The match field and version number for the next switch S_{n+1} will be piggybacked by the first data packet handled by the command message and carried to S_{n+1}. A new **WAIT** action is also defined in Attendre. The packets matching the entry with WAIT action type will be inserted into the switch buffer and a buffer ID will be associated to the entry. An entry with this WAIT action type will be inserted to the flow table when a packet-in is sent from a switch. A WAIT entry can also be inserted when a switch receives a data packet carrying the match field and version number information, and the corresponding command message has not been processed by the switch. The WAIT entry will be removed or replaced after the switch receives the corresponding command message. At the same time, the packets queued in the buffer associated with the being eliminated WAIT entry will be dequeued and processed according to the actions of the command.

3. BENEFITS TO VERIFICATION

We have integrated the Attendre mechanism into NICE by modifying its network component models accordingly. We compare the number of state transitions incurred by original OpenFlow and Attendre.

For simplicity, the network topology used is a simple chain. Two hosts, a sender and a replier, locate at the ends of the chain. The sender sends packets to the replier and will receive the same number of reply packets. The application on the controller is a "route flow" application that installs entries along the chosen route for the forwarding of the packets. The verification is done for different numbers of injected packets and switches along the path.

The numbers of state transitions generated during the model checking for different experiment configurations (upto eight switches and one million state transitions) are shown in Figure 2. When the number of packets or switches increases, the number of states transition increases exponentially. With one injected packet (square markers in Figure 2), Attendre could save the number of state transitions by about 50% if there are eight switches on the path. If there is more than one packet, the number of state transitions can be reduced by several orders of magnitude, since in this case, the following packets could queue at the switch which results in far more state transition reduction. For example, when there are six switches along the path and two packets injected (diamond markers in Figure 2), the state transitions could be reduced by two orders of magnitude. Note that NICE does not model the Type 2 race, it assumes a single-threaded controller. The results in Figure 2 for OpenFlow without Attendre are expected to become much worse if Type 2 race is accounted for. In other words, the actual benefit of Attendre is even larger than Figure 2 shows. It is not hard to see that if Attendre is adopted, far more complex network topologies and applications can be successfully verified for a given amount of computing resources.

4. REFERENCES

[1] M. Canini, D. Venzano, P. Perešíni, D. Kostić, and J. Rexford. A nice way to test openflow applications. In *NSDI'12: Proceedings of the 9th USENIX Symposium on Networked Systems Design and Implementation*, NSDI'12, pages 127–140, Berkeley, CA, USA, 2012. USENIX Association.

Dynamic Frequency Scaling Architecture for Energy Efficient Router[*]

Wenliang Fu, Tian Song[†], Shian Wang
School of Computer Science and Technology
Beijing Institute of Technology
{fuwenl, songtian, yiyun}@bit.edu.cn

Xiaojun Wang
School of Electronic Engineering
Dublin City University
wangx@eeng.dcu.ie

ABSTRACT

Recently, energy expenditures of the Internet have increased dramatically, raising energy issue of routers an urgent problem in relative research areas. In fact, much device surplus and redundancy are introduced during network planning for rarely appeared traffic peak hours and device failures, wasting energy most of the time. In this work, an energy-aware architecture is proposed for routers, which could trade system performance for energy savings while traffic is low by scaling frequencies of its inner components. We also explore multi-frequency modulation strategies to optimize the energy saving effect. The result shows that our prototype router could save about 40% of its peak power consumption.

Categories and Subject Descriptors

C.2.1 [**Computer-Communication Networks**]: Network Architecture and Design

General Terms

Design, Performance

Keywords

Energy efficiency, Frequency scaling, Router architecture

1. INTRODUCTION

Recently, many researches have been published on energy efficient technologies for networking devices, such as smart port sleeping, dynamic buffer adapting and green routing.

Maruti Gupta and Suresh Singh in [1] initially argued that network devices could save energy by putting idle ports to sleep. Later, their works on smart port sleeping led to the establishment of standard IEEE 802.3az. However, due to considerable time and energy costs for transitions, this method could only be effective while device utilization rate is under 10% [2].

A. Vishwanath *et al.* exam buffer usage of modern routers and develop scheme to dynamically turn on/off SRAM and

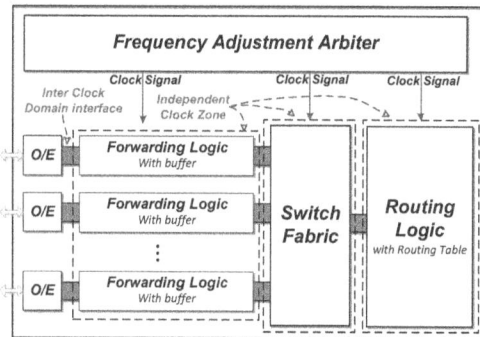

Figure 1: Block Diagram of Frequency Adjustment Router Architecture.

DRAM buffers according to traffic demand [3]. Their results show that up to 10% of the total energy consumption can be saved at the cost of negligible traffic congestion and latency.

Green routing mainly focuses on concentrating traffic in a subset of network links, and create more opportunities for energy efficient behaviors of single network devices.

Based on our initially work [5], we propose an energy-aware router architecture, which could dynamically adapt lower frequencies for energy savings while being underused. The frequency scaling scheme has advantages in the following aspects. First, frequency scaling zones cover nearly all functional components, which consume more then 50% of the overall energy, much more than other methods [4]. Second, multiple energy states with different trade-offs between performance and energy efficiency are more flexible for strategic switching. Besides, frequency scaling scheme is a supplement, not a replacement, for current green networking technologies.

2. FREQUENCY SCALING ARCHITECTURE

As is shown in Fig.1, we regard functional components as elementary units for frequency scaling. In order to release clock constraints, *inter clock domain interfaces* are inserted before targeted components and turn them into frequency irrelevant and adjustable units. Generally, *inter clock domain interfaces* is an asynchronous FIFO based one-way data path, which could relay signals between different clock zones and capable of buffering packets. Frequency switchings are triggered by buffer occupancy of directly attached interfaces, which is regarded as a indicator of gaps between incoming workload and current performance. As a central controller,

---

[*]This work was partially supported by the National Natural Science Foundation of China (No. 61272510, 60803002), Beijing Key Discipline Program.

[†]Tian Song is the corresponding author of this work.

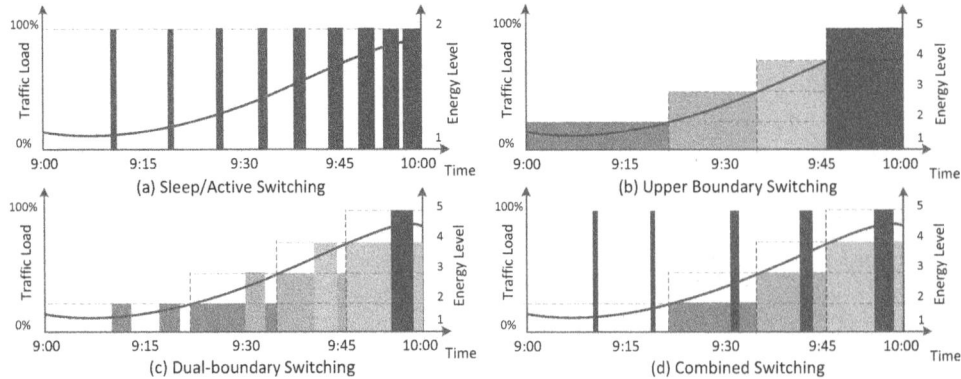

Figure 2: Energy State Switching Strategies.

frequency adjustment arbiter is introduced to monitor system information, such as buffer occupancies, module utilization rates and current energy states of each modules, and make frequency adjustment decisions.

3. DYNAMIC FREQUENCY SCALING STRATEGIES

Based on frequency scaling architecture, we propose several switching strategies distinguished by different buffer thresholds and preset cold down time between transitions to guarantee stay time in low energy states and reduce switches as follows:

Sleep/active switching: A two-frequency switch strategy as shown in Fig.2 (a) is proposed, which allows modules adopt very low frequency (as sleep) for energy savings, or active (in red) for packet processing. This scheme gains higher device utilization rate but causes frequent switching.

Upper Boundary switching: This scheme permits multi-frequency switching, and always adopts the higher frequency level that barely offering more capacity than necessary, which is shown in Fig.2 (b). Comparing to two-frequency switching, this method reduces frequency transitions at the cost of energy efficiency.

Dual-boundary switching: Fig.2 (c) describes a optimized strategy of upper boundary idea, which introduces a lower boundary to further reduce stay time in higher frequency levels with limited switches.

Combined switching: This strategy combines ideas of sleep/active and boundary switching, as shown in Fig.2 (d), treating low traffic time (such as nights) with sleeping strategy, and high traffic time with the aid of lower boundary for reducing transitions.

4. EXPERIMENT AND RESULT

We developed a software router prototype which supports 10-frequency adjustments range from 125 MHz to 12.5 MHz, corresponding to 1000Mbps to 100Mbps respectively in traffic capacity. We fed real traces for the experiments, which were captured from an office network of about 1000 users, spinning over a period of more than three hours and the loss rate was kept under 1% of the total traffic. The preliminary results are shown in Fig.3, by limiting switches with cold down time and processing buffered packets as fast as possible, our proposed combined strategy achieved the most significant savings.

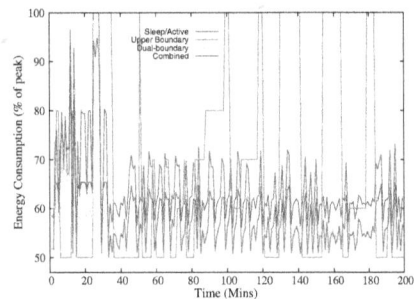

Figure 3: Frequency Scaling for Energy Efficiency.

5. CONCLUSION

In this work, we propose an energy efficient router architecture which allows inner components to adapt frequencies for energy savings. In addition, frequency modulation strategies for tuning various network environments are also introduced. Simulations on real traces show that our frequency adjustment router could save up to 40% of the total energy consumption.

6. REFERENCES

[1] M. Gupta and S. Singh. Greening of the internet. In *Proceedings of the 2003 conference on Applications, technologies, architectures, and protocols for computer communications*, pages 19–26. ACM, 2003.

[2] P. Reviriego, K. Christensen, J. Rabanillo, and J. Maestro. An initial evaluation of energy efficient ethernet. *Communications Letters, IEEE*, 1–3, 2011.

[3] A. Vishwanath, V. Sivaraman, Z. Zhao, C. Russell, and M. Thottan. Adapting router buffers for energy efficiency. In *Proceedings of the Seventh COnference on emerging Networking EXperiments and Technologies*, page 19. ACM, 2011.

[4] R.S. Tucker, R. Parthiban, J. Baliga, K. Hinton, R.W.A. Ayre, and W.V. Sorin. Evolution of wdm optical ip networks: A cost and energy perspective. *Journal of Lightwave Technology*, 27(3):243–252, 2009.

[5] W. Fu and T. Song. A frequency adjustment architecture for energy efficient router. In *Proceedings of the ACM SIGCOMM 2012 conference on Applications, technologies, architectures, and protocols for computer communication*, pages 107–108. ACM, 2012.

Extensible Hierarchical Simulation of Network Systems

Xinming Chen
Department of Electrical and Computer
Engineering
University of Massachusetts, Amherst, MA, USA
xinmingchen@ecs.umass.edu

Tilman Wolf
Department of Electrical and Computer
Engineering
University of Massachusetts, Amherst, MA, USA
wolf@ecs.umass.edu

ABSTRACT

The system architecture of a network system is of fundamental importance to the performance of computer networks. However, evaluating different architectural and algorithmic choices of network systems is usually difficult. This poster presents NetSysSim, a network system simulator implemented in SystemC. NetSysSim provides a hierarchical framework of different levels of abstraction. With SystemC's modular and timing based simulation, researchers can simulate any part of the system with the desired level of details. Our design of NetSysSim allows for performance evaluation of data plane architectures, including different queuing, scheduling or transmission schemes.

Categories and Subject Descriptors

C.2.6 [**Computer-Communication Networks**]: Internetworking—*Routers*

Keywords

system architecture, simulation, hierarchical abstraction

1. INTRODUCTION

Network systems such as switches and routers are at the core of data communication networks. The system architecture of these devices are of significant importance for network performance. However, so far there is no comprehensive method for network system evaluation. Researchers can (1) implement a prototype system with programmable devices such as NetFPGA or network processor, (2) implement a switch architecture based on existing network simulator such as NS2 [7] [6], or (3) write a dedicated simulation tool from the ground up, mostly using C/C++ language [3] [5].

The first method requires understanding of many details that may not be relevant to the research objective, which is costly in effort. The second method focuses on the design of network protocols and their operation, instead of the system architecture that implements networks. For example, network simulators like NS2 or GloMosim can not simulate how the memory bandwidth affects the performance of a packet processor in a router, because there is not enough detail of the router. The third method may face difficulty in describing timing, signal and interconnect of components, because

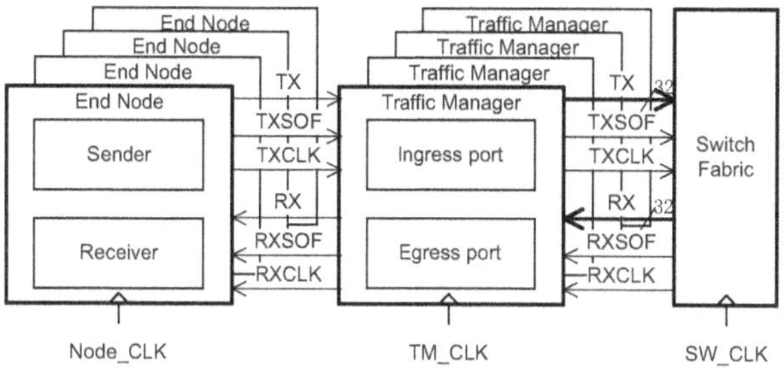

Figure 1: Architecture of NetSysSim.

standard C/C++ is not designed to describe hardware systems.

Our work, a hierarchical system-level simulator of network systems, fill this gap. Our NetSysSim is based on SystemC [1], a C++ class library and a methodology for system-level modeling. SystemC allows users to describe cycle-accurate algorithms and hardware architectures since it includes the concept of timing. It brings convenience to construct a framework that unify hardware and software. Based on SystemC, we build NetSysSim, which is capable for hierarchical system-level simulation.

NetSysSim simulates the core functionalities of network systems, such as switches and routers. Researchers can use the platform as tool for proof of concepts. It is different from network simulators because users can implement any level of details in it through the transistor level. And it is different from software routers like Click, because NetSysSim provides a unified, hardware independent simulation result.

2. ARCHITECTURE OF NETSYSSIM

The architecture of NetSysSim is shown in Figure 1. It consists of three parts:

- Switch Fabric: This component provides basic packet switching function for traffic managers. Existing switch architectures, such as point-to-point, shared memory and shared bus can be simulated. It can be configured as various queuing schemes such as input queuing and output queuing. It ensures in-order delivery from one specific ingress port to another egress port.

- Traffic Managers: These components provide main features and functions such as IP lookup, traffic shaping and security processing.

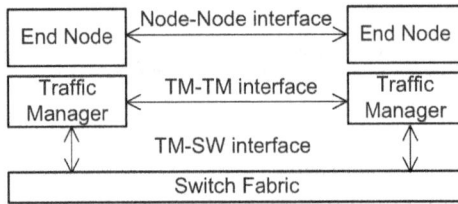

Figure 2: Three-layer interface.

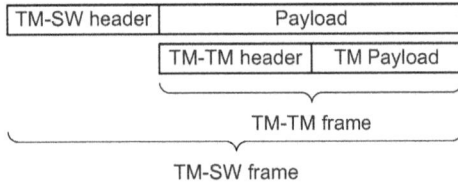

Figure 3: TM-SW frame format. It also illustrates TM-TM frame structure.

- End Nodes: These components send and receive Ethernet frames. Simulated traffic can either be generated or by replaying tcpdump traces.

These parts communicate in a packet based interface. Such interface is divided into three layers: Traffic Manager to Switch Fabric (TM-SW), Traffic Manager to Traffic Manager (TM-TM), and End Node to End Node (Node-Node), which is illustrated in Figure 2.

The TM-SW interface refers to the common switch interface (CSIX-L1) specification [2], some functions such as parity check is omitted. Control signals and payloads are packed (and fragmented if exceed maximal length) into packets. Figure 3 illustrates the frame format of TM-SW interface, it also illustrates the layered structure for TM-TM interface.

TM-TM interface contains TM-TM header and payload. The payload could be either Ethernet frames or control information between TMs. Traffic manager needs to fragment Ethernet frames and encapsulate them with TM-SW headers at the ingress port, and to reassemble packets at the egress port.

Node-Node interface uses standard Ethernet frames.

3. DEMONSTRATION

The first objective of NetSysSim is to simulate network systems with enough detail at certain specific aspect. It should behave the same as a real system on that desired aspect if correctly modeled. For example, if input queue and output queue structures are implemented in switch fabric, the saturation throughput should behave the same as the analysis in [4]. Figure 4 shows the simulation result of input queuing meets the analytical result. Another demonstration is the switch fabric queue length. Given a packet arrival time distribution and processing speed, the simulated queue length is the same as analytical result using queuing theory.

The second objective of NetSysSim is to enable hierarchical modeling. Researchers can use different level of abstraction to model each part of the system. For example, in research about IP lookup algorithms, other parts of the

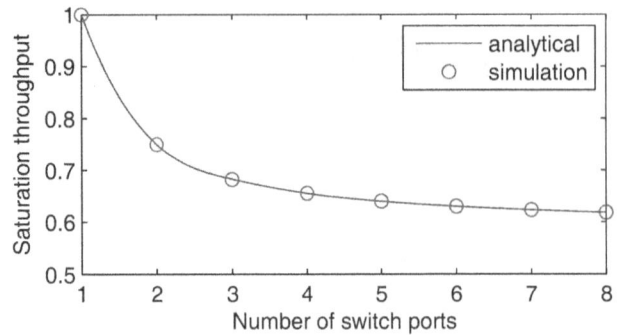

Figure 4: Simulation result for saturation throughput with input queuing scheme, consistent with the analysis in [4].

switch can be implemented in a minimal functional way. At different stages of research, different granularity can be used for the same part. In the previous example, one can first write an IP lookup module in a software implementation, then change it to hardware based model to observe its performance.

We believe that an accurate simulation tool for network systems, such as NetSysSim, is an important step toward effectively exploring design choices in network systems.

4. REFERENCES

[1] Approved ieee draft standard systemc language reference manual (superseded by 1666-2005). *IEEE Std P1666/D2.1.1*, 2005.

[2] CSIX. Csix-l1: Common switch interface specification-l1, aug. 2000.

[3] http://klamath.stanford.edu/tools/SIM/.

[4] M. Karol, M. Hluchyj, and S. Morgan. Input versus output queueing on a space-division packet switch. *Communications, IEEE Transactions on*, 35(12):1347 – 1356, dec 1987.

[5] M. Malgosa-Sanahuja, J. Castells-Cuscullola, and J. Garcia-Haro. An atm switch simulation tool based on the c++ object oriented programming language. In *Communications, Computers and Signal Processing, 1997. 10 Years PACRIM 1987-1997 - Networking the Pacific Rim. 1997 IEEE Pacific Rim Conference on*, volume 2, pages 972 –976 vol.2, aug 1997.

[6] F. Yang, Z. kai Wang, J. ya Chen, and Y. jie Liu. Design of parallel packet switch simulation system based on ns2. In *Wireless Communications, Networking and Mobile Computing, 2009. WiCom '09. 5th International Conference on*, pages 1 –4, sept. 2009.

[7] H. Zheng, Y. Zhao, and C. Chen. Design and implementation of switches in network simulator (ns2). In *Innovative Computing, Information and Control, 2006. ICICIC '06. First International Conference on*, volume 1, pages 721 –724, 30 2006-sept. 1 2006.

FlowOS: A Pure Flow-based Vision of Network Traffic

Abdul Alim[†], Mehdi Bezahaf[†], and Laurent Mathy[‡]
[†] Computing Dept., Lancaster University, United Kingdom
m.bezahaf,a.alim@lancaster.ac.uk
[‡] University of Liege
laurent.mathy@ulg.ac.be

ABSTRACT

The original Internet architecture lacked the concept of a flow, and considered each traffic as a set of packets. In this short paper, we rethink this concept inside middlebox-based platform and handle each traffic as a whole block instead of packets. We design a whole system where each input packet matching some criteria is placed in a specific structure which is shared between all processing modules that interact in a parallel manner with this flow. Thus, this new design improves flexibility of traffic and also increases the flow processing performances.

Categories and Subject Descriptors

C.2.1 [**Computer-Communication Networks**]: Network Architecture and Design—*Network communications*

Keywords

Network Flow Processing, Parallelisme, FlowOS

1. MOTIVATION

The use of middlebox processing services is increasingly popular as modern enterprises needs become more ubiquitous to improve security and performance of their networks. Nowadays, these middleboxes consist mainly in specialized expensive box devices plugged to the network and carry out one crucial network functionality at the time such as load balancing, packet inspection and intrusion detection. In the context of the european project CHANGE [1], the Flowstream platforms [1], which are a set of programmable switches and x86 servers, are deployed on the network and susceptible to receive input traffic at any time. These platforms can be along the path between the source and the destination as they can be off-path (traffic attraction). Received traffic must be processed by user-defined processing functionality running inside. We propose in this short paper an overview of FlowOS, an incremental deployable architecture based around the notion of flow processing platform. Using FlowOS, users can remotely instantiate different kind of processing in the same platform at the same time and on the same flow (Figure 1).

---

[1]http://www.change-project.eu

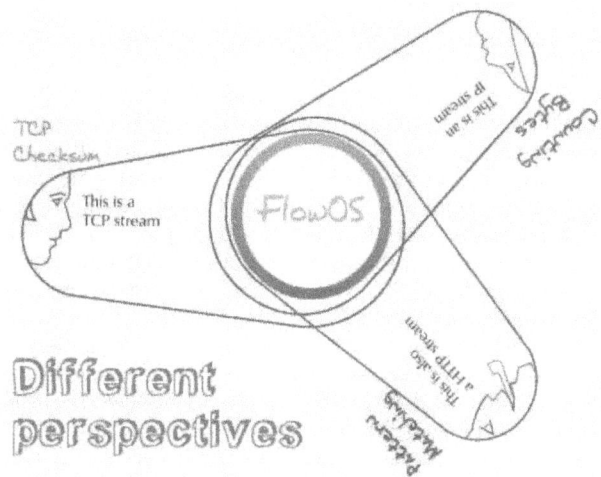

Figure 1: Different perspectives of FlowOS.

2. OVERVIEW OF FLOWOS

The original Internet architecture lacks the concept of a flow, treating traffic as independent packets. Moreover, Operating Systems naturally deal with packets, which are an artifact of the network - apart from the ease of moving them around in the network, there is nothing "logical" about packets: applications manipulate flows. As soon as we start reasoning about "applications" or anything else in the network, we need flows.

In current OSes, flows are really materialized at the socket interface in user space. That is not good for high-performance in network flow processing. FlowOS is based on new programming models and abstractions that are better suited to processing network traffic as flows. We design FlowOS with both an execution model that gives us much higher performance, and much greater flexibility as to the definition of what a flow is, than the socket interface.

Each input packet matching some criteria defined by the user is placed in a specific structure which is shared between all processing modules [2] that interact in a parallel manner with this flow. In fact, all these processing modules are placed inside a single pipeline named processing

---

[2]We define a processing module as one network functionality such as load balancing.

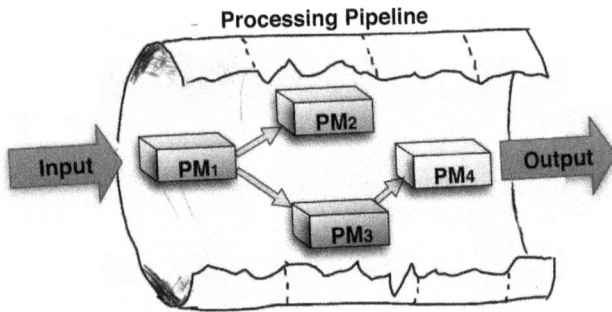

Figure 2: Example of processing pipeline, where PM_1 processes first received bytes and releases them to PM_2 and PM_3 that run concurrently. Finally, PM_4 is served by PM_3.

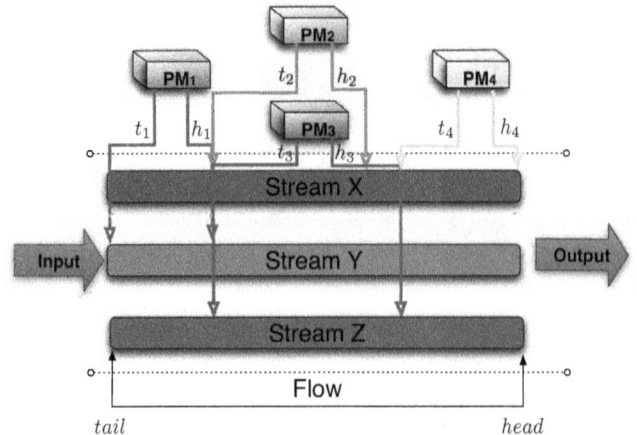

Figure 3: Pointers management inside the processing pipeline.

pipeline, where they can see the same flow as multiple independent flows, without having to bother about even the existence of other bytes (Figure 2). Thus, the potential performance benefits are twofold. First, it provides an easier parallelism through a better adapted programming model (each processing module in our execution pipes has a "window of flow data" within which it does what it wants, with the synchronization between neighbouring processing modules being done through "increasing/shrinking" those windows). Second, the packet-to-flow-to-packet translations is done efficiently by the system itself (in a packet oriented system, this must be done by the modules themselves, resulting in possibly things being done several times).

When a user defines a flow, FlowOS registers an RX handler to the input port to capture incoming IP packets. A *packet classifier* on the input side, which uses the same parameters defined on OpenFlow [2] to classify packets, is responsible for putting IP packets into appropriate flows. It comes with a set of protocol parsers, which are responsible for parsing IP packets and constructing virtual streams for a flow. In FlowOS, a flow contains one or more streams where each stream represents a protocol. On the output side, FlowOS has a *packetizer*, which is responsible for repacketizing data. If some processing modules changes the content of a stream, then the packetizer has to reflect these changes to packet headers since processing modules do not know about underlying packets. It then sends IP packets to the TX handler, which subsequently sends them to output interfaces.

FlowOS constructs flows in the shared kernel space in order to avoid the overhead of copying IP packets from kernel to user space. Each stream consists of a sequence of virtual buffers delimited by the `start` and `end` pointers in SK_BUFFs [3]. A Flow enters into the processing pipeline at one end called the `tail` and leaves it at the other end called the `head`. At a certain point in time, a flow is characterised by its `head` and `tail` pointers.

A flow is then fed to a *pipeline* of processing modules where streams get processed by one or more processing modules. A processing module (PM), which is also a kernel module, processes a specific stream of the flow (e.g., IP, TCP, application). Each PM runs as an independent kernel thread

---

[3]SK_BUFFs are the buffers in which the linux kernel handles network packets.

to process the stream of a flow. Note that a PM thread sleeps when there is no data to process in a stream.

FlowOS allows users to configure PMs to run concurrently or sequentially or a combination of them. Since streams are shared among PMs, it is important to synchronize PMs in order to access streams in right order. It is the administrator's responsibility to define the correct order and interdependency of PMs. In a processing pipeline, a PM i has access to a particular section "window" of the flow being processed. This window is delimited by h_i and t_i pointers. Figure 3 illustrates relationships among these PMs pointers inside the processing pipeline. When a set of PMs are processing concurrently (PM_2 and PM_3 in Figure 3), the last one of them to move its h_i pointer from a byte has to release it (PM_3 releases data to PM_4 in the illustration).

Different processing modules have been written and tested for FlowOS. The preliminary results show that these modules work perfectly in concurrent mode on FlowOS.

3. CONCLUSION AND FUTURE WORK

FlowOS is a novel OS that we have started to build from ground up with the purpose of supporting flow processing. After design and implementation phases, we have already implemented some processing modules that seems working perfectly in a parallel way. Next steps of the present work include performing a more thorough and rigorous analysis, using complex scenarios and comparing to the usual per packet processing. Furthermore, we plan to explore the inter-communication between different platforms, where a flow can migrate from one physical platform to another one.

4. REFERENCES

[1] A. Greenhalgh, F. Huici, M. Hoerdt, P. Papadimitriou, M. Handley, and L. Mathy. Flow processing and the rise of commodity network hardware. *ACM SIGCOMM Computer Communication Review*, 39(2):20–26, Mar. 2009.

[2] N. McKeown, T. Anderson, H. Balakrishnan, G. Parulkar, L. Peterson, J. Rexford, S. Shenker, and J. Turner. Openflow: enabling innovation in campus networks. *ACM SIGCOMM Computer Communication Review*, 38(2):69–74, Mar. 2008.

Modular Design of Data Vortex Switch Network

Brett Burley
Harvey Mudd College
Claremont, California
brett_burley@hmc.edu

Qimin Yang
Harvey Mudd College
Claremont, California
qimin_yang@hmc.edu

ABSTRACT
The Data Vortex switch architecture provides promising routing performance with scalability but requires complicated connections as its size increases. The modular design is proposed that allows a larger network to be formed from smaller Data Vortex units for flexible implementations. Routing performance is shown to be similar or improved.

Categories and Subject Descriptors
B.4.3 [**Input/Output and Data Communications**]: Interconnections (Subsystems) – *fiber optics, topology*.

Keywords
Switch network, Data Vortex, routing, modular.

1. INTRODUCTION
High performance computing and communication systems require switch networks of high throughput and low latency. Recent trends to optically implement such networks lead to new designs in switching devices and network architectures. Data Vortex (DV) architecture is designed to effectively combine optics and electronics and allows for very high throughput and low latency. Such network is scalable to thousands of I/O ports while maintaining its routing performance.

The DV network generally uses a cylindrical layout of routing nodes with angle A and height H. The number of cylinders is $C=log_2H + 1$ since each cylinder decodes one bit of the binary height address of packet. The last cylinder is added to provide angular resolution and optical buffering as it maintains its height before exit. A recent research shows that a A=1 network may achieve optimum performance, benefitting from no angular resolution [1].

As an example, Figure 1 shows the arrangement of network for A=1, H=4. The intra-cylinder paths loop back to the same angle instead of to the next angle as it would in multi-angle cylinders. The inter-cylinder paths are simply parallel paths to the same height. The control links are used regulate the traffic to maintain the single packet processing, which in combination of deflection routing allows for simple node without optical buffering [2]. The packet starts its routing process when injected at the outermost cylinder, and through each of the cylinders, where the intra-cylinder paths allow the packet to locate the right height group until it exits to the output at the innermost cylinder. Many of the physical implementation issues have been addressed in a previous study [3].

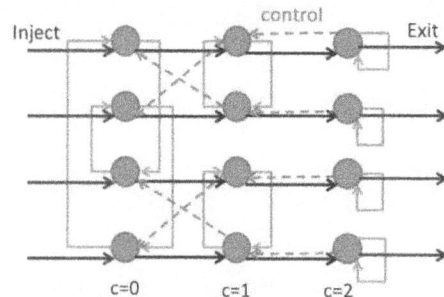
Figure 1. Data Vortex for A=1, H=4.

2. MODULAR DESIGN OF DV SWITCH
A single DV structure at much larger sizes requires complicated path patterns which lead to difficulty in fabrication. In addition, a single network does not allow flexibility in network operation or provide more dynamic service for different traffics. Making the DV network modular provides a solution to both issues. In particular, Clos network has been widely and successfully used in existing switch networks for such purpose [4]. Therefore, this study explores a combination of DV topology and Clos network to provide modular designs. Based on [1], we focus on using single angle modules where each sub-network is specified by HxH. Instead of decoding all log_2H bits in a single DV structure, we break it into multiple stages, each of which uses smaller DVs that decode a subset of the header bits. The full connections between stages follow similar pattern in Clos network. As an example, Figure 2 shows a 16x16 DV network built from the 4x4 modules in a two-stage arrangement, each decoding half of the header bits. A much larger network can be built in similar fashion or in multiple hierarchies. The number of stages determines the size of sub-modules and total number of elements required.

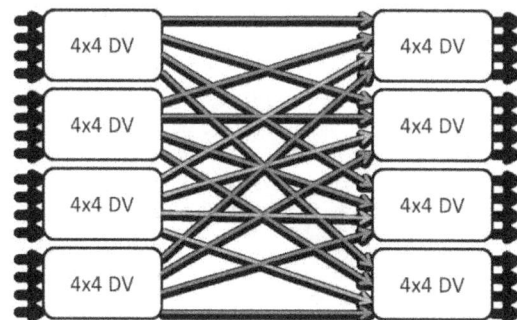
Figure 2. 16x16 DV built from 4x4 DVs in two stages.

Depending on whether the final buffering cylinder is included within each sub-DV, the cost of the modular network might differ from the original single DV. The cost is determined by both the number of required nodes and links necessary to fully connect the nodes. Cases for both with and without the buffer cylinder are considered for comparison purpose. An example network of 256x256 in modular and single DV is listed in Table 1.

Table 1. Hardware costs comparison for 256x256 DV

Data Vortex Type	Nodes	Links
Non-Buffered Single DV	2048	3840
Non-Buffered Modular DV	2048	3840
Buffered Single DV	2304	4352
Buffered Modular DV	2560	4864

The subset of header bits of the total $log_2 H$ header bits that are decoded in each stage is a parameter that can be varied to produce different modular DV configurations. In a 256x256 modular DV, there are 8 header bits, so each modular DV can be described by the number of bits that are decoded in each stage. For example, a 4-/4-bit modular DV would have 2 stages, each decoding 4 bits, made of 16x16 DVs.

3. PERFORMANCE EVALUATION

A simulator written in Java was used to compare the latency and throughput of a single 256x256 DV versus several variations of modular equivalents. Statistics in average latency in hops and throughput measured as successful injection rate were collected over sufficiently long simulation time. Random, uniform traffic was tested under loads of 10%, 50%, and 90%.

3.1 Non-Buffered Modular DV

With the buffering cylinder omitted, the cost of the non-buffered single DV and non-buffered modular DV is the same as shown in Table 1. Under the same traffic load, the routing performance of the single DV and modular DVs in two-stage configurations is found to be exactly the same for various subsets of header bits in each stage.

3.2 Buffered Modular DV

The inclusion of the buffering cylinder in the sub-DVs of a modular DV increases the number of nodes and links each by about 11% as listed in Table 1. Figure 3 (a) and (b) show the modular DV performance in comparison to the single DV in throughput and latency respectively. Overall, the performances are in a comparable range. For higher traffic loads of 50% and 90%, the throughput is shown to improve in the modular DV cases with a slight degradation of average latency.

Even though there is about an 11% cost increase in the buffered modular DV when compared to the buffered single DV, there is up to a 15% performance increase with respect to throughput in the symmetrical 4-/4-bit case shown in Figure 3(a). In all 5 tested configurations of the modular DV, there is a slight 6% performance decrease in the average latency under a 10% traffic load as compared in Figure 2(b). However, in the 5-/3-bit and 6-/2-bit configurations, there is a slight performance increase in average latency under heavier 50% and 90% traffic loads.

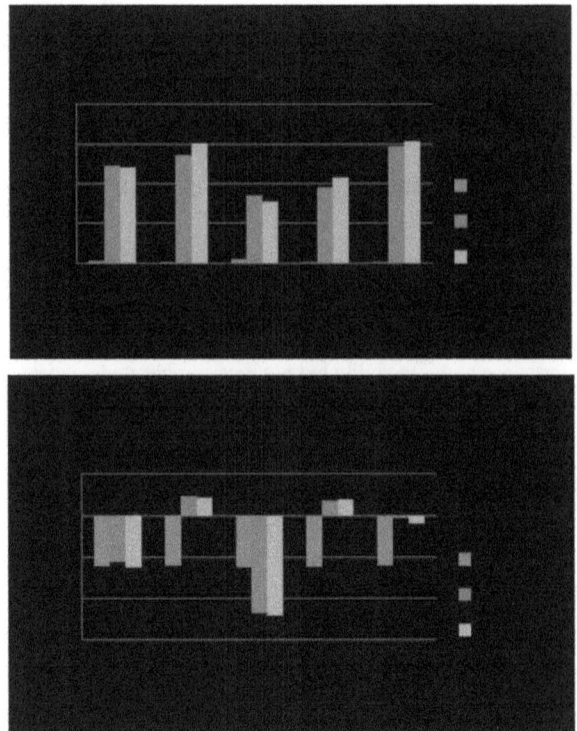

Figure 3. (a) Improvement of successful injection rate, and (b) improvement of average latency for modular DVs relative to performance of single, buffered DV.

4. CONCLUSIONS

The modular DV simplifies the connections typically required in a single, large DV and allows design flexibility within each sub-DV. The non-buffered modular DV exhibited the same routing performance as its non-buffered single DV counterpart in simulation with no cost addition. The buffered modular DV slightly increases the cost over the single buffered DV but has similar performance characteristics, and in some configurations under heavier traffic conditions improves performance.

5. ACKNOWLEDGMENTS

We thank the Rose Hill Research Grant for supporting the study.

6. REFERENCES

[1] Ilias Iliadis, etal, "Performance Evaluation of the Data Vortex Photonic Switch", *IEEE Journal of Selected Areas in Communications*, Vol. 25, No.6, pp. 20-35, Aug 2007.

[2] A. Shacham etal, "A Fully Implemented 12x12 Data Vortex Optical Packet Switching Interconnection Network," *Journal of Lightwave Technology*, vol. 23, No.10, pp. 3066-3075, Oct 2005.

[3] O. Liboiron-Ladouceur, etal, "Physical layer scalability of a WDM optical packet interconnection network," *Journal of Lightwave Technology*, vol. 24, pp. 262-270, Jan. 2006.

[4] Jan Cheyns, etal, "Clos Lives On in Optical Packet Switching", *IEEE Communication Magazine*, pp.114-121, Feb 2004.

PVNs: Making Virtualized Network Infrastructure Usable

Shufeng Huang, James Griffioen and Ken Calvert
Laboratory for Advanced Networking
Department of Computer Science
University of Kentucky
Lexington, KY 40506
{shufeng,griff,calvert}@netlab.uky.edu

ABSTRACT

Network virtualization is becoming a fundamental building block of future Internet architectures. Although the underlying network infrastructure needed to dynamically create and deploy custom virtual networks is rapidly taking shape (e.g., GENI), constructing and using a virtual network is still a challenging and labor intensive task, one best left to experts.

In this paper, we present the concept of a Packaged Virtual Network (PVN), that enables normal users to easily download, deploy and use application-specific virtual networks. At the heart of our approach is a PVN Hypervisor that "runs" a PVN by allocating the virtual network resources needed by the PVN and then connecting the PVN's participants into the network on demand. To demonstrate our PVN approach, we implemented a multicast PVN that runs on the PVN hypervisor prototype using ProtoGENI as the underlying virtual network, allowing average users to create their own private multicast network.

Categories and Subject Descriptors

C.2.1 [**Computer-Communication Networks**]: Network Architecture and Design

Keywords

Virtual Network, GENI

1. INTRODUCTION

While virtual machines (VMs) have been rapidly replacing physical machines, *virtual networks* have been slower to appear. Early examples include PlanetLab [7, 1] and Emulab [5, 2], which allowed users to reserve virtual machines ("slivers") for use as virtualized network routers. The emerging GENI [3] network offers a wide range of (virtualized) network resources that span the nation and enable wide-area virtual networks. In GENI terminology, each virtual network is a "slice" consisting of "slivers" connected via virtual links/channels. More recently, hardware vendors have begun to support network virtualization in commercial hardware. Juniper routers, for example, support "logical routers" [4], and a variety of other router vendors now support the *Open-Flow* [6] protocol and architecture. In short, it is becoming

clear that network virtualization is becoming a fundamental building block for future networks.

However, unlike a virtual machine, which has a relatively limited number of configuration options, a virtual network can be instantiated in an endless number of ways depending on the resources (virtual routers and links) selected for inclusion. Determining which virtual routers to include in the virtual network, and how those routers should be interconnected into a topology is a non-trivial process. Moreover, having identified and allocated the virtual network resources, one must load, configure, and initialize the software for all the resources that make up the virtual network—an even more challenging task. Today, creating a virtual network is a task best left for network experts or ISPs that see a niche market and decide to create a special-purpose network for a particular type of traffic.

To address this problem, we present a new model, called *Packaged Virtual Networks (PVNs)*. PVNs enable users to easily download, deploy, and use custom virtual networks designed for a specific purpose (e.g., video conferencing). Inspired by the concept of a *virtual appliance* [9], PVNs bundle together the network topology configuration, software, and network services needed to create and deploy a custom virtual network. The idea is that users download PVNs from PVN repositories and then "run" them on virtualized network infrastructure, in the same way users download and run virtual appliances on a virtual machine.

2. THE PVN ARCHITECTURE

A PVN can be thought of as a "virtual network image", much like a virtual appliance can be thought of as a "virtual machine image". Virtual appliances are run on virtual machines (VMs), while PVNs are run on *Virtual Network Infrastructure Providers (VNIPs)*. A VNIP typically owns and operates physical network infrastructure, but allows that infrastructure to be virtualized or "sliced". We expect that, like current ISPs, VNIPs would "sell" virtualized network resources for a fee. Much like there are many ISPs today, we envision there being many different VNIPs. In most cases, PVNs will want to create "slices" that span multiple VNIPs. Fig 1 illustrates the PVN architecture with various PVNs being build on top of several different VNIPs.

Much like a virtual appliance relies on a hypervisor to provide the interface to the underlying VM, the PVN abstraction relies on a *PVN hypervisor* to provide the interface between a PVN and the underlying VNIP hardware. The PVN hypervisor "executes" the PVN and, based on instructions from the PVN, will then identify , reserve and stitch

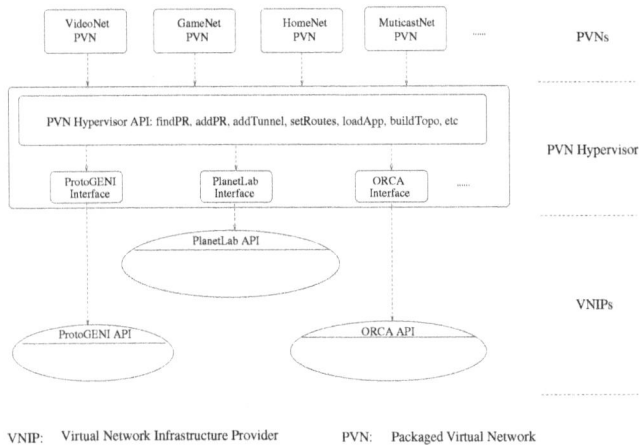

VNIP: Virtual Network Infrastructure Provider   PVN: Packaged Virtual Network

Figure 1: The PVN Architecture

together the VNIP resources needed to create the virtual network required by the PVN. To provide a standard platform for PVNs to run on, the PVN hypervisor exports a standard set of API calls that enable PVNs to create and manage virtual networks. Example API calls include:

- **findPR():** Find the virtual (programmable) router closest to a particular participant.

- **addPR(), addTunnel():** Reserve a virtual (programmable) router, or create a virtual link between two virtual routers and add them into the virtual network.

- **setAddress(), setRoute(), loadApp():** Set the address, routing table of a virtual router, or load a specific application onto a virtual router.

- **seeTopo(), seePath(), findCentralNode():** Explore the complete physical topology[1]; explore the shortest path between nodes; find the central node among a set of participants.

- **buildTopo():** Deploy the virtual network on VNIPs.

These are only a subset of the API calls available to the designers of PVNs. Each PVN offers a different application-specific virtual network, and will use these API calls to create a virtual network designed specifically for a particular network application and its participants.

Any number of PVN hypervisors can coexist. Hypervisors purchase resources and services in bulk from VNIPs and resell them to PVN users thereby avoiding the need for every PVN user to establish business relationships with every VNIP. As a result, the PVN hypervisor acts as a broker between users and VNIPs.

3. A PROTOTYPE PVN HYPERVISOR

To demostrate our PVN architecture, we implemented a prototype of the PVN Hypervisor using ProtoGENI as the VNIP. Our hypervisor implementation includes a VNIP interface which talks to ProtoGENI VNIP (via the GENI Aggregate Manager API[8]) to discover, manage and monitor ProtoGENI resources. The interface to the hypervisor is the

set of Hypervisor APIs described earlier. The heart of the hypervisor consists of three logical components: 1) an Information Base, 2) a Location Manager and 3) a Topology and Routing Server (TS/RS). The Information Base keeps a current view of the physical resources provided by the VNIPs, as well as all running PVNs. It pulls physical topology information from VNIPs (via the VNIP interface) and updates resource availability upon creation of new PVN virtual networks, tearing down existing ones. The Location Manager supports the **findPR()** API call. Each VNIP issues probe messages to determine the location of participants relative to that VNIP. The Location Manager then finds the closest virtual (programmable) router by collecting the probe results. The TS/RS explores the underlying physical topology. It gets updated topology information from the Information Base and runs routing algorithms to determine the shortest path, the central node, etc.

By using the hypervisor API calls, with less than 300 lines of java code, we built a "multicast PVN" that can be deployed by an average user to create a fully functional multicast virtual network. Our PVN package automatically reserves a tree topology, picks the central node of the tree as the rendezvous point, and loads and runs PIM Sparse Mode Multicast protocols on each tree node.

4. ACKNOWLEDGMENTS

This work supported in part by the National Science Foundation under grants CNS-0626918 and CNS-1111040.

5. REFERENCES

[1] https://www.planet-lab.org.

[2] http://www.emulab.net.

[3] GPO. Global Environment for Network Innovations - System Requirements Document. 2009.

[4] Juniper. Juniper M7I Router. http://www.juniper.net.

[5] W. D. Laverell, Z. Fei, and J. N. Griffioen. Isn't it Time You Had an Emulab? In *Proceedings of the 39th SIGCSE technical symposium on Computer science education*, SIGCSE '08, pages 246–250, New York, NY, USA, 2008. ACM.

[6] N. McKeown, T. Anderson, H. Balakrishnan, G. Parulkar, L. Peterson, J. Rexford, S. Shenker, and J. Turner. Openflow: Enabling innovation in campus networks. *SIGCOMM Comput. Commun. Rev.*, 38(2):69–74, 2008.

[7] L. Peterson, V. Pai, N. Spring, and A. Bavier. Using PlanetLab for Network Research: Myths, Realities, and Best Practices. Technical Report PDN–05–028, PlanetLab Consortium, June 2005.

[8] L. Peterson, S. Sevinc, J. Lepreau, R. Ricci, J. Wroclawski, T. Faber, S. Schwab, and S. Baker. Slice-Based Facility Architecture. 2009.

[9] vmware. Vmware Virtual Appliances. http://www.vmware.com/appliances.

---

[1]The actual topology exposed to the caller is controlled by the hypervisor.

Securing Multi-core Multi-threaded Packet Processors

Danai Chasaki
Department of Electrical and Computer Engineering
Villanova University, Villanova, PA, USA
{danai.chasaki}@villanova.edu

ABSTRACT

Modern routers use high-performance multi-core multi-threaded packet processing systems to implement protocol operations and to forward traffic. As the number of processor cores/threads increases, it becomes increasingly difficult to track their correct operation at runtime. In this paper, we discuss how to extent our existing monitoring scheme to support highly parallel environments.

Categories and Subject Descriptors

C.2.6 [**Computer-Communication Networks**]: Internetworking—*Routers*

Keywords

multi-core processor, attack, runtime monitoring

1. INTRODUCTION

Over the last few years, the advancement in the performance of general-purpose multi-core processors has enabled the development of routers with highly parallel, embedded MPSoCs [4] as integral components. To take it one step further in terms of performance, multithreading is currently discussed for next-generation packet processors. For example, virtualization demands larger numbers of parallel slices where multiple hardware threads share each processor core.

When the data path of routers was still ASIC-based, it did not present a potential target for attacks. With the use of programmable components in the data path, this premise has changed. In the data plane of the network, where the actual network traffic is transmitted between end-systems and routers, attackers may aim to eavesdrop on or intercept communications. In our recent work, we have shown that vulnerabilities in the protocol processing functions of routers can be exploited [3].

Defense mechanisms against such attacks in single-core processors have been proposed in our previous work [1]. In this paper, we extend our monitoring approach to address processing security in the context of multi-core and multi-threaded packet processors. We show that defense via monitoring in multi-core multi-threaded packet processing systems scales in a way that takes care of both security and performance constraints. This is achieved by sharing monitoring resources among threads and processors.

ANCS'12, October 29–30, 2012, Austin, Texas, USA.
ACM 978-1-4503-1685-9/12/10.

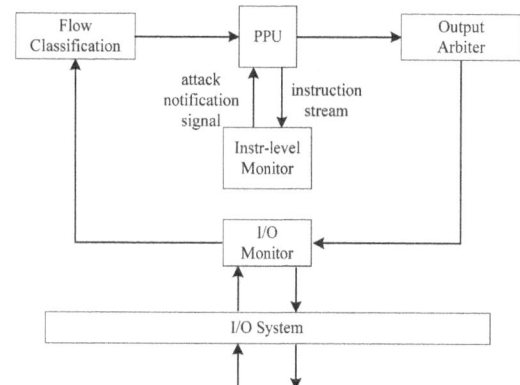

Figure 1: Security Monitors for Single-core.

2. SECURITY ON SINGLE CORE PACKET PROCESSORS

In [3] we have demonstrated the feasibility and effects of an in-network denial-of-service attack in the data plane. With just one malicious packet we exploit an integer vulnerability and manage to absorb the whole system bandwidth on a custom packet processor. This packet can propagate the effect to downstream routers as well.

A way to counter this type of attacks is to use a processing monitor which tracks the operations on the network processor, as we proposed in [1]. Figure 1 shows an overview of the instruction-level monitor working in parallel with the packet processing unit (PPU). The monitor detects an attack if the processor's operations deviate from the operations that are valid - as determined by offline analysis of the processing binary. Using the instruction-level monitor we can effectively monitor the program execution as a whole. Our resource utilization and performance results on a single-core processor in [1] show that the security monitor burdens the packet processing system with only 0.8 % additional hardware resources (LUTs) and does not slow down the core's speed [1].

Due to vulnerabilities in the data path, we can expect attacks at the protocol level as well. We can have a situation where valid processing instructions are executed on the network processor, but still the overall router behavior is abnormal. For this reason, we enhance our security mechanism with an extra monitor, the I/O monitor, which tracks the system operation as a whole. This monitor analyzes the processing of individual modules and validates data-path processing by profiling the processing time and delay that

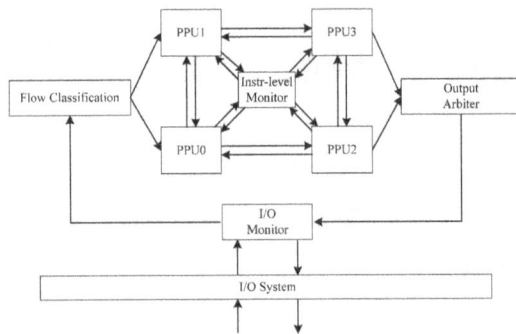

Figure 2: Security Monitors for Multi-core.

each packet encounters in every module. Such characteristics depend on the functionality of each module and help us determine if the system encounters any unusual delays while processing the packets [2]. A single I/O monitor is used for data flow validation of the whole packet processing system, be it a single-core or multi-core architecture.

3. SECURITY ON MULTI-CORE PACKET PROCESSORS

If we try to extrapolate the monitoring idea to a multi-core environment we could have one instruction-level monitor assigned to each core which validates each processor's behavior. This design clearly wastes hardware resources since we could have all four cores share the same monitor as shown in Figure 2. Our monitoring scheme is storing essential information for processing validation in on-chip memory. The level of monitor reusability depends on the characteristics of that memory. In our design, the block ram memory has two read ports, so we can read from both of them simultaneously. If we also clock the memory read at twice the speed of the monitoring logic we can achieve 4-way sharing of the same monitor. This means that we can have secure packet processing on a 4-core design by paying the same resource utilization penalty as in Section 2.

4. SECURITY ON MULTI-CORE MULTI-THREADED PACKET PROCESSORS

Multiple hardware threads are supported by modern packet processing systems. Multiple threads can share each processor, and processors have to maintain state for each thread in order to transition smoothly between thread executions. Hardware-supported multithreading has the objective of more efficient core utilization by avoiding processor stalls. In the same way that threads share a processor core, they can also share the hardware monitor that runs in parallel to the processor. The only limitation is that since threads have to read frequently from our monitor's dual-port memory, only four threads can share the same monitor given our memory architecture and our design's clock frequency.

Figure 3 shows the design for a 4-core system that supports 4-way multithreading. Each core's threads use the same security monitor. Alternatively, the security monitors can be shared by threads across cores. Table 1 compares the number of monitors required to secure several packet processing architectures. Since every monitor consumes very little hardware resources, the system would scale well even for

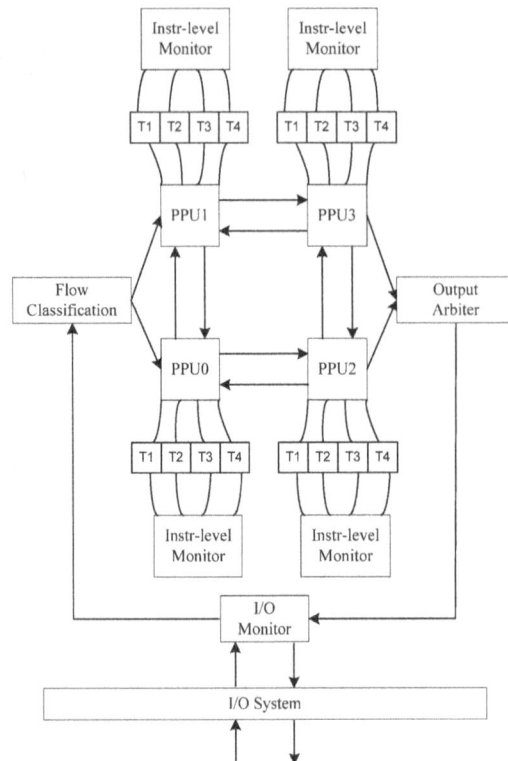

Figure 3: Security Monitors for Multi-core Multi-threaded.

Table 1: Sharing of Monitors in Different Packet Processing Systems

Configuration	Number of Monitors
4-Core	1
16-Core	4
64-Core	16
4-Core 2-way multithreading	2
4-Core 4-way multithreading	4

a 64-core configuration. However, for dense multi-threaded processing configurations, we would have to explore multi-ported memory architectures, that allow more than 8 memory reads simultaneously [5].

5. REFERENCES

[1] CHASAKI, D., AND WOLF, T. Design of a secure packet processor. In *Proc. of ACM/IEEE Symposium on Architectures for Networking and Communication Systems (ANCS)* (San Diego, CA, Oct. 2010).

[2] CHASAKI, D., WU, Q., AND WOLF, T. Inferring packet processing behavior using input/output monitors. In *ANCS 2011*.

[3] CHASAKI, DANAI, W. Q., AND WOLF, T. Attacks on network infrastructure. In *Proc. of IEEE International Conference on Computer Communications and Networks (ICCCN)* (Aug. 2011).

[4] CISCO SYSTEMS, INC. *The Cisco QuantumFlow Processor.* San Jose, CA, Feb. 2008.

[5] LAFOREST, C. E., AND STEFFAN, J. G. Efficient multi-ported memories for fpgas. In *Proceedings of the International Symposium on Field programmable gate arrays* (2010), FPGA '10.

A Hardware Spinal Decoder

Peter A. Iannucci, Kermin Elliott Fleming, Jonathan Perry, Hari Balakrishnan, and
Devavrat Shah
Massachusetts Institute of Technology
Cambridge, Mass., USA
{iannucci,kfleming,yonch,hari,devavrat}@mit.edu

ABSTRACT

Spinal codes are a recently proposed capacity-achieving rateless code. While hardware encoding of spinal codes is straightforward, the design of an efficient, high-speed hardware decoder poses significant challenges. We present the first such decoder. By relaxing data dependencies inherent in the classic M-algorithm decoder, we obtain area and throughput competitive with 3GPP turbo codes as well as greatly reduced latency and complexity. The enabling architectural feature is a novel "α-β" incremental approximate selection algorithm. We also present a method for obtaining hints which anticipate successful or failed decoding, permitting early termination and/or feedback-driven adaptation of the decoding parameters.

We have validated our implementation in FPGA with on-air testing. Provisional hardware synthesis suggests that a near-capacity implementation of spinal codes can achieve a throughput of 12.5 Mbps in a 65 nm technology while using substantially less area than competitive 3GPP turbo code implementations.

Categories and Subject Descriptors: B.4.1 [Data Communications Devices]: Receivers; C.2.1 [Network Architecture and Design]: Wireless communication

General Terms: Algorithms, Design, Performance

Keywords: Wireless, rateless, spinal, decoder, architecture

1. INTRODUCTION

At the heart of every wireless communication system lies a *channel code*, which incorporates methods for error correction. At the transmitter, an encoder takes a sequence of message bits (e.g., belonging to a single packet or link-layer frame) and produces a sequence of coded bits or coded symbols for transmission. At the receiver, a decoder takes the (noisy or corrupted) sequence of received symbols or bits and "inverts" the encoding operation to produce its best estimate of the original message bits. If the recovered message bits are identical to the original, then the reception is error-free; otherwise, the communication is not reliable and additional actions have to be taken to achieve reliability (these actions may be taken at the physical, link, or transport layers of the stack).

The search for good, practical codes has a long history, starting from Shannon's fundamental results that developed the notion of

channel capacity and established the *existence* of capacity-achieving codes. Shannon's work did not, however, show how to construct and decode practical codes, but it set the basis for decades of work on methods such as convolutional codes, low-density parity check (LDPC) codes, turbo codes, Raptor codes, and so on. Modern wireless communication networks use one or more of these codes.

Our interest is in *rateless codes*, defined as codes for which any encoding of a higher rate is a prefix of any lower-rate encoding (the "prefix property"). Rateless codes are interesting because they offer a way to achieve high throughput over *time-varying* wireless networks: a good rateless code inherently sends only as much data as required to communicate reliably under any given channel conditions. As conditions change, a good rateless code adapts naturally.

In recent work, we proposed and evaluated in simulation the performance of *spinal codes*, a new family of rateless codes for wireless networks. Theoretically, spinal codes are the first rateless code with an efficient (i.e., polynomial-time) encoder and decoder that essentially achieve Shannon capacity over both the additive white Gaussian noise (AWGN) channel and the binary symmetric channel (BSC).

In practice, however, polynomial-time encoding and decoding complexity is a necessary, but hardly sufficient, condition for high throughput wireless networks. The efficacy of a high-speed channel code is highly dependent on an efficient hardware implementation. In general, the challenges include parallelizing the required computation, and reducing the storage requirement to a manageable level.

This paper presents the design, implementation, and evaluation of a hardware architecture for spinal codes. The encoder is straightforward, but the decoder is tricky. Unlike convolutional decoders, which operate on a finite trellis structure, spinal codes operate on an exponentially growing tree. The amount of exploration the decoder can afford has an effect on throughput: if a decoder computes sparingly, it will require more symbols to decode and thus achieve lower throughput. This effect is shown in Figure 1. A naïve decoder targeted to achieve the greatest possible coding gain would require hardware resources to store and sort upwards of a thousand tree paths per bit of data, which is beyond the realm of practicality.

Our principal contribution is a set of techniques that enable the construction of a high-fidelity hardware spinal decoder with area and throughput characteristics competitive with widely-deployed cellular error correction algorithms. These techniques include:

1. a novel method to select the best B states to maintain in the tree exploration at each stage, called "α-β" incremental approximate selection, and

2. a method for obtaining hints to anticipate successful or failed decoding, which permits early termination and/or feedback-driven adaptation of the decoding parameters.

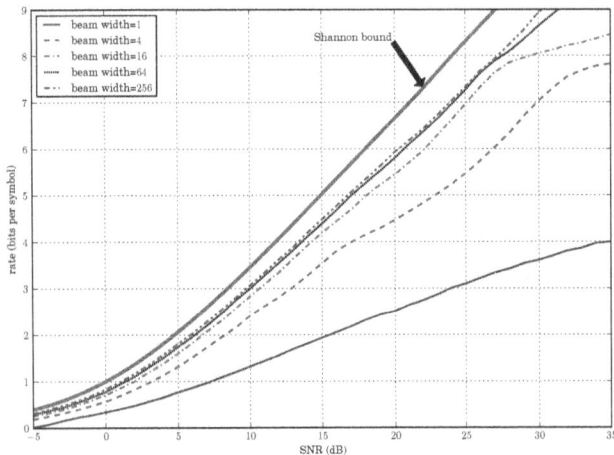

Figure 1: **Coding efficiency achieved by the spinal decoder increases with the width of the explored portion of the tree. Hardware designs that permit wide exploration are desirable.**

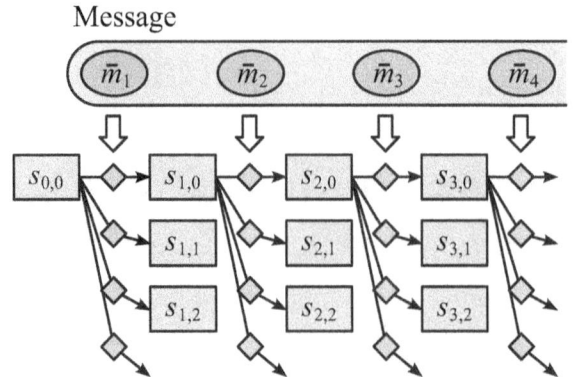

Figure 2: **Computation of pseudo-random words $s_{i,j}$ in the encoder, with hash function application depicted by a diamond. Each $\bar{m}_i$ is k message bits.**

We have validated our hardware design with an FPGA implementation and on-air testing. A provisional hardware synthesis suggests that a near-capacity implementation of spinal codes can achieve a throughput of 12.5 Megabits/s in a 65 nm technology while using substantially less area than competitive 3GPP turbo code implementations.

2. BACKGROUND & RELATED WORK

Wireless devices taking advantage of ratelessness can transmit at more aggressive rates and achieve higher throughput than devices using fixed-rate codes, which suffer a more substantial penalty in the event of a retransmission. Hybrid automatic repeat request (HARQ) protocols reduce the penalty of retransmission by puncturing a fixed-rate "mother code". These protocols typically also require the use of *ad hoc* channel quality indications to choose an appropriate signaling constellation, and involve a demapping step to convert I and Q values to "soft bits", which occupy a comparatively large amount of storage.

Spinal codes do not require constellation adaptation, and do not require demapping, instead operating directly on I and Q values. Spinal codes also impose no minimum rate, with encoding and decoding complexity polynomial in the number of symbols transmitted. They also retain the sequentiality, and hence potential for low latency, of convolutional codes while offering performance comparable to iteratively-decoded turbo and LDPC codes.

2.1 Spinal Codes

For context, we review the salient details of spinal codes [21, 20].

The principle of the spinal encoder is to produce pseudo-random bits from the message in a sequential way, then map these bits to output constellation points. As with convolutional codes, each encoder output depends only on a prefix of the message. This enables the decoder to recover a few bits of the message at a time rather than searching the huge space of all messages.

Most of the complexity of the encoder lies in defining a suitable sequential pseudo-random generator. Most of the complexity of the decoder lies in determining the heuristically best (fastest, most reliable) way to search for the right message.

Encoder. The encoder breaks the input message into k-bit pieces $\bar{m}_i$, where typically $k = 4$. These pieces are hashed together to obtain a pool of pseudo-random 32-bit words $s_{i,j}$ as shown in Figure 2. The

initial value $s_{0,0} = 0$. Note that each hash depends on k message bits, the previous hash, and the value of j. The hash function need not be cryptographic.

Once a certain hash $s_{i,j}$ is computed, the encoder breaks it into c-bit pieces and passes each one through a *constellation map* $f(\cdot)$ to get $\lfloor 32/c \rfloor$ real, fixed-point numbers. The numbers generated from hashes $s_{i,0}, s_{i,1} \ldots$ are indexed by ℓ to form the sequence $x_{i,\ell}$.

The $x_{i,\ell}$ are reordered for transmitting so that resilience to noise will increase smoothly with the number of received constellation points. Symbols are transmitted in *passes* indexed by ℓ. Within a pass, indices i are ordered by a fixed, known permutation [21].

Decoder. The algorithm for decoding spinal codes is to perform a pruned breadth-first search through the tree of possible messages. Each edge in this tree corresponds to k bits of the message, so the out-degree of each node is 2^k, and a complete path from the root to a leaf has N edges. To keep the computation small, only a fixed number B of nodes will be kept alive at a given depth in the tree. B is named after the analogy with beam search, and the list of B alive nodes is called the beam. At each step, we explore all of the $B \cdot 2^k$ children of these nodes and score each one according to the amount of signal variance that remains after subtracting the corresponding encoded message from the received signal. Lower scores (path metrics) are better. We then prune all but the B lowest-scoring nodes, and move on to the next k bits. With high probability, if enough passes have been received to decode the message, one of the B leaves recovered at the end will be the correct message. Just as convolutional codes can be terminated to ensure equal protection of the tail bits, spinal codes can transmit extra symbols from the end of the message to ensure that the correct message is not merely one of the B leaves, but the best one.

The decoder operates over received samples $y_{i,\ell}$ and candidate messages encoded as $\hat{x}_{i,\ell}$. Scores are sums of $(y_{i,\ell} - \hat{x}_{i,\ell})^2$. Formally, this sum is proportional to the log likelihood of the candidate message. The intuition is that the correct message will have a lower path metric in expectation than any incorrect message, and the difference will be large enough to distinguish if SNR is high or there are enough passes. "Large enough" means that fluctuations do not cause the correct message to score worse than B other messages.

To make this more concrete, consider the AWGN channel with $y = x + n$, where the noise n is independent of x. We see that $\text{Var}(y) = \text{Var}(x) + \text{Var}(n) = P \cdot (1 + \frac{1}{\text{SNR}})$, where P is the power of the received signal. If $\hat{x} = x$, then $\text{Var}(y - \hat{x}) = \frac{P}{\text{SNR}}$. Otherwise,

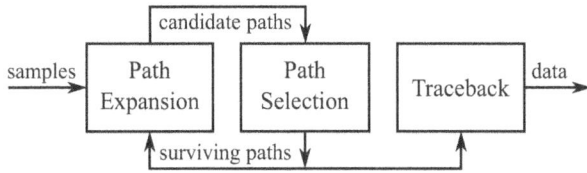

Figure 3: Block diagram of M-algorithm hardware.

$\mathrm{Var}(y - \hat{x}) = P \cdot (2 + \frac{1}{\mathrm{SNR}})$. The sum of squared differences is an estimator of this variance and discriminates between the two cases.

2.2 Existing M-Algorithm Implementations

The decoder described above is essentially the M-algorithm (MA) [2]. A block diagram of MA is shown in Figure 3. In our notation, the expansion phase grows each of B paths by one edge to obtain $B \cdot 2^k$ new paths, and calculates the path metric for each one. The selection stage chooses the best B of these, and the last stage performs a Viterbi-style [18] traceback over a window of survivor paths to obtain the output bits.

There have been few recent VLSI implementations of MA, in part because modern commercial wireless error correction codes operate on a small trellis [1]. It is practical to instantiate a full Viterbi [9] or BCJR [3] decoder for such a trellis in silicon. MA is an approximation designed to reduce the cost of searching through a large trellis or a tree, and consequently it is unlikely to compete with the optimal Viterbi or BCJR decoders in performance or area for such codes. As a result, the M-algorithm is not generally commercially deployed. Existing academic implementations [10] [22] focus on implementing decoders for rate 1/2 convolutional codes.

These works recognize that the sorting network is the chief bottleneck of the system, and generally focus on various different algorithms for achieving implementations. However, these implementations deal with very small values of B and k, for instance $B = 16$ and $k = 2$, for which a complete sorting network is implementable in hardware. Spinal codes on the other hand require B and k to be much larger in order to achieve maximum performance. Much of the novel work in this paper will focus on achieving high-quality decoding while minimizing the size of the sort network that must be constructed.

The M-algorithm implementation in [22] leverages a degree of partial sorting among the generated $B \cdot 2^k$ nodes at the expansion stage. Although our implementation does not use their technique, their work is, to the best of our knowledge, the first to recognize that a full sort is not necessary to achieve good performance in the M-algorithm.

The M-algorithm is also known as beam search in the AI literature. Beam search implementations do appear as part of hardware-centric systems, particularly in the speech recognition literature [15] where they are used to solve Hidden-Markov Models describing human speech. However, in AI applications, computation is typically dominated by direct sensor analysis, while beam search which appears at a high level of the system stack where throughput demands are much lower. As a result, there seems to have no attempt to create a full hardware beam search implementation in the AI community.

3. SYSTEM ARCHITECTURE

Our decoder is designed to be layered with an inner OFDM or CDMA receiver, so we are not concerned with synchronization or equalization. The decoder's inputs are the real and imaginary parts (I and Q) of the received (sub)carrier samples, in the same order that the encoder produced its outputs x_n. The first decoding step is to invert the encoder's permutation arithmetic and recover the matrix

$y_{i,\ell}$ corresponding to $x_{i,\ell}$. Because of the sequential structure of the encoder, $y_{i,\ell}$ depends on $\bar{m}_{1\ldots i}$, the first ik bits of the message. Each depth in the decoding tree corresponds to an index i and some number of samples $y_{i,\ell}$.

The precise number of samples available for some i depends on the permutation and the total number of samples that have been received. In normal operation there may be anywhere from 0 to, say, 24 passes' worth of samples stored in the sample memory. The upper limit determines the size of the memory.

To compute a score for some node in the decoding tree, the decoder produces the encoded symbols $\hat{x}_{i,\ell}$ for the current i (via the hash function and constellation map) and subtracts them from $y_{i,\ell}$. The new score is the sum of these squared differences plus the score of the parent node at depth $i - 1$. In order to reach the highest level of performance shown in Figure 1, we need to defer pruning for as long as possible. Intuitively, this gives the central limit theorem time to operate – the more squared differences we accumulate, the more distinguishable the correct and incorrect scores will be. This requires us to keep a lot of candidates alive (ideally $B = 64$ to 256) and to explore a large number of children as quickly as possible.

There are three main implementation challenges, corresponding to the three blocks shown in Figure 3. The first is to calculate $B \cdot 2^k$ scores at each stage of decoding. Fortunately, these calculations have identical data dependencies, so arbitrarily many can be run in parallel. The calculation at each node depends on the hash $s_{i-1,0}$ from its parent node, a proposal $\hat{m}_i$ for the next k bits of data, and the samples $y_{i,\ell}$. We discuss optimizations of the path expansion unit in §5.

The second problem is to select the best B of $B \cdot 2^k$ scores to keep for the next stage of path expansion. This step is apparently an all-to-all shuffle. Worse yet, it is in the critical path, since computation at the next depth in the decoding tree cannot begin until the surviving candidates are known. In §4 we describe a surprisingly good approximation that relaxes the data dependencies in this step and allows us to pipeline the selection process aggressively.

The third problem is to trace back through the tree of unpruned candidates to recover the correct decoded bits. When operating close to the Shannon limit (low SNR or few passes), it is not sufficient, for instance, to put out the k bits corresponding to the best of the B candidates. Viterbi solves this problem for convolutional codes using a register-exchange approach reliant on the fixed trellis structure. Since the spinal decoding tree is irregular, we need a memory to hold data and back-track pointers. We show in §6 how we keep this memory small and minimize the time spent tracing back through the memory, while also obtaining valuable decoding hints.

While we could imagine building $B \cdot 2^k$ path metric blocks and a selection network from $B \cdot 2^k$ inputs to B outputs, such a design is too large, occupying up to 1.2 cm^2 (for $B = 256$) in a 65 nm process. Worse, the vast majority of the device would be dark at any given time: data would be either moving through the metric units, or it would be at some stage in the selection network. Keeping all of the hardware busy would require pipelining dozens of simultaneous decodes, with a commensurate storage requirement.

3.1 Initial Design

The first step towards a workable design is to back away from computing all of the path metrics simultaneously. This reduces the area required for metric units and frees us from the burden of sorting $B \cdot 2^k$ items at once. Suppose that we have some number W of path metric units (informally, *workers*), and we merge their W outputs into a register holding the best B outputs so far. If we let $W = 64$, the selection network can be reduced in area by a factor of 78 and in latency by a factor of three relative to the all-at-once design, and workers also occupy 1/64 as much area. The cost is that 64 times

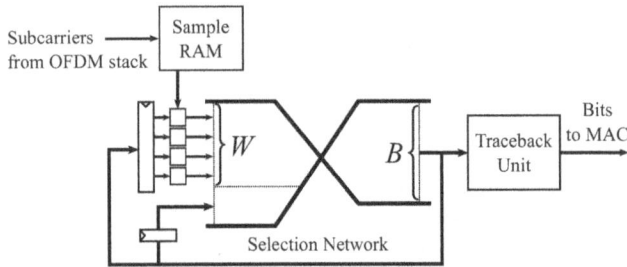

Figure 4: The initial decoder design with W workers and no pipelining.

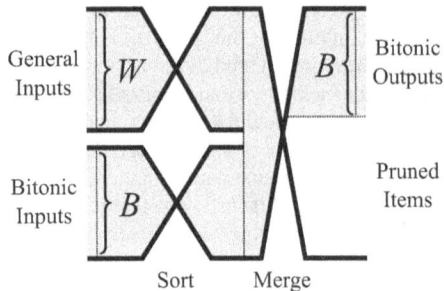

Figure 5: Detail of the incremental selection network.

Beam Width	8 Workers	16 Workers	32 Workers
8	14601		
16	22224	44898	
32	39772	61389	122575

Table 1: Area usage for various bitonic sorters in μm² using a 65 nm process. An 802.11g Viterbi implementation requires 120000 μm² in this process.

as many cycles are needed to complete a decode. This design is depicted in Figure 4. The procedure for merging into the register is detailed in §4.1.

4. PATH SELECTION

To address the problem of performing selection efficiently, we describe a series of improvements to the sort-everything-at-once baseline. We require the algorithm to be streaming, so that candidates are computed only once and storage requirements are minimal. Thus, during each cycle, the selection network must combine a number of fresh inputs with the surviving inputs from previous cycles, and prune away inputs which score poorly.

4.1 Incremental Selection

This network design accepts W fresh items and B old items, and produces the B best of these, allowing candidates to be generated over multiple cycles (Figure 5). While the W fresh items are in arbitrary order, it is possible to take advantage of the fact that the B old items are previous outputs of the selection network, and hence can be sorted or partially sorted if we wish. In particular, if we can get independently sorted lists of the best B old candidates and the best B new candidates, we can merge the lists in a single step by reversing one list and taking the pairwise min. The result will be in bitonic order (increasing then decreasing). Sorting a bitonic list is easier than sorting a general list, allowing us to save some comparators. We register the bitonic list from the merger, and restore it to sorted order in parallel with the sorting of the W fresh items. If $W \neq B$, a few more comparators can be optimized away. We use the bitonic sort because it is regular and parametric. Irregular or non-parametric sorts are known which use fewer comparators, and can be used as drop-in replacements.

4.2 Pipelined Selection

The original formulation of the decoder has a long critical path, most of which is spent in the selection network. This limits the throughput of the system at high data rates, since the output of the selection network is recirculated and merged with the next set of W outputs from the metric units. This dependency means that even if we pipeline the selection, we will not improve performance unless we find another way to keep the pipeline full.

Fortunately, candidate expansion is perfectly parallel and sorting is commutative. To achieve pipelining, we divide the $B \cdot 2^k$ candidates into α independent threads of processing. Now we can fill the selection pipeline by recirculating merged outputs for each thread independently, relaxing the data dependency. Each stage of the pipeline operates on an independent thread.

This increases the frequency of the entire system without introducing a dependency bottleneck. Registers are inserted into the pipeline at fixed intervals, for instance after every one or two comparators.

At the end of the candidate expansion, we need to eliminate $B(\alpha - 1)$ candidates. This can be done as a merge step after sorting the α threads at the cost of around $\alpha \log \alpha$ cycles of added latency. This may be acceptable if α is small or if many cycles are spent expanding candidates ($B \cdot 2^k \gg W$).

4.3 α-β Approximate Selection

We now have pipeline parallelism, which helps us scale throughput by increasing clock frequency. However, we have yet to consider a means of scaling the B and k parameters of the original design. An increase in k improves the maximum throughput of the design linearly while increasing the amount of computation exponentially, making this direction unattractive. For fixed k, scaling B improves decoding strength.

In order to scale B, we need to combat the scaling of sort logic, which is $\Theta(B \log B) + \Theta(W \log^2 W)$ in area and $\Theta(\max(\log B, \log^2 W))$ in latency. Selection network area can quickly become significant, as shown in Table 1. Fortunately, we can dodge this cost without a significant reduction in decoding strength by relaxing the selection problem.

First, we observe that if candidates are randomly assorted among threads, then on average $\beta \triangleq \frac{B}{\alpha}$ of the best B will be in each thread. Just as it is unlikely for one poker player to be dealt all the aces in a deck, it is unlikely (under random assortment) for any thread to receive significantly more than β of the B best candidates.

Thus, rather than globally selecting the B best of $B \cdot 2^k$ candidates, we can approximate by locally selecting $\beta = \frac{B}{\alpha}$ from each thread. There are a number of compelling reasons to make this trade-off. Besides eliminating the extra merge step, it reduces the width of the selection network from B to β, since we no longer need to keep alive the B best items in each thread. This decreases area by more than a factor of α and may also improve operating frequency. We call the technique α-β selection.

The question remains whether α-β selection performs as well as B-best selection. The intuition about being dealt many aces turns out to be correct for the spinal decoder. The candidates which are improperly pruned (compared with the unmodified M-algorithm) are certainly not in the top β, and they are overwhelmingly unlikely to be in the top $B/2$. In the unlikely event that the correct candidate

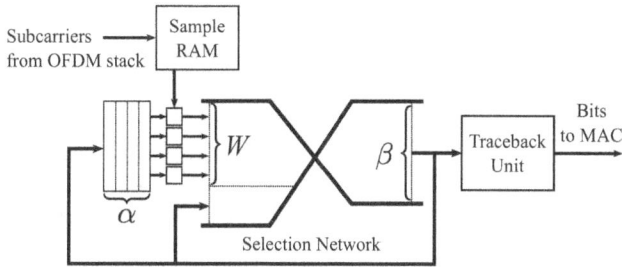

Figure 6: A parametric α-β spinal decoder with W workers. A shift register of depth α replaces the ordinary register in Figure 4. Individual stages of the selection pipeline are not shown, but the width of the network is reduced from B to $\beta = B/\alpha$.

is pruned, the packet will fail to decode until more passes arrive. A detailed analysis is given in §4.6.

Figure 6 is a block diagram of a decoder using α-β selection. Since each candidate expands to 2^k children, the storage in the pipeline is not sufficient to hold the B surviving candidates while their children are being generated. A shift register buffer of depth α placed at the front of the pipeline stores candidates while they await path expansion. We remark in passing that letting $\alpha = 1$, $\beta = B$ recovers the basic decoder described in §4.1.

4.4 Deterministic α-β Selection

One caveat up to this point has been the random assortment of the candidates among threads. Our hardware is expressly designed to keep only a handful of candidates alive at any given time, and consequently such a direct randomization is not feasible. We would prefer to use local operations to achieve the same guarantees, if possible.

Two observations lead to an online sorting mechanism that performs as well as random assortment. The first is that descendants of a common parent have highly correlated scores. Intuitively, the goal is not to spread the candidates randomly, but to spread them *uniformly*. Consequently, we place sibling candidates in different threads. In hardware this amounts to a simple reordering of operations, and entails no additional cost.

The second observation is that we can randomize the order in which child candidates are generated from their parents by scrambling the transmitted packet. The hash function structure of the code guarantees that all symbols are identically distributed, so the scores of incorrect children are i.i.d. conditioned on the score of their parents. This guarantees that a round-robin assignment of these candidates among the threads is a uniform assignment. The children of the correct parent are not i.i.d., since one differs by being the correct child. By scrambling the packet, we ensure that the correct child is assigned to a thread uniformly. The scrambler can be a small linear feedback shift register in the MAC, as in 802.11a/g.

The performance tradeoffs for these techniques are shown in Figure 9. Combining the two proposed optimizations achieves performance that is slightly better than a random shuffle.

4.5 Further Optimization

A further reduction of the α-β selection network is possible by concatenating multiple smaller selection networks as shown in Figure 8. This design has linear scaling with W. One disadvantage is that child candidates are less effectively spread among threads if a given worker only feeds into a single selection network. At the beginning of decoding, for instance, this would prevent the children of the root node from ever finding their way into the workers serving

the other selection networks, since no wires cross between the selection networks or the shift registers feeding the workers. A cheap solution is to interleave the candidates between the workers and the selection networks by wiring in a rotation by $\frac{W}{2\gamma}$. This divides each node's children across two selection networks at the next stage of decoding. A more robust solution is to multiplex between rotated and non-rotated wires with an alternating schedule.

4.6 Analysis of α-β Selection

We consider pipelining the process of selecting the B best items out of N (i.e. $B \cdot 2^k$). Our building block is a network which takes as input W unsorted items plus β bitonically presorted items, and produces β bitonically sorted items.

Suppose that registers are inserted into the selection network to form a pipeline of depth α. Since the output of the selection network will not be available for α clock cycles after the corresponding input, we will form the input for the selection network at each cycle as W new items plus the β outputs from α cycles ago. Cycle n only depends on cycles $n' \equiv n \pmod{\alpha}$, forming α separate threads of execution.

After N/W uses of the pipeline, all of the threads terminate, and we are left with α lists of β items. We'd like to know whether this is a good approximation to the algorithm which selects the $\alpha\beta$ best of the original N items. To show that it is, we state the following theorem.

THEOREM 1. *Consider a selection algorithm that divides its N inputs among N/n threads, each of which individually returns the best β of its n inputs, for a total of $N\beta/n$ results. We compare its output to the result of an ideal selection algorithm which returns precisely the $N\beta/n$ best of its N inputs. On randomly ordered inputs, the approximate output will contain all of the best m inputs with probability at least*

$$\mathbb{P} \geq 1 - \sum_{i=1}^{m} \sum_{j=\beta}^{n} \frac{\binom{n-1}{j}\binom{N-n}{i-j-1}}{\binom{N-1}{i-1}} \tag{1}$$

For e.g. $N = 4096$, $n = 512$, $\beta = 32$, this gives a probability of at least $1 - 3.1 \cdot 10^{-4}$ for all of the best 128 outputs to be correct, and a probability of at least $1/2$ for all of the best 188 outputs to be correct. Empirically, the probability for the best 128 outputs to be correct is $1 - 2.4 \cdot 10^{-4}$, so the bound is tight. The empirical result also shows that the best 204 outputs are correct at least half of the time.

PROOF. Suppose that the outputs are sorted from best to worst. Suppose also that the input consists of a random permutation of $(1, \ldots, N)$. For general input, we can imagine that each input has been replaced by the position at which it would appear in a list sorted from best to worst. Under this mapping, the exact selection algorithm would return precisely the list $(1, \ldots, N\beta/n)$. We can see that the best m inputs appear in the output list if and only if m is the mth integer in the list. Otherwise, some item $i \leq m$ must have been discarded by the algorithm. By the union bound,

$$\mathbb{P}(m^{\text{th}} \text{ output} \neq m) \leq \sum_{i=1}^{m} \mathbb{P}(i \text{ discarded})$$

An item i is discarded only when the thread it is assigned also finds at least β better items. So

$$\mathbb{P}(i \text{ discarded}) = \mathbb{P}(\exists \beta \text{ items} < i \text{ in same thread})$$

$$= \sum_{j=\beta}^{n} \mathbb{P}(\text{exactly } j \text{ items} < i \text{ in same thread})$$

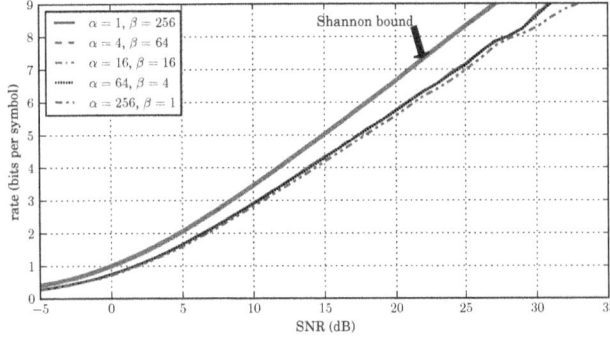

(a) Bits per Symbol

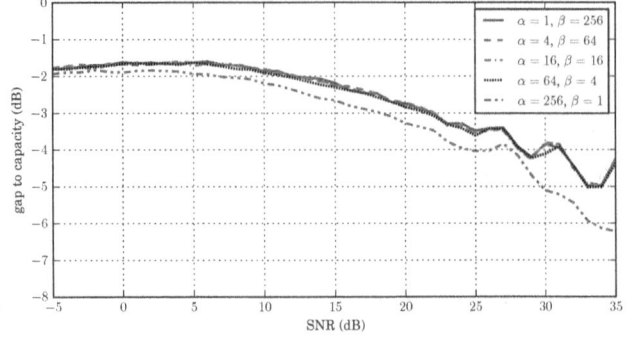

(b) Gap to Capacity

Figure 7: Decoder performance across α and β parameters. Even β=1 decodes with good performance.

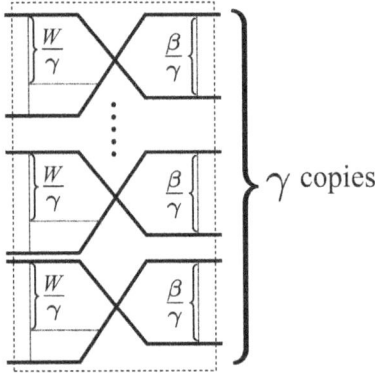

Figure 8: Concatenated selection network, emulating a β, W network with γ smaller selection networks.

What do we know about this thread? It was assigned a total of n items, of which one is item i. Conditional on i being assigned to the thread, the assignment of the other $n-1$ (from a pool of $N-1$ items) is still completely random. There are $i-1$ items less than i, and we want to know the probability that a certain number j are selected. That is, we want the probability of drawing exactly j colored balls in $n-1$ draws from a bucket containing $N-1$ balls, of which $i-1$ are colored. The drawing is without replacement. The result follows the hypergeometric distribution, so the number of colored balls is at least β with probability

$$\mathbb{P}(i \text{ discarded}) = \sum_{j=\beta}^{n} \frac{\binom{n-1}{j}\binom{N-n}{i-j-1}}{\binom{N-1}{i-1}}$$

Thus, we have

$$\mathbb{P}(\text{best } m \text{ outputs correct}) =$$

$$1 - \mathbb{P}(m^{\text{th}} \text{ output} \neq m) \geq 1 - \sum_{i=1}^{m}\sum_{j=\beta}^{n} \frac{\binom{n-1}{j}\binom{N-n}{i-j-1}}{\binom{N-1}{i-1}}$$

$\square$

Similarly, the expected number of the best m inputs which survive selection is

$$\mathbb{E}\left[\sum_{i=1}^{m} \mathbb{1}_{\{i \text{ in output}\}}\right] = \sum_{i=1}^{m} \mathbb{P}(i \text{ in output})$$

$$= m - \sum_{i=1}^{m}\sum_{j=\beta}^{n} \frac{\binom{n-1}{j}\binom{N-n}{i-j-1}}{\binom{N-1}{i-1}}$$

5. PATH EXPANSION

Thanks to the optimizations of §4, path metric units occupy a large part of the area of the final design. The basic worker is shown in Figure 10. This block encodes the symbols corresponding to k bits of data by hashing and mapping them, then subtracts them from the received samples and computes the squared residual. In the instantiation shown, the worker can handle four passes per cycle. If there are more than four passes available in memory, it will spend multiple cycles accumulating the result. By adding more hash blocks, we can handle any number of passes per cycle; however, we observe that in the case where many passes have been received and stored in memory, we are operating at low SNR and consequently low throughput. Thus, rather than accelerate decoding in the case where the channel and not the decoder is the bottleneck, we focus on accelerating decoding at high SNR, and we only instantiate one hash function per worker in favor of laying down more workers. We can get pipeline parallelism in the workers provided that we take care to pipeline the iteration control logic as well.

Samples in our decoder are only 8 bits, so subtraction is cheap. There are three major costs in the worker. The first is the hash function. We used the Jenkins one-at-a-time hash [13]. Using a smaller hash function is attractive from an area perspective, but hash and constellation map collisions are more likely with a weaker hash function, degrading performance. We leave a satisfactory exploration of this space to future work.

The second major cost is squaring. The samples are 8 bits wide, giving 9 bit differences and nominally an 18 bit product. This can be reduced a little by taking the absolute value first to give $8 \times 8 \rightarrow 16$ bits, and a little further by noting that squaring has much more structure than general multiplication. Designing e.g. a Dadda tree multiplier for squaring 8 bits gives a fairly small circuit with 6 half-adders, 12 full-adders, and a 10 bit summation. By comparison, an 8×8 general Dadda multiplier would use 7 half-adders, 35 full-adders, and a 14 bit summation.

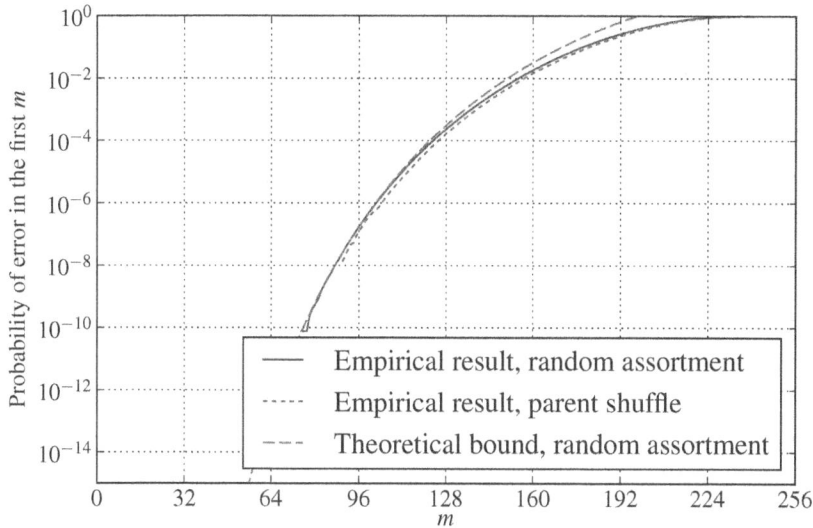

(a) Bound of Theorem 1 and empirical probability of losing one of the m-best inputs.

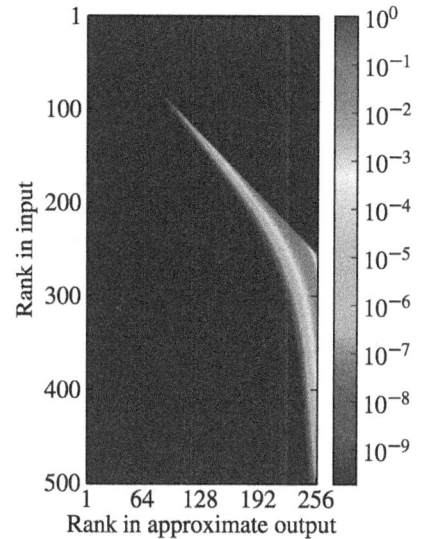

(b) Empirical log probability for each input position to appear in each output position.

Figure 9: Frequency of errors in α-β selection with $B \cdot 2^k = 4096$ and $\beta = 32$, $\alpha = 8$. The lower order statistics (on the left of each graph) are nearly perfect. Performance degrades towards the right as the higher order statistics of the approximate output suffer increasing numbers of discarded items. The graph on the left measures the probability mass missing from the main diagonal of the color matrix on the right. The derandomized strategy makes fewer mistakes than the fully random assortment.

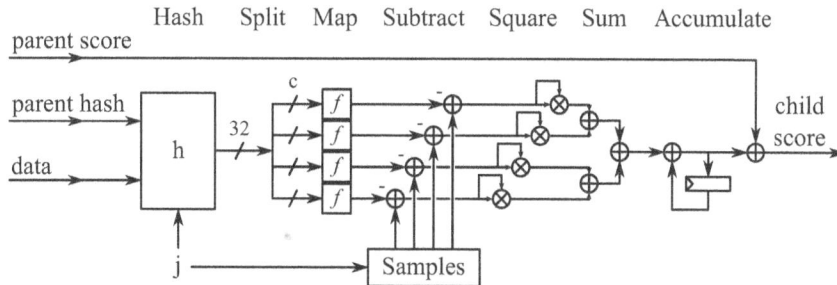

Figure 10: Schematic of the path metric unit. Over one or more cycles indexed by j, the unit accumulates squared differences between the received samples in memory and the symbols which would have been transmitted if this candidate were correct. Not shown is the path for returning the child hash, which is the hash for $j = 0$.

The third cost is in summing the squares. In Viterbi, the scores of all the live candidates differ by no more than the constraint length K times twice the largest possible log likelihood ratio. This is because the structure of the trellis is such that tracing back a short distance from two nodes always leads to a common ancestor. Thanks to two's complement arithmetic, it is sufficient to keep score registers that are just wide enough to hold the largest difference between two scores.

In our case, however, there is no guarantee of common ancestry, save for the argument that the lack of a recent common ancestor is a strong indication that decoding will fail (as we show in §6). As a consequence, scores can easily grow into the millions. We used 24 bit arithmetic for scores. We have not evaluated designs which reduce this number, but we nevertheless highlight a few known techniques from Viterbi as interesting directions for future work. First, we could take advantage of the fact that in low-SNR regimes where there are many passes and scores are large, the variance of

the scores is also large. In this case, the low bits of the score may be swamped with noise and rendered essentially worthless, and we should right-shift the squares so that we accumulate only the "good" bits.

A second technique for reducing the size of the scores is to use an approximation for the x^2 function, like $|x|$ or $\min(|x|, 1)$. The resulting scores will no longer be proportional to log likelihoods, so the challenge will be to show that the decoder still performs adequately.

6. ONLINE TRACEBACK

The final stage of the M-algorithm decoder is traceback. Ideally, at the end of decoding, traceback begins from the most likely child, outputting the corresponding set of k bits and recursing up the tree to the node's parents. The problem with this ideal approach is that it requires the retention of all levels of the beam search until the

157

end of decoding. As a result, traceback is typically implemented in an online fashion. For each new beam, a traceback of c steps is performed starting from the best candidate, and k bits are produced. The variable c represents the effective constraint length of the code: the maximum number of steps until all surviving paths converge on a single, likely ancestor. Beyond this point of convergence, the paths are identical. Because only c steps of traceback need to be performed to find this ancestor, only c beams' worth of data need to be maintained. For many codes, c is actually quite small. For example, convolutional code traceback lengths are limited to $\log_2 s$, where s is the number of states in the code. In spinal codes, particularly with our selection approximations, it is possible for bad paths to appear with very long convergence distances. However, in practice we find that convergence is usually quite rapid, on the order of one or two traceback steps.

Online traceback implementation are well-studied and appear in most implementations of the Viterbi algorithm. Viterbi implementations typically implement traceback using the register-exchange microarchitecture [7, 19]. However, spinal codes can have a much wider window of B live candidates at each backtrack step. Moreover, unlike convolutional codes wherein each parent may have only two children, in spinal codes, a parent may have 2^k children, which makes the wiring the register-exchange expensive. Therefore, we use the RAM-based backtrace approach [7]. Even hybrid backtrace/register-exchange architectures [5] are likely to be prohibitive in complexity. In this architecture, pointers and data values are stored in RAM and iterated over during the traceback phase. For practical choices of parameters the required storage is on the order of tens of kilobits. Figure 13 shows empirically obtained throughput curves for various traceback lengths. Even an extremely short traceback length of four is sufficient to achieve a significant portion of channel capacity. Eight steps represents a good tradeoff between decoding efficiency and area.

The traditional difficulty with traceback approaches is the long latency of the traceback operation itself, which must chase c pointers to generate an output. We note however, that c is a pessimistic bound on convergence. During most tracebacks, "good" paths will converge long before c. Leveraging this observation, we memoize the backtrack of the preceding generation, as suggested by Lin et al. [14]. If the packet being processed will be decoded correctly, parent and child backtracks should be similar. Figure 11 shows a distribution of convergence distances under varying channel conditions, confirming this intuition.

If, during the traceback pointer chase, we encounter convergence with the memoized trace, we terminate the traceback immediately and return the memoized value. This simple optimization drastically decreases the expected traceback length, improving throughput while simultaneously decreasing power consumption.

Figure 12 shows the microarchitecture of our backtrace unit. The unit is divided in half around the traceback RAM. The front half handles finding starting points for traceback from among the incoming beam, while the back half conducts the traceback and outputs values. The relatively simple logic in the two halves permits them to be clocked at higher frequencies than other portions of the pipeline. Our implementation is fully parameterized, including both the parameters of the spinal code and the traceback length.

7. EVALUATION

7.1 Hardware Platforms

We use two platforms in evaluating our hardware implementation. Wireless algorithms operate on the air, and the best way to achieve a high-fidelity evaluation is of wireless hardware is to measure its on-air performance. The first platform we use to evaluate the spinal

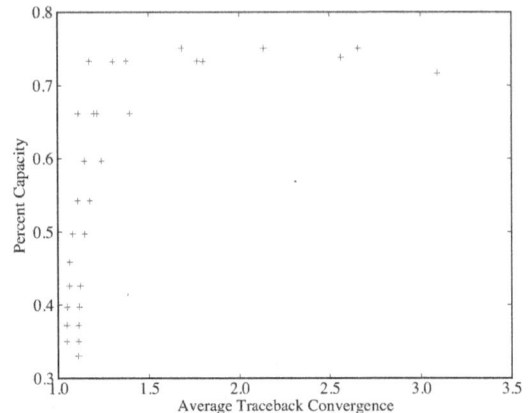

Figure 11: Average convergence distance between adjacent tracebacks, collected for various SNRs and numbers of passes. Near capacity, tracebacks begin to take longer to converge.

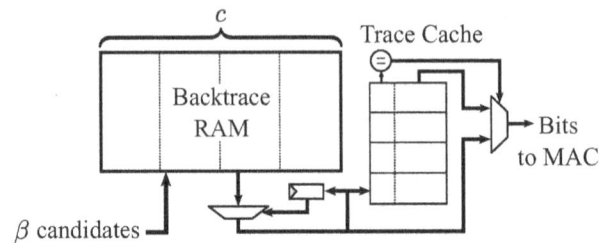

Figure 12: Traceback Microarchitecture. Some control paths have been eliminate to simplify the diagram.

decoder is a combination of an XUPV5 [23] and USRP2 [8]. We use the USRP2 to feed IQ samples to an Airblue [17]-based OFDM baseband implemented on the larger XUPV5 FPGA.

However, on-air operation is insufficient for characterizing and testing new wireless algorithms because over-air operation is difficult to control. Experiments are certainly not reproducible, and some experiments may not even be achievable over the air. For example, it is interesting to evaluate the behavior of spinal codes at low SNR, however the Airblue pipeline does not operate reliably at SNRs below 3dB. Additionally, from a hardware standpoint, some interesting decoder configurations may operate too slowly to make on-air operation feasible.

Therefore, we use a second platform for high-speed simulation and testing: the ACP [16]. The ACP consists of two Virtex-LX330T FPGAs socketed in to a Front-Side Bus. This platform not only offers large FPGAs, but also a low-latency, high-bandwidth connection to general purpose software. This makes it easy to interface a wireless channel model, which is difficult to implement in hardware, to a hardware implementation while retaining relatively high simulation performance. Most of our evaluations of the spinal hardware are carried out using this high-speed platform.

7.2 Comparison with Turbo Codes

Although spinal codes offer excellent coding performance and an attractive hardware implementation, it is important to get a feel for the properties of the spinal decoder as it compares to existing error correcting codes. Turbo codes [4] are a capacity-approaching code currently deployed in most modern cellular standards.

There are several metrics against which one might compare hard-

	3G Turbo (1dB)	3G Turbo(1dB) [6]	Spinal (1dB)	Spinal(-5dB)
Parity RAM	118 Kb	86kB		
Systemic RAM	92 Kb	25kB		
Interleaver RAM	16 Kb			
Pipeline Buffer RAM	27 Kb	12kB		
Symbol RAM			41Kb	135Kb
Backtrace RAM			8Kb	8Kb
Total RAM	253 Kb	123 kB	49Kb	143Kb

Table 2: Memory Usage for turbo and spinal decoders supporting 5120 bit packets. Memory area accounts for more than 50% of turbo decoder area.

ware implementations of spinal and turbo codes: implementation area, throughput, latency, and power consumption.

A fundamental difference between turbo and spinal decoders is that the former are *iterative*, while spinal decoders are sequential and thus can be streaming. This means that a turbo implementation must fundamentally use more memory than a spinal implementation since turbo decoders must keep at least one pass worth of soft, extrinsic information alive at any point in time. Because packet lengths are large and soft information is wide, this extra memory can dominate implementation area. On the other hand, spinal codes store much narrower symbol information. We therefore conjecture that turbo decoders must use at least twice the memory area of a spinal decoder with a similar noise floor. This conjecture is empirically supported by Table 2, which compares 3G-compliant implementations of turbo codes with spinal code decoders configured to similar parameters.

It is important to note that spinal decoder memory usage scales with the noise floor of the decoder since more passes must be buffered, while turbo codes use a constant memory area for any noise floor supported. If we reduce the supported noise floor to 1dB from -5dB, then the area required by the spinal implementation drops by around a factor of 4. This is attractive for short-range deployments which do not require the heavy error correction of cellular networks.

7.3 Performance of Hardware Decoder

Figure 13 shows the performance of the hardware decoder across a range of operational SNRs. Throughputs were calculated by running the full Airblue OFDM stack on FPGA and collecting packet error rates across thousands of packets, a conservative measure of throughput. The decoder performs well, achieving as much as 80% of capacity at relevant SNRs. The low SNR portion of the range is limited by Airblue's synchronization mechanisms, which do not operate reliably below 3dB.

Table 3 shows the implementation areas of various modules of our reference hardware decoder in a 65 nm technology. Memory area dominates the design, while logic area is attractively small. The majority of the area of the design is taken up by the score calculation logic. Individually, these elements are small. However there are β of them in our parameterized design. The α-β selection network requires one-fourth the design. In contrast, a full selection network for $B = 64$ requires around 360000 μm^2, much more than our entire decoder.

As a basis for comparison, state of the art turbo decoders [6] at the 65 nm node require approximately .3 mm^2 for the active portion of the decoder. The remaining area (also around .3 mm^2) is used for memory. Our design is significantly smaller in terms of area, using half the memory and around 80% the logic area. However, our design at 200 MHz, processes at a maximum throughput of 12.5 Mbps, which is somewhat lower than the Cheng et al., who approached 100 Mbps.

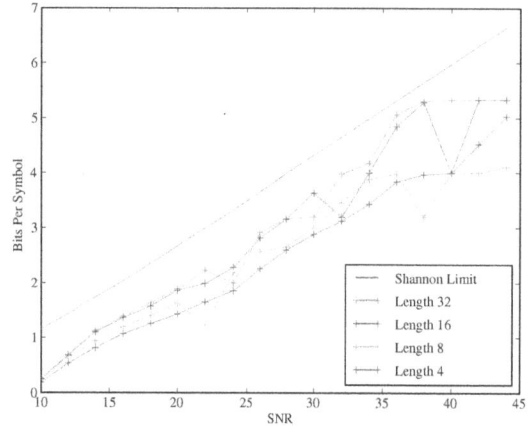

Figure 13: Throughput of the hardware decoder with various traceback lengths.

In our choice of implementation, we have attempted to achieve maximum decoding efficiency and minimum gap-to-capacity. However, maximum efficiency may not yield the highest throughput design. Should throughput be a priority, we note that there are several ways in which we could improve the throughput of our design. The most obvious direction is reducing B to 32 or 16. These decoders suffer slightly degraded performance, but operate 2 and 4 times faster. Figure 14 shows an extreme case of this optimization with $B = 4$. This design has low decoder efficiency, but much higher throughput. We note that a dynamic reduction in B can be achieved with relatively simple modifications to our hardware. A second means of improvement is optimizing the score calculators. There are three ways to achieve this goal. First, we can increase the number of score calculators. This is slightly unattractive because it also requires scaling in the sorting network. Second, the critical path of our design runs through the worker units and is largely unpipelined. Cutting this path should increase achievable clock period by at least a few nano-seconds. Related to the critical path is the fact that we calculate error metrics using Euclidean distance, which requires multiplication. Strength reduction to absolute difference has worked well in Viterbi and should apply to spinal as well. By combining these techniques it should be possible to build spinal decoders with throughputs greater than 100 Mbps.

159

Module	Total (μm^2)	Combinational (μm^2)	Sequential (μm^2)	RAM(Kbits)
Selection Network	60700	25095	35907	
Backtrack	8575	3844	4720	8
Score Calculator	10640	8759	1881	
SampleRAM	5206	2592	2613	41
Total	245526	181890	63703	49

Table 3: Area usage for modules with $B = 64$, $W = \beta = 16$, $\alpha = 4$. Area estimates were produced using Cadence Encounter with a 65 nm process, targeting 200 MHz operating frequency. Area estimates do not include memory area.

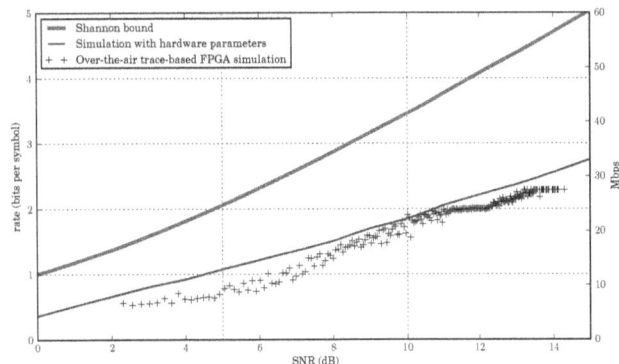

Figure 14: Performance of $B = 4$, $\beta = 4$, $\alpha = 1$ decoder over the air versus identically parameterized C++ model. Low code efficiency is due to the narrow width of the decoder, which yields a high throughput implementation.

7.4 On-Air Validation

The majority of the performance results presented in this paper were generated via simulation, either using an idealized, floating-point C++ model of the hardware or using an emulated version of the decoder RTL on an FPGA with a software channel model. Although we have taken care to accurately model both the hardware and the wireless channel, it is important to validate the simulation results with on-air testing.

Figure 14 show a preliminary on-air throughput curve obtained by using the previously described USRP set-up plotted against an identically parameterized C++ model. The performance differential between hardware and software across a wide range of operating conditions is minimal, suggesting that our simulation-based results have high fidelity.

7.5 Integrating with Higher Layers

Error correction codes do not exist in isolation, but as part of a complete protocol. Good protocols require feedback from the physical layer, including the error correction block, to make good operational choices. Additionally, the spinal decoder itself requires a degree of control to decide when to attempt a decode when operating ratelessly. Decoding too early results in increased latency due to failed decoding, while decoding too late wastes channel bandwidth. It is therefore important to have mechanisms in the decoder, like SoftPHY [12], which can provide fine-grained information about the success of decoding.

Traceback convergence in spinal codes, which bears a strong resemblance to confidence calculation in SOVA [11], is an excellent candidate for this role. As Figure 11 shows, a sharp increase in convergence length suggests being near or over capacity. By monitoring the traceback cache for long convergences using a simple filter, the hardware can terminate decodes that are likely to be incorrect early in processing, preventing significant time waste. Moreover, propagating information about when convergences begin to narrow gives upper layers an excellent measure of channel capacity which can be used to improve overall system performance.

8. CONCLUSION

Spinal codes are, in theory and simulation, a promising new capacity-achieving code. In this paper, we have developed an efficient microarchitecture for the implementation of spinal codes by relaxing data dependencies in the ideal code to obtain smaller, fully pipelined hardware. The enabling architectural features are a novel "α-β" incremental approximate selection algorithm, and a method for obtaining hints to anticipate successful or failed decoding, which permits early termination and/or feedback-driven adaptation of the decoding parameters.

We have implemented our design on an FPGA and have conducted over-the-air tests. A provisional hardware synthesis suggests that a near-capacity implementation of spinal codes can achieve a throughput of 12.5 Megabits/s in a 65 nm technology, using substantially less area than competitive 3GPP turbo code implementations.

We conclude by noting that further reductions in hardware complexity of spinal decoding are possible. We have focused primarily on reducing the number of candidate values alive in the system at any point in time. Another important avenue of exploration is reducing the complexity and width of various operations within the pipeline. Both Viterbi and Turbo codes operate on extremely narrow values using approximate arithmetic. It should be possible to reduce spinal decoders in a similar manner, resulting in more area-efficient and higher throughput decoders.

ACKNOWLEDGMENTS

We thank Lixin Shi for helpful comments. Support for P. Iannucci and J. Perry came from Irwin and Joan Jacobs Presidential Fellowships. An Intel Fellowship supported K. Fleming. Additional support for J. Perry came from the Claude E. Shannon Research Assistantship.

REFERENCES

[1] 3rd Generation Partnership Project. *Technical Specication Group Radio Access Networks, TS 25.101 V3.6.0*, 2001-2003.
[2] J. Anderson and S. Mohan. Sequential coding algorithms: A survey and cost analysis. *IEEE Trans. on Comm.*, 32(2):169–176, 1984.
[3] L. Bahl, J. Cocke, F. Jelinek, and J. Raviv. Optimal Decoding of Linear Codes for Minimizing Symbol Error Rate (Corresp.). *IEEE Trans. Info. Theory*, 20(2):284–287, 1974.
[4] C. Berrou, A. Glavieux, and P. Thitimajshima. Near Shannon limit error-correcting coding and decoding: Turbo-codes. 1. In *ICC*, 2002.
[5] P. Black and T.-Y. Meng. Hybrid Survivor Path Architectures for Viterbi Decoders. *Acoustics, Speech, and Signal Processing, IEEE International Conference on*, 1:433–436, 1993.
[6] C.-C. Cheng, Y.-M. Tsai, L.-G. Chen, and A. P. Chandrakasan. A 0.077 to 0.168 nJ/bit/iteration scalable 3GPP LTE turbo decoder with

an adaptive sub-block parallel scheme and an embedded DVFS engine. In *CICC*, pages 1–4, 2010.

[7]   R. Cypher and C. Shung. Generalized Trace Back Techniques for Survivor Memory Management in the Viterbi Algorithm. In *Proc. IEEE GLOBECOM*, Sec 1990.

[8]   Ettus Research USRP2. `http://www.ettus.com/products`.

[9]   G. J. Forney. The Viterbi Algorithm. In *Proceedings of the IEEE*, volume 61, pages 268–278. IEEE, Mar. 1973.

[10] L. Gonzalez Perez, E. Boutillon, A. Garcia Garcia, J. Gonzalez Villarruel, and R. Acua. VLSI Architecture for the M Algorithm Suited for Detection and Source Coding Applications. In *International Conference on Electronics, Communications and Computers*, pages 119 – 124, feb. 2005.

[11] J. Hagenauer and P. Hoeher. A Viterbi Algorithm with Soft-Decision Outputs and its Applications. In *Proc. IEEE GLOBECOM*, pages 1680–1686, Dallas, TX, Nov. 1989.

[12] K. Jamieson. *The SoftPHY Abstraction: from Packets to Symbols in Wireless Network Design*. PhD thesis, MIT, Cambridge, MA, 2008.

[13] B. Jenkins. Hash functions. Dr. Dobb's Journal, 1997.

[14] C.-C. Lin, C.-C. Wu, and C.-Y. Lee. A Low Power and High-speed Viterbi Decoder Chip for WLAN Applications. In *ESSCIRC '03*, pages 723 –726, 2003.

[15] E. C. Lin, K. Yu, R. A. Rutenbar, and T. Chen. A 1000-word vocabulary, speaker-independent, continuous live-mode speech recognizer implemented in a single FPGA. In *FPGA*, pages 60–68, 2007.

[16] Nallatech. Intel Xeon FSB FPGA Socket Fillers. *http://www.nallatech.com/intel-xeon-fsb-fpga-socket-fillers.html*.

[17] M. C. Ng, K. Fleming, M. Vutukuru, S. Gross, Arvind, and H. Balakrishnan. Airblue: A System for Cross-Layer Wireless Protocol Development. In *ANCS'10*, San Diego, CA, 2010.

[18] M. C. Ng, M. Vijayaraghavan, G. Raghavan, N. Dave, J. Hicks, and Arvind. From WiFI to WiMAX: Techniques for IP Reuse Across Different OFDM Protocols. In *MEMOCODE'07*.

[19] E. Paaske, S. Pedersen, and J. Sparsø. An Area-efficient Path Memory Structure for VLSI Implementation of High Speed Viterbi Decoders. *Integr. VLSI J.*, pages 79–91, Nov. 1991.

[20] J. Perry, H. Balakrishnan, and D. Shah. Rateless spinal codes. In *HotNets-X*, Oct. 2011.

[21] J. Perry, P. Iannucci, K. Fleming, H. Balakrishnan, and D. Shah. Spinal Codes. In *SIGCOMM*, Aug. 2012.

[22] M. Power, S. Tosi, and T. Conway. Reduced Complexity Path Selection Networks for M-Algorithm Read Channel Detectors. *IEEE Transactions on Circuits and Systems*, 55(9):2924 –2933, Oct. 2008.

[23] Xilinx. Xilinx University Program XUPV5-LX110T Development System. *http://www.xilinx.com/univ/xupv5-lx110t.htm*.

Fast Submatch Extraction using OBDDs[*]

Liu Yang
Rutgers University
lyangru@cs.rutgers.edu

Pratyusa Manadhata
HP Laboratories
manadhata@hp.com

William Horne
HP Laboratories
william.horne@hp.com

Prasad Rao
HP Laboratories
prasad.rao@hp.com

Vinod Ganapathy
Rutgers University
vinodg@cs.rutgers.edu

ABSTRACT

Network-based intrusion detection systems (NIDS) commonly use *pattern languages* to identify packets of interest. Similarly, security information and event management (SIEM) systems rely on pattern languages for real-time analysis of security alerts and event logs. Both NIDS and SIEM systems use pattern languages extended from regular expressions. One such extension, the *submatch* construct, allows the extraction of substrings from a string matching a pattern. Existing solutions for submatch extraction are based on non-deterministic finite automata (NFAs) or recursive backtracking. NFA-based algorithms are time-inefficient. Recursive backtracking algorithms perform poorly on pathological inputs generated by algorithmic complexity attacks. We propose a new approach for submatch extraction that uses *ordered binary decision diagrams* (OBDDs) to represent and operate pattern matching. Our evaluation using patterns from the Snort HTTP rule set and a commercial SIEM system shows that our approach achieves its ideal performance when patterns are combined. In the best case, our approach is faster than RE2 and PCRE by one to two orders of magnitude.

Categories and Subject Descriptors

C.2 [**Computer-Communication Networks**]: Network Operations

General Terms

Algorithms

Keywords

Regular expression, pattern matching, submatch, tagged-NFA, Ordered Binary Decision Diagram (OBDD)

---

[*]The first author completed parts of this work during an internship at HP Labs, Princeton, NJ.

1. INTRODUCTION

Regular expression-like pattern languages are widely used as building blocks of network security products. For example, network intrusion detection systems (NIDS) use thousands of patterns to describe malicious traffic. Security information and event management (SIEM) systems also use patterns to process event logs generated by hardware devices and software systems.

Pattern languages commonly used by NIDS are regular expressions extended with other features. One of the important features is the *capturing group*. A capturing group is a syntax used in modern regular expression implementations to specify a subexpression of a regular expression. Given a string that matches the regular expression, *submatch extraction* is the process of extracting the substrings corresponding to those subexpressions. In Snort 2012 rule set, more than 10% of `pcre` fields of the HTTP rules contain capturing groups. When a pattern containing a capturing group matches an input string, the submatch construct can identify parts of the input that are of interest to security administrators for analysis. For a regular expression like `username=(.*),hostname=(.*)` with an input string `username=Bob,hostname=Foo`, submatch construct can extract the two substrings `Bob` and `Foo` specified by the two capturing groups (the subexpressions wrapped by the two pairs of parentheses).

Likewise, SIEM systems, which perform real-time analysis of event logs and security alerts in enterprise networks, also make extensive use of submatch extraction. SIEM systems often collect data from a variety of hardware and software sensors, and must therefore normalize this data into a common format by extracting common fields from various data sources. SIEM systems use submatch extraction during data normalization and alert reporting. In a typical SIEM system, more than 90% of regular expressions used for data normalization contain capturing groups.

In both SIEM systems and NIDS, scalability of pattern matching and submatch extraction is key. NIDS are often deployed over high-speed network links, which require algorithms for pattern matching and submatch extraction be efficient enough to provide high throughput intrusion detection on large volume of network traffic. Similarly, a typical SIEM system collects logs from hundred of devices and applications, and must process terabytes of logs every day in enterprise networks.

There is plenty of prior work on making pattern matching for regular expressions time-efficient [1, 26, 6, 7, 15, 3, 13, 12, 18] and space-efficient [28, 2, 1, 23, 20, 21]. However, most of these works only considered regular expressions containing no capturing groups, i.e., they did not support submatch extraction. Existing solutions for submatch extraction are based on non-deterministic finite automata (NFAs) [14, 10] or recursive backtracking [16]. While NFAs are space-efficient and can extract submatches with a compact memory footprint, they are not time-efficient because they maintain a *frontier*, i.e., a set of states in which a NFA can be at any instant, that can contain $O(n)$ states where n is the NFA's number of states. This leads to an $O(n)$ operation time for the NFA for each input symbol. Google's RE2 package uses a combination of DFAs and NFAs to improve the time efficiency of submatch extraction [10]. RE2 constructs DFAs on demand (determination on the fly) and uses DFAs to locate a pattern's overall match location in an input string and then uses a NFA-based method to extract submatches. The time-efficiency of DFAs, however, often comes with a cost of state blow-up. RE2 can be very slow when the DFA construction fills up the limited state cache; it has to empty the state cache and restart the DFA construction process. Moreover, the actual submatch extraction of RE2 is performed using a NFA-based method, which is space-efficient, but not time-efficient. Tools such as Perl, PCRE, and Python use recursive backtracking for regular expression matching. The execution time of backtracking, however, can be exponential for certain types of regular expressions [9]; NIDS which employ backtracking suffer from algorithmic complexity attacks [19].

We present a novel approach to perform submatch extraction for regular expression-like pattern languages. Our approach is an extension of the NFA-OBDD work by Yang et al [26]. While both works employ the ordered binary decision diagram (OBDD) data structure, the NFA-OBDD approach in [26] did not consider the submatch construct, making it inapplicable to the 90% of regular expressions in a typical SIEM system. We extend the NFA-OBDD approach [26] in two ways: (1) we propose an approach to annotate capturing groups in regular expressions, and (2) present a new approach to perform submatch extraction. To demonstrate the feasibility of our approach, we evaluated our approach using patterns extracted from the Snort NIDS and a commercial SIEM product. Our experiments show that our approach achieves its ideal performance when patterns are combined. In the best case, our approach is faster than RE2 and PCRE by one to two orders of magnitude. In particular, we make the following contributions:

• We propose a new approach to tag capturing groups in a regular expression, and extend Thompson's NFA construction approach to convert a regular expression with capturing groups to a tagged-NFA.

• We present a novel and time-efficient technique (henceforth called *Submatch-OBDD*) to perform submatch extraction for regular expression-like pattern languages.

• We evaluated our approach's time efficiency and space efficiency by matching the patterns from the Snort system and a commercial SIEM system with network traces, synthetic traces, and enterprise event logs, and then compared our performance with two popular regular expression engines: RE2 and PCRE.

The remainder of the paper is organized as follows. We briefly describe ordered binary decision diagrams (OBDDs) as background knowledge in Section 2. After that, we present our design and implementation of Submatch-OBDD in Section 3, followed by our evaluation in Section 4. We discuss related work in Section 5 and conclude in Section 6.

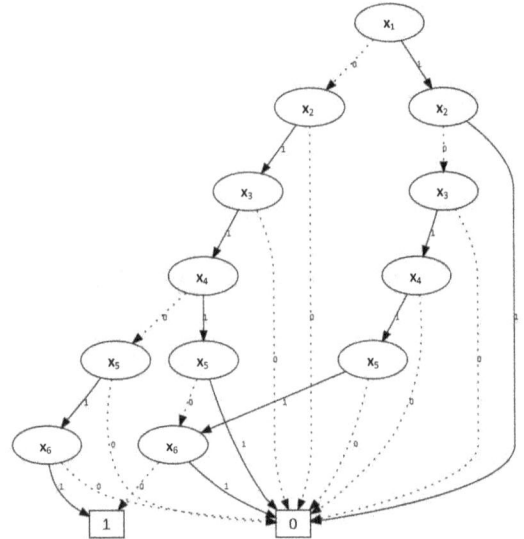

Figure 1: The ordered binary decision diagram of a Boolean function $f(x_1, x_2, x_3, x_4, x_5, x_6) = (\bar{x_1} \wedge x_2 \wedge x_3 \wedge \bar{x_4} \wedge x_5 \wedge x_6) \vee (\bar{x_1} \wedge x_2 \wedge x_3 \wedge x_4 \wedge \bar{x_5} \wedge \bar{x_6}) \vee (x_1 \wedge \bar{x_2} \wedge x_3 \wedge x_4 \wedge x_5 \wedge \bar{x_6})$ with ordering $x_1 \prec x_2 \prec x_3 \prec x_4 \prec x_5 \prec x_6$.

2. ORDERED BINARY DECISION DIAGRAMS

Bryant proposed ordered binary decision diagrams as a data structure for symbolically representing arbitrary Boolean functions [4]. OBDDs are widely used in digital logic design and testing, artificial intelligence, and model checking [5] to construct efficient algorithms for Boolean functions.

An OBDD represents a Boolean function, $f(x_1, x_2, \ldots, x_n)$, as a rooted directed acyclic graph (DAG), which has two types of nodes. A *non-terminal* node, v, is associated with an argument $\in \{x_1, x_2, \ldots, x_n\}$ and has two children, $low(v)$ and $high(v)$. A *terminal* node takes value $\in \{0, 1\}$. An OBDD is organized so that nodes along all paths from the root to terminal nodes follow a total order, $\prec$. An OBDD with root v denotes a function, f_v, which is recursively defined as:

• If v is a terminal node, then $f_v = value(v)$;

• If v is a non-terminal node and v is associated with argument x_i, then
$f_v(x_1, x_2, \ldots, x_n) = \quad \bar{x_i} \cdot f_{low(v)}(x_1, x_2, \ldots, x_n) \quad \vee \quad x_i \cdot f_{high(v)}(x_1, x_2, \ldots, x_n)$.

To evaluate a function with a set of argument values $x_1, x_2, \ldots, x_n$, start from the root, where if a node v is associated with x_i, then traverse to $low(v)$ if $x_i = 0$ and to $high(v)$ if $x_i = 1$, until a terminal node is reached. The function's value equals the value of the terminal node at the end of the traversal. Figure 1 shows an example Boolean function $f(x_1, x_2, x_3, x_4, x_5, x_6)$ and its OBDD representation in the order of $x_1 \prec x_2 \prec x_3 \prec x_4 \prec x_5 \prec x_6$.

OBDDs allow Boolean functions to be represented and manipulated efficiently. Testing satisfiability with OBDDs simply involves comparing a graph to that of a constant function **0**. Many operations of Boolean functions can be represented as two types of graph operations in OBDDs: APPLY and RESTRICT. APPLY allows Boolean operators such as $\vee$ and $\wedge$ to be applied to a pair of OBDDs. It takes a pair of OBDDs representing two functions f_1 and f_2, a binary operator OP, and produces a reduced graph representing function $\text{APPLY}(\text{OP}, f_1, f_2) = f_1 \text{OP} f_2$. The resulting OBDD has the same variable ordering as the input OBDDs. RESTRICT is a unary operator. It transforms a function f into one representing the function $f|_{x_i=b}$ for a specified argument value $x_i = b$. The resulting OBDD does not have any node associated with x_i. The complexity of APPLY and RESTRICT is polynomial in the size of input OBDDs.

One important operation used in our Submatch-OBDD design is *existential quantification*. In particular, $\exists x_i \cdot f(x_1, x_2, \ldots, x_n) = f(x_1, x_2, \ldots, x_n)|_{x_i=0} \vee f(x_1, x_2, \ldots, x_n)|_{x_i=1}$. Expressed by OBDD, we have $OBDD(\exists x_i \cdot f(x_1, x_2, \ldots, x_n))$
$= \text{APPLY}(\vee, \text{RESTRICT}(OBDD(f), 1 \leftarrow x_i),$
$\text{RESTRICT}(OBDD(f), 0 \leftarrow x_i))$. As a result, $OBDD(\exists x_i \cdot f(x_1, x_2, \ldots, x_n))$ will have no node associated with x_i.

OBDDs are extremely useful in obtaining concise representations of relations over finite domains. If R is a n-arity relation over $\{0, 1\}$, then R can be represented by an OBDD using its characteristic function $f_R(x_1, x_2, \ldots, x_n) = 1$ iff $R(x_1, x_2, \ldots, x_n)$. For example, a 3-ary relation $R = \{(1, 0, 1), (1, 1, 0)\}$ can be expressed by $f_R(x_1, x_2, x_3) = (x_1 \wedge \bar{x}_2 \wedge x_3) \vee (x_1 \wedge x_2 \wedge \bar{x}_3)$, which is a Boolean function and can therefore be represented using an OBDD.

A set of elements can also be expressed as an OBDD. If S is a set over a domain D, then we can define a relation $R_S(s) = 1$ if $s \in S$. Operations on sets can be expressed as Boolean operations and manipulated by OBDDs. For example, $\text{ISEMPTY}(S \cap T)$ is equivalent to checking the satisfiability of $OBDD(\text{APPLY}(\wedge, S, T))$. Our Submatch-OBDD design in Section 3 converts relations and sets to OBDDs to achieve time efficient operations.

3. DESIGN AND IMPLEMENTATION

We first give an overview of our approach before describing the technical details.

3.1 Solution Overview

A key observation underlying our approach is that adding a capturing group to a regular expression does not change the language defined by the regular expression. It is known that every language defined by a regular expression is also defined by a finite automaton [11]. However, traditional automata do not support capturing groups. We present an approach to annotate capturing groups in regular expressions and extend Thompson's approach to convert a regular expression with capturing groups to a NFA-like machine where transitions within capturing groups are tagged. We then present a novel approach to do submatch extraction using the tagged-NFAs. To improve the time efficiency of submatch extraction, we represent tagged-NFAs with symbolic Boolean functions, and manipulate the Boolean functions using ordered binary decision diagrams (OBDDs).

3.2 Tagging NFAs for Submatch

The syntax of regular expressions with capturing groups on an alphabet Σ is

$$E ::= \epsilon \cup a \cup EE \cup E|E \cup E* \cup (E) \cup [E]$$

where a stands for an element of Σ, and ϵ denotes for zero occurrence of a symbol. We use square brackets $[,]$ to group

terms in a regular expression that are not capturing groups, because the usual parentheses $(,)$ are reserved for marking capturing groups. If X and Y are sets of strings we use XY to denote $\{xy : x \in X, y \in Y\}$, and $X|Y$ to denote $X \cup Y$. We use $E*$ to denote the closure of E under concatenation.

We use tags to distinguish the capturing groups within a regular expression. Given a regular expression containing c capturing groups, we assign tags $t_1, t_2, \ldots, t_c$ to each capturing group in the order of their left parentheses as E is read from left to right. We denote the set of tags by $T = \{t_1, t_2, \ldots, t_c\}$. We use $tag(E)$ to refer to the resulting tagged regular expression. For example, if $E = ((a*)|b)(ab|b)$ then $tag(E) = ((a*)_{t_2}|b)_{t_1}(ab|b)_{t_3}$.

The language $L(F)$ for a tagged regular expression $F = tag(E)$ is a set of tagged strings, defined by $L(\epsilon) = \{\epsilon\}$, $L(a) = \{a\}$, $L(F_1 F_2) = L(F_1) \cdot L(F_2)$, $L(F_1|F_2) = L(F_1) \cup L(F_2)$, $L(F*) = L(F)*$, $L([F]) = L(F)$, and $L((F)_t) = \{\alpha_t : \alpha \in L(F)\}$, where $()_t$ denotes a capturing group with tag t and α_t denotes the string α tagged with t. A string α is tagged by t, if and only if each character in α is tagged by t. Substrings of α may be tagged by other tags. Since capturing groups can be nested, a character can be tagged by multiple tags. An example of tagged string for a tagged regular expression is: $ab_{t_1} b_{t_1} b_{t_1} \in L(a(b*)_{t_1})$.

Definition A *valid assignment of submatches* for a string α that matches regular expression E is a map $sub : \{t_1, t_2, \ldots t_c\} \rightarrow \Sigma^*$ such that there exists $\beta \in L(tag(E))$ satisfying the following:

(i) $\beta|_\Sigma = \alpha$, where $\beta|_\Sigma$ represents the projection of characters in β onto their corresponding values of Σ;

(ii) if t_i occurs in β then $sub(t_i)$ is the last consecutive sequence of characters that are assigned with tag t_i;

(iii) if t_i does not occur in β, then $sub(t_i) = \text{NULL}$;

For example, consider the regular expression $[(a|c)(b|d)]*$ with input string $abcd$. A valid submatch assignment satisfying the above conditions is $sub(t_1) = c$, $sub(t_2) = d$.

It is well known that a regular expression can be converted to an ϵ-NFA which defines the same language using Thompson's approach [11]. An ϵ-NFA can be reduced to an ϵ-free NFA through an ϵ-closure mechanism [11]. In this paper, we extend Thompson's algorithm in a way such that it can convert a regular expression containing capturing groups to a tagged ϵ-NFA defining the same language. A tagged ϵ-NFA can be described by a 7-tuple $A = (Q, \Sigma, T, \delta, \gamma, S, F)$, where Q is a finite set of states, Σ is a finite set of input symbols, T is a finite set of tags that each represents a capturing group, S is a set of start states, F is a set of accept states, δ is the transition function, and γ is a tag output function $\gamma : Q \times \Sigma \times Q \rightarrow 2^T$, which associates each transition with a tag set (which can be empty).

A tagged NFA can be constructed as follows, starting from the three base cases shown in Figure 2. Figure 2(a) is the NFA of expression ϵ, Figure 2(b) handles the empty regular expression, and Figure 2(c) gives the NFA of a single symbol a with a set of tags $\tau \in 2^T$ corresponding to capturing groups associated with the illustrated transition. More complex tagged NFAs can be constructed using the union, concatenation, and closure constructs, by combining smaller tagged NFAs as shown in Figure 3. A tagged NFA constructed using the above approach contains ϵ transitions. Such a tagged NFA can be converted to an ϵ-free NFA in

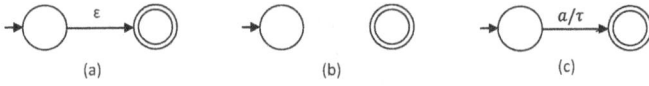

Figure 2: Constructing tagged NFAs for (a) NFA of ϵ; (b) NFA of an empty regular expression; (c) NFA of a symbol a wrapped by capturing groups denoted by τ.

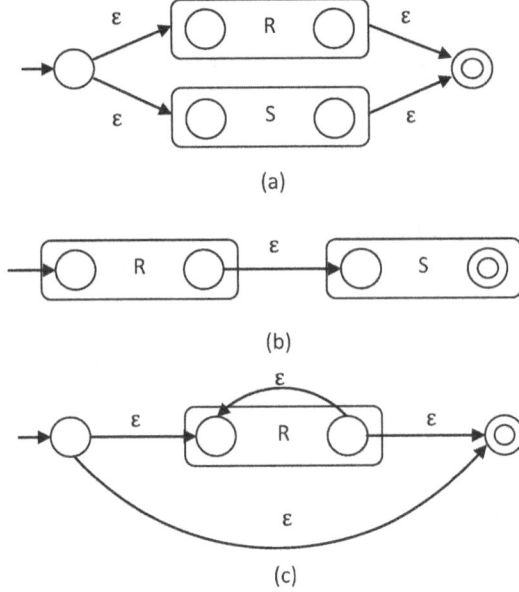

Figure 3: The (a) union $R|S$, (b) concatenation RS, and (c) closure constructs $R*$ of tagged NFA construction from a regular expression.

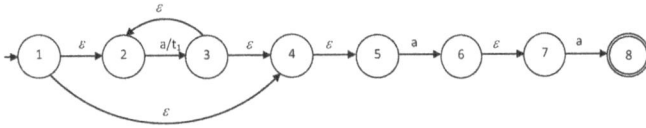

Figure 4: The tagged ϵ-NFA of $(a*)aa$, where the transition associated with the leftmost character a is tagged by t_1 because $a*$ is within a capturing group.

a manner akin to the standard ϵ-closure algorithm for standard NFAs. We denote the corresponding ϵ-free tagged NFA as $A_1 = (Q_1, \Sigma, T, \delta_1, \gamma_1, S_1, F_1)$, where the components of A_1 are defined in a manner akin to A (the tagged ϵ-NFA).

Example Consider an example regular expression $(a*)aa$. Figure 4 shows an the tagged ϵ-NFA, where the capturing group is tagged by t_1. Figure 5 shows the corresponding ϵ-free tagged-NFA.

3.3 Operations on Tagged NFAs

The transition function δ_1 and tag output function γ_1 of a tagged ϵ-free NFA can be represented by a four-column table denoted by $\Delta(x, i, y, \tau)$, which is a set of quadruples (x, i, y, τ) such that there is a transition labeled by input symbol i from state x to state y with a set of output tags τ. Table 1 shows the tagged transition table of the example NFA in Figure 5, where each tagged transition is represented

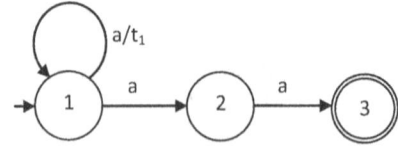

Figure 5: The tagged ϵ-free NFA of $(a*)aa$ after ϵ-elimination, where state numbers 1, 2, and 3 are obtained by renaming and merging states 2, 5, 7, and 8 in the ϵ-NFA during ϵ-closure calculation.

x	i	y	τ
1	a	1	$\{t_1\}$
1	a	2	ϕ
2	a	3	ϕ

Table 1: Transition table of the tagged NFA in Figure 5.

by a row in the table. $\Delta(x, i, y, \tau)$ allows us to perform two key operations on tagged NFAs — *match test* and *submatch extraction*, where the match test checks whether an input string is accepted by a tagged NFA; if so, the submatch extraction procedure returns a valid assignment of submatches of the input string.

3.3.1 Match Test

Testing whether an input string matches a regular expression with capturing groups can be done by operating its tagged NFA. The process is similar to operating a traditional NFA, except that we need to do bookkeeping to be used for submatch extraction. The match test of a tagged NFA for a given input string $a_1 a_2 \ldots a_l \in \Sigma^*$ is performed by consuming one input symbol at a time, and modifying the frontier of active states appropriately using the transition function δ_1. As we modify the frontier, we also record the transitions that the tagged-NFA makes by recording quadruples that store the states traversed by each transition, as well as the tags corresponding to those transitions. We denote these sets of transitions using Δ_1, Δ_2, and Δ_l, where each Δ_i is a set of quadruples of the form (x, i, y, τ) corresponding to a source state, an input symbol, a target state, and the corresponding tag.

After the last input symbol a_l is consumed, we check whether any state in the frontier set belongs to accept states F_1. If so, the input string $a_1 a_2 \ldots a_l$ is accepted by the tagged NFA A_1, i.e., the input string matches the regular expression defined by A_1.

Example Consider the example regular expression $(a*)aa$ in Figure 5, where its tagged transitions $\Delta(x, i, y, \tau)$ are shown in Table 1. For convenience, we denote the three quadruples in Table 1 by row_1, row_2, and row_3. Let's use $aaaa$ as an input string. For the ith input symbol, we use X_i to denote the current frontier set and Y_i to denote the next frontier set after the symbol is consumed. Start from the first input symbol a and start states $S_1 = \{1\}$, we have $\Delta_1 = \{row_1, row_2\}$, $Y_1 = \{1, 2\}$. Rename Y_1 to X_2 and follow the process described in the frontier derivation, we can obtain $\Delta_2 = \{row_1, row_2, row_3\}$, $X_3 = \{1, 2, 3\}$, $\Delta_3 = \{row_1, row_2, row_3\}$, $X_4 = \{1, 2, 3\}$, and $\Delta_4 = \{row_1, row_2, row_3\}$, $X_5 = \{1, 2, 3\}$. Figure 6 visualizes how the frontier set evolves after consuming each input symbol during the match test. An arrow between two

166

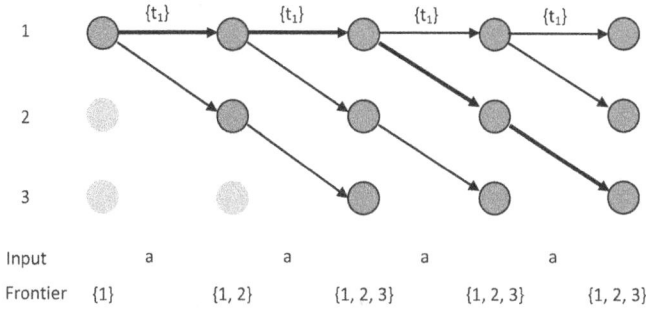

Figure 6: Example of frontier derivation for the tagged NFA in Figure 5 with input string $aaaa$. The dark circles of each column stands for the frontier states after consuming an input symbol. A light gray circle means that a state is not in a frontier state set. An arrow between two circles represent a transition. An arrow is labeled by a submatch tag if the denoted transition is within a capturing group.

nodes denotes a transition. If an arrow is tagged, it means that a transition is associated with one or more submatch tags, e.g., the $\{t_1\}$ above the arrow between states 1 and 1 indicate this transition is within a capturing group.

3.3.2 Submatch Extraction

If an input string is accepted by a regular expression which has capturing groups, the submatches of the input string need to be extracted. Recall that the NFA match test process described above actually considers all possible branches (transitions) when consuming each input symbol. If the input string is accepted by the tagged NFA, then there exists at least one path from a start state to an accept state, where edges of the path denote transitions between states and are sequentially associated with the individual symbols of the input string. An edge may be associated with one or more submatch tags, or no tag at all. For example, the bold arrows in Figure 6 shows a path from start state 1 to the accept state 3. Such a path allows us to perform submatch extraction.

In fact, *any path from a start state to an accept state during a match test on a tagged NFA generates a valid assignment of submatches*. A review of the match test process can help us to understand why: Since a path from a start state to an accept state is a number of sequential and valid operations of a tagged NFA on an input string, the assignment of submatch tags on each input symbol is also valid. The collections of the last consecutive sequences of symbols associated with the same tags that satisfy the conditions of the definition in Section 3.2 generate a valid assignment of submatches.

Path Finding Assume an input string $a_1 a_2 \ldots a_l$ is accepted by a tagged NFA A_1, and q_f is an accept state after consuming the last symbol a_l. We present a backward traversal approach to find a path which allows for submatch extraction. Starting from one of the accept states q_f, with the last input symbol a_l, perform a lookup on Δ_l for quadruples (x, i, y, τ) such that $y = q_f$ and $i = a_l$. Pick any quadruple (q_l, a_l, q_f, τ) which satisfies this condition, then q_l is a previous state which leads the automaton to q_f with the last input symbol a_l, and τ is the corresponding submatch tags associated with a_l. We note that τ can be empty. Using q_l, with input symbol a_{l-1}, perform a lookup on Δ_{l-1}

for quadruples (x, i, y, τ) such that $y = q_l$ and $i = a_{l-1}$. Such quadruples will allow us to find a previous state of q_l with input symbol a_{l-1}, along with submatch tags associated with a_{l-1} if there are any. Continue this process for $a_{l-2}, \ldots,$ and a_1. Finally, we will reach a start state q_1. Then $q_1, q_2, \ldots, q_l, q_f$ is a valid traversal path for input string $a_1 a_2 \ldots a_l$. During the backward path finding, each symbol in $a_1 a_2 \ldots a_l$ is assigned with zero or a set of submatch tags. Submatches of an accepted input string can be extracted by scanning the input strings and collecting the last consecutive sequence of symbols associated with the same submatch tags. Given a regular expression and a matching string, there might exist multiple paths from a start state to an accept state. Thus, there might exist multiple ways to assign valid submatches.

Example Figure 6 shows a traversal path of input string $aaaa$ on the tagged NFA shown in Figure 5. The path is marked by bold arrows. Along this path, we can see that the first two symbols of $aaaa$ are associated with tag t_1, and the last two symbols have no submatch tag. Thus, the submatch of $aaaa$ for regular expression $(a*)aa$ is the substring of the first two symbols, i.e., aa.

The match test and submatch extraction algorithms described in Section 3.3.1 and 3.3.2 are space efficient since the construction is based on NFA. However, they are not time efficient. During the match test, the number of states in a frontier is $O(|Q_1|)$ (the size of NFA). To derive the next frontier, states in the current frontier need to be processed one by one. Thus, the number of lookups at the transition table during the match test for an input string of length l is $O(|Q_1| \times l)$. Similarly, the number of table lookups performed during submatch extraction can be estimated as $O(|Q_1| \times l)$. If we can find an approach which allows us to derive frontiers (in match test) and previous states (in submatch extraction) more efficiently, then the time efficiency of the algorithms can be improved. Fortunately, we already have the data structures to do that. Our approach is to represent tagged NFAs, perform match testing and submatch extraction using Boolean functions (Section 3.4), and manipulate the Boolean functions using ordered binary decision diagrams (OBDDs) (Section 3.5).

3.4 Boolean Function Representation

For convenience, we discuss tagged NFAs in which ϵ transitions have been eliminated. The Boolean function of a tagged NFA $A_1 = (Q_1, \Sigma, T, \delta_1, \gamma_1, S_1, F_1)$ uses four vectors of Boolean variables, $\boldsymbol{x}, \boldsymbol{y}, \boldsymbol{i},$ and $\boldsymbol{t}$. Vectors $\boldsymbol{x}$ and $\boldsymbol{y}$ are used to denote states in Q_1, and they contain $\lceil \lg |Q_1| \rceil$ Boolean variables each. Vector $\boldsymbol{i}$ is used to denote symbols in Σ and it contains $\lceil \lg |\Sigma| \rceil$ Boolean variables. Vector $\boldsymbol{t}$ is used to denote submatch tags and it contains $\lceil |T| \rceil$ Boolean variables. We construct the following Boolean functions for the tagged NFA A_1.

• $\Delta(\boldsymbol{x}, \boldsymbol{i}, \boldsymbol{y}, \boldsymbol{t})$ denotes the tagged transition table of A_1. It is a disjunction of all tagged transition relations (x, i, y, t). As an example, the Boolean encoding of transition relations in Table 1 is shown in Table 2, where states are encoded by two bits, input symbol is encoded by one bit (since there is only one symbol a), and submatch tags are encoded by one bit. Specifically, states 1, 2, and 3 are encoded as 01, 10, and 11; symbol a is encoded as 1; and submatch tag t_1 is encoded as 1. The fifth column of Table 2 lists the function values for each set of Boolean encodings. The function value of

x	i	y	t	$\Delta(x,i,y,t)$
0 1	1	0 1	1	1
0 1	1	1 0	0	1
1 0	1	1 1	0	1

Table 2: Boolean encoding of transitions in Table 1.

Boolean encodings for tagged transitions is 1. The Boolean encoding in Table 2 can be symbolically translated to

$$\Delta(x,i,y,t) = (\bar{x}_1 \wedge x_2 \wedge i \wedge \bar{y}_1 \wedge y_2 \wedge t_1)$$
$$\vee (\bar{x}_1 \wedge x_2 \wedge i \wedge y_1 \wedge \bar{y}_2 \wedge \bar{t}_1)$$
$$\vee (x_1 \wedge \bar{x}_2 \wedge i \wedge y_1 \wedge y_2 \wedge \bar{t}_1)$$

$\Delta(x,i,y,t)$ is equivalent to the function f in Figure 1 if we rename variables i, y_1, y_2, and t_1 to x_3, x_4, x_5, and x_6 respectively.

- $\mathcal{I}_\sigma(i)$ stands for the Boolean representation of symbols in Σ. As an example, symbol a in Table 1 can be symbolically represented by $\mathcal{I}_a(i) = i$.
- $\mathcal{F}(x)$ is a Boolean function representing frontier states. In the tagged NFA shown in Figure 5, consider state $\{1\}$ with input symbol a, the new frontier has two states $\{1,2\}$, which can be symbolically represented by $\mathcal{F}(x) = (\bar{x}_1 \wedge x_2) \vee (x_1 \wedge \bar{x}_2)$.
- $\Delta_\mathcal{F}(x,i,y,t)$ is used to represent the intermediate transitions for frontier $\mathcal{F}(x)$ during a match test process.
- $\mathcal{A}(x)$ is used to define the Boolean representation of accept states of a tagged NFA. For the tagged NFA shown in Figure 5, the accept states is $\{3\}$, thus, $\mathcal{A}(x) = x_1 \wedge x_2$.

The Boolean functions described above can be automatically computed for any tagged NFA. We next describe how to perform the match test and submatch extraction described in Section 3.3 using these Boolean functions.

3.4.1 Match Test

The match test process is similar to that described in [26], except that we do book-keeping here to be used for submatch extraction. Suppose the frontier of a tagged NFA is $\mathcal{F}(x)$ at some instant of frontier derivation, and the next input symbol is σ, then the next frontier states can be computed using the following Boolean operations:

$$\mathcal{G}(y) = \exists \, x \cdot \exists \, i \cdot \exists \, t \cdot [\Delta_\mathcal{F}(x,i,y,t)] \quad (1)$$

where

$$\Delta_\mathcal{F}(x,i,y,t) = \mathcal{F}(x) \wedge \mathcal{I}_\sigma(i) \wedge \Delta(x,i,y,t) \quad (2)$$

We now explain why Equation (1) produces the new frontier states. Recall that $\Delta(x,i,y,t)$ is the disjunction of the tagged transitions of a NFA. The conjunctions of $\Delta(x,i,y,t)$ with $\mathcal{F}(x)$ and $\mathcal{I}_\sigma(i)$ on the right side of Equation (2) actually *selects* rows in the truth table of $\Delta(x,i,y,t)$ that correspond to outgoing transitions from the states in the current frontier $\mathcal{F}(x)$ labeled with symbol σ. These transitions are denoted by $\Delta_\mathcal{F}(x,i,y,t)$, which is a function of x,i,y, and t. The new frontier states are the target states of the selected transitions and are only associated with y. To extract the new frontier states, we existentially quantify x,i, and t using the existential quantification operator introduced in Section 2. We rename y to x to express the new frontier states in terms of x.

Consider the tagged NFA in Figure 5. Suppose the current frontier is $\{1\}$ and the next input symbol is a. Then

$$\mathcal{F}(x) \wedge \mathcal{I}_a(i) \wedge \Delta(x,i,y,t) = (\bar{x}_1 \wedge x_2 \wedge i \wedge \bar{y}_1 \wedge y_2 \wedge t_1)$$
$$\vee (\bar{x}_1 \wedge x_2 \wedge i \wedge y_1 \wedge \bar{y}_2 \wedge \bar{t}_1)$$

Apply existential quantification of x,i, and t on the above conjunctions we obtain $(\bar{y}_1 \wedge y_2) \vee (\wedge y_1 \wedge \bar{y}_2)$, which is the symbolic Boolean representation of the new frontier states $\{1,2\}$.

To check whether the automaton is in an accept state, simply check the satisfiability of the conjunction between $\mathcal{F}(x)$ and $\mathcal{A}(x)$. Rename the above example frontier $(\bar{y}_1 \wedge y_2) \vee (\wedge y_1 \wedge \bar{y}_2)$ to $(\bar{x}_1 \wedge x_2) \vee (\wedge x_1 \wedge \bar{x}_2)$ and do a conjunction with $\mathcal{A}(x) = x_1 \wedge x_2$. The result is not satisfiable, thus, the automaton is not in an accept state.

3.4.2 Submatch Extraction.

Now we discuss how to extract submatches using Boolean function operations. The process starts from the last symbol and one of the states where the input string is accepted. For convenience, we call the current state of a backward path finding a reverse frontier, which contains only one state because we are only interested in finding one path. Suppose at an instant of the path finding the reverse frontier representation is $\mathcal{F}_r(y)$, and the previous input symbol is σ. A previous state which leads the automaton to $\mathcal{F}_r(y)$ can be derived from the following Boolean function:

$$\Delta_r(x,i,y,t) = \mathcal{F}_r(y) \wedge \mathcal{I}_\sigma(i) \wedge \Delta_\mathcal{F}(x,i,y,t) \quad (3)$$

where $\Delta_\mathcal{F}(x,i,y,t)$ denotes the intermediate tagged transitions corresponding to symbol σ during the match test process. The conjunctions on the right side of Equation (3) *selects* tagged transitions (labeled by σ) from $\Delta_\mathcal{F}(x,i,y,t)$ where the target state is $\mathcal{F}_r(y)$. The previous states are associated with x in $\Delta_r(x,i,y,t)$. Since we are only interested in one path, we simply pick one row in the truth table of $\Delta_r(x,i,y,t)$ to find one previous state of $\mathcal{F}_r(y)$. If we denote the picked row as $\text{PICKONE}(\Delta_r(x,i,y,t))$, a previous state $\mathcal{G}(x)$ of $\mathcal{F}_r(y)$ can be derived by

$$\mathcal{G}(x) = \exists \, y \cdot \exists \, i \cdot \exists \, t \cdot \mathcal{H}(x,i,y,t) \quad (4)$$
$$\mathcal{H}(x,i,y,t) = \text{PICKONE}(\Delta_r(x,i,y,t)) \quad (5)$$

To obtain submatch tags $\tau(t)$ associated with σ, we existentially quantify x, i, and y on $\mathcal{H}(x,i,y,t)$.

$$\tau(t) = \exists \, x \cdot \exists \, i \cdot \exists \, y \cdot \mathcal{H}(x,i,y,t) \quad (6)$$

Consider the example in Figure 6. After consuming the fourth input symbol of $aaaa$, the automaton accepts and

$$\Delta_\mathcal{F}(x,i,y,t) = (\bar{x}_1 \wedge x_2 \wedge i \wedge \bar{y}_1 \wedge y_2 \wedge t_1)$$
$$\vee (\bar{x}_1 \wedge x_2 \wedge i \wedge y_1 \wedge \bar{y}_2 \wedge \bar{t}_1)$$
$$\vee (x_1 \wedge \bar{x}_2 \wedge i \wedge y_1 \wedge y_2 \wedge \bar{t}_1)$$

Starting from the accept state 3 ($\mathcal{F}_r(y) = y_1 \wedge y_2$) and the last symbol a ($\mathcal{I}_a(i) = i$), do a conjunction according to Equation (3) we get $\Delta_r(x,i,y,t) = (x_1 \wedge \bar{x}_2 \wedge i \wedge y_1 \wedge y_2 \wedge \bar{t}_1)$, which has only one tagged transition. Perform existential quantifications according to Equation (4) and (5) we obtain the Boolean representation of a previous state as $x_1 \wedge \bar{x}_2$, which translates to state 2. Do an existential quantifications according to Equation (6) we get $\tau(t) = \bar{t}_1$, which means that no tag is associated with the fourth symbol a. Applying the same approach on the 3rd, 2nd, and 1st symbols we obtain a path from state 1 to 3, where the 1st and 2nd

168

symbol a are assigned with submatch tag t_1. Thus, the submatch of $aaaa$ to $(a*)aa$ is $sub(t_1) = aa$.

A submatch assignment obtained by our approach is not necessarily the left most, longest submatch, which is required by POSIX. However, POSIX doesn't have a notion of "greedy" and "reluctant" closures, which give some control over the length of the submatch. Thus, POSIX is incomplete. Standard libraries like Java and PCRE have behaviors that are not POSIX compliant.

3.5 Submatch-OBDD

To improve the efficiency of the match test and submatch extraction, we represent and manipulate the Boolean functions defined in Section 3.4 using OBDDs. We call our model Submatch-OBDD. A Submatch-OBDD for a tagged NFA $A_1 = (Q_1, \Sigma, T, \delta_1, \gamma_1, S_1, F_1)$ is a 5-tuple $[OBDD(\Delta(\boldsymbol{x}, \boldsymbol{i}, \boldsymbol{y}, \boldsymbol{t})), \{OBDD(\mathcal{I}_\sigma | \forall \sigma \in \Sigma)\}, \{OBDD(\mathcal{T}_t | \forall t \in T)\}, OBDD(\mathcal{F}_{S_1}), OBDD(\mathcal{A})]$, where $\Delta(\boldsymbol{x}, \boldsymbol{i}, \boldsymbol{y}, \boldsymbol{t})$ is Boolean representation of tagged transitions, $\mathcal{I}_\sigma$ is the Boolean representation of a symbol $\sigma \in \Sigma$, $\mathcal{T}_t$ is Boolean representation of a tag $t \in T$, $\mathcal{F}_{S_1}$ is Boolean representation of start states, and $\mathcal{A}$ is the Boolean representation of accept states F_1.

To understand why OBDDs can improve the time-efficiency of tagged NFA operations, consider frontier derivation on a tagged NFA. To derive a new set of frontier states, the tagged transition table must be retrieved for each state in the current frontier $\mathcal{F}$, leading to $O(|\mathcal{F}|)$ operations for each input symbol. On the other hand, the time-complexity of using OBDDs to derive the next frontier is determined by the two conjunctions and one existential quantification in Equation (1) and (2). When the frontier set $\mathcal{F}$ is large, the cost of doing the two conjunctions and one existential quantification is often smaller than doing $|\mathcal{F}|$ lookups on the transition table. Using the same method, we can calculate that the time complexity of submatch extraction is the same as the match test process. For a tagged-NFA with n states, the size of frontier set $|\mathcal{F}|$ is $O(n)$. Thus, the cost to process an input string l bytes by our approach is between $O(l)$ and $O(nl)$. In other words, the time complexity of Submatch-OBDD is between a pure DFA and a pure NFA approach.

The space efficiency of Submatch-OBDD is comparable to tagged NFAs. The space cost of a Submatch-OBDD is dominated by $OBDD(\Delta(\boldsymbol{x}, \boldsymbol{i}, \boldsymbol{y}, \boldsymbol{t}))$, which needs a total of $2 \times \lceil \lg |Q_1| \rceil + \lceil \lg |\Sigma| \rceil + \lceil |T| \rceil$ Boolean variables. In the worst case, the size of the OBDD is $O(|Q_1|^2 \times |\Sigma| \times 2^{|T|})$, which is comparable to the size of transitions of a tagged NFA. We note that the OBDDs of intermediate transitions $\Delta_\mathcal{F}(\boldsymbol{x}, \boldsymbol{i}, \boldsymbol{y}, \boldsymbol{t})$ for all input symbols also take some space, mainly depending on the size of input string. We will show that such a cost is not a concern in practice in Section 4.

3.6 Implementation

We implemented Submatch-OBDD as a toolchain in C++. The toolchain has two offline components, RE2TNFA and TNFA2OBDD, and one online component, PATTERNMATCH. RE2TNFA accepts patterns as input and outputs tagged-NFAs that defines the same languages as the input patterns. TNFA2OBDD then generates the tagged-NFAs' OBDD representations. PATTERNMATCH then performs match test and submatch extraction on an input stream using the OBDD representations. Our implementation interfaces with the popular CUDD library [22] for OBDD construction and manipulation.

In comparison, both PCRE and RE2 are implemented in C++. PCRE uses a recursive backtracking approach: It compiles a pattern into a tree like structure and then uses recursive backtracking to match patterns and extract submatches. RE2 uses a combination of DFAs and NFAs for submatch extraction: Given a pattern and an input string, RE2 constructs and uses backward and forward DFAs to locate the pattern's overall match in the input string. It then uses NFA based approaches to find submatches in the overall match. For memory efficiency, RE2 doesn't construct entire DFAs. It creates DFA states on demand (determination on-the-fly) and stores them in a limited sized cache; when the cache gets full, RE2 empties the cache and restarts the DFA construction process.

4. EVALUATION

We evaluated the performance of our Submatch-OBDD implementation using patterns used in real systems. We measured Submatch-OBDD's time efficiency and space efficiency by matching the patterns with network traces, synthetic traces, and enterprise event logs, and then compared our performance with two popular regular expression engines: RE2 and PCRE. Our findings suggest that Submatch-OBDD achieves its ideal performance when patterns are combined. In the best case, Submatch-OBDD is faster than RE2 and PCRE by one to two orders of magnitude. All the performance numbers of Submatch-OBDD reported in this section were obtained based on the variable ordering of $\boldsymbol{i} \prec \boldsymbol{x} \prec \boldsymbol{y} \prec \boldsymbol{t}$.

4.1 Data Sets

We used three sets of patterns and trace files in our evaluation.

Snort-2009.

We extracted 115 patterns from a Snort 2009 HTTP rule set of 3078 patterns. All patterns were extracted from the pcre fields of the rules. Since our focus is submatch extraction, we excluded patterns containing no capturing groups and patterns containing back references as patterns with back references cannot be represented by regular languages. Each extracted pattern contains one to six capturing groups.

We used two network traces and one synthetic trace to evaluate the performance of our approach on the Snort-2009 pattern set.

- The first web trace was a 1.2GB network traffic collected using tcpdump from our department's web server. The average packet size of this trace is 126 bytes with a standard deviation of 271. The second web trace was a 1.3GB network traffic collected by crawling URLs that appeared on Twitter using a python script and recording the full length packets using tcpdump. The average packet size of the second trace is 1202 bytes with a standard deviation of 472.

- We also created a synthetic trace to observe how different implementations perform under the backtracking algorithmic complexity attack [19]. By reviewing the 115 patterns of the Snort-2009 pattern set, we found that several of them are vulnerable to the backtracking algorithmic complexity attack if a regular expression engine is implemented by backtracking, e.g., PCRE. We then crafted a 1MB trace which can exploit the backtracking behavior of a backtracking-based pattern matching engine. The average line length of the trace is 311 bytes with a standard deviation of 5.

Snort-2012.

We also evaluated our approach with the latest rules from the Snort system. We extracted 403 patterns (regular expressions with capturing groups) from a snapshot of the Snort-2012 HTTP rule set containing 3990 rules. All patterns were extracted from the `pcre` fields of the rules. Like the patterns of Snort-2009, we excluded patterns containing back references as they can not be represented by regular languages. Patterns containing no capturing group are also excluded as our focus was on submatch extraction. Each extracted pattern has one to ten capturing groups.

We used two web traces and one synthetic trace to evaluate the performance of different approaches on this pattern set. The two web traces are the same as those used in the Snort-2009 pattern set evaluation. The synthetic trace was created after reviewing the 403 patterns: We found that several of the 403 patterns are vulnerable to backtracking algorithmic attacks. We then crafted a 1MB trace which can exploit the backtracking behavior of a backtracking-based pattern matching engine and evaluated its effects on Submatch-OBDD, RE2, and PCRE. The average line length of this synthetic trace is 689 bytes with a standard deviation of 41.

Firewall-504.

We also obtained a set of 504 patterns used by a commercial SIEM system C to normalize logs generated by a commercial firewall, F. For commercial reasons, we do not disclose the names of the SIEM system and the firewall. Each pattern in the set has 1-22 capturing groups. We collected 87 MBs of firewalls logs generated by F in an enterprise setting and measured our performance on the logs. The logs consist of 1.01 million lines of text and the average line size is 87 bytes with standard deviation of 51. We did not create synthetic trace for this pattern set as firewall logs cannot easily be controlled by an attacker.

4.2 Experimental Setup

We conducted our experiments on an Intel Core2 Duo E7500 Linux-2.6.3 machine running at 2.93 GHz with 2 GB of RAM. We measure the time efficiency of different approaches in the average number of CPU cycles needed to process one byte of a trace file. We only measure pattern matching and submatch extraction time, and exclude pattern compilation time. Similarly, we measure memory efficiency in megabytes (MB) of RAM used during pattern matching and submatch extraction.

We measure the performance of each approach on a pattern set in two *configurations*. In one configuration, Conf.S, we match each pattern with the input stream sequentially. For example, we match each pattern in the Snort-2009 set with each packet in the network traces. Combining all patterns of a pattern set into one single pattern, however, allows us to match each packet with all patterns in one pass. This configuration, Conf.C, is also useful in the log normalization process of a SIEM system. The system can match an event log with all rules in one pass and extract all fields of interest instead of matching the logs with each rule sequentially.

Given a pattern set with n patterns and an input trace of M bytes, we measured performance of an approach in the following two configurations.

- **Conf.S (Sequential):** We compile each pattern individually and then match the compiled patterns with the trace sequentially. If the ith pattern's execution time for the M bytes trace is t_i cycles, then the time efficiency of an approach to the pattern set is $\frac{t_1+\cdots+t_n}{M}$ cycles/byte.

- **Conf.C (Combination):** We combine the n patterns together into one pattern using the UNION operation. We compile the combined pattern and match it with the input trace. If the combined pattern's execution time for the M bytes trace is t cycles, then an approach's time efficiency to the pattern set is $\frac{t}{M}$ cycles/byte. When an input string matches a specific pattern in the combined pattern, Submatch-OBDD emits the submatches, as well as the pattern that matches the input string.

4.3 Performance Results

4.3.1 Snort-2009

Table 3 shows the execution times (cycles/byte) and memory consumption of RE2, PCRE, and Submatch-OBDD for the Snort-2009 pattern set on the web traces and synthetic trace. We have the following observations:

- Submatch-OBDD achieves its ideal performance in Conf.C, i.e., when patterns are combined together for pattern matching and submatch extraction.

- Submatch-OBDD is the fastest approach among the three. For the web traces, Submatch-OBDD's best performance (in Conf.C) is an order of magnitude faster than the other approaches' best performance (in Conf.S).

- PCRE suffers from backtracking algorithmic complexity attacks, while Submatch-OBDD and RE2 don't. With the web traces, the best time efficiency of PCRE was 3.67×10^4. However, PCRE was slowed down by two orders of magnitude when the synthetic trace was used, as is shown in Table 3(b). The reason is that the synthetic trace caused PCRE to perform heavily backtracking for some patterns.

- In Conf.C, the memory consumption of Submatch-OBDD and RE2 are comparative, while PCRE consumes the least memory. We do not report the memory requirements in Conf.S as the three approaches use very little memory for simple patterns.

We note that in Conf.S, RE2 is faster than Submatch-OBDD. This is because many patterns did not fill up the DFA state cache and hence did not trigger the DFA reconstruction process. In the case of simple patterns, the cost of OBDD operations, e.g., frontier derivation and existential quantification, is higher than the cost of several lookups on NFA transition table because the frontier size is often very small. Thus, Submatch-OBDD performs slower than RE2 in such situations. The cost of OBDD operations will be paid off when the frontier size of a tagged-NFA is large.

We recommend that Submatch-OBDD to be used in cases where a group of patterns are combined together. The performance boost of Submatch-OBDD is due to the *redundancy elimination*: The OBDD representation eliminates the redundancy in the Boolean representation of tagged-NFAs.

Method	Conf.S Exec-time	Conf.C	
		Exec-time	Memory (MB)
RE2	2.31×10^4	1.21×10^5	7.3
PCRE	3.67×10^4	1.13×10^6	1.2
OBDD	8.76×10^4	3.63×10^3	9.4

(a) Performance numbers with the web traces

Method	Conf.S Exec-time	Conf.C	
		Exec-time	Memory (MB)
RE2	8.20×10^4	2.22×10^5	7.6
PCRE	1.44×10^6	1.40×10^6	1.0
OBDD	2.12×10^5	2.20×10^4	7.0

(b) Performance numbers with the synthetic trace

Table 3: Execution time (cycles/bytes) and memory consumption for the Snort-2009 data set with (a) the web traces and (b) the synthetic trace. In both traces, Submatch-OBDD's best execution time (Conf.C) is much shorter than RE2's and PCRE's best execution times (Conf.S).

4.3.2 Snort-2012

Table 4 shows the performance of RE2, PCRE, and Submatch-OBDD on the 403 patterns from Snort-2012 rule set.We have the following observations:

- Submatch-OBDD achieves its ideal time efficiency in Conf.C, i.e., when patterns are combined together for matching test and submatch extraction.

- For the web traces, Submatch-OBDD is faster than RE2, but slower than PCRE. While for the synthetic trace, Submatch-OBDD is faster than both RE2 and PCRE.

- Like in the Snort-2009 data set, PCRE suffers from the backtracking algorithmic attack performed by the synthetic trace. PCRE's time efficiency under the synthetic trace is two to three orders of magnitude than under the web traces.

- In Conf.C, the memory consumption of Submatch-OBDD and RE2 are comparative.

Although we observed that PCRE performed better for the web traces in Table 4, PCRE is still not recommended to be used as pattern matching engine for a network intrusion detection system (NIDS). The main reason is that it is easy for attackers to craft network traffic performing backtracking algorithmic attacks on PCRE, as was shown by Smith et al. in [19]. Our experimental results in Table 3 and Table 4 also demonstrated that PCRE is easily to be slowed down by hundreds of times with carefully crafted synthetic traces.

4.3.3 Firewall-504

Table 5 shows the three approaches' performance on the Firewall-504 data set. Submatch-OBDD is the fastest approach on this data set. In Conf.C, Submatch-OBDD is orders of magnitude faster than RE2 and PCRE. Also, Submatch-OBDD's best performance (in Conf.C) is 62% faster than RE2's best performance (in Conf.S). In memory usage, PCRE is most space compact. Submatch-OBDD consumes slightly more memory than RE2.

4.4 Discussion

During our evaluation, we found a small number regular expressions from the Snort 2009 and 2012 rule sets that can

Method	Conf.S Exec-time	Conf.C	
		Exec-time	Memory (MB)
RE2	4.79×10^4	2.09×10^6	15.0
PCRE	7.70×10^4	2.69×10^3	1.0
OBDD	3.83×10^5	1.08×10^4	6.3

(a) Performance numbers with the web traces

Method	Conf.S Exec-time	Conf.C	
		Exec-time	Memory (MB)
RE2	2.92×10^5	8.21×10^6	15.0
PCRE	1.47×10^6	7.64×10^5	1.0
OBDD	4.70×10^5	1.10×10^5	15.3

(b) Performance numbers with the synthetic trace

Table 4: Execution time (cycles/bytes) and memory consumption for the Snort-2012 data set with (a) the web traces and (b) the synthetic trace.

Method	Conf.S Exec-time	Conf.C	
		Exec-time	Memory (MB)
RE2	2.04×10^5	2.20×10^7	21.0
PCRE	6.88×10^5	1.60×10^6	1.1
OBDD	6.31×10^5	1.25×10^5	30.0

Table 5: Execution time (cycles/bytes) and memory consumption for the Firewall-504 data set.

cause either PCRE or RE2 to perform poorly. For example, if we use PCRE to match

```
.*\x2F[^\s]*\.(dat|xml)\?[^\s]*v=[^\s]*t=[^\s]*c=
```

with input string
```
/;/;/;.dat?;.dat?;.dat?;v=;v=;v=;t=;t=;t=;c.
```
Then PCRE will perform $O(3 \times 3 \times 3 \times 3)$ backtracking evaluations before eventually concluding that the string does not match the pattern. The evaluation time of PCRE will increase exponentially if we increase the number of repetitions of the /;, .dat?, v=, and t=; in the input string. We observed that when these substrings were repeated 20 times, the execution time of PCRE for this regular expression was in the order of 10^6 cycles/byte. Details on how to create pathological traces to exploit the backtracking behavior of PCRE can be found in [19].

RE2 can perform poorly under the case when the DFA states of a regular expression blow up. The blow-up will cause the limited state cache be filled quickly and RE2 has to empty the cache and restart the DFA construction. In our experiments, we have observed an individual regular expression from Snort-2009 where the time efficiency of RE2 is an order of magnitude slower than Submatch-OBDD, which does not suffer from state blow up as it is a NFA-based approach. We also found eight patterns from the SIEM system which cause RE2 to blow up in its DFA construction. For these patterns, the time efficiency of RE2 is an order of magnitude slower than Submatch-OBDD. For commercial reasons, we do not disclose these patterns in the paper.

Please note that RE2 and PCRE are mature and popular engines and their code bases are heavily optimized. We have not devoted significant time to try to optimize Submatch-OBDD. We believe that Submatch-OBDD's performance can be further improved with better optimization.

5. RELATED WORK

Regular expressions are extensively used to construct attack signatures in NIDS and to process event logs in SIEM systems. Finite automata are natural representations for regular expressions. DFAs are fast, but suffer from state blow-up for certain types of regular expressions. NFAs are compact, but slow in operation. Many techniques have been proposed to improve DFAs' space efficiency: compression [2], determinization on-the-fly [23], building multiple DFAs (MDFA) from a group of signatures [28], extending DFAs with scratch memory (XFAs) [20, 21], and constructing DFA variants with hardware implementations [3, 13, 15]. Similarly, many techniques have been proposed to improve NFAs' time efficiency: hardware based parallelism [7, 15, 6, 12, 18] and software based speedup [26, 27]. Hybrid finite automata [1] combines the benefits of NFAs and DFAs.

Submatch extraction, however, has not received much attention from the research community. Pike implemented a submatch extraction approach in the sam text editor [17] using a straightforward modification of Thompson's NFA simulation [24]. Google's RE2 tool also uses the modified NFA simulation approach. Laurikari proposed TNFA, an NFA-based approach for submatch extraction, where an NFA is augmented with tags to represent capturing groups [14]. Our approach also uses tags, but we associate tags with non-ϵ transitions whereas TNFA associates tags with ϵ transitions. We use OBDDs to represent and operate on tagged NFAs to achieve time efficiency. We did not include TNFA in our experiments as we already compared with a mature NFA-based approach, RE2.

Java, PCRE, Perl, Python, Ruby, and many other tools implement pattern match and submatch extraction using recursive backtracking, where an input string may be scanned multiple times before a match is found. The backtracking approach's worst case performance is exponential running time [9]. These tools use backtracking to efficiently handle backreference, a non-regular construct that improves the pattern language's expressive power. In contrast, our Submatch-OBDD approach is an NFA-based technique and does not suffer from exponential running time.

Google's RE2 is an open source automata based pattern matching tool that supports submatch extraction [10]. RE2 employs a DFA approach to test whether an input string matches a pattern. If a pattern contains capturing groups, RE2 uses a DFA approach to find the pattern's overall match in an input string and then runs an NFA approach to extract the submatches in the overall match. Similar to RE2, our Submatch-OBDD is NFA-based. We, however, use OBDDs to perform NFA operations and hence improve time efficiency. Submatch-OBDD performs better than RE2 when patterns are combined. Both RE2 and Submatch-OBDD do not support backreferences.

Yang et al.'s NFA-OBDD model [26, 27] is the most relevant work to Submatch-OBDD. A commonality between NFA-OBDD and our Submatch-OBDD is the use of "implicit state enumeration" by means of OBDDs [8, 25]. NFA-OBDD, however, does not support submatch extraction. In our work, we propose a new approach to tag capturing groups in a regular expression, and extend Thompson's NFA construction to support capturing groups. We propose a novel submatch extraction approach using OBDDs.

6. CONCLUSION

We present Submatch-OBDD, which allows fast submatch extraction in regular expression-like pattern matching. We propose a new approach to tag capturing groups in a regular expression, and extend Thompson's NFA construction approach to support regular expressions with capturing groups.

We present a novel technique to perform submatch extraction. Our use of OBDDs improves the time efficiency of match test and submatch extraction. We evaluated our Submatch-OBDD implementation using patterns used in the Snort NIDS and a commercial SIEM system. Our experiments on real network traces, synthetic traces, and enterprise event logs show that Submatch-OBDD achieves its ideal performance when patterns are combined. In the best case, our approach is faster than RE2 and PCRE by one to two orders of mangintude.

Acknowledgments

This work was supported in part by NSF grants CNS-0952128 and CNS-1117711. We thank the anonymous ANCS reviewers for helpful comments on an earlier draft of this paper.

7. REFERENCES

[1] M. Becchi and P. Crowley. A hybrid finite automaton for practical deep packet inspection. In *Proceedings of the 2007 ACM CoNEXT conference*, CoNEXT '07, pages 1:1–1:12, New York, NY, USA, 2007. ACM.

[2] M. Becchi and P. Crowley. An improved algorithm to accelerate regular expression evaluation. In *Proceedings of the 3rd ACM/IEEE Symposium on Architecture for networking and communications systems*, ANCS '07, pages 145–154, New York, NY, USA, 2007. ACM.

[3] B. C. Brodie, D. E. Taylor, and R. K. Cytron. A scalable architecture for high-throughput regular-expression pattern matching. In *Intl. Symp. Computer Architecture*, pages 191–202. IEEE Computer Society, 2006.

[4] R. E. Bryant. Graph-based algorithms for boolean function manipulation. *IEEE Trans. Comput.*, 35:677–691, August 1986.

[5] J. R. Burch, E. M. Clarke, K. L. McMillan, D. L. Dill, and J. Hwang. Symbolic model checking: 10^{20} states and beyond. In *Symp. on Logic in Computer Science*, pages 401–424. IEEE Computer Society, 1990.

[6] D. Chasaki and T. Wolf. Fast regular expression matching in hardware using nfa-bdd combination. In *Proceedings of the 6th ACM/IEEE Symposium on Architectures for Networking and Communications Systems*, ANCS '10, pages 12:1–12:2, New York, NY, USA, 2010. ACM.

[7] C. R. Clark and D. E. Schimmel. Scalable pattern matching for high-speed networks. In *Symp. on Field-Programmable Custom Computing Machines*, pages 249–257. IEEE Computer Society, 2004.

[8] O. Coudert, C. Berthet, and J. C. Madre. Verification of synchronous sequential machines based on symbolic execution. In *Proceedings of the international workshop on Automatic verification methods for finite state systems*, pages 365–373, New York, NY, USA, 1990. Springer-Verlag New York, Inc.

[9] R. Cox. Regular expression matching can be simple and fast. http://swtch.com/~rsc/regexp/regexp1.html, 2007.

[10] R. Cox. Implementing regular expressions. http://swtch.com/~rsc/regexp/, Last retrieved in August 2011.

[11] J. E. Hopcroft, R. Motwani, and J. D. Ullman. *Introduction to automata theory, languages, and computation.* Addison Wesley, 2001.

[12] B. L. Hutchings, R. Franklin, and D. Carver. Assisting network intrusion detection with reconfigurable hardware. In *Symp. on Field-Programmable Custom Computing Machines*, pages 111–120. IEEE Computer Society, 2002.

[13] S. Kumar, S. Dharmapurikar, F. Yu, P. Crowley, and J. Turner. Algorithms to accelerate multiple regular expressions matching for deep packet inspection. In *ACM SIGCOMM Conference*, pages 339–350. ACM, 2006.

[14] V. Laurikari. NFAs with tagged transitions, their conversion to deterministic automata and application to regular expressions. In *SPIRE'00*, September 2000.

[15] C. Meiners, J. Patel, E. Norige, E. Torng, and A. X. Liu. Fast regular expression matching using small TCAMs for network intrusion detection and prevention systems. In *19th USENIX Security Symposium*, August 2010.

[16] PCRE. The Perl compatible regular expression library. http://www.pcre.org.

[17] R. Pike. The text editor sam. *Softw. Pract. Exper.*, 17:813–845, November 1987.

[18] R. Sidhu and V. Prasanna. Fast regular expression matching using FPGAs. In *Symp. on Field-Programmable Custom Computing Machines*, pages 227–238. IEEE Computer Society, 2001.

[19] R. Smith, C. Estan, and S. Jha. Backtracking algorithmic complexity attacks against a NIDS. In *Annual Computer Security Applications Conf.*, pages 89–98. IEEE Computer Society, 2006.

[20] R. Smith, C. Estan, and S. Jha. XFA: Faster signature matching with extended automata. In *Symp. on Security and Privacy*, pages 187–201. IEEE Computer Society, 2008.

[21] R. Smith, C. Estan, S. Jha, and S. Kong. Deflating the Big Bang: Fast and scalable deep packet inspection with extended finite automata. In *SIGCOMM Conference*, pages 207–218. ACM, 2008.

[22] F. Somenzi. CUDD: CU decision diagram package, release 2.4.2. Department of Electrical, Computer, and Energy Engineering, University of Colorado at Boulder. http://vlsi.colorado.edu/$\sim$fabio/CUDD.

[23] R. Sommer and V. Paxson. Enhancing byte-level network intrusion detection signatures with context. In *CCS'03*, pages 262–271. ACM, 2003.

[24] K. Thompson. Programming techniques: Regular expression search algorithm. *Commun. ACM*, 11:419–422, June 1968.

[25] H. J. Touati, H. Savoj, B. Lin, R. K. Brayton, and A. Sangiovanni-Vincentelli. Implicit state enumeration of finite state machines using bdd's. In *IEEE International Conference on Computer-Aided Design*, pages 130–133, Santa Clara, CA, 1990. IEEE.

[26] L. Yang, R. Karim, V. Ganapathy, and R. Smith. Improving nfa-based signature matching using ordered binary decision diagrams. In *RAID'10*, volume 6307 of *Lecture Notes in Computer Science (LNCS)*, pages 58–78, Ottawa, Canada, September 2010. Springer.

[27] L. Yang, R. Karim, V. Ganapathy, and R. Smith. Fast, memory-efficient regular expression matching with nfa-obdds. *Computer Networks*, 55(15):3376–3393, October 2011.

[28] F. Yu, Z. Chen, Y. Diao, T. V. Lakshman, and R. H. Katz. Fast and memory-efficient regular expression matching for deep packet inspection. In *ACM/IEEE Symp. on Arch. for Networking and Comm. Systems*, pages 93–102, 2006.

LEAP: Latency- Energy- and Area-optimized Lookup Pipeline

Eric N. Harris[†]   Samuel L. Wasmundt[†]   Lorenzo De Carli[†]
Karthikeyan Sankaralingam[†]   Cristian Estan[‡]
[†]University of Wisconsin-Madison   [‡]Broadcom Corporation
enharris@uwalumni.com   {wasmundt,lorenzo,karu}@cs.wisc.edu   cestan@broadcom.com

ABSTRACT

Table lookups and other types of packet processing require so much memory bandwidth that the networking industry has long been a major consumer of specialized memories like TCAMs. Extensive research in algorithms for longest prefix matching and packet classification has laid the foundation for lookup engines relying on area- and power-efficient random access memories. Motivated by costs and semiconductor technology trends, designs from industry and academia implement multi-algorithm lookup pipelines by synthesizing multiple functions into hardware, or by adding programmability. In existing proposals, programmability comes with significant overhead.

We build on recent innovations in computer architecture that demonstrate the efficiency and flexibility of dynamically synthesized accelerators. In this paper we propose LEAP, a latency- energy- and area- optimized lookup pipeline based on an analysis of various lookup algorithms. We compare to PLUG, which relies on von-Neumann-style programmable processing. We show that LEAP has equivalent flexibility by porting all lookup algorithms previously shown to work with PLUG. At the same time, LEAP reduces chip area by $1.5\times$, power consumption by $1.3\times$, and latency typically by $5\times$. Furthermore, programming LEAP is straight-forward; we demonstrate an intuitive Python-based API.

Categories and Subject Descriptors

B.4.1 [**Data Communication Devices**]: Processors; C.1 [**Computer Systems Organization**]: Processor Architectures

Keywords

Network processing, Lookups, TCAM, Dynamically-specialized datapath

1. INTRODUCTION

Lookups are a central part of the packet processing performed by network switches and routers. Examples include forwarding table lookups to determine the next hop destination for the packet and packet classification lookups to determine how the given packet is to be treated for service quality, encryption, tunneling, etc. Software based table lookups [17], lookup hardware integrated into the packet processing chip [18], and dedicated lookup chips [3, 4] are different implementations with the latter two being the preferred industry approach.

We observe that there is an increasing sophistication in the lookup processing required and reducing benefits from technology scaling. Borkar and Chien show that energy efficiency scaling of transistors is likely to slow down, necessitating higher-level design innovations that provide energy savings [9]. This paper's goal is to investigate a new class of flexible lookup engines with reduced latency, energy consumption, and silicon area that ultimately translate into cost reductions or more aggressive scaling for network equipment as described below.

1. Latency: Lookup engine latency affects other components on the router interface. The exact nature of the savings depends on the line card architecture, but it can result in a reduction in the size of high-speed buffers, internal queues in the network processor, and the number of threads required to achieve line-speed operation. A major reduction in the latency of the lookup engine can indirectly result in important area and power savings in other chips on the line card.

2. Energy/Power: Reducing the power consumption of routers and switches is in itself important because the cost of electricity is a significant fraction of the cost of operating network infrastructure. Even more important, reducing power improves scalability because the heat dissipation of chips, and the resulting cooling challenges, are among the main factors limiting the port density of network equipment. Our design, LEAP, demonstrates energy savings for lookups through architectural innovation.

3. Area: Cutting-edge network processors and stand-alone lookup engines are chips of hundreds of square millimeters. Reducing the silicon area of these large chips results in a super-linear savings in costs.

The architecture we propose is called LEAP. It is meant to act as a co-processor or lookup module for network processors or switches, not to implement the complete data plane processing for a packet. It is a latency- energy- and area-optimized lookup pipeline architecture that retains the flexibility and performance of earlier proposals for lookup pipeline architectures while significantly reducing the overheads that earlier proposals incur to achieve flexibility. Instead of programmable microengines, we use a dynamically configurable data path. We analyze seven algorithms for forwarding lookups and packet classification to determine a mix of functional units suitable for performing the required processing steps. We have designed, implemented in RTL, and verified one instance of the LEAP architecture and synthesized it to a 55nm ASIC library. PLUG[13] is an earlier proposal for a tiled smart memory architecture that can perform pipelined lookups. At the same technology node, LEAP achieves the same throughput as PLUG and supports the same lookup algorithms but has $1.5\times$ lower silicon area, and $1.3\times$ lower energy. Latency savings depend on the lookup algorithms used: we observe between $1.7\times$ and $6.5\times$, with typical savings exceeding $5\times$.

The remainder of this paper is organized as follows. We first present motivating background and related work in Section 2. Section 3 presents a characterization of lookup processing, Section 4 presents the LEAP architecture, Section 5 discusses programming mechanisms and our Python API. Section 6 presents quantitative evaluation of LEAP based on a diverse set of seven lookup algorithms. Section 7 concludes the paper.

2. BACKGROUND AND MOTIVATION

We first describe background to place our work in context. Two main approaches exist for implementing lookups, namely, run-to-completion (RTC) architectures and dedicated lookup-based engines. The RTC paradigm exploits packet-level parallelism: the core in charge of each packet fully performs the lookup based on the data in the packet. Lookup data structures are maintained by the network processor on a globally shared memory (across the processor's cores). Examples include Cisco's Silicon Packet Processor [18]. An alternative is dedicated lookup engines interfaced with the network processor. These engines are organized as a pipeline where, at each pipeline step (which is a hardware block), relevant fields from packet headers are moved close to relevant lookup data structures. The goal is to minimize the amount of data moved around. In principle, both approaches are equally valid. In this work we focus on systems that belong to the latter class of dedicated lookup-based engines.

Within this domain, two main approaches exist for implementing lookups: i) relying on massive bit-level hardware parallelism and ii) using algorithms.

2.1 Bit-level hardware parallelism

In this approach, typically a ternary content-addressable memory (TCAM) is employed to compare a search key consisting of packet header fields against all entries of the table in parallel. Due to low density of TCAM storage and power challenges, much research effort has focused on the second approach of finding good RAM-based algorithmic solutions.

2.2 Algorithmic Lookup Engine Architectures

The main challenge for lookup algorithms is minimizing the amount of memory required to represent the lookup table while keeping the number of memory references low and the processing steps simple as described in a mature body of work [20, 31, 37, 6]. The common characteristics across different types of lookups are the following: performance is dominated by frequent memory accesses with poor locality, processing is simple and regular, and plentiful parallelism exists because many packets can be processed independently. Here each lookup is partitioned into individual tasks organized in a pipeline. Multiple packets can be concurrently processed by a single lookup engine by pipelining these tasks across the packets. Figure 1 plots lookup engines classified according to flexibility and efficiency and we elaborate on this design space below.

Fixed-Function: Specialized approaches where a chip is designed for each type of lookup provide the most efficiency and least flexi-

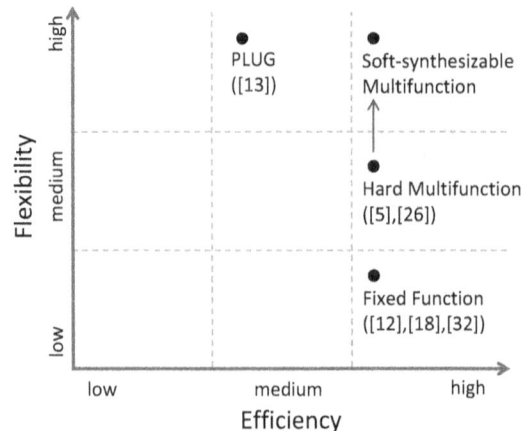

Figure 1: Design space of Lookup Engines

bility. In the FIPL architecture [32], an array of small automata are connected to a single memory to provide an efficient specialized architecture for the IP forwarding lookup problem. By placing the processing close to memory and using specialized hardware, high efficiency is achieved. Such specialized algorithmic lookup engines are widely used both as modules inside network processors, for example, in QuantumFlow [12] and as stand-alone chips [3, 4] acting as a coprocessors to network processors or other packet processing ASICs. The fixed-function approach is often implemented with partitioned memories accompanied by their own dedicated hardware modules to provide a pipelined high-bandwidth lookup engine.

Hard Multi-Function: Versatility can be increased without compromising efficiency by integrating different types of fixed-function processing that all use the same data-paths for memories and communication between pipeline stages [26]. Baboescu *et al.*[5] propose a circular pipeline that supports IPv4 lookup, VPN forwarding and packet classification. Huawei is building tiled Smart Memories [28] that support 16 functions including IP forwarding lookup, Bloom filters that can be used in lookups [16, 15], sets and non-lookup functions such as queues, locks, and heaps. As is common for industry projects, details of the internals of the architecture and programming model are not disclosed.

Specialized Programmable: Hard multi-function lookup pipelines lack the flexibility to map lookups whose functionality has not been physically realized at manufacture time. The PLUG [13] lookup pipeline achieves flexibility by employing principles from programmable von-Neumann style processors in a tiled architecture. Specifically, it simplifies and specializes the general purpose processor and processing is achieved through 32 16-bit microcores in each tile that execute processing steps of lookup algorithms based on instructions stored locally. Communication is provided by six separate 64-bit networks with a router in each tile. These networks enable non-linear patterns of data flow such as the parallel processing of multiple independent decision trees required by recent packet classification algorithms such as Efficuts [37].

While PLUG attains high flexibility, it lacks high efficiency. It is instructive to understand why it has inefficiencies. Using the Mc-PAT [30] modeling tool, we modeled a simple 32-bit in-order core similar to a PLUG microcore. In Figure 2, we show the percentage contribution of energy consumption of this processor due to its four main primitive functions: instruction-fetch and decode, register-file read, execute, and write-back (includes register write, pipeline staging etc.). Events other than execute are *overhead* yet they account for more than 80% of the energy. While it may seem surprising,

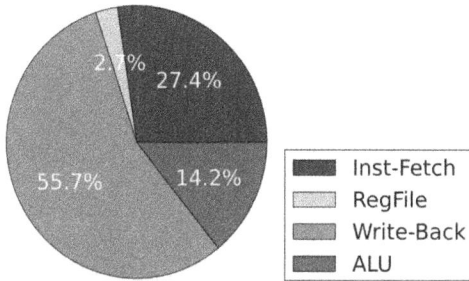

Figure 2: Programmable processor energy breakdown.

Context	Approach/Key data structures
Ethernet forwarding	D-left hash table [8, 10, 39]
IPv4 forwarding	Compressed multi-bit tries [14]
IPv6 forwarding	Compressed multi-bit tries + hash tables [25]
Packet classification	Efficuts with parallel decision trees [37, 36]
DFA lookup	DFA lookup in compressed transition tables[27]
Ethane	Parallel lookup in two distinct hash tables [11]
SEATTLE	Hash table for cached destinations, B-tree for DHT lookup [24]

Table 1: Characterization of lookup algorithms.

considering that a single pipeline register consumes 50% of the energy of an ALU operation and staging uses three pipeline registers for PLUG's 3-cycle load, these overheads soon begin to dominate. We are not the first to make this observation. In the realm of general purpose processors, Hameed *et al.*[22] recently showed how the von-Neumann paradigm of fetching and executing at instruction granularity introduces overheads.

2.3 Obtaining Efficiency and Flexibility

With over 80% of energy devoted to overhead, it appears the von-Neumann approach has far too much inefficiency and is a poor starting point. Due to fundamental energy limits, recent work in general purpose processors has turned to other paradigms for improving efficiency. This shift is also related to our goal and can be leveraged to build efficient and flexible *soft-synthesizable multi-function* lookup engines. We draw inspiration from recent work in hardware specialization and hardware accelerators for general purpose processors that organizes coarse-grained functional units in some kind of interconnect and dynamically synthesizes new functionality at run-time by configuring this interconnect. Some examples of this approach include: DySER [19], FlexCore [35], Qs-Cores [38], and BERET [21]. These approaches by themselves cannot serve as lookup engines because they have far too high latencies, implicitly or explicitly rely on a main-processor for accessing memories, or include relatively heavyweight flow-control mechanisms and predication mechanisms internally. Instead, we leverage their insight of dynamically synthesizing functionality, combine it with the unique properties of lookup engine processing and develop a stand-alone architecture suited for lookup engines.

FPGAs provide an interesting platform and their inherent structure is suited for flexibility with rapid and easy run-time modification. However, since they perform reconfiguration using fine-grained structures like 4- or 8-entry lookup tables, they suffer from energy and area efficiency and low clock frequency problems. Also, a typical FPGA chip has limited amounts of SRAM storage. While they may be well suited in some cases, they are inadequate for large lookups in high-speed network infrastructure.

3. TOWARD DYNAMIC MULTI-FUNCTION LOOKUP ENGINES

We now characterize the computational requirements of lookups. We assume a dataflow approach where processing is broken up in steps and mapped to a tile based pipeline similar to PLUG. We focus on the processing to understand how to improve on existing flexible lookup approaches and incorporate the insights of hardware specialization.

3.1 Description of Lookup Algorithms

We examined in detail seven lookup algorithms. Each algorithm is used within a network protocol or a common network operation. Table 1 summarizes the application where each lookup algorithm originated, and the key data structures being used. Much of our

analysis was in understanding the basic steps of the lookup processing and determining the hardware requirements by developing design sketches as shown in Figure 3. We describe this analysis in detail for two algorithms, Ethernet forwarding and IPv6 and the next section summarizes overall findings. These two serve as running examples through this paper, with their detailed architecture discussed in Section 4.4 and programming implementation discussed in Section 5.3.

Ethernet forwarding

Application overview: Ethernet forwarding is typically performed by layer-II devices, and requires retrieving the correct output ports for each incoming Ethernet frame. Our implementation uses a lookup in a hash table that stores, for every known layer-II address (MAC), the corresponding port.

Ethernet lookup step: This step, depicted in Figure 3a, checks whether the content of a bucket matches a value (MAC address) provided externally. If yes, the value (port) associated with the key is returned. To perform this, a 16-bit *bucket id* from the input message is used as a memory address to retrieve the bucket content. The bucket key is then compared with the key being looked up, also carried by the input message. The bucket value is copied to the output message; the message is sent only if the two keys match. The input is also forwarded to other tiles – this enables multiple tiles to check all the entries in a bucket in parallel.

IPv6 lookup

Application overview: The IPv6 forwarding approach discussed here is derived from PLUG, and is originally based on the "Lulea" algorithm [14]. This algorithm uses compressed multibit tries with fixed stride sizes. The IPv6 lookup algorithms extends it by using two pipelines operating in parallel. Pipeline #1 covers prefixes with lengths up to 48 bits, and is implemented as a conventional multibit trie with five fixed-stride stages. Pipeline #2 covers prefixes of length between 48 and 128 bits, and is implemented as a combination of hash tables and trie. If both pipelines return a result, pipeline #2 overrides #1.

In multibit tries, each node represents a fixed number of prefixes corresponding to each possible combination of a subset of the address bits. For example, a stage that consumes 8 bits covers 256 prefixes. In principle, each prefix is associated with a pointer to a forwarding rule and a pointer to the next trie level. In practice, the algorithm uses various kinds of compressed representations, avoiding repetitions when multiple prefixes are associated with the same pointers. In the following we describe one of the processing steps for computing a rule pointer in IPv6.

IPv6 rule lookup step: This step, represented in Figure 3b, uses a subset of the IPv6 address bits to construct a pointer into a forwarding rule table. Specifically, it deals with the case where the 256

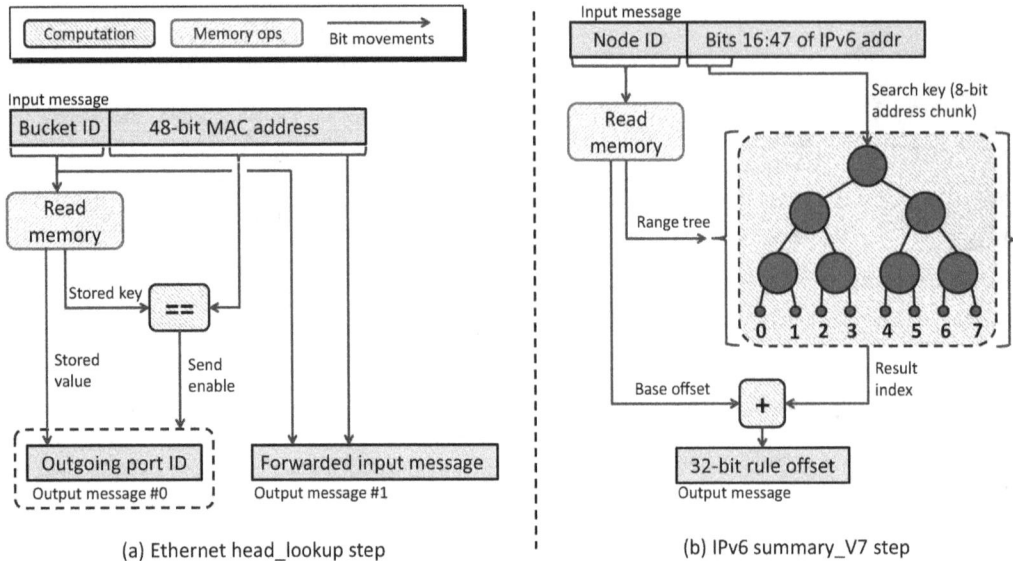

(a) Ethernet head_lookup step   (b) IPv6 summary_V7 step

Figure 3: Rule offset computation in IPv6

prefixes in a trie node are partitioned in up to seven ranges, each associated with a different rule. In the rule table, the seven rules are stored sequentially starting at a known base offset. The goal of this step is to select a range based on the IPv6 address being looked up and construct a pointer to the corresponding rule.

Initially, 16 bits from the input message (the *node id*) are used as an address to retrieve a trie node from memory. The node stores both the base offset for the rules, and the seven ranges in which the node is partitioned. The ranges are represented as a 3-level binary search tree. Conceptually, the lookup works by using 8 bits from the IPv6 address as a key, and searching the range vector for the largest element which does not exceed the key. The index of the element is then added to the base offset to obtain an index in the rule table. Finally, the result is forwarded to the next stage.

3.2 Workload Analysis

Similar to the above two, we analyzed many processing steps of several lookup algorithms. This analysis revealed common properties which present opportunities for dynamic specialization and for eliminating von-Neumann-style processing overheads. We enumerate these below, tying back to our discussion in Section 2.3 and conclude with the elements of an abstract design.

1. Compound specialized operations: The algorithms perform many specific bit-manipulations on data read from the memory-storage. Examples include bit-selection, counting the bits set, and binary-space partitioned search on long bit-vector data. Much of this is efficiently supported with specialized hardware blocks rather than through primitive instructions like add, compare, or, etc. For example, one bstsearch instruction that does a binary search through 16-bit chunks of a 128-bit value, like used in [29], can replace a sequence of cmp, shift instructions. There is great potential for reducing latency and energy with simple specialization.

2. Significant instruction-level parallelism: The lookups show opportunity for instruction-level parallelism (ILP), i.e. several primitive operations could happen in parallel to reduce lookup latency. Architectures like PLUG which use single-issue in-order processors cannot exploit this.

3. Wide datapaths and narrow datapaths: The algorithms perform operations on wide data including 64-bit and 128-bit quanti-

ties, which become inefficient to support with wide register files. They also produce results that are sometimes very narrow: only 1-bit wide (bit-select) or 4-bits wide (bit count on a 16-bit word) for example. A register file or machine-word size with a fixed width is over-designed and inefficient. Instead, a targeted design can provide generality and reduced area compared to using a register file.

4. Single use of compound specialized operations: Each type of compound operation is performed only once (or very few times) per processing step, with the result of one operation being used by a *different compound operation*. A register-file to hold temporary data is not required.

5. Many bit movements and bit extractions: Much of the "processing" is simply extracting bits and moving bits from one location to another. Using a programmable processor and instructions to do such bit movement among register file entries is wasteful in many ways. Instead, bit extraction and movement could be a hardware primitive.

6. Short computations: In general, the number of operations performed in a tile is quite small - one to four. De Carli et al[13] also observe this, and specifically design PLUG to support "code-blocks" no more than 32 instructions long.

These insights led us to a design that eliminates many of the overhead structures and mechanisms like instruction fetch, decode, register-file, etc. Instead, a lookup architecture can be realized by assembling a collection of heterogeneous functional units of variable width. These units communicate in arbitrary dynamically decided ways. Such a design transforms lookup processing in two ways. Compared to the fixed-function approach, it allows dynamic and lookup-based changes. Compared to the programmable processor approach, this design transforms long multi-cycle programs to a single-cycle processing step. In the next section, we describe the LEAP architecture, which is an implementable realization of this abstract design.

4. LEAP ARCHITECTURE

In this section we describe the architecture and implementation of the LEAP lookup engine. First, we present its organization and execution model and discuss its detailed design. We then walk

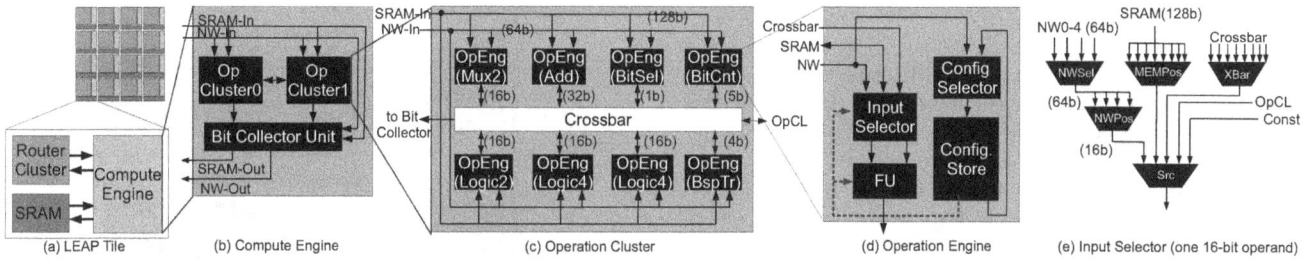

Figure 4: LEAP Organization and Detailed Architecture

through an example of how lookup steps map to LEAP. We conclude with a discussion of physical implementation and design trade-offs. The general LEAP architecture is flexible enough to be used to build substrates interconnected through various topologies like rings, buses, meshes etc. and integrated with various memory technologies. In this paper, we discuss in detail a mesh-based chip organization and integration with SRAM.

4.1 Hardware Organization

For clarity we first present LEAP assuming all computation steps are one cycle. Section 4.5 relaxes this assumption.

Organization: Figure 4 presents the LEAP architecture spanning the coarse-grained chip-level organization showing 16 tiles and the detailed design of each tile. We reuse the same tiled design as PLUG ([25]) in which lookups occur in steps, with each step mapped to a tile.

Each tile consists of a LEAP compute engine, a router-cluster, and an SRAM. We first summarize the chip-level organization before describing the details of LEAP. At the chip-level, the router cluster in each tile is used to a form a mesh network across the tiles. We mirror the design of PLUG in which tiles communicate only to their immediate neighbors and the inter-tile network is scheduled at compile time to be conflict-free. PLUG used six inter-tile networks, but our analysis showed only four were needed. Each network is 64 bits wide. Lookup requests arrive at the top-left tile on the west interface, and the result is delivered on the east output interface of the bottom-right tile. With four networks, we can process lookups that are up to 256 bits wide. Each SRAM is 256KB and up to 128 bits can be read or written per access.

Each LEAP compute engine is connected to the router cluster and the SRAM. It can read and write from any of the four networks, and it can read and write up to 128 bits from and to the SRAM per cycle. Each compute engine can perform computation operations of various types on the data consumed. Specifically, we allow the following seven types of primitive hardware operations decided by workload analysis: *select, add, bitselect, bitcount, 2-input logical-op, 4-input logical-op, bsptree-search* (details in Table 2 and Section 4.2). For physical design reasons, a compute engine is partitioned into two identical operation clusters as shown in Figure 4c. Each of these communicate with a bit-collection unit which combines various bits and sends the final output message. Each operation cluster provides the aforementioned hardware operations, each encapsulated inside an operation engine. The operation engine consists of the functional unit, an input selector, and a configuration-store.

Execution model: The arrival of a *message* from the router-cluster triggers the processing for a lookup request. Based on the *type* of message, different processing must be done. The processing can consist of an arbitrary number of primitive hardware operations performed serially or concurrently such that they finish in a single cycle (checked and enforced by the compiler). The LEAP compu-

Functional Unit	Description
Add	Adds or subtracts two 32-bit values. It can also decrement the final answer by one.
BitCnt	Bitcounts of all or a subset of a 32-bit input
BitSel	Can shift logically or arithmetically and select a subset of bits
BSPTr	Performs a binary space tree search comparing an input to input node values.
Logic2	Logic function "a op b" where op can be AND,OR,XOR,LT,GT,EQ,etc
Logic4	Operates as "(a op b) op (c op d)" or chooses based on an operation: "(a op b) ? c : d"
Mux2	Chooses between 2 inputs based on a input

Table 2: Functional Units Mix in the operation engines

tation engine performs the required computation and produces results. These results can be written to the memory, or they can result in output messages. In our design, a single Compute Engine was sufficient. With this execution model, the architecture sustains a throughput of one lookup every cycle. As described in Section 4.6, our prototype runs at 1 GHz, thus providing a throughput of 1 billion lookups per second.

Pipeline: The high-level pipeline abstraction that the organization, execution-model, and compilation provides is a simple pipeline with three stages (cycles): *memory-read (R)*, *compute (C)*, and *memory-write/network-write (Wr)*.

For almost all of our lookups, a simple 3-stage (3-cycle) pipeline of *R, C, Wr* is sufficient in every tile. This provides massive reductions in latency compared to the PLUG approach. In the case of updates or modifications to the lookup table, the *R* stage does not do anything meaningful. We support coherent modifications of streaming reads and writes without requiring any global locks by inserting "write bubbles" into the lookup requests[7]. The R stage forms SRAM addresses from the network message or through simple computation done in the computation engine (but this computation must not conflict with any configuration of computation in the C stage). Our analysis showed this sufficient.

For packet classification, the SRAM was *logically* enhanced to handle strided access. The config store sequences the SRAM so one 128-bit value is treated as multiple addresses. The only enhancement to the SRAM is an added external buffer to hold the 128 bits and logic to select a subset based on configuration signals.

Compilation: Lookup processing steps can be specified in a high-level language to program the LEAP architecture. Specifically, we have developed a Python API and used it to implement several lookups (details in Section 5). As far as the programmer is concerned, LEAP is abstracted as a sequential machine that executes one hardware operation at a time. This abstraction is easy for programmers. The compiler takes this programmer-friendly abstraction and maps the computation to the hardware to realize the single-cycle compute-steps. The compiler uses its awareness of the hardware's massive concurrency to keep the number of compute steps low. The compiler's role is threefold: i) data-dependence analysis between the hardware operations, ii) hardware mapping to the hardware functional units, and iii) generation of low-level configuration signals to orchestrate the required datapath patterns to accomplish the processing. The end result of compilation is simple: a set of configuration bits for each operation engine in each operation cluster and configuration of the bit-collection unit to determine which bits from which unit are used to form the output message, address, and data for the SRAM. This compilation is a hybrid between programmable processors that work on serial ISAs and hardware synthesis.

4.2 Design

Tile (Figure 4(a)): A single tile consists of one LEAP compute-engine, interfaced to the router-cluster and SRAM.

Compute engine (Figure 4(b)): The compute engine must be able to execute a large number of primitive hardware operations concurrently while allowing the results from any hardware unit to be seen by any other hardware unit. To avoid introducing excessive delay in the forwarding path, it must perform these tasks at low latency. Our workload characterization revealed that different lookups require different types of hardware operations, and they have large amounts of concurrency ranging up to four logical operations in IPv6 for example. Naively placing four copies of each of the eight hardware operations on a 32-wide crossbar would present many physical design problems. To overcome these, we build a clustered design with two identical *operation clusters*, allowing one value to be communicated between the clusters (to limit the wires and delay).

Different lookups combine bits from various operations to create the final output message or the value for the SRAM. To provide this functionality in as general a fashion as possible, the compute engines are interfaced to a bit collector, which receives the operation engine results being fed to it. This unit includes a bit-shifter for the input coming from each operation engine, one level of basic muxing and a 4-level OR-tree that combines all of the bits to produce 64-bit messages, 128-bit value, and 32-bit address for outgoing network messages and SRAM value/address respectively.

Operation cluster(Figure 4(c)): The operation cluster combines eight operation engines communicating with each other through a crossbar. It also receives inputs from and outputs to all four networks and the SRAM. It receives one input from the neighbor operation cluster and produces outputs to the bit collector. Depending on compiler analysis, the crossbar is configured into different datapaths as shown by the two examples in Figure 5. Based on our workload analysis, we found the *4-input logical-op* unit was used the most, hence we provide two of them in each cluster.

Operation engine(Figure 4(d)): The core computation happens in each operation engine, which includes a configuration store, an input selector, and the actual hardware functional unit like an adder or a comparator. We provide seven types of hardware functional units as described in Table 2. The main insight behind the operation engine is a throwback to micro-controlled machines which encode the control signals into a micro-control store and sequence operations. In LEAP, we effectively have loosely distributed concurrent micro-controlled execution across all the operation engines. Each operation engine must first select its inputs from one of the four networks, values from the SRAM, values from any other operation engine (i.e. the crossbar), or values from a neighboring operation cluster. This is shown by the selection tree in Figure 4(e). Fur-

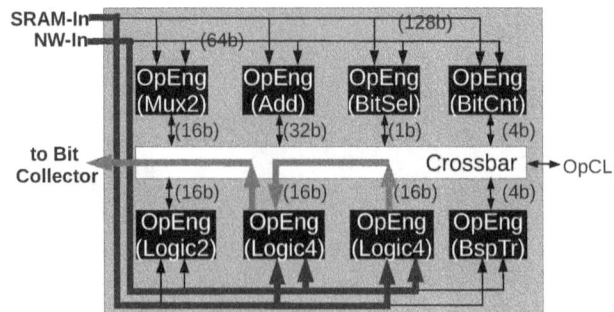

(a) Example Datapath for Ethernet Forwarding mapping a 48 bit equality to 2 Logic4's

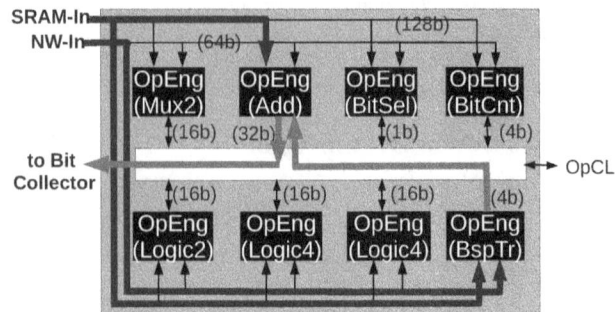

(b) Example Datapath for IPv6
Figure 5: Dynamically created datapaths.

thermore, the actual inputs delivered to the functional unit can be a subset of bits, sign- or zero-extended, or a constant provided by the configuration store. A final selection step decides this and provides the proper input to the functional unit. The result of the functional unit is sent to the crossbar. The configuration store includes the control signals for all elements in the operation engine. Each operation engine has a different sized configuration vector, depending on the number and type of operands, but most 16-bit operands each require: Src (3 bits), NWPos (2 bits), NWSel (2 bits), MEMPos (3 bits), XBar (3 bits), and a Const (16 bit). These configuration bits correspond to the input selection shown in Figure 4(e). Section 5 provides a detailed example showing the Python-API and its compiler-generated configuration information.

Reading the configuration-store to control the operation-engine proceeds as follows. Every cycle, if a message arrives its bits are used to index into the configuration store and decide the configuration to load the controls signals for the operation engine. The compiler is aware of the timing of each operation engine and only chains operation engines together in paths that fit within the single cycle compute-step. An important optimization and insight is the use of such pre-decoded control information as opposed to instruction fetch/decode like in von-Neumann processing. By using configuration information, we eliminate all decoding overhead. More importantly, if successive messages require the same compute step, no reconfiguration is performed and no additional dynamic energy is consumed. Further application analysis is required to quantify these benefits, and our quantitative estimates do not account for this.

4.3 Implementation

Based on workload analysis, we arrived at the mix of functional units and the high-level LEAP design. We have completed a prototype implementation of the compute engine in Verilog. We have also synthesized the design along with the associated SRAM memory to a 55nm technology library using the Synopsys design compiler. Since the compute engine occupies a very small area, we use high-performance transistors which provide high-frequency and low-latency operations - their leakage power is not a large concern. Using these synthesis results we obtain the energy consumed by a lookup access, which we then use to compute power. For SRAM, we consider memories built with low-standby power transistors. The partitioned design restricts a single operation cluster to eight operations and meets timing with a clock period of 1ns. The SRAMs dictate final frequency.

In Section 2.3 we argued that instruction fetch, decode and register files incur area and energy overheads. LEAP eliminates many of these overhead structures. Area and energy costs are now dominated by computation. Detailed quantitative results are in Section 6.

4.4 Mapping lookups to LEAP's architecture

To demonstrate how lookup steps map to LEAP, we revisit the examples introduced in Section 3.1. Figure 5 shows how the steps shown in Figure 3 are configured to run on LEAP.

In our example Ethernet forwarding step, the *R-stage* reads a bucket containing a key (MAC) and a value (port). The *C-stage* determines if the 48-bit key matches the key contained in the input message. If it matches, the bit collector sends the value out on the tile network during the *memory-write/network-write (Wr) stage*. In order to do a 48-bit comparison, two Logic4 blocks are needed. The first Logic4 can take four 16 bit operands and is fed the first 32 bits (2 operands of 16 bits) of the key from SRAM and the first 32 bits of the key from the input message. This Logic4 outputs the logical AND of two 16-bit equality comparisons. The second Logic4 ANDs the output of the first Logic4 with the equality comparison of the remaining pair of 16 bits to check. The result is sent to the bit collector, which uses the result to conditionally send. Since data flows freely between the functional units, computation completes in one cycle (as it also does in the shown IPv6 example). If we assume SRAM latency is 1 cycle and it takes 1 cycle to send the message, LEAP completes both the Ethernet forwarding step and IPv6 step in Figure 5 in 3 cycles. The equivalent computation and message formation on PLUG's von-Neumann architecture would take 10 cycles for the Ethernet forwarding step and 17 cycles for the IPv6 step. With LEAP, computation no longer dominates total lookup delay. These examples are just one step; to complete the lookup the remaining steps are mapped to other tiles in the same manner.

4.5 Multi-cycle compute step

LEAP can easily be extended to handle sophisticated lookups requiring multiple *C* stages. The functional units are augmented with a data-store that allows buffering values between compute steps. In the interest of space and clarity we defer a more detailed description to [23].

4.6 Discussion of Tradeoffs

Functional unit mix: Based on our analysis from Section 3.1, we implemented in Verilog specialized hardware designs to determine an appropriate functional-unit mix(details in Section 6). We found that various lookups use a different mix of a core set of operations, justifying a dynamically synthesized lookup engine. Table 3 presents a sample across different applications showing the use of different operation engines by listing the critical path in terms of functional units serially processed in a single compute step.

Bit selection: From the fixed-function implementation, we observed that a commonly used primitive was to select a subset of bits

Algorithm	Critical Path
Ethernet Forwarding	Logic4→Logic4
	Add
IPv4 forwarding	BitCnt→Add→Mux2
	BRPTr→Mux2→Add
IPv6 forwarding	Logic4→Logic2→Logic2→Mux2
	BitCnt→Add→Mux2
	BSPTr→Add→Mux2
Packet Classification	BitSel→Logic4→Mux2→Mux2→Add
	Mux2→Logic4→Add→BitSel
	Logic4→Logic→Mux2
DFA lookup	BSPTr→BitSel→Add→BitSel
	Logic2
Ethane	Logic4→Logic4→Logic→Mux2
SEATTLE	Logic4→Logic2→Mux2

Table 3: Examples of processing steps' critical paths.

produced by a previous operation. In a programmable processor like PLUG this is accomplished using a sequence of shifts and or's, which uses valuable cycles and energy. To overcome these overheads, every functional unit is preceded by a bit-selector which can select a set of bits from the input, and sign- or zero- extend it. This is similar to the shift mechanisms in the ARM instruction sets [1].

Crossbar design: Instead of designing a "homogenous" cross-bar that forwards 16 bits across all operation engines, we designed one that provides only the required number of bits based on the different functional units. For example *bitcount*, *bitselect*, and *bsptree-search* produce 5 bits, 1 bit, and 4 bits of output respectively. This produces savings in latency and area of the crossbar.

A second piece of the crossbar's unusual design is that its critical path dynamically changes based on the configuration. We have verified through static timing analysis that any four serial operations can be performed in a single cycle. This would change if our mix of functional units changed.

Scalability: A fundamental question for the principles on which LEAP is constructed is what ultimately limits the latency, throughput, area, and power. This is a sophisticated multi-way tradeoff, denoted in a simplified way in Figure 6. With more area, more operation engines can be integrated into a compute engine. However, this will increase the latency of the crossbar, thus reducing frequency and throughput. If the area is reduced, then specialized units like the *bsptree-search* must be eliminated and their work must be accomplished with primitive operations, increasing latency (and reducing throughput). A faster clock speed cannot make up the processing power lost because the cycle time is lower-bounded by the SRAM. Increasing or reducing power will cause a similar effect. For the architecture here, we have proposed an *optimized* and *balanced* design for a target throughput of 1 billion lookups-per-second and overall SRAM size of 256KB (split across four 64KB banks). With a different target, the type and number of elements would be different.

5. PROGRAMMING LEAP

We now describe the programmer's abstract machine model view of LEAP, a specific Python API we have implemented, and outline an example in detail showing final translation to LEAP configuration bits. The API provides a familiar model to programmers despite our unique microarchitecture.

5.1 Abstract machine model

The abstract machine model of LEAP hides the underlying concurrency in the hardware. Specifically, the programmer assumes a

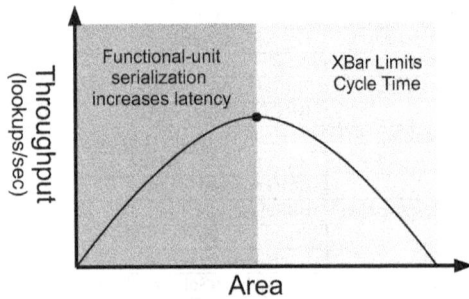

Figure 6: Design scalability tradeoff

serial machine that can perform one operation at a time. The only types of operations allowed are those implemented by the operation engines in the hardware. The source for all operators is either a network message, a value read from memory, or the result of another operator. The native data-type for all operators is bit-vector and bit range selection is a primitive supported in the hardware. For example, `a[13:16]` selects bits 13 through 16 in the variable `a`, and comes at no cost in terms of latency.

This machine model, while simple, has limitations and cannot express some constructs. There is no register file, program counter, control-flow, stack, subroutines or recursion. While this may seem restrictive, in practice we found these features unnecessary for expressing lookups.

Lack of control-flow may appear to be a significant limitation, but this is a common choice in specialized architectures. For example, GPU programming did not allow support for arbitrary control flow until DirectX 8 in 2001. The LEAP machine model does allow two forms of conditional execution. Operations that alter machine state – stores and message sends – can be executed conditionally depending on the value of one of their inputs. Also, a special select primitive can dynamically pick a value from a set of possible ones, and return it as output. Both capabilities are natively supported by LEAP, and we found that they were flexible enough to implement every processing step we considered.

5.2 Python API for LEAP

We have developed a simple Python API to make programming LEAP practical. In our model, programmers express their computational steps as Python sub-routines using this API. We chose Python because it is simple to import our API as a module, but we do not rely on any Python-only functionality. Alternatively we could have chosen to implement our API in a non-interpreted language such as C.

Given the simplicity of LEAP units and the abundance of functionality (e.g. bitvectors) in Python, a software-only implementation of LEAP API calls is trivial. Developers simply run the code on a standard Python interpreter to verify syntactic and semantic correctness. After this debugging phase, our compiler converts this Python code into binary code for the configuration store.

The functionality provided by every functional unit is specified as a Python subroutine. An entire compute step is specified as a sequential set of such calls. Recall the compiler will extract the concurrency and map to hardware. Table 4 describes the most important calls in our API. In most cases a call is mapped directly to a LEAP functional unit; some calls can be mapped to multiple units, for example if the operands are larger than a word. The common case of a comparison between two long bitvectors is optimized through the use of the LOGIC4 unit, which can perform two partial comparisons, and AND them together.

```
1  # In Msg:   0:15: Previous rule
2  #          16:31: Pointer to target node
3  #          32:63: Bytes 16:47 of IPv6 address
4  # Out Msg0: 0:23: Bits 24:47 of IPv6 address
5  #          24:39: Pointer to new rule
6  #          40:55: Previous rule
7  # Out Msg1: 0:31: Pointer to child (unused)
8  def cb0(nw, nw_out):
9    data = read_mem(nw[0].body[16:31])
10   ret = bspsearch3_short(nw[0].body[32:39],
11        data[32:87],data[88:95],0x053977)
12   offset = add(data[0:31], ret)
13   nw_out[0].head[8:17] = offset[2:11]
14   nw_out[0].body[0:23] = nw[0].body[40:63]
15   nw_out[0].body[24:39] = offset[16:31]
16   nw_out[0].body[40:55] = nw[0].body[0:15]
17   send(nw_out[0], 1)
18   nw_out[1].head[8:17] = offset[2:11]
19   nw_out[1].body[0:31] = 0xFFFFFFFF
20   send(nw_out[1], 1)
```

(a) Python code

Unit	ADD (a)	ADD (b)	BSPTr (val)	BSPTr (vec)	BSPTr (cfg)	BSPTr (res)	RAM RD	RAM WR
Src	001 (MEM)	002 (XBar)	000 (NW)	001 (MEM)	001 (MEM)	003 (CFG)	00 (NW)	X
NWPos	X	X	11	X	X	X	1	
NWSel	X	X	00	X	X	X	00	X
Mask								X
MemPos	00	X	X	0x4	0xB	X		
Const	X	X	X	X	X	0x53977	X	
XbarPos			X			X		
Xbar	X	111 (BSPTr)	X	X	X	X	X	X
OpCl	X	X	X	X	X	X	X	X

(b) LEAP configuration bits

Figure 7: IPv6 example compute step

5.3 Implementing Lookups in LEAP

To give a sense practical application development, we briefly discuss the implementation of the IPv6 processing step introduced in Section 3.1.

As previously discussed, this compute step works by searching a binary tree using 8 bits of the IPv6 address as the key. The result represents the relative offset of a forwarding rule corresponding to the address. The absolute offset is then obtained by adding the result of the search to a base offset. Both the binary tree and the base offset are obtained by retrieving a 96-bit word from memory, consisting of:

1) Bits 0-31: Base rule offset
2) Bits 32-87: Binary tree, stored as 7-entry vector
3) Bits 88-95: Vector bitmask (specifies valid entries)

Figure 7(a) depicts the LEAP implementation of this step, together with input and output messages. The node is retrieved from memory in line #9. Lines #10-11 perform the search in the range vector. In line #12 the result of the search (index of the range where the prefix lie) is added to the base rule offset to generate a rule index for the next stage. The rest of the code prepares outgoing messages.

The API calls in Figure 7(a) are translated to the configuration bits shown in Figure 7(b). The rows in Figure 7(b) correspond to the generalized input selection shown in Figure 4(f). X's are "don't cares" in the figure and a grayed box indicates those config bits are not present for the particular operand. Each operand in the table has different available config bits because of the different widths and possible sources for each operand.

6. EVALUATION

We now present our evaluation. After describing the methodology for our evaluations, we discuss characteristics of the lookups

182

API Call	Functional unit	Description
		Access functions
read_mem(addr)	SRAM	Load value from memory
write_mem (addr)	SRAM	Store value to memory
send (value, enable)	Network	Send `value` on a on-chip network if `enable` is not 0.
		Operator functions
select (v0, v1, sel, width=16|32|64)	Mux2	Selects either `v0` or `v1` depending on the value of `sel`
copy_bits (val[a:b])	BitSel	Extracts bits between position a and position b from `val`
bitwise_and(a, b)	Logic2	bitwise AND
bitwise_or(a, b)	Logic2	bitwise OR
bitwise_xor(a, b)	Logic2	bitwise XOR
eq(a, b)	Logic2	Comparison (returns 1 if a == b, 0 otherwise)
cond_select(a, b, c, d, "logic-function")	Logic4	Apply the logic-function to a and b and select c or d based on the result.
add(a, b)	Add	Sum a to b
add_dec(a, b)	Add	Sum a to b and subtracts 1 to the result
sub(a, b)	Add	Subtract b from a
bitcount(value, start_position)	BitCnt	Sum bits in `value` from bit `start_position` to the end
bsptree3_short(value, vector, cfg, res)	BSP-Tree	Perform a binary-space-tree search on vector. `value` is an 8-bit value; `vector` is a 64-bit vector including 7 elements, each 8 bits; `cfg` is an 8-bit configuration word (1 enable bit for each node) `res` is a 24-bit value consisting of 8 result fields, each 3 bits wide.
bsptree3_long(value, vector, cfg, res)	BSP-Tree	Perform a binary-space-tree search on 128-bit vector with 16-bit value.

Table 4: Python API for programming LEAP

Algorithm	Lookup data
Ethernet forwarding	Set of 100K random addresses
IPv4 forwarding	280K-prefix routing table
IPv6 forwarding	Synthetic routing table [40]
Packet classification	Classbench generated classifiers [33]
DFA lookup	Signature set from Cisco [2]
Ethane	Synthetic data based on specs [11]
SEATTLE	Synthetic data based on specs [24]

Table 5: Datasets used

implemented on LEAP and a performance and sensitivity analysis. To quantitatively evaluate LEAP, we compare it to an optimistic model we construct for a fixed-function lookup engine for each lookup. We also compare LEAP to PLUG which is a state-of-art programmable lookup engine.

6.1 Methodology

We have implemented the seven algorithms mentioned in Table 1 using our Python API including the multiple algorithmic stages and the associated processing. For this work, we manually translated the code from the Python API into LEAP configuration bits, since our compiler work is on-going and describing it also is beyond the scope of one paper. For each lookup, we also need additional data like a network traffic trace or lookup trace and other dataset information to populate the lookup data structures. These vary for each lookup and Table 5 describes the data sets we use. In the interest of space, omitted details can be found in [23]. To be consistent with our quantitative comparison to PLUG, we picked similar or equivalent traces and datasets. For performance of the hardware we consider parameters from our RTL prototype implementation: clock frequency is 1 GHz and we used the 55nm Synopsys synthesis results to determine how many compute steps lookup processing took for each processing step at each algorithmic stage.

Modeling of other architectures: To determine how close LEAP comes to a specialized fixed-function lookup engine (referred to as FxFu henceforth), we would like to consider performance of a FxFu hardware RTL implementation. Recall that the FxFu is also combined with an SRAM like in PLUG and LEAP. We implemented them for three lookups to first determine whether such a detailed implementation was necessary. After implementing FxFu's for Ethernet forwarding, IPv4, and Ethane, we found that the they easily operated within 1 ns, consumed *less than 2%* of the tile's area, and the contribution of processing to power consumption was always

less than 30%. Since such a level of RTL implementation is tedious and ultimately the FxFu's contribution compared to the memory is small, we did not pursue detailed fixed-function implementations for other lookups and adopted a simple optimistic model: we assume that processing area is fixed at 3% of SRAM area, power is fixed at 30% of total power, and latency is always 2 cycles per-tile (1 for memory-read, 1 for processing) plus 1 cycle between tiles.

We also compare our results to the PLUG design by considering their reported results in [25] which includes simulation- and RTL-based results for area, power, and latency. For all three designs we consider a tile with four 64KB memory banks. With 16 total tiles, we can get 4MB of storage thus providing sufficient storage for all of the lookups.

Metrics: We evaluate latency per lookup, worst-case total power (dynamic + static), and area of a single tile. Chip area is tile area multiplied by the number of tiles available on the chip plus additional wiring overheads, area of IO pads, etc. The fixed-function engines may be able to exploit another source of specialization in that the SRAM in tiles can be sized to exactly match the application. This requires careful tuning of the physical SRAM sub-banking architecture when algorithmic stage sizes are large along with a design library that supports arbitrary memory sizes. We avoid this issue by assuming FxFu's also have fixed SRAM size of 256 KBs in every tile. Finally, when SRAM sizes are smaller than 64KB, modeling tools like CACTI [34] overestimate. Our estimate of the FxFu area could be conservative since it does not account for this memory specialization.

6.2 Implementing Lookups

First, we demonstrate that LEAP is able to flexibly support various different lookups. Table 6 summarizes code statistics to demonstrate the effectiveness of the Python API and ease of development for the LEAP architecture. As shown in the second and third columns, these applications are relatively sophisticated, require accesses to multiple memories and perform many different types of processing tasks. The fourth column shows that all these algorithmic stages can be succinctly expressed in a few lines of code using our Python API. This shows our API provides a simple and high-level abstraction for high-productive programming. All algorithmic stages in all applications except Packet classification are ultimately transformed into single-cycle compute steps.

Algorithm	Total Algorithmic Stages	Total Compute Steps	Avg. Lines per Compute Step
Ethernet forwarding	2	6	9.5
IPv4	8	42	10.8
IPv6	26	111	12.1
Packet classification	3	3	98
DFA matching	3	7	9.5
Ethane	5	22	11.5
SEATTLE	4	19	9.3

Table 6: Application Code Statistics

Algorithm	FxFu	PLUG	LEAP
Ethernet forwarding	6	18	6
IPv4 forwarding	24	90	24
IPv6 forwarding	42	219	42
Packet classification	23	130	75
DFA matching	6	37	6
Ethane	6	39	6
SEATTLE	9	57	9

Table 7: Latency Estimates (ns)

Result-1: LEAP and its programming API and abstraction are capable of effectively implementing various lookups.

6.3 Performance Analysis

Tables 7-9 compare the fixed-function optimistic engine (FxFu), PLUG and LEAP along the three metrics. All three designs execute at a 1 GHz clock frequency and hence have a throughput of 1 billion lookups per second on all applications except Packet-classification.

Latency: Table 7 shows latency estimates. For FxFu, latency in every tile is the number of SRAM accesses plus one cycle of compute plus one cycle to send. Total latency is always equal to the tile latency multiplied by number of tiles accessed. For LEAP, all lookup steps except Packet classification map to one 1ns compute stage. The latencies for PLUG are from reported results. For FxFu and LEAP the latencies are identical for all cases except Packet-classification since compute-steps are single cycle in both architectures. The large difference in packet classification is because our FxFu estimate is quite optimistic - we assume all sophisticated processing (over 400 lines of C++ code) can be done in one cycle, with little area or energy. For PLUG, the latencies are universally much larger, typically on the order of 5× larger, for two reasons. First, due to its register-file based von-Neumann design, PLUG spends many instructions simply assembling bits read-from/written-to the network. It also uses many instructions to perform operations like bit-selection which are embedded into each operation engine in LEAP. A second and less important factor is that LEAP includes the *bsptree-search* unit that is absent in PLUG.
Result-2: LEAP matches the latency of fixed-function lookups and outperforms PLUG by typically 5×.

Energy/Power: Since all architectures operate at the same throughput, energy and power are linearly related; we present our results in terms of power. For FxFu and LEAP, we estimate power based on the results from RTL synthesis and the power report from Synopsys Power Compiler, assuming its default activity factors. For PLUG, we consider previously reported results also at 55nm technology. Peak power of a single tile and the contribution from memory and processing are shown in Table 8.
Result-3: LEAP is 1.3× better than PLUG in overall energy efficiency. In terms of processing alone, LEAP is 1.6× better.
Result-4: Fixed-function designs are a further 1.3× better than LEAP, suggesting there is still room for improvements.

	FxFu	PLUG	LEAP
Total	37 mWatts	63 mWatts	49 mWatts
Memory %	70	42	54
Compute %	30	58	46

Table 8: Power Estimates

	FxFu	PLUG	LEAP
Total	2.0 mm^2	3.2 mm^2	2.1 mm^2
Memory %	97	64	95
Compute %	3	36	5
Small Memory Tile 64 KB			
Total	0.3 mm^2	0.88 mm^2	0.36 mm^2
Memory %	98	35	83
Compute %	2	65	17

Table 9: Area Estimates

Area: We determined tile area for FxFu and LEAP from our synthesis results and use previously reported results for PLUG. These are shown in Table 9. The network and router area is small and is folded into the memory percentage.
Result-5: LEAP is 1.5× more area efficient than PLUG overall. In terms of processing area alone, it is 9.4× better.
Result-6: LEAP is within 5% of the area-efficiency of fixed-function engines, overall.

7. CONCLUSION

Data plane processing in high-speed routers and switches has come to rely on specialized lookup engines for packet classification and various forwarding lookups. In the future, flexibility and high performance are required from the networking perspective and improvements in architectural energy efficiency are required from the technology perspective.

LEAP presents an architecture for efficient soft-synthesizable multifunction lookup engines that can be deployed as co-processors or lookup modules complementing the packet processing functions of network processors or switches. By using a dynamically configurable data path relying on coarse-grained functional units, LEAP avoids the inherent overheads of von-Neumann-style programmable modules. Through our analysis of several lookup algorithms, we arrived at a design based on 16 instances of seven different functional units together with the required interconnection network and ports connecting to the memory and on-chip network. Comparing to PLUG, a state-of-art flexible lookup engine, the LEAP architecture offers the same throughput, supports all the algorithms implemented on PLUG, and reduces the overall area of the lookup engine by 1.5×, power and energy consumption by 1.3×, and latency typically by 5×. A simple programming API enables the development and deployment of new lookup algorithms. These results are comprehensive, promising, and show the approach has merit.

A complete prototype implementation with an ASIC chip or FPGA that runs protocols on real live-traffic is on-going, and future work will focus on demonstrating LEAP's quantitative impact in product and deployment scenarios. By providing a cost-efficient way of building programmable lookup pipelines, LEAP may speed up scaling and innovation in high-speed wireline networks enabling yet-to-be-invented network features to move faster from the lab to the real network.

Acknowledgments

We thank the anonymous reviewers for their comments. Support for this research was provided by NSF under the following grants: CNS-0917213. Any opinions, findings, and conclusions or recommendations expressed in this material are those of the authors and do not necessarily reflect the views of NSF or other institutions.

8. REFERENCES

[1] Arm instruction set reference https://silver.arm.com/download/download.tm?pv=1199137.

[2] Cisco intrusion prevention system. http://www.cisco.com/en/US/products/ps5729/Products_Sub_Category_Home.html.

[3] Cypress delivers industry's first single-chip algorithmic search engine. http://www.cypress.com/?rID=179, Feb. 2005.

[4] Neuron and neuronmax search processor families. http://www.cavium.com/processor_NEURON_NEURONMAX.html, Aug. 2011.

[5] F. Baboescu, D. Tullsen, G. Rosu, and S. Singh. A tree based router search engine architecture with single port memories. In *ISCA '05*.

[6] F. Baboescu and G. Varghese. Scalable packet classification. In *SIGCOMM '01*.

[7] A. Basu and G. Narlikar. Fast incremental updates for pipelined forwarding engines. In *IEEE INFOCOM '03*.

[8] F. Bonomi, M. Mitzenmacher, R. Panigraphy, S. Singh, and G. Varghese. Beyond Bloom filters: From approximate membership checks to approximate state machines. In *SIGCOMM '06*.

[9] S. Borkar and A. A. Chien. The future of microprocessors. *Commun. ACM*, 54(5):67–77, 2011.

[10] A. Broder and M. Mitzenmacher. Using multiple hash functions to improve IP lookups. In *INFOCOM '01*.

[11] M. Casado, M. J. Freedman, J. Pettit, J. anying Luo, N. McKeown, and S. Shenker. Ethane: taking control of the enterprise. In *SIGCOMM '07*.

[12] Cisco Public Information. The cisco quantumflow processor: Cisco's next generation network processor. http://www.cisco.com/en/US/prod/collateral/routers/ps9343/solution_overview_c22-448936.html, 2008.

[13] L. De Carli, Y. Pan, A. Kumar, C. Estan, and K. Sankaralingam. Plug: Flexible lookup modules for rapid deployment of new protocols in high-speed routers. In *SIGCOMM '09*.

[14] M. Degermark, A. Brodnik, S. Carlsson, and S. Pink. Small forwarding tables for fast routing lookups. In *SIGCOMM '97*.

[15] S. Dharmapurikar, P. Krishnamurthy, T. Sproull, and J. Lockwood. Deep packet inspection using parallel bloom filters. In *IEEE Micro*, pages 44–51, 2003.

[16] S. Dharmapurikar, P. Krishnamurthy, and D. E. Taylor. Longest prefix matching using bloom filters. In *SIGCOMM '03*.

[17] M. Dobrescu, N. Egi, K. Argyraki, B.-G. Chun, K. Fall, G. Iannaccone, A. Knies, M. Manesh, and S. Ratnasamy. Routebricks: exploiting parallelism to scale software routers. In *SOSP '09*.

[18] W. Eatherton. The push of network processing to the top of the pyramid. Keynote, ANCS '05.

[19] V. Govindaraju, C.-H. Ho, and K. Sankaralingam. Dynamically specialized datapaths for energy efficient computing. In *HPCA '11*.

[20] P. Gupta, S. Lin, and N. Mckeown. Routing lookups in hardware at memory access speeds. In *INFOCOM '98*.

[21] S. Gupta, S. Feng, A. Ansari, S. Mahlke, and D. August. Bundled execution of recurring traces for energy-efficient general purpose processing. In *MICRO '1*.

[22] R. Hameed, W. Qadeer, M. Wachs, O. Azizi, A. Solomatnikov, B. C. Lee, S. Richardson, C. Kozyrakis, and M. Horowitz. Understanding sources of inefficiency in general-purpose chips. In *ISCA '10*.

[23] E. Harris. Leap: Latency- energy- and area-optimized lookup pipeline. Master's thesis, The University of Wisconsin-Madison, 2012.

[24] C. Kim, M. Caesar, and J. Rexford. Floodless in SEATTLE: A scalable ethernet architecture for large enterprises. In *SIGCOMM '08*.

[25] A. Kumar, L. De Carli, S. J. Kim, M. de Kruijf, K. Sankaralingam, C. Estan, and S. Jha. Design and implementation of the plug architecture for programmable and efficient network lookups. In *PACT '10*.

[26] S. Kumar, M. Becchi, P. Crowley, and J. Turner. CAMP: fast and efficient IP lookup architecture. In *ANCS '06*.

[27] S. Kumar, S. Dharmapurikar, F. Yu, P. Crowley, and J. Turner. Algorithms to accelerate multiple regular expressions matching for deep packet inspection. In *SIGCOMM '06*.

[28] S. Kumar and B. Lynch. Smart memory for high performance network packet forwarding. In *HotChips*, Aug. 2010.

[29] H. Le and V. Prasanna. Scalable tree-based architectures for ipv4/v6 lookup using prefix partitioning. *IEEE Trans. Comp.*, PP(99):1, '11.

[30] S. Li, J. H. Ahn, R. D. Strong, J. B. Brockman, D. M. Tullsen, and N. P. Jouppi. McPAT: an integrated power, area, and timing modeling framework for multicore and manycore architectures. In *MICRO 42*.

[31] S. Singh, F. Baboescu, G. Varghese, and J. Wang. Packet classification using multidimensional cutting. In *SIGCOMM '03*.

[32] D. Taylor, J. Turner, J. Lockwood, T. Sproull, and D. Parlour. Scalable ip lookup for internet routers. *Selected Areas in Communications, IEEE Journal on*, 21(4):522 – 534, may 2003.

[33] D. E. Taylor and J. S. Turner. Classbench: A packet classification benchmark. In *IEEE INFOCOM '05*.

[34] S. Thoziyoor, N. Muralimanohar, J. H. Ahn, and N. P. Jouppi. Cacti 5.1. Technical Report HPL-2008-20, HP Labs.

[35] M. Thuresson, M. Sjalander, M. Bjork, L. Svensson, P. Larsson-Edefors, and P. Stenstrom. Flexcore: Utilizing exposed datapath control for efficient computing. In *IC-SAMOS '07*.

[36] N. Vaish, T. Kooburat, L. De Carli, K. Sankaralingam, and C. Estan. Experiences in co-designing a packet classification algorithm and a flexible hardware platform. In *ANCS '11*.

[37] B. Vamanan, G. Voskuilen, and T. N. Vijaykumar. Efficuts: optimizing packet classification for memory and throughput. In *SIGCOMM '10*.

[38] G. Venkatesh, J. Sampson, N. Goulding, S. K. V, S. Swanson, and M. Taylor. Qscores: Configurable co-processors to trade dark silicon for energy efficiency in a scalable manner. In *MICRO '11*.

[39] B. Vöcking. How asymmetry helps load balancing. In *IEEE-FOCS '99*.

[40] K. Zheng and B. Liu. V6gene: A scalable IPv6 prefix generator for route lookup algorithm benchmark. In *AINA '06*.

Floating Ground Architecture: Overcoming the One-Hop Boundary of Current Mobile Internet

Hajime Tazaki
NICT, Japan
tazaki@nict.go.jp

Rodney Van Meter
Keio University, Japan
rdv@sfc.wide.ad.jp

Ryuji Wakikawa
TOYOTA InfoTechnology
Center, U.S.A., Inc.
ryuji@us.toyota-itc.com

Noriyuki Shigechika
RCA Co., Ltd., Japan
nazo@rca.co.jp

Keisuke Uehara
Keio University, Japan
kei@wide.ad.jp

Jun Murai
Keio University, Japan
jun@wide.ad.jp

Abstract

We propose the *Floating Ground Architecture (FGA)* for network mobility and ad hoc network convergence. Various factors, including excessive dependence on intelligence in the fixed network, result in the Internet having a de facto logical boundary one hop from the fixed network. To reduce these dependencies, FGA introduces a new logical layer, called *Floating Ground*, between the fixed network infrastructure and the mobile network, aiming to bridge these different types of network systems. Thanks to the effect of this buffer layer, the architecture: 1) optimizes routes in a deeply nested mobile router arrangement, 2) simplifies mobility event handling under frequent movement of the nodes, and 3) transparently introduces additional functionality with no additional intelligence on the infrastructure side. Through evaluation of our proposed architecture using an actual software implementation running via Direct Code Execution simulation, optimized routes are confirmed with three possible mobility scenarios, demonstrating the handoff duration is dramatically reduced in the short-distance movement scenario, which happens in 78.4%, at maximum, of the handoff events under actual taxi cabs movement in real world. Qualitative analysis of FGA shows it minimizes modification of the network components and existing standardized protocols, and is therefore more suitable for self-organized, distributed network extension than competitive approaches.

Categories and Subject Descriptors

C.2.1 [**Network Architecture and Design**]: Distributed networks

Keywords

Floating Ground, MANEMO, ad hoc network, network mobility, architecture design

1. INTRODUCTION

Self-organizing mobile networks supporting universal reachability of nodes are required for emergency use, spanning the communication black hole that arises immediately following a huge disaster, such as Japan's March 2011 earthquake and tsunami, or the earlier Chuetsu earthquake. Normal data communication between those in the disaster area and those outside is cut off [6], although some communication devices may be available. Reinstating communication with the outside world in such events has been considered a primary use case for self-organized systems. However, it is difficult to accomplish this without a lot of support from the network operators.

The current network architecture and proposed solutions depend on the intelligence of routers on both sides of the infrastructure/user border, impeding the ability of networks to self-organize. Meeting the goal of providing universal reachability to all nodes normally requires careful planning, including management of global address blocks and routing to mitigate the impact of complex local topologies on the global routing system. These problems are especially acute in the case of mobile networks with rapidly-changing topologies, and require the intervention of a seasoned human operator. Self-organizing mobile networks therefore are rarely connected to the Internet. These constraints result in the current mobile Internet having a de facto boundary *one hop away* from the fixed network that prevents integration of mobile networks and fixed networks, as shown in Figure 1. This brittle boundary is a problem when we are urgently trying to restore communications in a post-disaster scenario. If users, rather than network administrators, want to extend the network by adding wireless routers to an existing network, they must use Network Address Translation (NAT) [22] or a tunneled approach such as NEMO (NEtwork MObility) [9]. However, using NAT results in broken application transparency and NEMO or related tunnel protocols could provide global reachability for additional routers, but the nested NEMO problem [15] such as ping-pong routing (shown in the right of Figure 1) occurs when the distance from the fixed network is more than one hop. Existing work such as Unified MANEMO Architecture (UMA) [16] also addressed the integration of self-organized network and the Internet by carefully adapting the extension of NEMO, but still requires interaction with nodes in the fixed network.

In this paper, we propose the *Floating Ground Architec-*

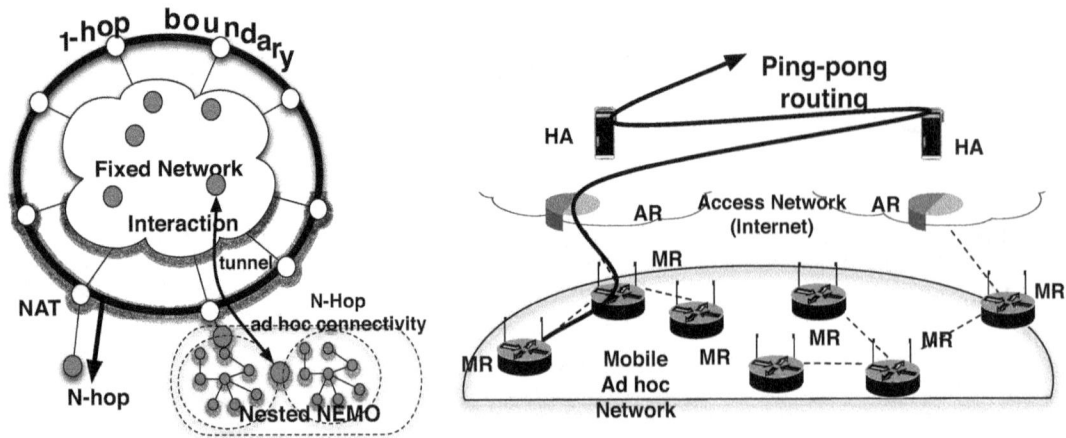

Figure 1: The classical Internet limits the node to logical one-hop only extension at the mobile network (left); NAT prevents application transparency, while NEMO introduces the nested NEMO problem (right) that requires network intelligence to solve.

ture (FGA) to overcome the above one-hop boundary and maximize the possibility of the Internet connectivity. In order to reduce our dependency on the intelligence of the infrastructure, we take as our mantra *Smart Core, Dumb End hosts, and* **Intelligent Floating Routers**, promoting the injection of intelligence on the user side and avoiding replacing the end-node functionality since the large installed base of the Internet is difficult to touch. We condense all of the newly required intelligence into the *Floating Router (FR)* to encourage the deployment into the current Internet, as shown in Figure 2. FGA exploits Mobile Ad Hoc Network (MANET) [8] for NEMO (MANEMO) [26], the combination of local and global mobility management to widen the reachability of IP networks, as a basic design element of our architecture. By carefully considering the issues of MANEMO and providing solutions for its shortcomings, we redesign the last one-hop network for the mobile Internet with Floating Ground that provides a new logical component between fixed network infrastructure and mobile networks.

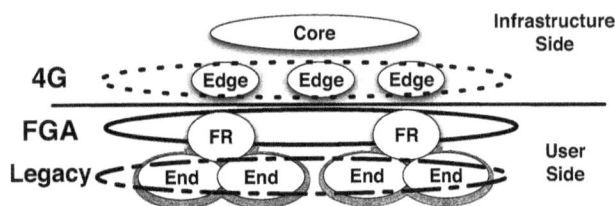

Figure 2: Different architectures require intelligence in different places: the legacy Internet requires intelligence at end nodes, 4G requires intelligent edges, while **FGA** requires Floating Routers (**FR**) on the user side of the boundary.

The contributions of this paper are as follows.

- We propose a network architecture, *Floating Ground Architecture*, to promote the integration of mobile network and fixed network.

- We demonstrate the optimized communication with FGA under nested router arrangement, in the best

case improving the handoff duration to almost zero on short-range movement, which is the most common movement (78.4% of movements at maximum) in urban taxi mobility trace dataset.

- We show that the architecture achieves a simplified topology construction in the mobile ad hoc network to relax the complexity of the protocol deployment.

The rest of this paper is organized as follows. In Section 2, we summarize the current difficulties of the mobile Internet and present the motivation of this paper, then define the problems that we focus on. Then, we present an overview of our proposed architecture, Floating Ground Architecture, in Section 3. In Section 4, we present our design of the FGA, then show the implementation in Section 5. We show the evaluation on the optimized communication and mobility event handling efficiency in Section 6. In Section 7, we summarize the related work and explain why existing work does not solve the problems.

2. PROBLEM OF THE INTEGRATION OF MOBILE NETWORK AND FIXED NETWORK

Extending the network, adding a network segment by adding router nodes, is a common event on the Internet, combining any two networks to scale up the size of the network. This extensibility is well planned at the core of the current Internet, but adding a new network to the user side negatively affects the state of the fixed network. MANET is the extreme case, as its dynamic topology alteration generates a bunch of routing and mobility events by movement that must be propagated throughout the Internet. Thus, NAT is the most common technique applied at the user side network or mobile network because it can be done without any planning and interaction with the original network. However, this address translation introduces the issues of the difficult implementation of application level gateway services, and limits the application transparency at the end node as a result.

188

Although Mobile IP [12] and its extended protocols (e.g., NEMO [9]) were originally introduced to provide addressing at the mobile environment, they relax the concerns with the above NAT issue by hiding the topologically incorrect address (e.g., private address) using tunneling. However, redundant paths (as shown in Figure 1) would occur as a result of a nested router arrangement [15] and require the assistance of the node on the infrastructure side to optimize the path introduced by these protocols. This dependency on the infrastructure side inhibits the flexibility and diversity of the network extension.

Although the nature of the Internet itself is "Inter Networking", combining two systems as different as mobile and fixed networks is not straightforward, as seen in the above examples. Corson *et al.* [8] defined a three-layer model for the integration of the Internet and MANET, as shown in Figure 3. The focal point of this model is the mobile router layer, which can be located multiple hops away from the fixed network and requires additional intelligence for the integration. Unified MANEMO Architecture (UMA) [16] is an example implementation of the mobile router layer, in which NEMO provides global connectivity and the UMA gateway provides addressing with optimized routing.

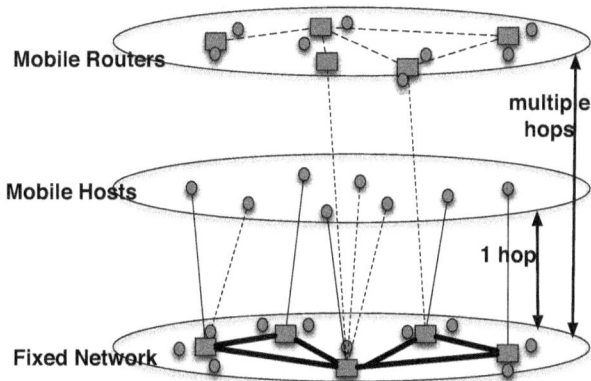

Figure 3: Reference model of the MANET and fixed network integration by Corson *et al.* [8]: the mobile routers layer is distinct from the mobile host layer, which is located a single hop away from the fixed network.

Improving the abilities and functionalities of the fixed network are possible approaches to promoting the integration of mobile and fixed networks. Better performance, availability of effectively unconstrained resources, and better security protection of *wired* routers as compared to *wireless* routers have high potential to provide universal connectivity to everyone. However, the fixed network infrastructure sometimes is not available in the case of the destruction of facilities such as a natural disaster; the network is required to be resilient in any situation even though wired networks are not available.

Considering the above difficulties seen in the integration of the Internet and mobile networks, we defined the following problems that this paper aims to solve.

P1) **The complexity of mobile network dynamics** prevents integration with a managed system (i.e., the Internet).

P2) **Sub-optimal route under nested NEMO:** Using tunneling protocols such as NEMO at the mobile router layer conceals the topology complexity of the MANET, but causes communication delay because of redundant routes when the Mobile Routers form a nested topology.

P3) **Dependencies on the intelligence of fixed network:** The required interactions and dependencies on the intelligence of the node in the fixed network increase deployment costs, and decrease the flexibility and diversity of the network environment. As a result, it reduces the ability of the network to self-organize.

These are the source of the one-hop boundary mentioned in Section 1 and solving these problems simultaneously will give us an implementation of the model of Corson *et al.*

3. THE ARCHITECTURE OVERVIEW

In this section, we present a new mobile network architecture, the Floating Ground Architecture (FGA), to tackle the problems defined in Section 2. FGA's principal philosophy is to give all Mobile Routers the appearance of being one hop away from the fixed Internet.

Figure 4 shows the network after the adaptation of FGA, based on Network Mobility (NEMO) for providing a set of global unique IP address. The key component of the FGA is the *Floating Ground*, the logical component constructed between the network on the infrastructure side (access and core networks) and the mobile networks. This logical network provides 1) a stable, optimized routing path for the node inside the mobile network, concealing the complexity of the status of mobile nodes, and 2) seamless concatenation of the access network and the mobile network without any modifications of the components in the infrastructure side. To construct the Floating Ground seamlessly, our design extends a routing protocol in order to exchange messages among Mobile Routers; we call such a Mobile Router a Floating Router. The nodes located at the border between access router and mobile network announce their roles as *root Floating Routers*. If a node is not attached directly to an access router, it is considered a *normal Floating Router*. Once a Floating Router gets the information announced by a root Floating Router, it uses this information in order to join the Floating Ground. When a Floating Router registers its binding information to its anchor point (a Mobile IP and NEMO operation), the Floating Router uses the information of the Floating Ground in order to eliminate redundant paths between the several Home Agents resulting in multiple encapsulations inside the mobile network. In our design, we have chosen to have the root Floating Router translate the Care-of-Address (CoA) of the Floating Router (using NAT). Thus, all the Floating Routers and root Floating Routers are logically located under the access router directly, as shown in Figure 4 (details will be discussed in Section 4.2). As a result, *all the Floating Routers move up (float) to the direct link of the access router*, and redundant paths caused by a nested router arrangement are optimized since the arrangement of Floating Router is logically flat. Thanks to the transparent existence of the node inside Floating Ground, any components in the infrastructure side are not aware of the transformation of Floating Ground. Thus, route optimization under NEMO or any other new functionality in the mobile network side is done transparently to the fixed network side.

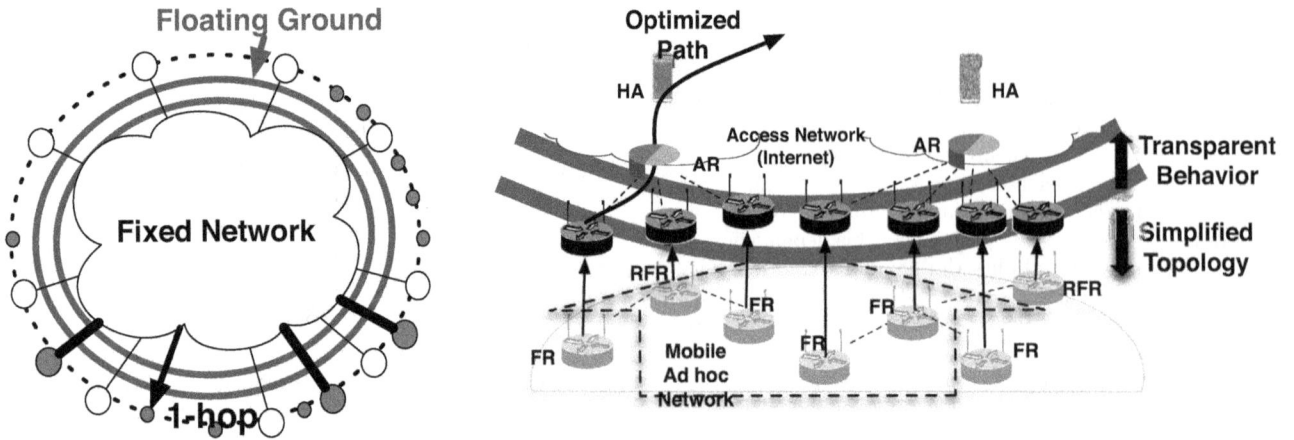

Figure 4: Conceptual model of the Floating Ground Architecture (FGA), where *Floating Routers (FR)* create Floating Ground. The buffer layer between the access network and the mobile network promotes the integration of these two networks. FRs are the only new element necessary to instantiate FGA.

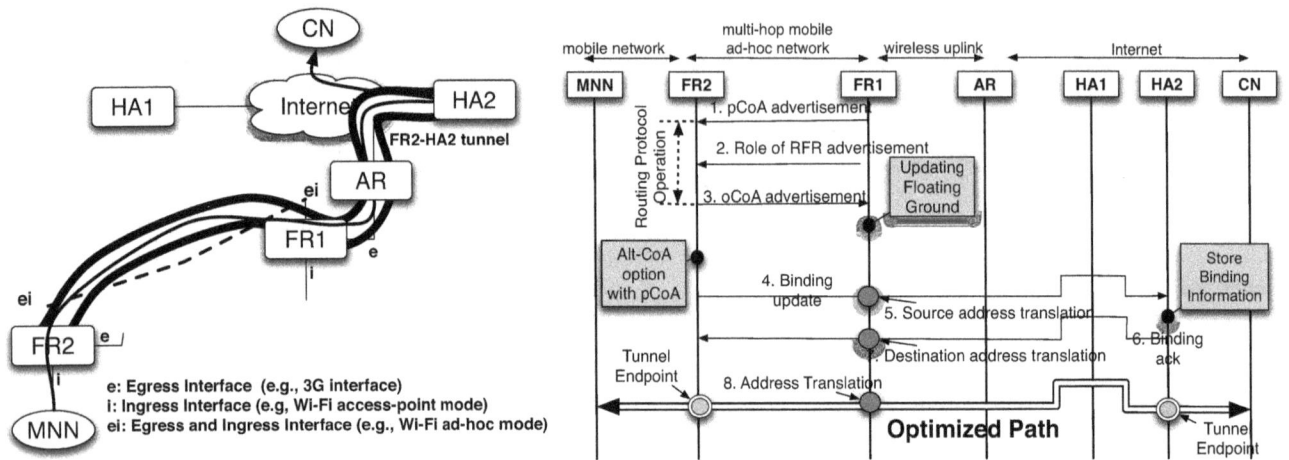

Figure 5: Sequence of the Floating Ground registration (right) using NAT-MANEMO for the network on the left. The dog-leg hop over HA1 indicates that packets bypass HA1 during normal packet forwarding.

4. ARCHITECTURE DESIGN

The main part of FGA's design is the creation of a logical, overlay network (i.e., Floating Ground) and the optimization of the communication at mobile networks. We design a routing protocol, Extended MANEMO Tree Discovery, and a method for mapping overlay, NAT-MANEMO [23], to achieve the goal of FGA. In this section, we briefly review these two protocols.

4.1 Signaling among Mobile Routers

In order to design the signaling protocol for FGA, we choose Tree Discovery protocol (TDP) and Network in Node Advertisement (NINA) [17] as a basic protocol set since scalability in MANEMO scenario shows good performance [24].

As with original TDP/NINA, Extended MANEMO Tree Discovery is a simple distance vector protocol built on the Router Advertisement (RA) message of the Neighbor Discovery protocol. The basic information of this protocol is carried in the Tree Information Option (TIO) extension of the RA message to create the tree topology, and the prefix information is encoded with Network In Node Option

(NINO) to build up the routing information based on the tree topology. In addition, Extended MANEMO Tree Discovery constructs Floating Ground by using an extended TIO and NINO message. This additional information is distributed among the Floating Routers and is used to map physical and logical location to join the Floating Ground.

4.2 Mapping Overlay for Floating Ground Registration

Figure 5 shows an example network configuration and the sequence of Floating Ground registration, conducted as a route optimization method, NAT-MANEMO, and also used as mapping overlay. The communication between Mobile Network Node (MNN) and Correspondent Node (CN) is performed with the interaction of the root Floating Router (FR1), FR2, and Home Agents based on the basic NEMO functionality.

Whenever a Floating Router receives an RA message on its egress interface (the upstream interface, e.g., 3G data interface) and recognizes the advertiser as a normal access router, the router becomes a root Floating Router in a sin-

190

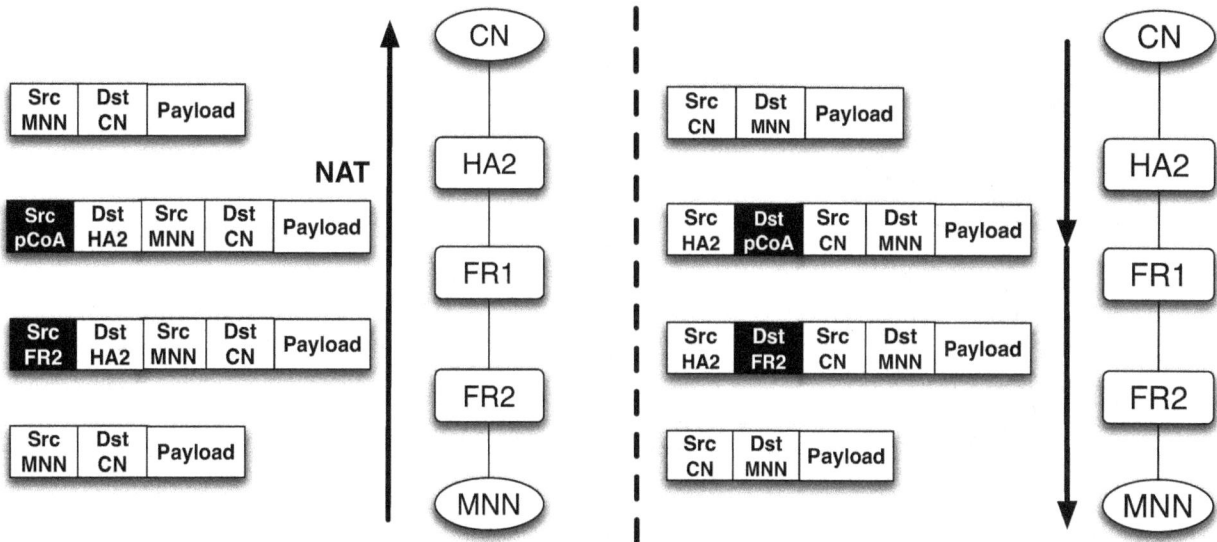

Figure 6: Packet processing at root Floating Router. Left: Outbound processing, Right: Inbound processing.

gle Floating Ground, and starts to advertise its role as root Floating Router with the public Care-of-Address (pCoA) obtained from the access router. The root Floating Router (FR1) also manages an address list of Floating Routers in order to recognize the addresses that should be translated. FR-HA mapping entries with the Floating Router's original CoA (oCoA, "original" meaning before translation) are added when the oCoA is announced by a Floating Router with its home address (HoA). This mapping table is used for translating packets exchanged between Home Agent and Floating Router, and the entry is valid only while Floating Router is in the routing table of FR1, as in normal routing protocol operation.

After FR2 detects the root Floating Router, it transmits a Binding Update (BU) packet to its Home Agent (HA2) using the alternate CoA option encoded with pCoA, which is obtained from FR1. Whenever a BU packet traverses FR1, FR1 rewrites the source address of this packet. Figure 6 shows the transformation of the packet between MNN and CN. The root Floating Router (FR1) performs the packet translation as follows.

- Outbound packet processing (left in Figure 6)
  A root Floating Router performs NAT when it receives a packet sourced from a Floating Router's oCoA. If the source address field of the packet is in the NAT table, the root Floating Router translates it to the pCoA.

- Inbound packet processing (right in Figure 6)
  When a root Floating Router receives a packet from the Home Agent bound to a pCoA, it translates the destination address of this packet to the oCoA. Thus the IP address of MNN is globally reachable even though oCoA is not globally reachable.

Note that this approach does not perform deep packet inspection: since only the tunnel wrapper IP header is affected, the transform engine is carefully constrained, and the impact on packet processing performance is limited. This limited NAT preserves both present and future application transparency and does not require NAT traversal techniques.

4.3 Address Configuration of Mobile Router over Ad Hoc Network

The Mobile Routers in a mobile network are capable of running over an ad hoc interface, but ad hoc address configuration is not compatible with classical IP links [2]. In the MANEMO architecture, the only address that is required to be globally reachable is the CoA of the Mobile Router: any other address is reachable via a home agent. In FGA, this CoA does not need to be globally reachable since Floating Ground provides the reachability.

If mobile network prefixes are preserved and configured statically, the CoAs of Mobile Routers are also configured statically (e.g., using home address): any node in the Internet does not need to be aware of the CoAs since Floating Ground concatenates two distinct topologies for the reachable address.

5. IMPLEMENTATION

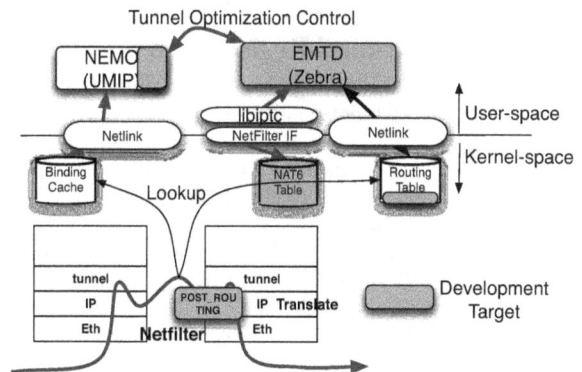

Figure 7: Software module construction for Floating Ground Architecture implementation. The wavy line at the bottom is the packet forwarding path.

In this section, we present our implementation of FGA on Linux. Note that we only modified the Mobile Router

191

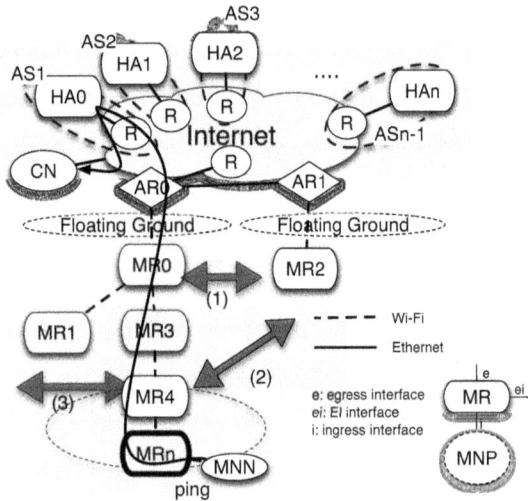

Figure 8: Network configuration and the simulation parameters for the Floating Ground experiment. Mobile Routers are configured with three Wi-Fi interfaces and belong to the different Home Agents located in each AS.

functionality, as shown in Figure 7 of the overview of the software structure.

5.1 Signaling

We implemented the Extended MANEMO Tree Discovery as an extension to the Zebra routing software[1]. All of the messages are generated and received by a user-mode application using a socket, while the routing information is stored in the kernel. In addition, this Zebra extension can interact with USAGI-patched Mobile IPv6 for Linux (UMIP)[2] for optimizing forwarding under nested NEMO environment.

NAT table management by signaling is extended functionality that the original TDP/NINA does not have. To manage the node joining/leaving the Floating Ground, the root Floating Router performs NAT according to the topology in the mobile network. Once the original Care-of-Address (oCoA) information is notified by child routers via Extended MANEMO Tree Discovery, the root Floating Router stores this oCoA with a translated address, defined by the public Care-of-Address (pCoA), into its NAT table.

5.2 Floating Ground Registration

In order to manage Floating Ground membership, we implemented IPv6 NAT functionality in the Linux net-next-2.6 kernel[3] as an extension of the netfilter functionality. This NAT implementation achieves stateless address translation with simplified higher 64-bit prefix translation (i.e., if oCoA is 1:2:3:4:a:b:c:d, it will be translated to 5:6:7:8:a:b:c:d if the prefix of pCoA is 5:6:7:8::) and Neighbor Discovery Proxies [25]. Thus, NAT devices do not have to remember the session state for address translation.

For the NEMO functionality, we used a modified version of UMIP and Linux net-next-2.6 kernel as-is. We implemented the interaction with Zebra and UMIP in order to encode

the alternate CoA option learned by Extended MANEMO Tree Discovery. Our modifications are limited to the Mobile Router; a Home Agent for NEMO can be used unmodified.

6. EVALUATION

Our evaluation demonstrates optimized communication under the nested router arrangement, simplified mobility event handling under various movement scenarios, and the analysis of the mobility pattern in actual vehicular movement to reveal the Floating Ground operation in the real world. We also show how the additional intelligence required remains architecturally transparent, hewing close to the original Internet philosophy.

6.1 Experimental Setup

Using all of the software described in Section 5, we ran the experiments over ns-3 Direct Code Execution (DCE) [14] with our enhancements[4]. This DCE environment allows us to use Zebra, UMIP, and the Linux kernel in our experiments on a network simulator without any modification, provides easy control of experiments, and achieves good agreement between network simulation and actual deployment since unmodified protocol implementation can be used.

Figure 8 shows the network configuration for our experiment and the configured parameters. We conducted 15 simulations with different random seeds. Mobile Routers are configured with three interfaces and form Egress/Ingress and Egress/Ingress (EI-EI) attachments [26] to discover the gateway in the nested NEMO clouds. We use the unnumbered addressing model for address configuration in each Mobile Router for the Floating Ground experiment; the address of the ad hoc interface is borrowed from the address of the ingress interface, which is in the range of the Mobile Network Prefix and is used as the node's oCoA with a 128 bit prefix length. Although this CoA is not reachable from the Internet without conducting a binding registration, Floating

---

[1]Our extension is available at http://web.sfc.wide.ad.jp/~tazaki/hg/zebra-manemo/

[2]USAGI-patched Mobile IPv6 for Linux: http://www.umip.org/, downloaded Jul 7 2010 version.

[3]http://git.kernel.org/?p=linux/kernel/git/davem/net-next.git, downloaded Aug 19 2010 version.

[4]The complete source code of this simulation is available: http://code.nsnam.org/thehajime/ns-3-dce-quagga-umip/

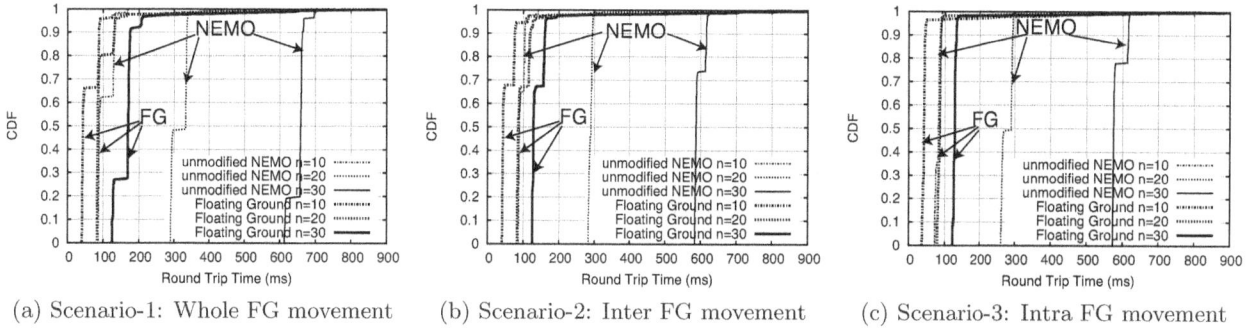

(a) Scenario-1: Whole FG movement (b) Scenario-2: Inter FG movement (c) Scenario-3: Intra FG movement

Figure 9: CDF of round trip time (RTT) between MNN and CN. Floating Ground achieves low latency communication as a result of route optimization.

Ground provides the connectivity since this CoA would be translated at the root Floating Router. For the unmodified NEMO experiment, we used a managed mode interface for the egress interface (marked 'e' in Figure 8) since unmodified NEMO is not able to utilize an ad hoc mode interface for its egress interface because of the Neighbor Discovery protocol limitation [1].

In this experiment, all of the Mobile Routers belong to different Home Agents, which are connected to the node in the topology of Exodus (AS3967) generated from a Rocketfuel [21] dataset. Although the original dataset represents the intra-Autonomous System (AS) topology of the Internet service provider, we use this topology as an inter-AS topology in which each AS operates a single Home Agent service using the Mobile IPv6/NEMO protocol. All the interfaces in this topology set are configured as Ethernet links with IPv6 addresses based on route information exchanged by Zebra `bgpd` routing daemon.

In the Floating Ground experiments, all of the Mobile Routers are configured with Extended MANEMO Tree Discovery to join Floating Ground attaching to access routers (AR0, AR1). By using these Mobile Routers, a group of nodes moves to change the topology as in the following three scenarios, as shown by the corresponding arrow in Figure 8:

Scenario-1 Whole topology movement to reconstruct Floating Ground,

Scenario-2 Partial group nodes movement to another existing Floating Ground, and

Scenario-3 Movement inside the same the Floating Ground.

Scenario-1 represents the most inefficient scenario for the mobile nodes, which involves reconstruction of the whole logical overlay caused by root Floating Router movement to attach to another access router (MR0 moves with MR1, MR3, MR4, MRn, and MNN in the case of Figure 8). Considering the arrangement of two access routers that provides different network prefixes via a wireless system (e.g., mobile cell-phone base station), this happens when the root Floating Router moves a long distance. This operation involves child nodes in the topology and requires signaling to rejoin to the Floating Ground after the movement. Scenario-2 is also a high-cost movement from the viewpoint of Floating Ground operation. The nodes change the Floating Ground to which they belong as a result of their own movement (MR4 moves with MRn and MNN to attach under MR2 in the

case of Figure 8). Compared to other scenarios, Scenario-3 highlights the good effects of Floating Ground since this movement is almost transparent to any node outside the Floating Ground; the topology change is concealed and unseen from the infrastructure side (the group of MR4 moves to attach under MR1 in Figure 8). Further, there is no interaction with Home Agent during this movement. Since this scenario occurs as a result of short-distance movements, it is the most common (details will be discussed in Section 6.4).

During these scenarios, we measure round trip time (RTT) and handoff duration with the different number of nodes. Since the handoff duration in Floating Ground operation dominates the convergence time of the routing protocol, the handoff performance with large numbers of Mobile Routers should be evaluated to show the overhead of Floating Ground.

6.2 Optimized Route under Nested Router Arrangement

Figure 9 represents the CDF of the RTT between MNN and CN, with an interval of 1 second and 64 byte packets as the number of Mobile Routers is changed. In this measurement, MR0 is attached to AR0 as shown in Figure 8 before the movement and moves as each scenario described in Section 6.1. Each result of unmodified NEMO has bigger delay than FGA because the path between MNN and CN is $MNN \rightarrow (MRn) \rightarrow \cdots \rightarrow MR3 \rightarrow MR2 \rightarrow MR1 \rightarrow MR0 \rightarrow AR0 \rightarrow HA0 \rightarrow HA1 \rightarrow HA2 \rightarrow HA3 \rightarrow \cdots \rightarrow (HAn) \rightarrow CN$, where each Ethernet link at access routers and Home Agents adds a 20 millisecond delay. On the other hand, the optimized path created by Floating Ground results in smaller delay because it bypasses $(n-1)$ Home Agents as a result of route optimization.

This shows that Floating Ground improves performance in terms of packet delivery because of the optimized path and the possibility of the implementation through the network experiment.

6.3 Simplified Mobility Event Handling

Figure 10 shows the handoff duration as the number of nodes (Mobile Routers) changes under the three scenarios described in Section 6.1. From all of the results in Figure 10, we can see the handoff duration of unmodified NEMO is almost the same in all of the scenarios. This is because only the binding registration of the moved node happens in these movements. On the other hand, reconstruction of Floating Ground as seen in Scenario-1 requires message exchange

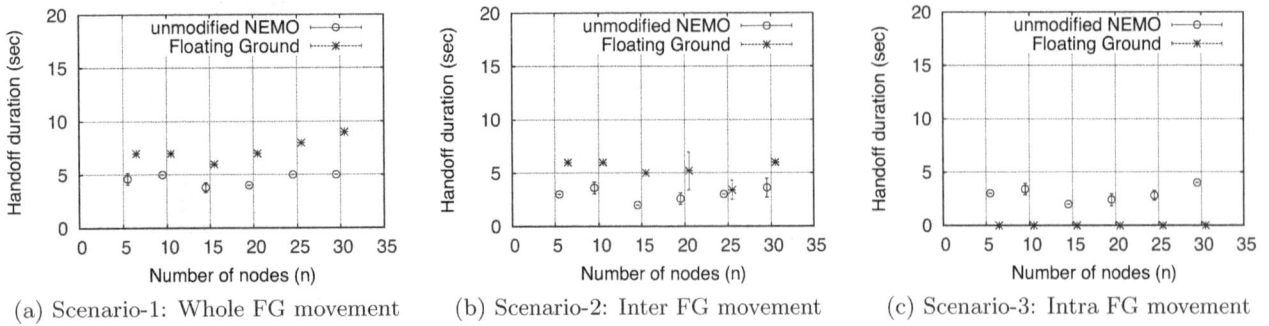

(a) Scenario-1: Whole FG movement (b) Scenario-2: Inter FG movement (c) Scenario-3: Intra FG movement

Figure 10: Mean handoff duration as a function of the number of nodes with standard deviation from 15 replications. Scenario-3 shows effectively zero handoff duration for Floating Ground, while unmodified NEMO stays at $3-5$ seconds duration.

among the nodes that belong to the same Floating Ground as shown in Figure 10(a). This causes longer handoff duration not only due to binding registration, but also increasing the convergence time of Extended MANEMO Tree Discovery caused by increasing numbers of Floating Routers in the mobile network. The result of Scenario-2 also indicates that unmodified NEMO outperforms Floating Ground as shown in Figure 10(b). In the last case, Figure 10(c) shows that handoff duration is negligible in Floating Ground operation while unmodified NEMO still requires the same handoff duration: there are no binding registrations in this movement because home agents are not aware of this movement. Instead, Floating Routers exchange route information and replace the NAT entry of a moved node. This clearly displays the most positive effect of the Floating Ground, concealing the user side movement from the network components.

6.4 Floating Ground Scenario Analysis from Real-world Activity

In this section, we give the analysis of the three possible Floating Ground scenarios discussed in previous sections, from the actual behavior of mobile vehicles. The objective of this analysis is showing the impact of Floating Ground deployment in a real environment and discussing the results obtained in Section 6.3.

Table 1: Simulation parameters for movement analysis.

Parameters	Value
Wi-Fi	802.11b Ad hoc mode, 1Mbps
# of root FRs	10,50,100,250,500
# of total routers	536
simulation time	550000 sec (about 6.3 days)
mobility trace	CRAWDAD epfl/mobility
RA interval	5 sec

We use the CRAWDAD epfl/mobility dataset [19] as a movement scenario of actual mobile network. This dataset contains mobility traces of taxi cabs in the San Francisco Bay Area[5] obtained over 20 days from 536 taxis equipped with GPS devices. The location information (timestamp, coordinates, and identifier) was recorded from May 17th 2008

to Jun 10 2008. The interval between two location records is less than 60 seconds allowing us to reproduce the environment.

In order to observe the frequency of each movement scenario mentioned in Section 6.1 from this mobility trace, we operated Extend MANEMO Tree Discovery under the Wi-Fi equipped cabs movement in network simulation. We also used the same software implementation of Zebra with ns-3 Direct Code Execution (Section 5). 536 nodes in the simulation were equipped with a single Wi-Fi 802.11b ad hoc mode interface running Extended MANEMO Tree Discovery to construct Floating Ground, and moving as described in the CRAWDAD dataset, which was imported using extended mobility functionality of ns-3.

We chose several nodes as manually configured root Floating Routers that have direct connectivity to access routers via always available 3G connectivity, for example. Then we observed the relative frequency of the three Floating Ground movement scenarios from the signaling behavior of Extended MANEMO Tree Discovery, which is available by parsing the log output. The parameters for this simulation are listed in Table 1.

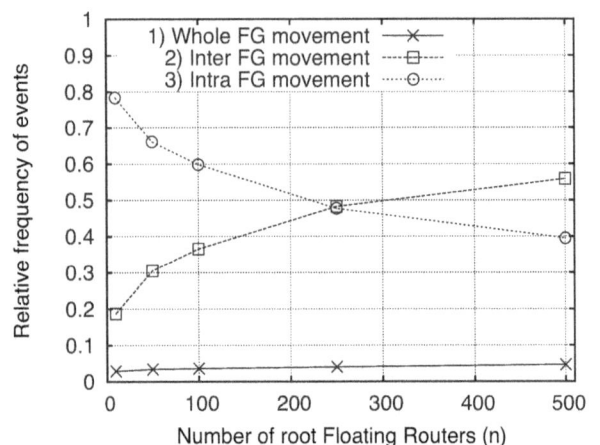

Figure 11: Relative frequency of Floating Ground events in San Fransisco cab mobility movement.

Figure 11 depicts the relative frequencies of each Floating Ground event as a result of varying the number of root

---

[5]http://cabspotting.org/

Table 2: Impact on initial deployment involvement.

	Access Network involvement	Home Agent replacement	Network Component involvement
(unmodified) NEMO			
MIRON	∨		
UMA		∨	
WINMO	∨		∨ (BGP router)
FGA			

Table 3: Impact on mobility signaling caused by node movement.

	Mobile Router	Anchor Point	Access Router	Global Routing Table
(unmodified) NEMO	∨	∨ (HA)		
MIRON	∨	∨ (HA)	∨	
UMA	∨	∨ (HA extension)		
WINMO	∨	∨ (BGP Aggregation Router)		∨
FGA	∨	∨ (HA)		

Floating Routers from 10 nodes to 500 nodes in the whole network. When the number of root Floating Routers is relatively small in the network, less than 250 in this case, Scenario-3 is the most common event among the three scenarios (78.4% of total events at the maximum), since movement between different Floating Ground is not likely to happen frequently. However, if the number of root Floating Routers is large, the event of Scenario-2 is more likely: 55.9% of all events are Scenario-2 when the number of root Floating Routers is 500, causing binding re-registration to their home agents and longer handoff durations.

From the specific mobility trace with the given Floating Ground configuration, we can see the third mobility scenario, Scenario-3 Intra Floating Ground movement, is the common case when the number of root Floating Router is less than 250: at maximum 78.4% of handoff events are concealed because of the effect of Floating Ground. This fact supports the result of Section 6.3 that the common handoff scenario does not affect the fixed network side. On the other hand, if the number of root Floating Routers is large, Scenario-2 is more likely, however, they also may have direct connectivity to access routers since the large number of routers are root Floating Routers, and do not handoff because they already may have the global connectivity directly. The Floating Ground Architecture works efficiently when the number of root Floating Routers is small: that is the exact case of the infrastructure destruction after a huge disaster when the existing network infrastructure is lost and the global connectivity is only partially available.

Although further study is of course required to ensure the mobility analysis, this example dataset shows the feasibility of our proposed architecture, which achieves less impact on the fixed network infrastructure when the nodes move around at the wireless network constructing ad hoc connectivity to the whole network.

6.5 Transparent Behavior

In this section, we present the transparent behavior of FGA with the analysis of existing solutions and FGA. The details of the compared protocols will be discussed in Section 7.

Table 2 summarizes the cost for the initial installation of each functionality. As FGA takes as its design concept *Smart Core, Dumb End hosts, and Intelligent Floating Routers*, the required installation into existing components is limited to the mobile network side. NEMO extensions (UMA [16], MIRON [5]) require additional functional cooperation from existing components, while WINMO [10] introduces different component behavior with their extensions. On the other hand, FGA only replaces Mobile Router functionality of NEMO and does not require the cooperation of the access network. All of the Home Agents used for unmodified NEMO are usable without any modifications. This is a big advantage when considering the initial deployment cost of FGA.

In the summary of the mobility signaling influence in Table 3, we see that WINMO is different. Being based on BGP to handle network mobility, it requires the state of mobile networks to be incorporated into the global routing table, which is a critical problem if the number of mobile networks is large. MIRON requires signals to be exchanged with network side components (i.e., access router). UMA and FGA basically influence mobile nodes and anchor point similar to unmodified NEMO.

In the analytical evaluation we see that FGA is a less disruptive solution to the network infrastructure compared to any existing solutions. Mobility functionality and its optimization are done by the mobile entities themselves, without the support of additional network resources and state consumption on the infrastructure side to achieve transparent integration of MANET and the Internet.

7. RELATED WORK

Floating Ground Architecture is most related to three areas of work: ad hoc network and Internet integration, NEMO route optimization, and mobility event simplification techniques.

Mobile ad hoc network (MANET) and the existing Internet integration has been the challenging topic in this area, as discussed in Section 2. McCarthy *et al.* proposed the Unified MANEMO Architecture (UMA) [16] in order to provide efficient local communication under MANET with global reachability, and showed the possibility of providing structured Authentication, Authorization, and Accounting (AAA), and multi-homing functionality in this architecture. MIPMANET [13] is another solution by Jönsson *et al.* using AODV and Mobile IPv4, trying to achieve a similar goal to UMA, using both dynamic tunneling (i.e., MIP or NEMO) and a MANET routing protocol to fill the gap between the mobile network and the fixed network. Meisel *et al.* [18] recently proposed an insightful architecture by adopting Named Data Networking [11] in an ad hoc network. Their approach is categorized as a clean-slate approach, redesigning the architecture from scratch to solve the current problem. In contrast, FGA tries to achieve the same goal as seen in these three proposals, while focusing on not requiring additional intelligence in the fixed network as shown in the evaluation in Section 6.5. This is a different motivation from existing solutions.

NEMO route optimization is one of the significant pieces of functionality of FGA. There are plenty of solutions to tackle this nested NEMO problem [15]; Light-NEMO [20] uses tunnel concatenation, MIRON [5] configures CoA with address delegation cooperate with access router, HAHA [28] migrates distributed home agents at the infrastructure side, WINMO [10] uses BGP infrastructure to distribute optimized prefix of mobile network, Correspondent Router [27] introduces new router functionality at the correspondent node side to bypass the redundant path. While these existing solutions introduce additional functionality to the non-mobile network side (which is not the main entity involved by the movement), FGA only involves mobile entities to achieve route optimization. This meshes smoothly with the concept of MANET, autonomous network construction performed by the mobile devices themselves. This new capability is one of the key differences and advantages of FGA.

Mobility event simplification is also a target of FGA goal to relax the integration with the fixed network. Several approaches, such as regional anchor point distribution (e.g., Anchor Chain Scheme [4]) and hierarchical mobility management (e.g., Hierarchical Mobile IPv6 [7]), have been tried to improve the binding overhead performance. However, the new functionality introduced by these and existing operational components are not transparently interoperable with current IETF standards, while FGA is carefully designed to slide in quietly among the existing operational components.

8. CONCLUSION

In this paper, we have proposed the *Floating Ground Architecture (FGA)* in order to complement the functionality of the current Internet architecture and fill the gaps in support for ad hoc networks. Various factors prevent the current mobile Internet from providing distributed network extension, limiting the Internet to a *one-hop boundary* between the fixed network and the mobile network. To tackle the problems created by this boundary, we introduce Floating Ground to promote the integration of mobile networks by creating a logical network between mobile network and fixed network infrastructure. This transparent architecture achieves optimized communication with less signaling overhead when the distance of the movement is short and simplifies the address configuration of the MANET nodes.

This paper could be a guideline for the future mobile network, reviving the original potential of the distributed communication network as envisioned by pioneers such as Baran [3]. By redesigning the last one-hop network for the mobile Internet with Floating Ground, this incrementally deployable architecture alleviates the brittleness of the current mobile Internet without requiring the network to be rebuilt from scratch.

9. REFERENCES

[1] BACCELLI, E., CLAUSEN, T. H., AND JACQUET, P. Ad Hoc Networking in the Internet: A Deeper Problem Than It Seems. Research Report RR-6725, INRIA, 2008.

[2] BACCELLI, E., AND TOWNSLEY, M. IP Addressing Model in Ad Hoc Networks. RFC 5889 (Informational), Sept. 2010.

[3] BARAN, P. On distributed communications networks. *IEEE Transactions on Communications Systems 12*, 1 (Mar. 1964), 1–9.

[4] BEJERANO, Y., AND CIDON, I. An anchor chain scheme for IP mobility management. In *Proceedings of the Nineteenth Annual Joint Conference of the IEEE Computer and Communications Societies.* (Mar. 2000), vol. 2 of *INFOCOM 2000*, IEEE, pp. 765–774.

[5] BERNARDOS, C. J., BAGNULO, M., AND CALDERÓN, M. MIRON: MIPv6 route optimization for NEMO. In *Proceedings of 4th Workshop on Applications and Services in Wireless Networks.* (Aug. 2004), ASWN 2004, IEEE, pp. 189–197.

[6] CABINET OFFICE, GOVERMENT OF JAPAN. Report on Niigata Prefecture Chuetsu-oki Earthquake in 2007 (in Japanese, online). `http://www.bousai.go.jp/kinkyu/080107jishin_niigata/jishin_niigata34.pdf` (accessed 2010-12-19), Oct. 2009.

[7] CASTELLUCCIA, C. HMIPv6: A hierarchical mobile IPv6 proposal. *ACM SIGMOBILE Mobile Computing and Communications Review. 4* (Jan. 2000), 48–59.

[8] CORSON, M. S., MACKER, J. P., AND CIRINCIONE, G. H. Internet-Based Mobile Ad Hoc Networking. *IEEE Internet Computing 3*, 4 (July 1999), 63–70.

[9] DEVARAPALLI, V., WAKIKAWA, R., PETRESCU, A., AND THUBERT, P. Network Mobility (NEMO) Basic Support Protocol. RFC 3963 (Proposed Standard), Jan. 2005.

[10] HU, X., LI, L., MAO, Z., AND YANG, Y. Wide-Area IP Network Mobility. In *Proceedings of the 27th Conference on Computer Communications.* (Apr. 2008), INFOCOM 2008, IEEE, pp. 951–959.

[11] JACOBSON, V., SMETTERS, D. K., THORNTON, J. D., PLASS, M. F., BRIGGS, N. H., AND BRAYNARD, R. L. Networking named content. In *Proceedings of the 5th international conference on Emerging networking experiments and technologies* (Dec. 2009), CoNEXT '09, ACM, pp. 1–12.

[12] JOHNSON, D., PERKINS, C., AND ARKKO, J. Mobility Support in IPv6. RFC 3775 (Proposed Standard), June 2004.

[13] JÖNSSON, U., ALRIKSSON, F., LARSSON, T., JOHANSSON, P., AND MAGUIRE, JR., G. Q. MIPMANET: Mobile IP for Mobile Ad Hoc Networks. In *Proceedings of the 1st ACM international*

symposium on Mobile ad hoc networking & computing (Aug. 2000), MobiHoc '00, IEEE Press, pp. 75–85.

[14] LACAGE, M. *Experimentation Tools for Networking Research.* PhD thesis, Universite De Nice-Sophia Antipolis, 2010.

[15] LIM, H.-J., KIM, M., LEE, J.-H., AND CHUNG, T. Route Optimization in Nested NEMO: Classification, Evaluation, and Analysis from NEMO Fringe Stub Perspective. *IEEE Transactions on Mobile Computing 8*, 11 (Nov. 2009), 1554–1572.

[16] MCCARTHY, B., EDWARDS, C., AND DUNMORE, M. Using NEMO to Support the Global Reachability of MANET Nodes. In *Proceedings of the 28th Conference on Computer Communications.* (Apr. 2009), INFOCOM 2009, IEEE, pp. 2097–2105.

[17] MCCARTHY, B., JAKEMAN, M., EDWARDS, C., AND THUBERT, P. Protocols to efficiently support nested NEMO (NEMO+). In *Proceedings of the 3rd international workshop on Mobility in the evolving internet architecture* (New York, NY, USA, Aug. 2008), MobiArch '08, ACM, pp. 43–48.

[18] MEISEL, M., PAPPAS, V., AND ZHANG, L. Ad hoc Networking via Named Data. In *Proceedings of the fifth ACM international workshop on Mobility in the evolving internet architecture* (2010), MobiArch '10, ACM, pp. 3–8.

[19] PIORKOWSKI, M., SARAFIJANOVIC-DJUKIC, N., AND GROSSGLAUSER, M. CRAWDAD data set epfl/mobility (v. 2009-02-24). Downloaded from http://crawdad.cs.dartmouth.edu/epfl/mobility, Feb. 2009.

[20] SABEUR, M., JOUABER, B., AND ZEGHLACHE, D. Light-NEMO+: Route Optimzation for Light-NEMO Solution. In *Proceedings of the 14th IEEE International Conference on Networks* (Sept. 2006), vol. 2 of *ICON '06*, IEEE, pp. 1–6.

[21] SPRING, N., MAHAJAN, R., WETHERALL, D., AND ANDERSON, T. Measuring ISP topologies with rocketfuel. *IEEE/ACM Transactions on Networking (TON) 12*, 1 (Feb. 2004), 2–16.

[22] SRISURESH, P., AND EGEVANG, K. Traditional IP Network Address Translator (Traditional NAT). RFC 3022 (Informational), Jan. 2001.

[23] TAZAKI, H., VAN METER, R., WAKIKAWA, R., UEHARA, K., AND MURAI, J. NAT-MANEMO: Route Optimization for Unlimited Network Extensibility in MANEMO. *Journal of Information Processing 19* (2011), 118–128.

[24] TAZAKI, H., VAN METER, R., WAKIKAWA, R., WONGSAARDSAKUL, T., KANCHANASUT, K., AMORIM, M., AND MURAI, J. MANEMO Routing in Practice: Protocol Selection, Expected Performance, and Experimental Evaluation. *IEICE Transactions on Communications 93*, 8 (Aug. 2010), 2004–2011.

[25] THALER, D., TALWAR, M., AND PATEL, C. Neighbor Discovery Proxies (ND Proxy). RFC 4389 (Experimental), Apr. 2006.

[26] WAKIKAWA, R. *Vehicular Networks: Techniques, Standards, and Applications.* Auerbach Publications, 2009, ch. 11 Mobile Ad Hoc NEMO, pp. 309–329.

[27] WAKIKAWA, R., KOSHIBA, S., UEHARA, K., AND MURAI, J. ORC: optimized route cache management protocol for network mobility. In *Proceedings of the 10th International Conference on Telecommunications* (Feb. 2003), vol. 2 of *ICT 2003*, IEEE, pp. 1194–1200.

[28] WAKIKAWA, R., VALADON, G., AND MURAI, J. Migrating home agents towards internet-scale mobility deployments. In *Proceedings of the 2006 ACM CoNEXT conference* (2006), CoNEXT '06, ACM, pp. 1–10.

ECOS: Leveraging Software-Defined Networks to Support Mobile Application Offloading

Aaron Gember, Christopher Dragga, Aditya Akella
University of Wisconsin, Madison
{agember,dragga,akella}@cs.wisc.edu

ABSTRACT

Offloading has emerged as a promising idea to allow resource-constrained mobile devices to access intensive applications, without performance or energy costs, by leveraging external computing resources. This could be particularly useful in enterprise contexts where running line-of-business applications on mobile devices can enhance enterprise operations. However, we must address three practical roadblocks to make offloading amenable to adoption by enterprises: (i) ensuring privacy and trustworthiness of offload, (ii) decoupling offloading systems from their reliance on the availability of dedicated resources and (iii) accommodating offload at scale. We present the design and implementation of ECOS, an enterprise-centric offloading framework that leverages *Software-Defined Networking* to augment prior offloading proposals and address these limitations. ECOS functions as an application running at an enterprise-wide controller to allocate resources to mobile applications based on privacy and performance requirements, to ensure fairness, and to enforce security constraints. Experiments using a prototype based on Android and OpenFlow establish the effectiveness of our approach.

Categories and Subject Descriptors

C.2.4 [**Computer-Communication Networks**]: Distributed Systems—*Client/Server*

General Terms

Algorithms, Management, Performance, Security

Keywords

Offloading, Mobile devices, Energy savings, Enterprise network

1. INTRODUCTION

Mobile devices such as smartphones and tablets are being increasingly recognized as critical business tools, and enterprises are targeting both specialized and common mobile applications to these platforms [23]. Unfortunately, the complexity and overhead of the applications [15, 17], and the accompanying security issues, are seen as major impediments to full-fledged deployment on mobile devices [4].

Application-independent offloading has long been recognized as an important mechanism for enabling smartphone users to access resource-intensive applications without incurring energy and performance costs [9, 13, 15, 22, 25]. As mobile devices become the primary platforms for some employees, we believe mobile application offloading will be essential for running resource-intensive enterprise applications—e.g., modeling and analysis tools, handwriting and speech recognition, etc.—with suitable performance and energy usage. Moreover, this need will persist for the foreseeable future as device demands continue to outstrip battery capabilities [8]. However, two key roadblocks currently prevent enterprise adoption of mobile application offloading.

1. Privacy and trust: Enterprise applications frequently operate on data with strict privacy requirements, requiring the use of trusted resources (e.g., servers in a local data center) for application execution. The majority of offloading systems ignore such privacy requirements, selecting compute resources solely based on connectivity characteristics and processing capabilities [14]. Even systems which are capable of limiting execution to specific compute resources [9] are insufficient, as they overly restrict offloading opportunities and may unnecessarily impose energy and latency costs.

2. Resource sharing and churn: Enterprises may have thousands of employees using mobile devices, all of which may desire offloading simultaneously. While existing systems address *what* and *how* to offload from a single device [15, 22], no attention has been given to the effects of many devices with different objectives simultaneously offloading to the same compute resources. The energy and latency benefits of offloading assumed by some frameworks to be fixed [14, 15] will, at enterprise scale, be quite dynamic. This dynamism increases even more when considering the range of potential compute resources available—idle desktops, local servers, public clouds—and the changes in capacity and availability that accompany this diversity.

We present an enterprise-centric offloading system (ECOS) that can be coupled with existing offloading frameworks to address the above roadblocks. ECOS is based on two observations: (i) There are plenty of idle resources available in enterprise networks; our unique measurements of resource availability in a campus network confirm this (§4). (ii) Tight administrative control over compute and network resources in enterprises provides the means for mobile applications to access trusted resources, enabling natural mechanisms to ensure privacy and trust. Thus, the central design guideline in ECOS is to allow many mobile application offloads to opportunistically leverage idle compute resources, while tightly controlling the locations where specific applications are offloaded depending on trust, privacy and performance constraints of different users and applications.

ECOS leverages *Software-Defined Networking (SDN)* to meet this guideline. To the best of our knowledge, this is the first attempt at using SDN to better meet the demands of mobile applications. ECOS functions as an application running at an enterprise-wide controller that orchestrates all mobile application offloads. The controller application: (*i*) enforces trust and privacy constraints—specified using a simple, expressive policy language—by tightly controlling the flow of traffic between mobile devices and selected compute resources and by triggering additional higher-layer security mechanisms as necessary; and (*ii*) uses fine-grained resource management algorithms to exercise control over an enterprise's network, desktop, cloud, and mobile device resources and *guarantee* the desired benefit in terms of latency improvement, energy savings, or both. While our centralized framework is an extreme point in the design space, we claim that simultaneously meeting the privacy, trust, and resource constraints identified above, while optimally supporting mobile offloading at scale in enterprises, necessitates this choice.

Key challenges arise in designing ECOS. First, we show that securing offload adds non-trivial energy and latency during both connection setup and state transfer. Hence, careful choices must be made in deciding whether to offload an application that requires privacy and also in designing offload schemes to control the overhead. Second, because a limited number of compute resources are shared by a variety of mobile applications with differing performance, energy and security requirements, we must design clever allocation algorithms that (*i*) adapt quickly to diverse application demands and changing resource availability, (*ii*) ensure applications see equitable and substantial benefits, and (*iii*) control the impact on regular desktop applications. Third, our approach should minimize the amount of work mobile devices undertake and shift a majority of the decision-making from the devices to the controller. Finally, the controller, where the algorithms run, must offer high offload request throughput and low latency. We describe our solutions in §3 and §4.

We have prototyped ECOS using OpenFlow [21] and Android [5]. We evaluate our prototype using two mobile applications that are representative of enterprise workloads. Using 12 phones and up to 6 desktops, we measure the benefits ECOS can provide in a small enterprise setting where phones have varying goals and privacy constraints. In all cases, application latency improves by as much as 94% and energy savings can be up to 47%. In addition, the amount of execution state applications need to send can be reduced by up to 98% for some applications by employing resource affinity and maintaining execution state on compute resources, further improving benefits.

We summarize our contributions as follows:

- We analyze the overhead of transport-layer encryption and categorize the risks associated with enterprise data. Based on these observations, we design (*i*) a simple, expressive policy language that captures privacy constraints of applications/devices and trust levels of resources, (*ii*) a decision process for applying encryption, and (*iii*) network-level policy enforcement mechanisms.

- Using measurements of resource availability from an enterprise-like setting, and our policy language, we design algorithms for allocating resources and managing offloading state at these resources in a way that provides equitable and desirable benefits.

- We prototype our system using Android and OpenFlow/NOX. We conduct several experiments using our prototype, illustrating that application latency improves by up to 94% and energy savings can be up to 47%.

2. BACKGROUND

In this section, we discuss: (*i*) prior proposals for offloading and how ECOS augments them, (*ii*) when ECOS can most help enterprise applications, and (*iii*) the design requirements to ensure ECOS is practical and useful. We conclude the section with an overview of ECOS.

2.1 Prior Offloading Proposals

In offloading, parts of a mobile application are run on a different compute resource to offer improved performance, lower energy usage, and/or higher utility to a mobile device user. Many offloading systems have been developed in the past decade, with somewhat different goals. AIDE dynamically partitions memory-demanding mobile Java applications, minimizing the required communication between the mobile device and the compute resource [22]. Chroma uses developer-specified execution strategies (i.e., tactics) to divide execution of code modules with varying complexity and accuracy between local and remote resources; a tactic is selected at runtime based on currently available mobile device and server resources. [9, 18]. MAUI offloads methods from .NET applications to a remote runtime environment based on a history of energy consumption [15]. CloneCloud uses function inputs and an offline model of runtime costs to dynamically partition Android applications between a weak device and the cloud, with the goal of increasing performance or improving failure resiliency [14, 13].

None of the proposals directly addresses the privacy requirements of offloaded applications. Chroma provides some notion of resource trust [9], but with limited flexibility. Alternative methods of augmenting a mobile device's capabilities require the use of specialized APIs [26, 32] or complex trust establishment schemes [25]. Moreover, existing proposals focus on *what* and *how* to offload from a single mobile device and do not consider the effects of multiple offloads sharing compute resources. ECOS can be coupled with any of the offload mechanisms above to overcome these limitations; our implementation (§5) extends a hybrid of Chroma and CloneCloud. However, the benefits ECOS offers may be different from prior systems, especially when applied at scale and when privacy is considered. We explain this next.

2.2 Application Benefits

Four properties allow both current and future mobile applications to potentially benefit from ECOS.

(P1) Significant computation. Offloading requires processing time on mobile devices to capture execution state and time to transfer it over the network. This overhead must be small enough to not offset the speedup from offloading. Thus, applications with significant compute blocks are the most likely candidates to observe *latency benefits*.

(P2) Small amounts of state exchange. The amount of execution state transferred from the mobile device should be small so the *energy cost* of wireless state transfer, which is known to be significantly more expensive than CPU usage [10], does not exceed the energy savings from offload.

The precise computation size and state size required for offloading to be beneficial depends on the quality of the network link and the processing speed of the compute resource, both of which are considered in existing offloading models [13, 20]. However, in ECOS, the security sensitivity of applications and the level of multiplexing on compute resources also influence whether offloading is beneficial.

(P3) Security sensitivity. Security-insensitive applications can run on any compute resource (idle desktop, local server, or public cloud) and require no higher-layer security; hence their performance is driven by **P1** and **P2**. In contrast, some applications are limited in the resources they can leverage—e.g., applications where data should not leave the enterprise premises due to legal issues can never run on public clouds—or applications may require additional security mechanisms—e.g., applications computing on private user data should always use encrypted communications. If no available enterprise resources provide high enough trust for such applications, then offloading is not possible. Similarly, if the latency and energy overhead of encryption—a result of additional CPU cycles and increases in state size—is too high, then the applications cannot benefit from ECOS.

(P4) Resourcing multiplexing. Because ECOS opportunistically leverages compute resources, unlike prior systems, it is possible that offloading saves energy but does not improve latency. This can happen, e.g., when *multiple* apps each satisfying **P1** and **P2** are offloaded to the same desktop. Of course, like prior systems, ECOS can also result in both latency and energy benefits, or latency benefits alone (for apps satisfying **P1** but not **P2**, e.g., face recognition).

Speech-to-text is a compelling example enterprise application that satisfies these properties. (P1) Analyzing the audio stream requires significant computation. (P2) The audio data is limited in size. (P3) Dictations may be confidential, requiring the data and its processing to remain within the enterprise. (P4) Reasonable amounts of delay can be tolerated.

2.3 ECOS Design Requirements

In considering the latency, energy, and security constraints of mobile applications in a large-scale enterprise setting, the ECOS framework must, ideally: (*i*) Know for a given application if the user is expecting energy or latency savings from offload; (*ii*) Identify if applications are security sensitive, pick candidate compute resources accordingly, and provide encrypted channels for such applications; (*iii*) Assign resources so the overall energy and latency benefits are significant, and benefits are equitably distributed across mobile devices and usage scenarios (e.g., privacy-sensitive vs not); (*iv*) Adapt dynamically to changing compute resources without impacting offloaded applications; (*v*) Require minimal decision-making involvement from mobile devices.

2.4 ECOS Overview

ECOS orchestrates all offloads using an SDN application running atop an enterprise-wide controller.

Mobile applications desiring offload contact the controller to request resources. ECOS determines what privacy level the mobile application requires and subsequently decides if the choice of compute resources needs to be limited and if data needs to be secured in transit between the mobile device and a compute resource (§3). The controller considers the costs of security relative to expected user benefits to ensure ECOS provides latency and/or energy savings despite the overhead of ensuring privacy. The challenge lies in creating an expressive policy language that allows administrators to exercise tight control over privacy and trust levels for applications, devices and resources.

The SDN application carefully selects a compute resource—among desktops, local servers and public clouds—based on gathered utilization information (§4). If a mobile user wants performance improvements, the controller assigns a resource with plenty of idle CPU time. Knowing exactly which resources to select is complicated by the fact that resource availability changes over time. ECOS prefers to use the same resource for subsequent offloads but can flexibly switch resources as necessary. Another challenge is ensuring fairness in resource allocation and overall efficiency.

While the actual offload takes places, ECOS enforces its security and resource decisions by installing and manipulating network forwarding state (§3.3). Programmable switches give ECOS tight control over communications and allow the mobile device to move within the network during offload. Dealing with rogue offloads is an additional challenge, which ECOS addresses using a network with default-off behavior.

3. SECURITY

Normally, all execution and associated data stays within a mobile device unless an application explicitly communicates with a third party. Offloading introduces the possibility for data to leave the confines of the mobile device without an application's explicit actions. Thus, privacy and trust (we use the term security to refer to both at once), which are paramount in enterprises, become important concerns.

Two issues arise when accommodating security: (*i*) when to invoke security mechanisms, and (*ii*) if an application needs security, how to determine if it should be offloaded and then, how to secure it. In addressing these issues, ECOS uses two insights. First, not all enterprise applications desire strict privacy; in some cases, it is sufficient if application data is kept within an enterprise boundary. Second, tight administrative control over compute and network resources in enterprises, through SDN, provides a direct way for mobile device users to access trusted resources for offloading, alleviating the need for complex trust establishment schemes.

3.1 Security Risks and Overheads

Mobile applications utilize many different types of data. For example, an image recognition application may work with photos taken at an office party, while an optical character recognition application may operate on patient medical records. Some of this data needs to remain confidential, while no risk is posed if other data is viewed by a third party.

Most enterprise data falls into one of three basic categories. *User-private* data should only be accessed by specific users, e.g., a person's medical information can only

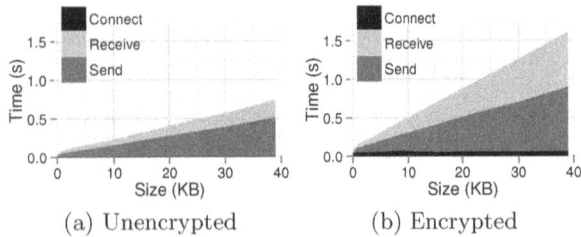

(a) Unencrypted   (b) Encrypted

Figure 1: Execution state transfer latency

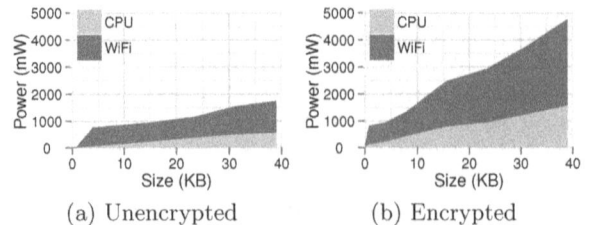

(a) Unencrypted   (b) Encrypted

Figure 2: Average power consumed for state transfer

be accessed by individuals working with the patient [6]. *Enterprise-private* data should not be leaked outside the enterprise, e.g., intellectual property such as code and internal memos. *No-private* data can be viewed by anyone, e.g. news releases.

The same privacy restrictions that apply to these categories of data must also be applied to offloads, since an application's execution state likely includes data the application has obtained (or generated). This means security mechanisms must be applied to offloading communications and executions to avoid compromising privacy. Unfortunately, these mechanisms, e.g., transport-layer encryption, can have significant costs.

3.1.1 Latency and Energy Overhead of Encryption

To understand the extent to which basic security mechanisms can influence the benefits of offloading, we measure the time and energy overhead of applying TLS encryption to offloading communications. We compare the overhead both with and without encryption for varying amounts of execution state. Our measurements use an Android emulator [5] and a 2GHz dual-core desktop running an x86 version of Android in a virtual machine [7]. The emulator and VM both run our modified Dalvik runtime environment capable of capturing and loading execution state. (We show in §6.2 that an emulator is reasonable approximation of an actual Android phone.) Each test consists of a single method call which includes several arguments and objects in its execution state but performs no computation.

Latency. The latency overhead of offload with and without encryption is shown in Figure 1. We observe, in both cases, that connection overhead occupies a noticeable portion of the time spent offloading, though it becomes less significant as the size of the offloaded state increases. This can be alleviated by reusing connections. More importantly, the costs of sending and receiving execution state are approximately 40% higher with encryption. Encryption introduces significantly more latency because of the additional control mechanisms, the need to spend processing time encrypting the data, and the higher net volume of encrypted data. For the chess application used in our evaluation (§6), the send (receive) time per-offload increases from ~0.2s (~0.1s) without encryption to ~0.35s (~0.3s) with encryption.

Energy. Physical energy measurements are difficult due to the limited availability of device schematics and the frequent interactions between individual components. Instead, we take advantage of an existing power model [31] that takes as input fine-grained measurements of CPU utilization, CPU frequency, and the number of packets and bytes sent and received wirelessly. We only include the CPU and Wi-Fi energy components of the model because the other components on a device are typically not impacted by offloading.

Figure 2 shows the results of our power estimates. Energy consumption grows steadily with the size of the offload state for both encrypted and unencrypted connections. CPU energy usage grows roughly linearly, due to increasing costs for deconstructing and reassembling the increasing amounts of state. The overhead from encryption nearly doubles these costs. The increasing Wi-Fi energy usage results from the increasing amount of time required to offload larger state sizes, which grew from a median of 0.1s or less when no state is sent to a median of 1.5s and 0.7s for encrypted and unencrypted connections, respectively, when sending 39 KB of state. During this time, the Wi-Fi interface remains in its high power state, causing it to consume more energy. Conversely, with small amounts of state (< 1KB) transfered over unencrypted connections, the power consumption is very low because the device is able to stay in low power mode throughout the operation.[1] For the chess application used in our evaluation (§6), the energy per-offload used for state transfer increases from ~950mW without encryption to ~2470mW with encryption.

Summary. These observations have key implications for the design of ECOS. First, security mechanisms that use mobile device resources, e.g., TLS encryption, should be used only when necessary. Second, the connection setups and data transfers required for offload should be minimized.

3.2 Security Policy

ECOS ensures private data remains secure, while maximizing offloading opportunities and benefits, through the use of a simple, yet expressive, security policy. This policy allows ECOS to minimize the use of resource-intensive TLS encryption and maximize the use of SDN—to exercise tight network control and carefully select compute resources.

The security policy (e.g., Listing 1) conveys (i) the privacy level of devices and applications and (ii) the trust level of compute resources. Mobile devices and compute resources are identified based on their MAC address, IP address, or primary user. Applications are identified based on a cryptographic hash derived from the executable. ECOS could be adapted to use pre-distributed certificates or user credentials for identifying and authenticating devices and applications.

Mobile devices and applications are labeled as utilizing *enterprise-private*, *user-private*, or *no-private* data. By default, mobile devices are no-private. Devices can be further classified based on who they belong to: e.g. the CEO's device is user-private or all company-owned devices are enterprise-private. Users may also request their devices to be registered as user-private; it is up to administrators to honor such re-

---

[1]While the additional power consumption in high power mode is significant, this will not be visible if the interface is being actively used by other tasks, causing it to already be in high power mode.

Listing 1 Sample enterprise security policy

```
### Assign privacy levels to mobile devices ###
mobile alice = 00:0F:89:B1:C3:D5 enterprise;
mobile bob = 00:0F:71:6A:17:DF user;
### Assign privacy levels to applications ###
app chess = <8232afd556a9fc56c68cc13113c3f2f5> none;
app speech = <beef35481503415c65555ea068c07ac5> user;
### Assign trust levels to resources ###
resource carol = 192.168.1.10 enterprise;
resource dave = 192.168.1.20 enterprise;
resource cloud = 10.0.0.50 none;
```

quests. Applications are assigned privacy levels based on the most private data they are expected to access. Only if an application has *zero* likelihood of accessing private data is it classified as no-private, e.g., a map application. Applications that are likely to access personal information are labeled as user-private, and all other applications are labeled as enterprise-private (the default label for applications).

We acknowledge that the granularity of these labels is quite coarse. Ideally, an application should be labeled based on the specific data objects it is accessing at a given point in time. To do this, it may be necessary to track the flow of private information on the mobile device [16, 24]. Data with known privacy requirements—e.g., data coming from specific servers, email senders, or file locations—can be tracked, enabling dynamic knowledge of the data contained in an offload. However, approaches to track information flow impose performance penalties (TaintDroid imposes 14% CPU overhead [16]) and are still evolving. Our coarser-grained approach, in contrast, takes a conservative view of privacy and trust. Privacy levels are assigned based on the strictest level of privacy an application may ever require, irrespective of its current data usage. The advantage is that it is far simpler and efficient to implement, and it subsumes privacy constraints based on individual data items and bytes. As will become clear after the description of the security mechanisms (§3.3), the downside of using coarse labels is that it may unnecessarily prevent offloading of some applications.

In ECOS, compute resources are assigned trust levels that mirror the privacy levels: *enterprise-trust, user-trust*, and *no-trust*. Resources internal to the enterprise—personal desktops and laptops and local data centers—are labeled *enterprise-trust*. External resources—public clouds—are labeled *no-trust*. *User-trust* resources are special because they are trusted by a specific user or subset of users. For example, if Alice and Bob are both friends, they may consider each others' desktops to provide *user-trust*. Our paper does not address how such trust relationships among users are derived and used to specify user-trust resources for a specific user, but there are several ways for an admin to do so: e.g., configuration by hand, based on observed communication patterns among users [27, 29], or based on existing user groups [28].

3.3 Security Mechanisms

One way to ensure data remains secure is to always encrypt offloading communications and always limit the compute resources used for offload, but this adds unnecessary overhead and artificially restricts offloading opportunities. In contrast, ECOS utilizes the nature of the mobile device and application in determining whether encryption and/or limiting the choice of compute resources are really necessary. ECOS bases its decision on the *strictest level* of privacy required by a device and application pair. Moreover, ECOS

examines the costs to determine if offloading should even be performed. In all cases, ECOS enforces its decisions using tight network control. We describe each mechanism below.

Require Encryption. All data and state belonging to the offload must be encrypted between the mobile device and the compute resource when the device or application is *user-private*. ECOS achieves this using TLS encryption at the transport layer. Since the added time and energy overhead (§3.1.1) can significantly decrease the number of situations where offloading is beneficial, we seek to reduce transfer sizes and reuse connections through resource affinity (§4.2).

Limit Choice of Compute Resources. When the application or mobile device is *enterprise-private*, offload can only happen to *enterprise-* or *user-trust* compute resources, and in the *user-private* case only to *user-trust* resources. This severely limits the number of resources available for these offloads, placing them at a potential disadvantage. Our resource management algorithm (§4.2) must therefore ensure that these offloads receive a fair amount of resources.

Control Network Flows. ECOS enforces its encryption and resource selection decisions using SDN. Enforcement is crucial because we must ensure that: (*i*) the offload system cannot be abused to inflict attacks on network links and compute resources or to attack other offloads, and (*ii*) the system cannot be used to compromise privacy and trust constraints. We realize these guarantees by relying on the ECOS SDN application to implement a default-off network, tightly control the path taken by individual offload flows from mobile devices to compute resources[2], and constantly monitor network/compute resources and traffic and filter abusive flows.

In a default-off network, network elements (e.g., switches, wireless APs, hosts, etc.) have no forwarding state by default. Forwarding state is installed by an SDN application on the basis of traffic engineering directives [19, 30] and/or reachability policies [11, 21]. Moreover, forwarding is controlled at fine-granularity, in terms of protocols, ports, VLANs, and source and destination IP and MAC addresses.

ECOS's use of SDN to implement a default-off network and control forwarding of traffic at fine-granularity offers key advantages. First, we can ensure that data is encrypted when privacy is desired and compromises of a device cannot lead to leakage of private information. To do this, the SDN application establishes forwarding state only for the appropriate flows: e.g., when a device or application is user-private, only packets destined for port 8443, corresponding to offloading with TLS in our implementation, are allowed to enter the network from the mobile device; all other traffic from the device is dropped at the first hop router. For enterprise-private devices or applications, the SDN application ensures that traffic does not leave the confines of the enterprise. The SDN application can also perform other checks, e.g., periodically routing offload traffic through a middlebox to check for exfiltration of confidential data.

Second, applications can be prevented from making unauthorized access to compute resources during offload. When unauthorized access is detected, e.g., based on unexpected resource consumption activity on a desktop, the controller can simply delete the corresponding forwarding entries in the network, terminating the offload. Rogue applications that aim to steal resources from legitimate offloaded applications

---

[2]The rest of the network could use other forms of routing.

203

can also be easily thwarted: ECOS limits the potential impact of rogue applications on performance seeking offloads by providing the latter a minimum share of CPU cycles. For offloads seeking energy savings, care must be taken to ensure rogue applications do not over consume network resources and inflate the communication overheads incurred by legitimate offloads. The controller constantly monitors for usage spikes on compute resources and network links to detect rogue applications; the controller installs drop rules in network elements to "terminate" these applications.

Finally, SDN provides mobility advantages [11]: As mobile devices move through the enterprise wireless network, paths are updated to keep devices and resources connected.

4. SELECTING RESOURCES

Existing offloading frameworks address *what* and *how* to offload from a single device [13, 15, 22], but they do not consider how multiplexing offloads from several devices on a single compute resource will impact performance and energy benefits, nor do they consider the effects of changes in resource capacity and availability over time. In contrast, ECOS is designed to *opportunistically leverage* multiple available enterprise resources to provide benefits to a network of mobile devices while honoring privacy constraints. In this section, we show that plenty of idle desktops exist in enterprise settings to serve application offloads, and we examine the volatility of these resources to understand the implications for resource scheduling in ECOS. We then show how to leverage these resources. The challenge lies in *optimally and fairly multiplexing* offloaded applications amongst these resources to meet the energy, latency and privacy needs of each mobile device. A further complication arises from the fact that we must assign resources without knowing a priori when other offloads will need specific types of resources.

4.1 Resource Availability

Enterprise networks contain a large supply of compute resources that can be leveraged for offloading. These resources tend to be under tight administrative control, making them relatively secure. Some of these resources can be dedicated specifically for offloading (e.g. servers), while others should only be utilized for offloading when they are not serving their primary purpose (e.g. personal desktops and laptops).

To understand idle desktop availability in enterprise settings, we tracked the resource usage of 325 machines (used by ∼600 users) in the Computer Science department network at a large university; this roughly corresponds to a deployment in a modest-sized enterprise. We queried the CPU load of both user desktops and dedicated servers every 5 minutes over a period of 10 days in April 2010.

At any given point in time, over half of the machines have the majority of their CPU idle. Figure 3 shows the median, average, and 90th percentile user load (normalized to one CPU core) per machine throughout the measurement period. The average user load (i.e. fraction of CPU time used by user processes) usually ranges from 15% to 30%, with the median usually ranging from 0% to 5%. Thus, there are suitable volumes of resources already existing in enterprises to serve the offloading needs of mobile devices.

We also measure how user load changes over time. For each 5 minute interval, we calculate the absolute change in user load (normalized to one CPU core) per machine. Figure 4 shows a cumulative distribution of the absolute

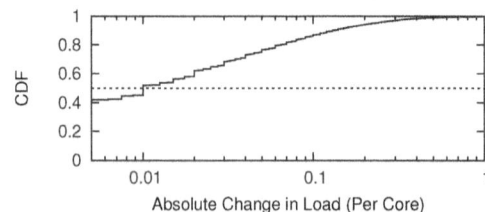

Figure 4: Change in used fraction of processing capacity (per core) over all 5 minute intervals and all machines

changes for all five minute intervals and all machines. We observe that about 42% of the time there is no change in user load between 5 minute intervals, and, over half of the time, the change is less than 0.01. Therefore, we conclude that the change in resource availability is usually small.

Assuming desktop CPUs are 4-5X faster than mobile device CPUs [2], the above results show that a medium-sized enterprise may have enough idle resources to support offload requests for a few hundred mobile applications. Moreover, resource capacity does not need to be tracked at small timescales to make optimal scheduling decisions as the resources are fairly stable; the mechanism for scheduling offloads must mainly worry about churn in the mobile device population.

4.2 Allocating Resources

We now describe how ECOS schedules compute resources to meet application and mobile device security, performance and energy needs in a fair and efficient manner.

In an environment where there are always more suitable compute resources (trusted desktops, servers, etc.) than mobile applications desiring offloading, we can use a simple approach: When a mobile application wants to offload it is assigned a dedicated resource which is not used by other offloaded applications; after an application's offloaded execution completes, a different application can be assigned to the resource. We call this approach *one-to-one scheduling*. Unfortunately, this simple approach is likely to be of limited use. First, we expect that in the future there will be orders of magnitude more mobile applications desiring offloading. The above approach causes some applications to be denied resources and forces them to run on the mobile device. Second, the approach cannot accommodate more complex offloading policies, e.g., parallel offload of application threads [12].

4.2.1 Multiplexing Applications in ECOS

A better alternative when there are limited compute resources is to assign multiple mobile applications to the same resource, an approach we call *multiplexed scheduling*. ECOS relies on a simple heuristic for pairing offload requests with resources in a manner that likely offers the desired benefits.

Input. The input to the heuristic for a given code block b includes the energy savings without encryption n_b, the energy savings with encryption e_b, the estimated execution time on the mobile device t_b, and the mobile device's CPU speed s. §5.2 explains how ECOS obtains these values.

Candidate resources. If an application is seeking energy benefits and the application is deemed to require encryption, then the controller verifies if the energy savings with encryption is positive; otherwise, no resource is assigned and offload is prevented. For applications that either (i) do not require encryption, (ii) have energy savings with

204

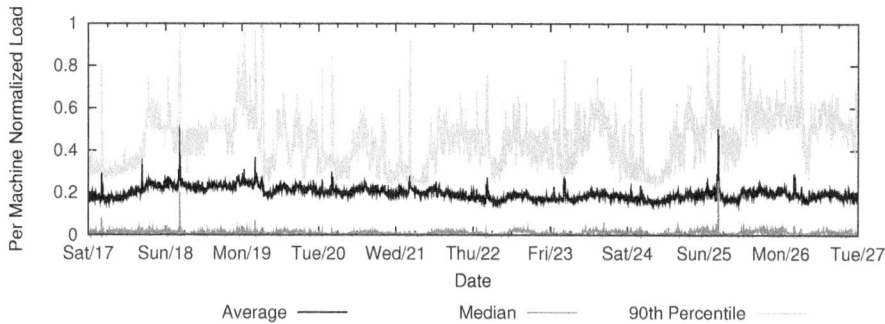

Figure 3: Fraction of CPU used per machine

encryption, or (*iii*) do not care about energy use, a decision must be made whether a resource with sufficient capacity is available.

The time taken, w_b, to execute a code block on a desktop, with or without encryption, depends on the code and the current load on the desktop. For applications seeking performance improvements, we must assign a resource r such that $w_b < t_b$. We can verify this will hold by ensuring the unassigned CPU cycles on r are at least the CPU speed of the mobile device. The available CPU time $a_r = s_r - \sum_m s_m$, where s_r is r's total idle CPU capacity and $\sum_m s_m$ is the sum of the mobile CPU speeds for the applications currently assigned to r. We know $w_b < t_b$ will hold if $s \leq s_r - \sum_m s_m$.

If an application does not seek performance improvements, w_b can be arbitrarily large, meaning that, b can be run anywhere and the application would still see a net energy benefit from b's execution. But care must be taken to ensure that running b does not impact other offloaded applications running on the same resource which care about latency.

Allocation (at the start of offload). If an application is not seeking performance benefits, it is assigned to a resource with other applications only seeking energy benefits. For applications seeking performance benefits, we assign the application to the resource which has the largest a_r at the time the application makes its first offload request. If no $a_r \leq s$, no resources are available and the application is unable to offload; however, it could ask for resources again the next time it desires to offload a particular codeblock. We could assign both performance and energy seeking applications to the same resource, but we separate the two types to avoid the need for complex scheduling mechanisms on the compute resources. If an application seeks both latency and energy savings, we follow the allocation policy for a performance seeking offload, as this places stricter bounds on execution time.

As stated in §3.3, the offload destination depends on the trust level of the application/device: e.g., user-private applications are only offloaded to user-trust machines, whereas no-private applications can be offloaded anywhere. By assigning performance seeking offloads to the machine with the largest available CPU time a_r, we spread the offloads amongst all available resources and increase the likelihood that a sufficiently trusted candidate resource will have available CPU time to serve the privacy constrained offload. In the case of energy seeking offloads, we can assign as many offloads as necessary to a resource, so there is a low likelihood that a sufficiently trusted resource will be unavailable.

Resource affinity. A key issue in allocation pertains to whether the resource used for offload should be changed during the course of an application's execution. Changing means that consecutive codeblocks can be offloaded to different resources. The cost of this is that it requires application state to be re-instantiated at the new resource and new transport connections to be set up. The benefit is that it allows fine-grained scheduling of resources across portions of mobile applications to dynamically adapt to demand and resource availability over time. However, our results in §3.1.1 show that the costs to the mobile device for sending state are significant, and are likely to outweigh the benefits of changing resources. Thus, ECOS avoids this overhead and instead tries to ensure *resource affinity* as much as possible: that is, ECOS seeks to always offload an application to the same resource every time it requests resources. This allows us to keep a partially loaded runtime on the resource to quickly serve future offloads from the same application.

Churn in available resources. Resource affinity relies upon the assigned resource having a constant amount of capacity available during the lifetime of the application, which may not always be true. Conceivably, the user of a desktop could start some resource-intensive task that requires the resources that have been provided to the phone. Our heuristic should not impede the user of a desktop, so we cannot allow the mobile device to continue to offload in such a case. We have several options for how to proceed:

• Deny future offloads from the device until the desktop resources free up. This is simple, but it may not adequately leverage idle resources that may exist in the network and may prevent the device from offloading for a long period.

• Assign the device a new desktop and require it to resend its state. As mentioned before, this could be quite costly to the device, but it may improve the desktop user's experience by allowing state space to be immediately reclaimed.

• Assign the device to a new desktop and have the original desktop migrate the state over the Ethernet the next time the device offloads. This is ideal from the perspective of the mobile device, since it only incurs the cost of establishing a new connection. However, state migration significantly complicates the protocol and may impact the user of the desktop.

• Store state in a network-wide cache after an offload completes. This avoids the need for resource affinity, since any desktop can retrieve the previous state from the cache without the involvement of the mobile device or the original desktop. However, this requires dedicated cache resources.

The best behavior depends on the characteristics of the network. If machines frequently become busy for long periods of time, it may be best to have the mobile device reestablish the state or have the desktops handle state migration. In contrast, if desktops are known to only become busy for brief periods, it may be best to keep the device associated with the desktop and deny its requests until the machine has resources. The controller can track prior usage patterns of a desktop to determine the right behavior.

5. ECOS IMPLEMENTATION

Our prototype implementation of ECOS consists of the following components: (*i*) Compute resource monitors provide resource availability information for offloading resources; compute resources also run a virtual machine (VM) in which offloaded applications are executed; (*ii*) A modified runtime environment (RE) on the phone passes metadata about candidate applications to the controller and performs the actual offload as directed by the controller; (*iii*) An SDN application that runs atop an enterprise-wide SDN controller, orchestrating all offloads and carefully managing the network paths connecting mobile devices and compute resources.

5.1 Compute Resources

Desktops, servers, and remote clouds capable of executing offloaded applications announce their availability and provide an environment for executing the applications. Compute resources provide updates on available CPU cycles. Resource monitors establish a connection to the controller at startup and, based on measurements in §4.1, send XML messages every 5 minutes with the available CPU capacity a_r measured using *mpstat*.

All compute resources also run a smartphone VM to provide an environment for executing offloaded applications. Specifically, we run a native version of Google Android [5]—the phone platform used in our system prototype—in a VirtualBox [7] VM. A *restore agent* in the Android environment manages the offloaded applications running in the VM. Each incoming socket connection from a phone is accepted and passed to a new RE instance which receives the name of the application and all associated execution state from the phone. We assume the application executable already exists on the compute resource; alternatively, applications could be downloaded from the Internet on first offload. The restore agent informs the controller when execution completes.

ECOS gives user applications running on the desktop priority over offloaded applications by over-provisioning for desktop applications: if the collective CPU utilization of desktop applications is u, then ECOS ensures CPU usage of offloaded applications doesn't exceed $a_r = 1 - (u + \delta)$, leaving δ amount of additional compute resources for desktop applications should they need them. We conservatively set $\delta = max(0.1, 0.5u)$.

5.2 Phone Runtime Environment (RE)

The RE on the phone is responsible for (*i*) selecting which parts of an application should be offloaded and (*ii*) transferring an application's execution state to a compute resource.

The RE considers estimated CPU usage, security overhead, and the availability of resources when considering what to offload. We build on ideas from CloneCloud [13] and Chroma [9]. Applications that want to take advantage of ECOS provide a list of computationally intensive methods with the estimated execution time and runtime state size for each method, similar to the specification of tactics in

Chroma. The RE uses the measurements from §3.1.1 and the provided state size estimate to approximate the time/energy overhead of offloading with and without encryption. Similarly, the phone approximates the time/energy required to execute the method on the phone, based on the provided execution time estimate t_b and an energy model [31].

If the overhead without encryption is less than executing on the phone, then the phone's *offload agent* asks the controller for resources. The agent connects to the controller when the phone joins the wireless network. An XML message sent to the controller specifies the device (MAC address), a hash of the application binary, energy savings without encryption n_b, energy savings with encryption e_b, estimated execution time t_b, and phone CPU speed s. It also specifies whether the phone desires performance benefits, energy savings, or both, based on user preference, the current battery level, or other factors.

When a method is selected for offload, the method signature, argument values, and any referenced objects must be transferred to the compute resource. Knowing exactly what state to send and where this state resides is challenging. We take a conservative approach to transferring objects, like in MAUI [15]: we transfer objects that are arguments and objects that are referenced by the arguments. After method execution on the compute resource completes, the phone RE must reconcile any changes between the old RE state and the RE state received from the compute resource. A reconciling process compares objects and updates them appropriately, allocating new objects as necessary. Based on the overhead measurements in §3.1.1 and the state similarity measurements in §6.5, ECOS preserves the RE on the compute resource, sending only a delta of the state across different offload points. Computing the delta incurs no more CPU overhead than sending the full state and requires fewer packets. While sending deltas requires additional memory on the phone to store hashes, and preserving state on the compute resource potentially limits the ability of a phone to opportunistically take advantage of available resources, we find that this is a reasonable strategy to use in practice (§6).

5.3 SDN Application

Based on the algorithms described earlier, the ECOS SDN application running on an SDN controller sends an XML message with the IP address of the assigned resource. The phone RE opens a socket (or secure) connection to the resource on port 8400 (or 8443). We implement our SDN application on top of NOX atop a bed of OpenFlow switches [21]. The forwarding state in network elements expires and is flushed out after 1s. Thus, phones with active offloaded applications have to periodically ask the controller for a route (every 0.5s in ECOS) to keep the path to the offload destination alive.

6. EVALUATION

In this section, we evaluate the advantages of using ECOS with a range of experiments to ultimately establish the viability of using ECOS to support enterprise applications on mobile devices. We study the following issues: (*i*) Can ECOS support enterprise applications with different latency, energy and security needs? What benefits do different application classes observe, and what are the costs? (§6.3) (*ii*) To what extent do resource affinity and the ability of ECOS to multiplex several applications on each secondary resource help? Are they ever detrimental in enterprise settings? (§6.4) (*iii*) To what extent do various optimizations to control ECOS's offload overhead, such as preserving state across different code block executions and transferring state under resource churn, help? (§6.5) We describe our setup in §6.1 and §6.2.

6.1 Methodology

We were unable to obtain real enterprise applications due to licensing issues and the lack of source code. Instead, we use two representative applications which can benefit from ECOS: chess, which we use as an (admittedly artificial) stand-in for an compute-intensive AI-based enterprise application that does not seek user privacy (e.g., non-linear decision-making and customer relationship management apps [1]), and speech recognition, which is a stand-in for a speech-to-text enterprise transcriber. The chess game features an AI engine, configured to use three decision iterations. We modified the game to play against itself for 50 moves, with a 10 second delay every other move to simulate the delay of a human. Speech recognition is a computationally and memory intensive application. Unfortunately, Android lacks some crucial audio libraries to run speech recognition, such as CMU Sphinx [3], directly on the phone. Therefore, we model the state size, CPU usage, and memory usage of speech recognition with a mock application to still show the benefits ECOS could provide if phones added the necessary audio support in the future. The application is configured to perform 20 recognitions, with a 10 second delay between tasks. We consider speech recognition to be user-private.

We measure ECOS for a small enterprise setting using a set of Android emulators and desktops. Phone emulators are used in our experiments because we only had access to a single Android developer phone. We show experiments in §6.2 to confirm that the emulators we use are a reasonable substitute for actual phones. Each phone emulator's CPU frequency is scaled to match the behavior of an Android phone in typical under-clocked use. The desktops we use for computational resources are 2.4GHz Intel quad core machines with 4GB of RAM, representative of a typical modern workstation. Our controller runs on a separate machine whose specs are the same as the desktops. The emulated phones and desktops communicate over wired Ethernet. In a real setting, phones would communicate using wireless links of lower speeds; thus, our latency measurements are likely to be an underestimate, but we do not expect the difference to be significant as the amount of state transferred during offload in the above applications is quite small. To estimate power consumption, we measure the number of packets and bytes sent, CPU usage and wireless NIC usage and plug them into energy models [31], similar to our study in §3.1.1.

6.2 Emulator/Phone Comparison

We first perform a set of small-scale experiments to affirm that Android emulators are a reasonable substitute for an actual Android phone. Our setup consists of a single desktop and a single emulator or phone. We run both example applications with and without offloading, measuring the total energy usage and total runtime.

The execution time of the chess and speech recognition applications without offloading are shown in Figure 5 for both the Android emulator and an Android developer phone. The applications run about 20% faster on the emulator, but the

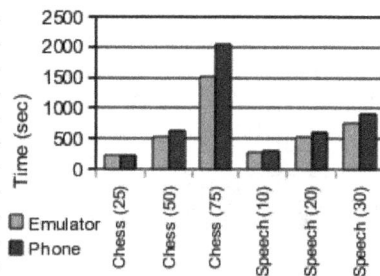

Figure 5: Comparison of application execution times without offloading using a phone versus an emulator

Phone	Application	Goal	Trusted Resources
P1,P2,P3	Chess	Latency	All
P4,P5,P6	Chess	Energy	All
P7	Speech Recognition	Latency	D1, D2, D6
P8	Speech Recognition	Latency	D2, D3, D6
P9	Speech Recognition	Latency	D1, D3, D6
P10	Speech Recognition	Energy	D1, D2, D6
P11	Speech Recognition	Energy	D2, D3, (D6)
P12	Speech Recognition	Energy	D1, D3, (D6)

Table 1: Configuration for applications used

scaling is consistent for both applications with varying numbers of moves/recognitions. We polled the CPU and memory utilization during application execution in all cases at frequent intervals. The CPU ran at 100% utilization whenever the application was busy executing in the case of both the phone and the emulator. Memory usage was also identical. We found the CPU frequency to be similar, although the emulator was running at a slightly higher speed.

We then examined how offload would function. We compared the packet stream to and from the remote desktop when using the phone versus the emulator, and found them to be identical in terms of the number of bytes and packets. Also, in the case of the phone, we did not observe a significant change in the bitrate during the course of any offload. The only difference was a longer offload state transmission time on the phone due to the slower link speed and higher loss rates of a wireless link: for 50 chess moves the total time spent transmitting state over wireless is 10s versus 2.9s over wired, and for 20 speech recognitions the total transmit time is 12s versus 2.3s over wired. We omit the detailed results for brevity.

In effect, these measurements show that the emulator provides a reasonable approximation of execution and offload on a physical phone. The data we collect on the emulator to help estimate phone power drain also provides a qualitatively similar estimate to real power drain on the phone (modulo the accuracy of the model in [31]).

6.3 Full System Analysis

We present an analysis of ECOS for a small enterprise setting consisting of 12 phones (P1-P12) and 4-6 desktops (D1-D6). A mix of both chess and speech recognition run on the phones, with varying goals and privacy levels. Table 1 shows the application specification for each phone. The constraint of only having 4 desktops to serve 12 phones stresses ECOS, but we find it still benefits most applications. With lesser contention when 6 desktops are available, the benefits from ECOS become more significant and equitable.

In these experiments, we assume the desktops are not running any other applications during the entire test du-

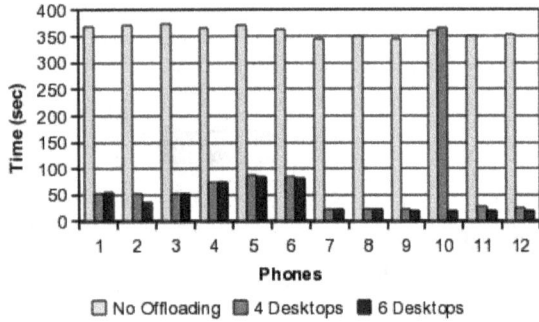

Figure 6: Comparison of application execution times

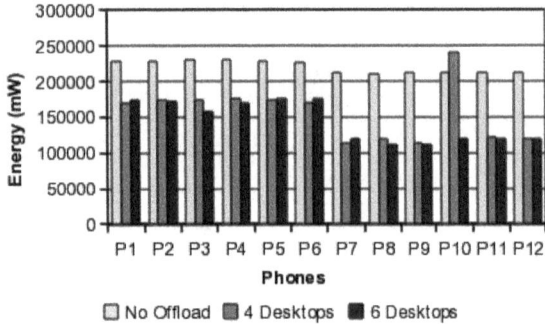

Figure 7: Comparison of application energy usage

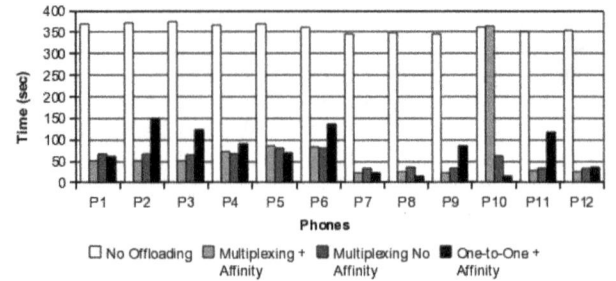

Figure 8: Total execution time using alternative resource allocation approaches

ration. Furthermore, the controller multiplexes assignments and enforces resource affinity. Later in this section, we consider other approaches to assigning resources—one-to-one scheduling and no-resource-affinity—in §6.4.

Latency. The execution time (excluding delay between moves or recognitions) for each phone is shown in Figure 6. Without offloading, all applications take approximately 350s to execute. Multiplexing the applications amongst four computational resources significantly reduces the execution time to between 22s and 87s for all phones except P10. The speech recognition application on P10 receives no performance benefit because there is no desktop it trusts that is available to serve applications seeking energy savings. Furthermore, phones P4-6 see nearly 50-60% higher latency than P1-3 although they are running the same chess application. This is because P4-6 requested energy savings, while P1-3 requested latency improvement.

Using 6 desktops provides enough resources leading to all phones seeing better latency than without offloading and 10-20% better net latency compared to using 4 desktops. Furthermore, P10 is able to offload to a user-trusted resource, and, with lower resource contention per desktop, P10's speech recognition application is able to run in as little as 20s.

Energy. Our experiments show that ECOS also offers energy benefits. Figure 7 shows the total energy used by the applications for the same scenario. The energy savings for both performance seeking and energy seeking applications ranges from 24% to 44% with 4 desktops and 23% to 47% with 6 desktops. Again, with 4 desktops, resources are constrained and ECOS is unable to provide energy benefits to P10. Increasing the number of available compute resources allows P10 to attain energy savings equivalent to its peers.

Although not present in these scenarios, in some cases ECOS imposes energy cost for applications that have explic-

itly requested low latency. Likewise, some applications desiring energy savings may see a degradation in latency compared to no-offload when there is high resource contention. Since our system is opportunistic, the presence of this cost is highly dependent on the application workload.

6.4 Resource Allocation Efficiency

Multiplexing applications on compute resources is one approach to assigning compute resources. However, §4.2 also discussed one-to-one scheduling—assigning a single application to a compute resource at a time—as an alternative method for assigning resources. We compare these two scheduling approaches, both with and without resource affinity. We use the same experimental setup of phones and desktops that was described in §6.3. We measure the total time and energy saved by applications for one-to-one-scheduling with affinity, multiplexed scheduling with affinity, and multiplex scheduling without affinity.

Figure 8 shows the execution time for each phone with the three approaches. First, we observe that one-to-one scheduling results in less offloading opportunities and higher execution latency for two-thirds of the phones. In some cases, e.g. P2, execution takes more than twice as long. This behavior results from the inability to serve more than 4 applications (as many desktops as we have) at any given time. At the same time, applications typically take less time to execute a given offload instance since compute resources are not shared with other offloaded applications. P10, for example, executes the fastest with one-to-one scheduling because it gets full use of the CPU when it is assigned to a desktop.

Second, we observe that for most phones, the resource allocation approaches that use affinity result in lower total latency. This decrease in latency stems directly from the decrease in execution state that must be transferred as a result of preserving execution state at the same desktop. However, avoiding affinity can help provide a fair sharer of benefits when the number of compute resources are limited: e.g., P10 receives significant benefit when using multiplexing with no-affinity as there is more churn in assignments and a greater opportunity for being allocated resources.

6.5 Preserving State Across Offloads

As discussed in section §5.1, ECOS preserves state on compute resources between offloads, decreasing the amount of state that must be transferred during subsequent offloads. This can be especially beneficial when security requirements force data to be encrypted.

We analyze the feasibility of preserving state by measuring how much the state changes between subsequent offloads of the chess application. Figure 9 shows the fraction of state

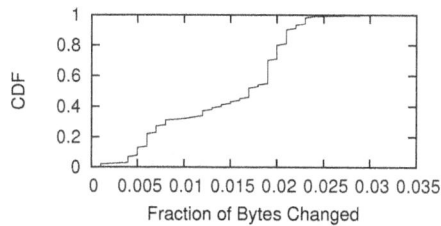

Figure 9: CDF of fraction of state changed between subsequent offloads of chess AI

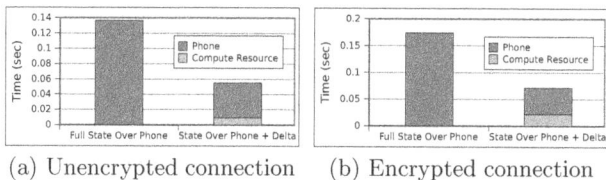

(a) Unencrypted connection   (b) Encrypted connection

Figure 10: Time to transmit state

that changes between subsequent offloads for the first 100 offloaded moves. At most 3.5% of the approximately 37 KB of state changes between offloads. The majority of offloads have only a 0.5% to 2% difference between the current and previous state. The amount of state variation is highly dependent on the specific application, but these results show that preserving state for some applications can significantly reduce the state that must be sent.

Preserving state between offloads can significantly reduce offloading overhead, but it also requires assigning applications to the same compute resource every time. If a compute resource no longer has idle capacity for offloading, the state must either be (i) sent in full from the phone or (ii) transfered from one compute resource to another. We already measured the overhead of sending the state from the phone in §3.1.1. Figure 10 compares the cost of resending all state from the phone versus transferring state between compute resources and the phone only sending a state delta for subsequent offloads for unencrypted and encrypted connections. For both types of connections, the latter method is more than twice as fast, with the bulk of the overhead occurring on the phone due to connection set up and state transfer. Thus, it is best to handle state transfer between compute resources directly.

In summary, we find that ECOS can support multiple applications with different performance and security needs, offering significant, equitable benefit at low cost. Our design choices have proven crucial for good overall performance, especially when opportunistically leveraging resources, and under encryption: Resource affinity appears to be essential to avoid the overhead of state transfer and connection cost, and ensure good offload performance especially for encryption. For similar reasons, preserving state across offloads and transferring state across connected desktops (when moving offloaded applications) is crucial. Multiplexing several smartphone applications on each desktop is important to ensure multiple applications observe equitable benefits.

Our controller offers high throughput for computing resource allocations, similar to what was observed in [11]. The latency between the time an offload request is made until the time the offload actually begins (which depends on computation of the resource assignment, computing a network

path and establishing forwarding state along the path) is negligible. We omitted these benchmarks for brevity.

7. CONCLUSION

We presented ECOS, an enterprise-based offloading system designed to address the security needs of mobile applications and opportunistically leverage available compute resources. ECOS extends the offloading decision process to take into account security requirements and costs, and in doing so, ECOS leverages the unique advantages that Software-Defined Networking (SDN) provides. In ECOS an enterprise-wide controller assigns (trusted) compute resources to applications based on resource availability, administrator specified security policies, and the performance or energy savings goals of mobile devices; ECOS also strictly enforces the security constraints through careful control of the network. We showed that ECOS provides both latency and energy benefits to both no-private and user-private applications.

The main contributions of our work are to show: (i) how to accommodate trust and privacy considerations in offloading without resorting to complex and error prone trust schemes and (ii) how to scale offloading to many mobile devices and compute resources, and opportunistically leverage these resources. By addressing these issues using SDN, we believe that we have paved the way for wider-spread adoption of offloading to assist current and future mobile applications in the enterprise context.

8. ACKNOWLEDGEMENTS

We would like to thank our shepherd Ripduman Sohan and the anonymous reviewers for their insightful feedback. This research was supported by the National Science Foundation under grants CNS-1050170, CNS-1017545 and CNS-0746531.

9. REFERENCES

[1] Ai enters the mainstream. http://www.domain-b.com/infotech/itfeature/20070430_Intelligence.htm/.
[2] Apple's iPhone 4: Thoroughly reviewed. http://anandtech.com/show/3794/the-iphone-4-review/12.
[3] Cmu sphinx. http://cmusphinx.sourceforge.net.
[4] Developing enterprise applications for mobile devices remains way too hard. http://zdnet.com/blog/gardner.
[5] Google android. http://android.com.
[6] Health information privacy. http://hhs.gov/ocr/privacy.
[7] Oracle virtualbox. http://virtualbox.org.
[8] Why your smartphone battery sucks. http://pcworld.com/article/228189.
[9] R. K. Balan, M. Satyanarayanan, S. Y. Park, and T. Okoshi. Tactics- based remote execution for mobile computing. In *MobiSys*, 2003.
[10] N. Balasubramanian, A. Balasubramanian, and A. Venkataramani. Energy consumption in mobile phones: implications for network applications. In *IMC*, 2009.
[11] M. Casado, M. J. Freedman, J. Pettit, J. Luo, N. McKeown, and S. Shenker. Ethane: taking control of the enterprise. In *SIGCOMM*, 2007.

[12] B. Chun and P. Maniatis. Augmented smartphone applications through clone cloud execution. In *HotOS*, 2009.

[13] B.-G. Chun, S. Ihm, P. Maniatis, M. Naik, and A. Patti. CloneCloud: elastic execution between mobile device and cloud. In *EuroSys*, 2011.

[14] B.-G. Chun and P. Maniatis. Dynamically partitioning applications between weak devices and clouds. In *MCS*, 2010.

[15] E. Cuervo, A. Balasubramanian, D. ki Cho, A. Wolman, S. Saroiu, R. Chandra, and P. Bahl. MAUI: Making Smartphones Last Longer with Code Offload. In *MobiSys*, 2010.

[16] W. Enck, P. Gilbert, B.-G. Chun, L. P. Cox, J. Jung, P. McDaniel, and A. N. Sheth. TaintDroid: an information-flow tracking system for realtime privacy monitoring on smartphones. In *OSDI*, 2010.

[17] L. Fiering and K. Dulaney. iPads: Not notebook replacements, but still useful for business. *Gartner, Inc.*, 2010.

[18] J. Flinn, S. Park, and M. Satyanarayanan. Balancing performance, energy, and quality in pervasive computing. In *ICDCS*, 2002.

[19] A. Greenberg, G. Hjalmtysson, D. Maltz, A. Myers, J. Rexford, G. Xie, H. Yan, J. Zhan, and H. Zhang. A clean slate 4D approach to network control and management. *ACM SIGCOMM CCR*, 2005.

[20] K. Kumar and Y.-H. Lu. Cloud computing for mobile users: Can offloading computation save energy? *Computer*, 99, 2010.

[21] N. McKeown, T. Anderson, H. Balakrishnan, G. Parulkar, L. Peterson, J. Rexford, S. Shenker, and J. Turner. OpenFlow: Enabling innovation in campus networks. *ACM SIGCOMM CCR*, 2008.

[22] A. Messer, I. Greenberg, P. Bernadat, D. Milojicic, D. Chen, T. Giuli, and X. Gu. Towards a distributed platform for resource-constrained devices. In *ICDCS*, volume 22, 2002.

[23] S. D. Nelson and D. A. Willis. Separating enterprise tablet applications from consumer apps. *Gartner, Inc.*, 2011.

[24] A. Ramachandran, Y. Mundada, M. B. Tariq, and N. Feamster. Securing enterprise networks using traffic tainting. Technical Report GT-CS-09-15, GaTech, October 2009.

[25] M. Satyanarayanan, V. Bahl, R. Caceres, and N. Davies. The case for vm-based cloudlets in mobile computing. *IEEE Pervasive Computing*, 2009.

[26] S. Smaldone, B. Gilbert, N. Bila, L. Iftode, E. de Lara, and M. Satyanarayanan. Leveraging smart phones to reduce mobility footprints. In *MobiSys*, 2009.

[27] J. R. Tyler, D. M. Wilkinson, and B. A. Huberman. Email as spectroscopy: automated discovery of community structure within organizations. *Communities and technologies*, pages 81–96, 2003.

[28] T. Whalen, D. Smetters, and E. F. Churchill. User experiences with sharing and access control. In *CHI*, 2006.

[29] A. Wu, J. M. DiMicco, and D. R. Millen. Detecting professional versus personal closeness using an enterprise social network site. In *CHI*, 2010.

[30] H. Yan, D. A. Maltz, T. S. E. Ng, H. Gogineni, H. Zhang, and Z. Cai. Tesseract: A 4D network control plane. In *NSDI*, 2007.

[31] L. Zhang, B. Tiwana, Z. Qian, Z. Wang, R. Dick, Z. Mao, and L. Yang. Accurate online power estimation and automatic battery behavior based power model generation for smartphones. In *CODES+ISSS*, 2010.

[32] X. Zhang, J. Schiffman, S. Gibbs, A. Kunjithapatham, and S. Jeong. Securing elastic applications on mobile devices for cloud computing. In *Workshop on Cloud computing security*, 2009.

On Pending Interest Table in Named Data Networking

Huichen Dai[†] Bin Liu[†] Yan Chen[§] Yi Wang[†]
[†] Tsinghua National Laboratory for Information Science and Technology
[†] Dept. of Computer Science and Technology, Tsinghua University, China
[§] Northwestern University, USA
dhc10@mails.tsinghua.edu.cn, liub@tsinghua.edu.cn, ychen@cs.northwestern.edu, wy@ieee.org

ABSTRACT

Internet has witnessed its paramount function transition from host-to-host communication to content dissemination. Named Data Networking (NDN) and Content-Centric Networking (CCN) emerge as a clean slate network architecture to embrace this shift. Pending Interest Table (PIT) in NDN/CCN keeps track of the Interest packets that are received but yet un-responded, which brings NDN/CCN significant features, such as communicating without the knowledge of source or destination, loop and packet loss detection, multipath routing, better security, etc.

This paper presents a thorough study of PIT for the first time. Using an approximate, application-driven translation of current IP-generated trace to NDN trace, we firstly quantify the size and access frequencies of PIT. Evaluation results on a 20 Gbps gateway trace show that the corresponding PIT contains 1.5 M entries, and the lookup, insert and delete frequencies are 1.4 M/s, 0.9 M/s and 0.9 M/s, respectively. Faced with this challenging issue and to make PIT more scalable, we further propose a Name Component Encoding (NCE) solution to shrink PIT size and accelerate PIT access operations. By NCE, the memory consumption can be reduced by up to 87.44%, and the access performance significantly advanced, satisfying the access speed required by PIT. Moreover, PIT exhibits good scalability with NCE. At last, we propose to place PIT on (egress channel of) the outgoing line-cards of routers, which meets the NDN design and eliminates the cumbersome synchronization problem among multiple PITs on the line-cards.

Categories and Subject Descriptors

C.2.1 [**COMPUTER-COMMUNICATION NETWORKS**]: Network Architecture and Design

Keywords

PIT, Size, Frequency, Encoding

1. INTRODUCTION

The functionality of Internet has evolved substantially, though it was originally designed for host-to-host communication, it now mostly serves content-centric applications. Therefore, researchers arrive at a widely recognized agreement that content should have a more central role in future network architectures than it does in the current Internet's host-centric conversation model [5, 7]. The research community addresses this problem with a paradigmatic shift – Content-Centric Networking [10] (CCN), which focuses on content dissemination based on content identifiers rather than content hosts. Named Data Networking [15] (NDN) is an instance of the general CCN paradigm.

NDN communication is requester-driven. A requester sends out an *Interest* packet, which carries a name – the identifier – that specifies the desired data. PIT in an intermediate router remembers the Interest name and from which interface the Interest comes in. Content providers respond to the Interest by sending back a *Data* packet that carries both the name and desired content. Once Data packet arrives at an intermediate router, the router looks up its name in the PIT to obtain the interface from which the requesting Interest comes, and deletes that name from PIT. Therefore, a router inserts every incoming Interest into PIT, and removes each received Data packet from PIT. Intuitively, due to the high-speed packet arrival rate, PIT will have a large size and demand extremely high access (lookup, insert and delete) frequency, which has caused wide debate on the feasibility of PIT. The features that PIT brings (Section 2.3) makes it indispensable to NDN/CCN. Therefore, the PIT issue becomes a knotty obstacle that hinders practical and scalable implementation of NDN.

Though current researches on NDN/CCN are in full swing, the study on PIT is quite exiguous. To the best of our knowledge, we are the first to conduct measurements on exact PIT size and access frequency, and further propose solutions to address this untouched issue, i.e., shrinking PIT size and improving PIT access performance. To unfold this problem, we are faced with the following challenges:

1. How to evaluate and quantify the PIT size and frequency while NDN/CCN has NOT been really deployed?
2. Different from merely lookup a name, how to well address the insert and delete operations?
3. How to support PIT scalable to large name sets while still sustaining high PIT access performance?

In this paper, we propose a real trace-translation/mapping method to measure the size and access frequency required by PIT. We captured a one-hour trace from a 20 Gbps gateway link in the China Education and Research Network (CERNET) for our experiment. Afterwards, a Name Component Encoding (NCE) solution is put forward to reduce PIT's memory consumption and promote the access performance. At last, we also present a scheme on where to place PIT in NDN routers when actually implementing PIT. Especially, we make the following contributions:

1. We emulate NDN applications' working paradigms by transferring the existing IP applications to the NDN platform. Then, by translating/mapping our captured IP trace to NDN scenario at the perspective of applications, we quantify the size and access (lookup, insert and delete) frequencies required by PIT. Experimental results on a trace collected from

a 20 Gbps gateway link show that the corresponding PIT has 1.5 M entries, and its lookup, insert and delete frequencies are 1.4 M/s, 0.9 M/s and 0.9 M/s, respectively. These results imply that directly storing and accessing PIT entries as character strings in commodity memories is not scalable and incurs great challenges, especially for insert and delete operations.

2. We continue to adopt an encoding-based idea (NCE) and make important improvements to shrink the PIT size and satisfy the access frequency requirement. The encoding idea was first proposed in our previous work [14]. Experimental results demonstrate that by our solution, the PIT size can be be reduced by up to 87.44%, and the lookup, insert and delete frequencies can achieve 3.27 M/s, 2.93 M/s and 2.69 M/s respectively on an Intel 2.27 GHz CPU, satisfying the access frequency requirement of the studied PIT.

3. Moreover, in combination with the router architecture, we design an ingenious PIT residence scheme that places PIT on packets' outgoing line-cards (egress channel), avoiding the cumbersome synchronization problem among multiple PITs on line-cards.

The remainder of the paper is organized as follows. Section 2 provides NDN background information and our motivation. Section 3 designs working paradigms on the NDN platform for existing IP network applications, and measures the size and access frequency of PIT. We propose an encoding-based solution to reduce the PIT size and promote PIT access performance in Section 4. Section 5 proposes a PIT residence solution on router outgoing line-cards. We evaluate our solutions in Section 6, Section 7 is the comparison of previous work, and Section 8 concludes this paper.

2. BACKGROUND AND MOTIVATION

2.1 NDN Introduction

NDN, a specific instance of the CCN paradigm, is a novel network architecture proposed by [15] recently. Different from current Internet practice, it makes content ("what") as its central role, rather than "where" content is located. A critical distinction from IP is that, every piece of content in NDN network has an assigned name, and packets are routed/forwarded by names, rather than IP addresses.

NDN names are application-dependent and opaque to the network, but they all share common characteristics – hierarchically structured and composed of explicitly delimited *components*. The delimiters, usually slash ('/') or dot ('.'), are not part of the name. The naming system is an important piece in the NDN architecture and is now still under active research. For the purpose of early exploring the properties of PIT before the naming specification finally goes to standard, in this paper, we temporarily use *hierarchically* reversed domain names as NDN names. For example, the scholar service provided by Google – *scholar.google.com* is hierarchically reversed to *com/google/scholar*, and *com*, *google*, *scholar* are three components of the name. For an HTTP URL, we hierarchically reverse its host name, and concatenate the rest part, such as absolute path, as an NDN name. For instance, URL *name.example.com/path/to/content* is transferred to an NDN name of *com/example/name/path/to/content*. The hierarchical structure, which resembles IP addresses, enables name aggregation and allows fast name lookup of Longest Prefix Match (LPM), and will be leveraged by PIT lookup.

NDN adopts a brand new data requester-driven communication mechanism. To receive data, a requester sends out an *Interest* packet, which carries a name – the identifier – that specifies the desired data. An intermediate router remembers the name and the interface from which the Interest comes in the PIT, and then forwards the Interest by looking up its name in the Forwarding Information Base (FIB).

When the Interest arrives at a node that serves the requested data, a *Data* packet that carries both the name and the content is returned. Once the Data packet reaches a router, the router looks up its name in the PIT to obtain the interface from which its corresponding Interest comes in, and then forwards Data to that interface. Therefore, Data travels back to the requester by taking the same path of the Interest, but in the reverse direction, i.e., symmetric routing. Moreover, the Data packet is strategically cached by a router's Content Store (CS) to serve subsequent Interests.

2.2 Packet Lookup and Forwarding in NDN

The NDN packet lookup and forwarding processes are a bit complicated than that of IP. To better comprehend the packet lookup and forwarding process, keep in mind that PIT keeps track of *pending* Interests, i.e., received but yet un-responded Interests. The specific lookup and forwarding processes of Interest and Data packet are shown in Figure 1(a) and Figure 1(b), respectively. Figure 1(a) shows that once an Interest packet arrives at interface i of an NDN router R, R:

1. consults CS if the desired content is present and returns a copy in Data packet via i,
2. if not, looks up PIT to see if PIT has an entry for this Interest. If so, adds i to that entry, and discards this Interest packet,
3. otherwise, creates a PIT entry for this Interest and add i to this entry, and
4. forwards Interest to the next-hop interface by looking up FIB.

When Data packet returns, Figure 1(b) shows that R:

1. forwards the Data packet over all the requesting interfaces in the corresponding PIT entry and deletes this entry,
2. caches the Data packet in the CS based on policies.

2.3 Motivation

We have presented the role that PIT plays in NDN, and PIT brings NDN significant features:

1. PIT enables Interest and Data packets be routed without specifying a source or destination address, which means PIT makes NDN communication concentrates on the content itself, rather than where the content locates or who are exchanging data.
2. The feature above inherently supports anonymous communication, making attacks difficult to launch and communication more secure.
3. PIT prevents Interests loop persistently, because Interest name plus a random nonce, which is stored in the PIT entry, can effectively identify duplicates to discard. Data packets do not loop since they take the reversed paths of Interests.
4. The property above also enables NDN to inherently support multipath routing, because NDN router can send out an Interest via multiple interfaces without worrying about loops.
5. PIT supports Data packet multicast when multiple Interests received by the router apply for the same content.
6. At last, PIT can detect Data packet losses if an Interest has not been responded beyond a time threshold.

All these significant features that PIT brings to NDN enable NDN/CCN to be an information/content-centric network, and therefore motivate us to conduct a thorough study on PIT.

3. PIT SIZE AND ACCESS FREQUENCY

There has been a debate on PIT since the proposal of PIT, because each Interest looks up, inserts and updates PIT, and each Data packet looks up and removes PIT entries, intuitively resulting in extremely large size and high access frequency of PIT. Interests are triggered

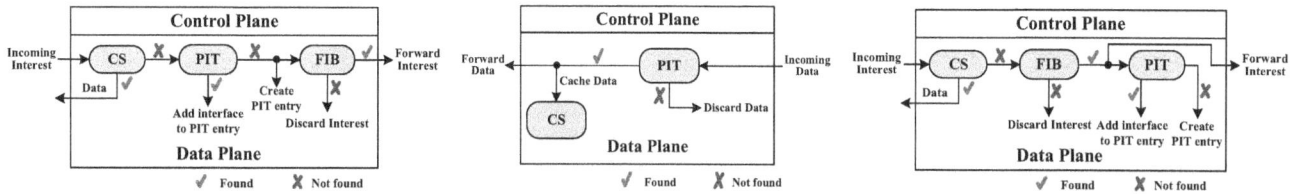

(a) Interest lookup and forwarding process. (b) Data lookup and forwarding process. (c) Interest lookup and forwarding process when placing PIT on the outgoing line-cards.

Figure 1: Packet lookup and forwarding process.

by applications for data communication. Each application has its own way to send out Interest and Data packets, and the mechanism of sending out packets influences the size and access frequency of PIT. In other words, each application has its own way to construct and destruct PIT entries. Therefore, we measure the size and access frequency of PIT at the perspective of application, rather than at the network layer.

3.1 Applications from IP to NDN platform

Researchers propose that Internet architecture switch from IP to NDN to better satisfy the requirements of users, such as browsing web pages, sharing files, and watching videos, etc. Therefore, the network applications demanded by users will not change, but require different implementations on the NDN platform. Consequently, we design working paradigms for the major applications in current Internet – HTTP, FTP, P2P, Email, Online Game, Streaming media, Instant messaging – based on NDN communication model[1], i.e., transfer applications from IP platform to NDN platform.

Fundamentally, NDN's requester-driven communication model *pulls* data from remote host to local host – a one-way data flow service. NDN has to solve the problem of bootstrapping two-way communications on top of a fundamentally one-way service.

3.1.1 HTTP

The HTTP protocol is a request/response protocol. A client sends a request to the server in the form of a request Method, request-URI (Uniform Resource Identifier), etc., over a connection to a server. The server responds with a message, which includes a success or error code, followed by entity meta information and possible entity-body content.

HTTP communication is initiated by a client sending out a request to be applied to a resource on some origin server. In the simplest case, this may be accomplished via a single connection between the client and the origin server. Therefore, HTTP accords with the NDN requester-driven communication model. The HTTP requests can be divide into different categories because of the different methods they take with them. The *Method* token of an HTTP request indicates the method to be performed on the resource identified by the Request-URI. Method includes:

1. OPTIONS: allows the client to determine the options and/or requirements associated with a resource, or the capabilities of a server.
2. GET: retrieves whatever information identified by the Request-URI.
3. HEAD: identical to GET except that the server MUST NOT return a message-body in the response, but only the metain-formation about the resource.
4. POST: requests that the origin server accept the data enclosed in the request as a new subordinate of the resource identified by the Request-URI.

5. PUT: requests that the enclosed data in the request be stored under the supplied Request-URI.
6. DELETE: requests that the origin server delete the resource identified by the Request-URI.
7. TRACE: invokes a remote, application-layer loop-back of the request message.
8. CONNECT: (reserved).

NDN only contains two kinds of packets: Interest and Data. Interest packet requests for a specific piece of content, i.e., *pulls* desired content, which is similar to the HTTP OPTIONS, GET and HEAD method. Thus, these three methods can be directly implemented by Interest packet, with the Request-URI being Interest name, and the method encapsulated in the Interest body. Their corresponding HTTP responses are implemented by Data packets. However, in order to implement the integrated HTTP protocol in NDN, we should extend the functionality of Interest packet to accommodate the rest methods. Let's examine them one by one.

For request with POST or PUT method, it tries to *push* data to the server, which violates NDN's pull communication model. To implement the pushing HTTP request, Interest packet should be *extended* to not only contain the name of the requested content, but also include a data block that contains the content to be pushed to the server. By this means, requests of POST and PUT method can be implemented by Interests as well, with their request-URIs also being Interest names. Their responding Data packets notifies the clients of the status of these requests, such as SUCCESSFUL, NOT ALLOWED, NOT IMPLEMENTED, etc. The rest two methods, DELETE and TRACE, neither pulls or pushes data, but invokes a function on the server. Now that we have extended the Interest from pull to push communication model, it is no harm that we further extend Interest to function calls (delete, echo-back, etc.). In this way, DELETE and TRACE can be implemented by the Interest packet as well. For DELETE, the Interest specifies the name of the to be deleted resource by Request-URI, and Data packet returns the status of the deletion. For TRACE method, the Interest is echoed-back by the server in Data packet.

It's worth pointing out that, though NDN is a brand new network-layer architecture, it is still built on current technologies. The data link layer evolves separately from network layer, and the current dominant layer-two standard is Ethernet. We believe that NDN still employs the data transmitting services that current data link layer technologies provide. Therefore, due to the Maximum Transmission Unit (MTU) of the data link layer, the HTTP response message, when being transmitted back to the client, may be segmented into multiple IP packets. Similarly, in NDN, a HTTP response message is likely to be segmented into multiple Data packets as well, rather than a single Data packet.

Subsequently, we examine the life time of PIT entries created by NDN HTTP connections. An NDN HTTP connection v is initialized by a client sending out an Interest packet. Assume that this Interest packet enters router R via line-card i and leaves via line-card j, line-card j (not line-card i, see Section 5) creates an PIT entry when receives it, and keeps this entry until all the responding Data

---

[1]NDN does not have a separate transport layer, thus all the NDN applications build directly on the request-driven communication model.

213

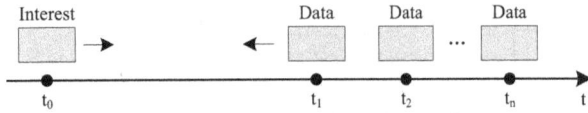

Figure 2: The life time of this PIT entry is $t_n - t_0$.

packets of this NDN HTTP response message have been returned, i.e., as the last Data packet arrives at line-card j, it removes the corresponding PIT entry. Therefore, one NDN HTTP request/response pair corresponds to one PIT entry, and the life time of this PIT entry is the time interval between the Interest packet and last Data packet observed by a router, as shown in Figure 2.

3.1.2 FTP

Similar to HTTP, we implement FTP on the NDN platform by assigning more roles to Interest and Data packet to accommodate the FTP requests and responses.

FTP is built on a client-server architecture and uses separate control and data connections between the client and the server, as depicted in Figure 3. The client's Protocol Interpreter (PI) initiates the control connection, then standard FTP commands are generated by the client PI and transmitted to the server process via the control connection. Standard replies with status codes are sent from the server PI to the user PI over the control connection in response to the commands. The FTP commands specify the parameters for the data connection (data port, transfer mode, representation type, etc.). Data connection can either be opened by the server (active mode) from its default port to a negotiated client port, or by the client (passive mode) from an arbitrary port to a negotiated server port as required to transfer file data.

The control connection is in an interactive mode, which can make direct use of NDN's request-driven communication model, with Interest packets implementing commands and Data packets acting as replies or acknowledgements. A command is encapsulated in the Interest packet, and now the Interest's name is not the name of desired data, but the FTP service name on the FTP server. Data packets are replies or acknowledgements of the commands and contain the status code of the command executing result.

IP FTP data connection may run in active or passive mode, which determines how the data connection is established. The difference between them is whether the client or the server opens the data connection. In active mode, the server initializes the data connection. In situations where the client is behind a firewall and unable to accept incoming connections, passive mode is used and the client initiates the data connection to the server. However, in whatever mode the data connection is initiated, a control connection must be initiated by the client first!

No matter established by active mode or passive mode, the data connection of IP FTP is duplex and bidirectional, i.e., the client and server can actively send data to each other over this very data connection. However, the NDN connection is unidirectional, the data can only flow from the host possessing the data to the host sending out Interests. Consequently, in NDN, the way how the data connection established alters and depends on whether the client C wants to upload file to or download file from the server S. If client C wants to download file from server S, it has to actively send out an Interest to initiate the data connection, which corresponds to the passive mode (initiated by client) of IP FTP. To implement this, client C first negotiates with server S over control connection by sending out an PASV command, then server S replies its information such as FTP service name. Next, client C uses this information to establish a data connection by sending out an Interest with S's FTP service name to S, now server S can send file to client C by sending back Data packets. This process is shown in Figure 4(a).

However, if client C wants to upload file, which corresponds to the active mode (initialized by server) of IP FTP, client C first sends

its information (FTP service name) to server S, and server S acknowledges with a Data packet. Then server S sends client C an Interest to establish a data connection. After receiving the Interest, client C can upload file to server S, which is illustrated in Figure 4(b). Till now, we have designed the working paradigm on NDN platform for both FTP control connection and data connection.

It's known that, each Interest creates a PIT entry and the PIT entry lasts until the responding Data packet returns. Because the control connection is interactive, a corresponding PIT entry is inserted and removed over and over again. At each snapshot, there can be at most one PIT entry that corresponds to the control connection, but control connection brings high PIT access frequency. For data connection, because the size of a file can be relatively large and will be segmented, the PIT entry will last until all the Data packets are returned, which is very like that of an HTTP connection.

3.1.3 P2P

P2P is no doubt a major application in current Internet and contributes vast traffic. To implement P2P on the NDN platform, we should know how it works. Fortunately, it also obeys the request-driven paradigm.

P2P network has two constructing ways: unstructured and structured. In unstructured P2P networks, a node sends out query packets by flooding or smarter algorithms [11, 12] to search for nodes serving desired files, and those nodes possess the requested files will send content back to the requester. In structured P2P networks, a node first consults its local Distributed Hash Table (DHT) about which nodes have the desired content, and then directly send requests to those nodes. No matter unstructured or structured P2P network, P2P's working paradigm is the closest to that of NDN. The query packets or the requests can be directly implemented by Interest packets "as-is", as well as the replies implemented by Data packets without any modification.

The corresponding PIT entries are created and removed in the way similar to that of HTTP.

3.1.4 Email

Implementing Email service in NDN will encounter some difficulties and calls for an a new mechanism. Because hosts *push* emails out to servers, which is fundamentally different from NDN's communication *pull* model.

Email servers and other email transfer agents use Simple Mail Transfer Protocol (SMTP) to send and receive email messages, user-level client email applications typically only use SMTP for sending messages to a email server for relaying. For receiving messages, client applications usually use either the Post Office Protocol (POP) or the Internet Message Access Protocol (IMAP).

Firstly, we consider sending emails. In IP, email is submitted by a mail client (MUA, mail user agent) to a mail server (MSA, mail submission agent) using SMTP. The messages can be directly sent to the server without any knowledge of the server in advance (except the email address). In however, in NDN, we need a "initiating" stage. Assume that a Gmail client (logged in with email address example@gmail.com) tries to send an email to "van@parc.com". It initiates the transaction by sending out an NDN Interest with the name: "parc.com/email_service/van/example@gmail.com/123". ("parc.com/email_service/" is a routable prefix for the email service of PARC, and 123 is a nonce.) After receiving the Interest, the PARC server responds to the client's Interest with an "OK" Data packet as acknowledgement. This process is called the "initiating" stage. Then the PARC server then sends out an Interest with the name: "gmail.com/email_service/example/van@parc.com/456" ("gmail.com/email_service" is the routable prefix to the Gmail server, and 456 is also a nonce) to pull down the incoming email. By this means, the Gmail mail client can send out an email via

214

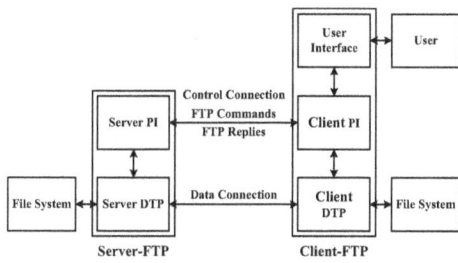

Figure 3: Model for FTP.

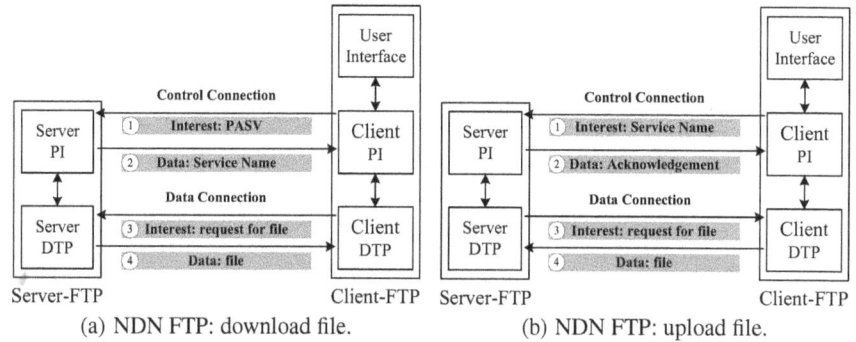

(a) NDN FTP: download file.

(b) NDN FTP: upload file.

Figure 4: Upload and download file process of NDN FTP.

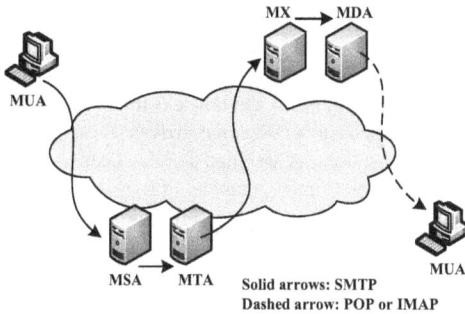

Solid arrows: SMTP
Dashed arrow: POP or IMAP

Figure 5: Email.

the responding Data packets. After the email is delivered to the MSA, it is forwarded to the Mail Transfer Agent (MTA) and Mail Delivery Agent (MDA) in the same way (Solid arrows in Figure 5). Therefore, for the NDN email sending process, it takes a little longer than that of IP for about a RTT time due to the initiating step.

Secondly, we consider the mail delivery procedure (Dashed arrows in Figure 5). Once delivered to the local mail server, the mail is stored for retrieval by authenticated email clients (MUAs), using POP or IMAP. When using POP, clients typically connect to the email server briefly, only as long as it takes to download new emails. Therefore, for POP, email clients actively send Interests to email servers when they want to receive emails, and mail server responds with emails by Data packets.

For IMAP, clients often stay connected as long as the user interface is active and download emails on demand. In other words, IMAP pushes emails to mail clients. Therefore, we adopt the same method for the email sending procedure. Let mail server send an Interest to mail client, then mail client acknowledges this Interest by a Data packet, and next sends another Interest to pull down the emails.

3.1.5 Online games

In current Internet online games, there must be a consistent connection between the client and server since an account logs in until it logs out, which is demanded by the application requirements. In NDN, this requirement does not alter, which reveals bidirectional communications between game client and game server. Therefore, the game client sends out an Interest with the user name and password to login to the server. (The Interest's name is set to a routable prefix of the game server.) This Interest is extended to a special kind of Interest that its corresponding PIT entries will not be removed while the user is logged in, in order to support the consistent connection requirement. By these PIT entries, a consistent channel from the server to the client is created. The server can arbitrarily send Data packets to the client. However, this unidirectional channel is not enough since the client also needs to send messages such as mouse clicks and keystrokes to the server. Consequently, after receiving the Interest from a client, the server sends an Interest to the client as well, which is also a special Interest that its PIT entries will

not be removed until the client logs out. In this way, a channel from the client to the server is also established, and the two unidirectional channels form a bidirectional connection between the game client and server.

3.1.6 Streaming media

Due to the entertainments brought by on-line videos, streaming media has become one of the major Internet applications, such as the Youtube website. To watch a video, a client (e.g., Adobe flash player) sends an Interest to Youtube website, and then Youtube sends back the video segments to the client in continuous Data packets, like a stream. If P2P accelerating techniques are used, the Interest is sent from a peer to another, and another peer sends back the video encapsulated in Data packets. Refer to *HTTP* and *P2P* for how the corresponding PIT entries are created and removed.

3.1.7 Instant messaging

Instant messaging (IM) has been integrated into people's lives. It also *pushes* out messages like sending emails, which violates the the NDN's communication model. In current Internet, IM mostly uses UDP to transport packets and thus provides best-effort service. Therefore, we design a mechanism for IM in NDN that consumes no PIT entry, which offers best-effort service as well.

In our design, an IM client logs into the IM server by sending an Interest to the server. The Interest's name is the IM service name of the server (a routable prefix to the server), and payload includes an account's user name, password, and local IM service name (a routable prefix to the client). The server acknowledges with a Data packet, which also includes the remote IM service names of the account's friends, indicating which friends are online. If the user wants to initiate a conversation with a friend, he sends out an Interest, with the friend's remote service name as the Interest's name, and the message that he wants to deliver to his friend as the Interest's payload. This kind of Interest is classified as another special kind of Interests that intermediate routers will NOT create PIT entry for it. Moreover, when the Interest reaches the destination client, the client will NOT return a Data packet either, providing best-effort service.

By now, we have designed working paradigms for the most major Internet applications on the NDN platform. In next subsection we will evaluate and quantify the size and access frequency of PIT.

3.2 Measuring PIT size and access frequency

We captured a one-hour trace from a 20 Gbps link in China Education and Research Network (CERNET). By mapping or translating this IP-generated trace to an NDN trace at the perspective of application, we quantify the PIT size and access frequency. Videlicet, to transmit the same data encapsulated in the payload of the IP trace via NDN (with the data link layer technologies remain the same), how many PIT entries and how much access frequency of PIT is required. For example, for a pair of HTTP request and response, a PIT entry will last since the arrival of the Interest packet until the arrival of the last Data packet of the response message. For an SMTP connection, we should add an initiating stage right before it starts, contributing a PIT entry for about a RTT time. We first parsed out the biflows[2] (connections for TCP and bidirectional flows for UDP) of each application discussed above. The parsing process takes advantage of multiple tools, including TIE [3, 6], Tstat [4, 8], l7-filter [2], etc., and we parsed our 88.76% of the total trace. The numbers of biflows of each application at each snapshot[3] are shown in Table 1 (except the second column is the # of yet un-responded HTTP requests).

From Table 1, as well as our designed working paradigms of NDN applications, we translate this trace from the IP scenario to the NDN scenario. E.g., the second column "Un-responded HTTP requests" presents the number of HTTP requests that have not been responded. Because HTTP/1.1 [1] adopts consistent connection and multiple HTTP packets can be sent via the same connection in a pipelined manner, but in NDN, each un-responded Interest packet has an associated entry in PIT, thus the number of PIT entries is related to the un-responded requests at a snapshot, rather than the number of HTTP connections. Another example is Online Game, as aforesaid, two unidirectional NDN channels realize the functionality of an IP connection, therefore, each Game connection corresponds to 2 PIT entries. In this way, by translating all the major applications, we derive that the number of PIT entries is around 1.5 million. We only consider the traffic of network applications, thus the traffic of network management (e.g., ICMP) and services (such as DNS, which is no more needed in NDN) are not counted (the last two columns in Table 1). IM belongs to the Conference group, which is not counted, either.

Moreover, the read frequency of PIT depends on how many Data packets (excluding the last Data of a request/response pair) arrive per second, the insert frequency is determined by the number of emerging biflows (new Interest arrive), and delete frequency the number of disappearing biflows (last Data packets arrive). These statistics of the first 10 seconds (out of 3,600 seconds) of the studied trace are provided in Table 2. It's worth pointing that we ignore all the packets without payload when calculating the PIT access frequency. From Table 2, the read frequency of PIT is around 1.4 M/s, and the insert and delete frequencies are both around 0.9 M/s.

4. ENCODING-BASED FAST PIT ACCESS AND SIZE SHRINKING

In this section, we propose an encoding-based approach to shrink PIT size, and accelerate the lookup, delete and update performance.

As the NDN names are composed of components and are relatively long compared to IPv4/v6 addresses. Directly storing them as character strings in a table and search for a match is not a wise idea. The hierarchical structure of NDN names enable some of them to share the same name prefixes, thus the names can be organized in Name Prefix Trie (NPT), which is very likely to the IP Prefix trie [9, 13] structure. NPT (middle part in Figure 7) makes the PIT organized and more manageable, but it does not help much in shrinking PIT size and accelerating PIT access frequency.

We propose to assign a code (an integer) to each component of the name, i.e., the Name Component Encoding (NCE) method, and use the codes to build the NPT, called the Encode Name Prefix Trie (ENPT): The codes are utilized to conduct lookup, delete and update on ENPT (rightmost part in Figure 7) to make PIT access faster. We should keep in mind the following problems when adopting NCE:

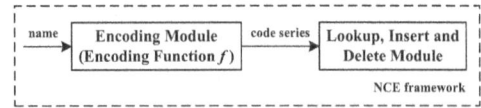

Figure 6: The NCE framework.

1. High-speed Longest Prefix Match (LPM) lookup, insert and delete, which are the major objectives of the NCE mechanism.
2. Fast component encoding. When a packet arrives, the name's components must be assigned codes before starting the longest prefix matching. Therefore the speed of component encoding should be no slower than the lookup, insert and delete speed.
3. Low memory cost. An effective encoding-based method should reduce the total memory cost of PIT as well.

The encoding-based solution was first introduced in our previous work [14]. In this paper, we only adopt the *encoding* idea, the ways to allocate codes, lookup, insert and delete is totally different. Based on our solution, when an NDN name arrives, its components are extracted and each of them is assigned a code. Then the code series are used to lookup, delete and update the PIT, as shown by Figure 6.

4.1 Comparing With Pattern Matching

Intuitively, our problem is to look for a string among a set of strings, which is very similar to the pattern matching problem at the first glance. However, there are some fatal differences.

1. Pattern matching checks if a string contains a pattern (or more than one patterns), beginning at arbitrary position of the string. However, we lookup the NDN names according to the LPM rule, which begins at the first character of the input string. Moreover, pattern matching resolves at character granularity, while our problem is of component granularity.
2. The patterns in conventional pattern matching problems are fixed or update at a very low frequency. While in our problem, the names in the PIT are all patterns, and the patterns are inserted and removed at a very high frequency, which will lead to frequent reconstruction of the Deterministic Finite Automaton (DFA) if adopted.

Therefore, traditional pattern matching solutions may not apply here.

4.2 Name Prefix Trie for Name Lookup

An NDN name is composed of explicitly delimited components, its hierarchical structure inspire us that it can be represented by Name Prefix Trie (NPT), a data structure very similar to IP Prefix Trie, but it is not necessarily a binary tree. A sample PIT with 9 names is shown in the leftmost part Figure 7, its corresponding NPT is illustrated by the middle part of Figure 7, with its edges standing for name components and nodes representing lookup states. The NPT is of component granularity, rather than character or bit granularity, since the longest name prefix lookup of NDN names can only match a complete component at once, i.e., no match happens in the middle of a component.

Name prefix lookups always begin at the root. When an Interest packet arrives, its name is extracted and the Longest Prefix Match lookup starts on NPT. The process is as follows: we first check if the name's first component matches one of the edges originated from the root node, i.e., the level-1 edge. If so, the transfer condition holds and then the lookup state transfers from the root node to the pointed level-2 node. The subsequent lookup process proceeds iteratively. When the transfer condition fails to hold or the lookup state reaches one of the leaf nodes, the lookup process terminates and outputs the index that the last state corresponds to.

---

[2]Each bifow is identified by the five tuple of $< src\_ip, dst\_ip, src\_port, dst\_port, protocol >$.

[3]We take 60 snapshots and show the first 20 of them.

Snapshot	Un-responded HTTP requests	P2P (TCP)	P2P (UDP)	Multimedia (TCP)	Multimedia (UDP)	Email (SMTP, POP, IMAP)	Online Game	FTP	Conference	Network Management	Services
1	296964	290259	802510	24124	10981	41680	23815	505	32160	1383	245229
2	297903	300391	802879	24159	10920	42447	23382	510	32214	1363	244204
3	295908	306238	806185	23602	10959	42102	22314	504	31872	1416	233500
4	293035	306394	803680	23719	11073	39462	22128	508	31246	1383	234829
5	289425	313038	814347	24478	11238	39529	23014	505	30771	1422	239554
6	296794	314074	802497	24540	11241	40849	23306	505	30528	1383	238147
7	290364	315705	796327	24280	11094	42456	22051	505	30970	1405	238524
8	292213	317899	794001	24691	11227	43039	22536	507	30711	1350	229399
9	289825	318622	796744	23895	11275	42036	22099	504	31290	1414	227809
10	289100	319768	792193	24174	11113	41097	21579	507	30385	1407	227809
11	291704	320269	800449	24066	11295	41310	22426	507	30120	1375	233416
12	284372	315984	808876	24435	11199	39768	22207	510	30375	1401	226281
13	281394	316593	800140	23904	11259	38079	21801	511	28992	1339	221280
14	285902	317544	785419	24022	11140	37252	20238	511	28993	1321	226810
15	282457	317049	798048	24118	11376	37377	21867	519	29445	1365	229740
16	281691	315820	789139	24081	10525	39309	22384	510	29296	1396	223627
17	284215	318627	794436	24280	10450	38898	22311	517	29481	1389	217819
18	283341	321657	809278	24046	10674	39459	22500	511	29532	1434	225202
19	283813	324276	817171	24126	10788	39729	22462	517	29967	1471	225826
20	281163	322878	800040	24111	10702	38428	22774	513	30255	1503	210875

Table 2: Lookup, insert and delete frequency.

Time (s)	1	2	3	4	5	6	7	8	9	10
Read frequency	1,409,082	1,409,477	1,411,578	1,406,273	1,409,961	1,408,013	1,409,447	1,409,731	1,410,433	1,409,555
Insert frequency	90,278	88,513	89,561	90,597	90,376	88,821	90,457	89,828	90,856	91,079
Delete frequency	91,422	92,264	90,296	89,132	90,280	90,614	90,805	90,303	90,475	90,394

4.3 Name Component Encoding (NCE)

NPT makes names in the PIT well organized and manageable, but it cannot contribute much to reducing PIT size or greatly improving PIT access performance.

In this section, we propose the NCE solution to solve the challenges. Each name component is encoded as an integer (code), and the bits allocated for a code is dependent on the amount of integers used. In this paper, a 32-bit integer for each code is sufficient. Therefore, the NPT will be encoded as Encode Name Prefix Trie (ENPT), whose edges stand for codes of name components, as shown by the rightmost part of Figure 7. ENPT is a logical structure and we then construct the State Transition Arrays (STA) – a data structure similar to Adjacency List that stores a graph, but more concise – to implement the ENPT, which significantly shrinks the size of NPT and enables fast lookup, insert and delete. Once an NDN names arrives, it is encoded to a series of codes, and the name is looked up against, inserted to, or deleted from PIT based on these codes. Moreover, a Code Allocation Mechanism is also designed to assign each component a (dynamic) code, which eliminates the potential size explosion problem of PIT.

As Figure 7 shows, the given 9 names can be organized as an NPT with 14 nodes. Different components (edges) leaving the same node should be encoded differently to distinguish themselves. A straightforward method is to assign unique codes to all the components in NPT, ranging from 1 to N, N is the number the edges in the NPT. However, in our solution these unique codes will not help accelerate the PIT access speed. Moreover, since the amount of edges in an NPT can be very large, unique codes will lead to codes of large numerical values and require more bits to store them.

We define the edges in the NPT leaving from the same node as a Code Allocation Set (CAS). which are illustrated by the dotted ellipse on the NPT in Figure 7. We propose that we allocate continuous unique codes within each CAS separately, as depicted by RULE 1.

RULE 1. *Assign each name component in a CAS a unique code. The codes should be as small and continuous as possible within each CAS.*

By default, the codes start from 1 within each CAS. After encoding each CAS, we arrive at the ENPT, which is shown by the rightmost part of Figure 7. Suppose that a CAS is composed of edges originated from node i, we denote this CAS as CAS i. By this method, components of the same level but in different CASes may have the same code, e.g., component "yahoo" in CAS 2 and "baidu" in CAS 9 share a common code 1. And the same component in different CASes may be assigned different codes, e.g., CAS

2 and CAS 9 both contain the component "google", but "google" in CAS 2 is assigned code 2 while "google" in CAS 9 is encoded as 3. These two cases, which can be called code assignment collisions, will not bring any negative effects to the name lookup, however. The latter case is in fact how to allocate codes to components, and will be discussed in Section 4.6. Now we assume that each component has been assigned a code based on RULE 1. We then prove the first case will not lead to lookup conflict. (A conflict arises when matching different level-i components of two different names, the lookup states transfer to the same state/node.) The proof is by contradiction.

PROOF. Given two names $C_1 C_2 \cdots C_i \cdots C_n$ and $C'_1 C'_2 \cdots C'_i \cdots C'_n$ with n level components, and they are encoded to $E_1 E_2 \cdots E_i \cdots E_n$ and $E'_1 E'_2 \cdots E'_i \cdots E'_n$, respectively. Their corresponding lookup paths are $N_0 N_1 N_2 \cdots N_i \cdots N_n$ and $N_0 N'_1 N'_2 \cdots N'_i \cdots N'_n$ (N_i represents a node in the ENPT, and N_0 is the root). Component $C_i \neq C'_i$, and are assigned the same component, i.e., $E_i = E'_i$. Assume that there is a conflict after matching C_i and C'_i, i.e., the lookup states both transfer to the same state/node, thus $N_i = N'_i$. And because (allowing for) $E_i = E'_i$, then $N_{i-1} = N'_{i-1}$. Therefore component C_i and C'_i belongs to the same CAS $i-1$. According to our code allocation algorithm, each component within a CAS is assigned a unique code, due to $E_i = E'_i$, we get $C_i = C'_i$, which contradicts previous assumption $C_i \neq C'_i$. □

4.4 State Transition Arrays for Encoded Name Prefix Trie

ENPT is a logical data structure, and is implemented by State Transition Array (STA), as depicted in Figure 8. (CAS i corresponds to Transition$_i$ of STA.) STA is composed of Transition arrays, each array of them, say Transition$_i$, stands for a state i and its children in the ENPT, as well as edges originated from state i. If there is a name prefix match at state i, a pointer to the corresponding PIT entry is included in Transition$_i$.

Figure 8 also shows the process of looking up "cn/google/maps", which is encoded as "2/3/1". The lookup process always begins at the root node, i.e., Transition$_1$. Searching in Transition$_1$ for code 2, the next state, 9, is obtained. Then we transfer to Transition$_9$ and continue to search for the second code – 3. The lookup process iterates like this and finally reaches Transition$_B$, where the pointer to the PIT is stored. By now, we successfully finds the PIT entry and the lookup process terminates. This lookup process make use of the codes (integers) to find a valid transfer from one state to another, compared to matching a component (character string) of variable length, matching an integer is much more easier.

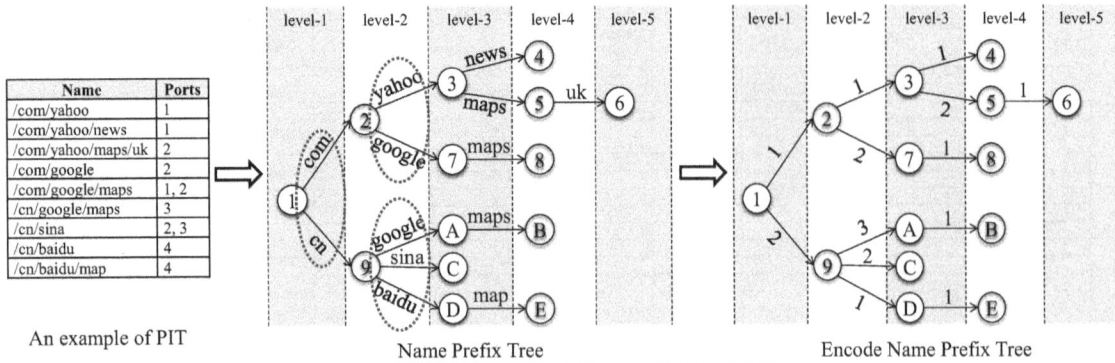

Figure 7: PIT, NPT and Encode NPT (ENPT).

An example of PIT

Name	Ports
/com/yahoo	1
/com/yahoo/news	1
/com/yahoo/maps/uk	2
/com/google	2
/com/google/maps	1, 2
/cn/google/maps	3
/cn/sina	2, 3
/cn/baidu	4
/cn/baidu/map	4

Name Prefix Tree

Encode Name Prefix Tree

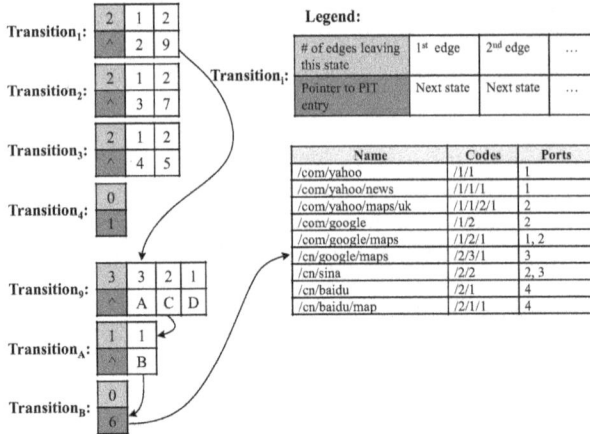

Figure 8: The STA data structure.

Name	Codes	Ports
/com/yahoo	/1/1	1
/com/yahoo/news	/1/1/1	1
/com/yahoo/maps/uk	/1/1/2/1	2
/com/google	/1/2	2
/com/google/maps	/1/2/1	1, 2
/cn/google/maps	/2/3/1	3
/cn/sina	/2/2	2, 3
/cn/baidu	/2/1	4
/cn/baidu/map	/2/1/1	4

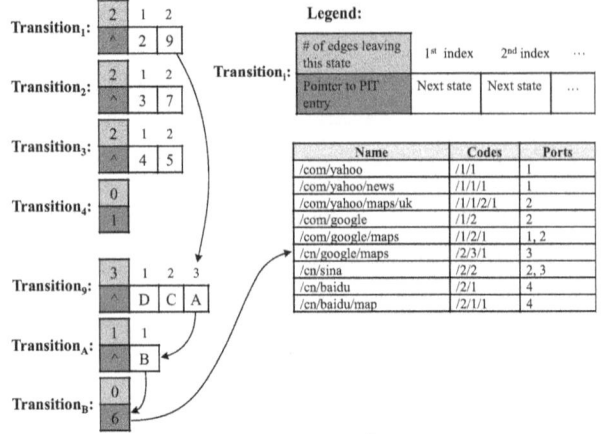

Figure 9: The Simplified STA (S^2TA) data structure.

Name	Codes	Ports
/com/yahoo	/1/1	1
/com/yahoo/news	/1/1/1	1
/com/yahoo/maps/uk	/1/1/2/1	2
/com/google	/1/2	2
/com/google/maps	/1/2/1	1, 2
/cn/google/maps	/2/3/1	3
/cn/sina	/2/2	2, 3
/cn/baidu	/2/1	4
/cn/baidu/map	/2/1/1	4

However, this matching method still requires linearly searching a code in a Transition$_i$, which brings low frequency. To relieve this problem, we further propose to directly locate the next state by the code, videlicet, taking code as an index of the Transition array. By this scheme, the STA data structure is simplified to what is shown by Figure 9, and we name it Simplified STA (S^2TA). The codes are no longer stored in the STA, but act as the indexes to locate the next lookup state. The lookup process now can simply be implemented by sequentially accessing four Transition elements: Transition$_1$[2], Transition$_9$[1], Transition$_A$[1] and the *Pointer to PIT entry* field of Transition$_B$. Significant advantages of this scheme include: 1) no need to move data when inserting and deleting names, 2) no complicated memory management involved. By this means, the required storage by S^2TA is reduced compared to the STA in Figure 8, and simultaneously the access frequency is markedly improved! Therefore, PIT is conceptually transferred to ENPT, and eventually implemented by S^2TA.

However, this method calls for strict requirements on the codes, which should be as continuous as possible and starting from a code of a numerical value as small as possible. Otherwise the memory consumption of PIT can be enormous. We address this problem in next subsection.

4.5 Dynamic Code – A solution to potential PIT explosion

We have demonstrated the benefits that NCE and S^2TA bring, but there is still one problem before the actual deployment of PIT. Because PIT is quite dynamic, though the number of PIT entries is relatively stable, the names are inserted (Interest arrives) and deleted (Data arrives) at a high frequency, which has been shown by the evaluation results in Section 3.2. Names arrive disorderly, therefore as well as the name components since names are composed of components. RULE 1 implies that, within each CAS, the code assigned

to a specific name component is unique, as well as *consistent*. Assume that a name component C_m of CAS i arrives, and C_m has a consistent code 1, then Transition$_i$[1] in the S^2TA will be occupied by the *next state* of C_m. Immediately following C_m, another component C_n in CAS i also arrives, which has a consistent code 1000, and Transition$_i$[1000] will also be occupied by the *next state* of C_n. Therefore, elements from Transition$_i$[2] to Transitions$_i$[999] are all wasted. The worst case is, names are all composed of new components, if we assign a consistent and unique code to each component, the numerical value of codes will increase to be extremely large. Because we utilize codes as indexes of the Transitions in S^2TA, after a specific name is deleted, the corresponding elements in S^2TA can not be used by other names, cumulatively the actual memory consumed by PIT will be quite huge!

Due to the above situation, though PIT contains 1.5 M valid entries, the memory actually consumed by the S^2TA can be extremely large. Therefore, only using RULE 1 is impractical to deploy. To address this problem, we propose assigning *dynamic* codes to components, which is summarized as RULE 2.

RULE 2. *Assign each name component in a CAS an available code. An available code means this code has not been assigned to a name component, or is freed by a leaving name component.*

RULE 2 also ensures the code assigned to each name component in a CAS is unique. RULE 2 proposes that we assign dynamic (and maybe different) codes to the same component at same level in the ENPT while they arrive at different time. In fact, we can view each CAS as a code pool. When an Interest packet arrives, we encode its name by selecting the available codes in the CAS. Assume that the encoding function is f, which takes component as one of its parameters and returns a code. (f will be discussed in detail in Section 4.6.) Each time f is called for each component of the **Interest name**, it picks up the smallest available code for this component and writes the <component, value> pair to a hash table. As the responding **Data** packet returns, f returns the same code series for its name. After the Data packet name is looked up and deleted from the S^2TA based on its code series, all the codes are freed and denoted as available. These codes will be reused for subsequent Interest and Data packets. An example is as follows (refer to Figure 7): an Interest with name "cn/sina" comes, and its name is encoded to "2/2". After the responding Data packet returns, the second level code (2) in CAS 9 is freed (assume that the first level code, 2, is occupied by another names, such as "cn/google/maps"). When another Interest name comes, e.g., "cn/yahoo", "cn" is still encodes as 2. For "yahoo", if consistent code allocation is adopted, it should be encode to 4. However, by dynamic code method, we find that in CAS 9, code 2 is an available code, and consequently assign 2 to "yahoo". It's worth pointing out that, a freed code also indicates that the corresponding element of the CAS in S^2TA is vacant, and thus this element can be reused to save memory consumption. Evaluation results show that the dynamic code method effectively reduces the amount of codes within each CAS and makes the codes as continuous as possible. Therefore, the numerical value of the largest is reduced, as well as the memory actually consumed by the STA.

Adopting RULE 2 involves searching for an available code for a component within a CAS, which will bring extra time cost compared to consistent code allocation. Assume that the largest code of a CAS is N, then the worst time complexity of searching a CAS is $O(N)$. However, evaluation results show that 75% search operations successfully return after only one try (edges to leaf nodes in ENPT), and 10% search operations successfully return with less than 5 tries. At last the average time complexity is approximately $O(\frac{N}{4})$.

4.6 Code Allocation Function f

Previously, we assume that the codes, either consistent or dynamic code, are correctly assigned to components. In Section 4.3, we have found that identical component may be assigned multiple codes. However, given a specific component in a specific name, only one code is correct for that component. In this subsection, we will present how to allocate correct codes to components, videlicet, how to implement the function f mentioned in Section 4.5.

Based on the fact that components of domains are separated by special delimiters, we can get which level a given component belongs to. We define a function $f(component, level, preceding\_code)$ that maps a component to its appropriate code by a hash function, which is borrowed from Python. f takes three parameters, the first is the component that is to be assigned a code, the second is its level in the whole name, and the third is the code of its preceding component. f returns the correct code of current component by hashing. If current component is the first component, $preceding\_code$ is set to 0. Note that f behaviors differently for Interest name and Data name. For Interest name, f assigns available codes to components within each CAS separately. As a result, identical component in different CASes may have different codes. For Data name, f is responsible for finding the correct code for a component that has multiple codes. An example of code assignment is as follows. Suppose that the looked up name is still "cn/google/maps" (refer to Figure 7). First we take "cn/google/maps" as an Interest name. "cn" is encode as 2 by invoking $f("cn", 1, 0)$, then $f("google", 2, 2)$ is invoked to encode "google", the second argument, 2, indicates that this is a second level component, and the third argument, also 2, indicates the branch or subtree that current component belongs to, and further figures out which CAS this component belongs to. In this example, "google" belongs to CAS 9. Assume that the code 3 is available within CAS 9, thus f returns 3 as the code of "google", and writes 3 to the appropriate entry of the hash table. Similarly, "maps" is

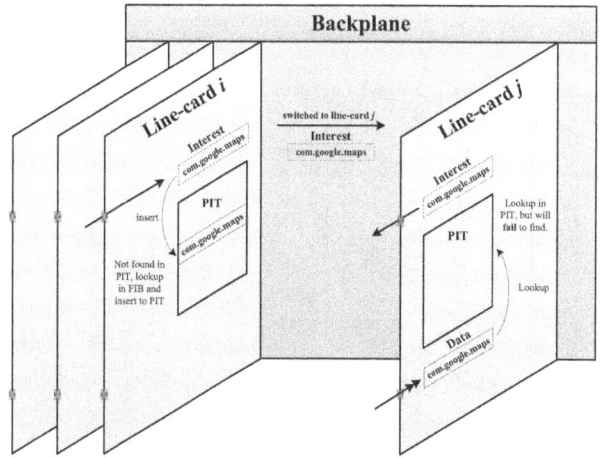

Figure 10: Place PIT on incoming line-cards.

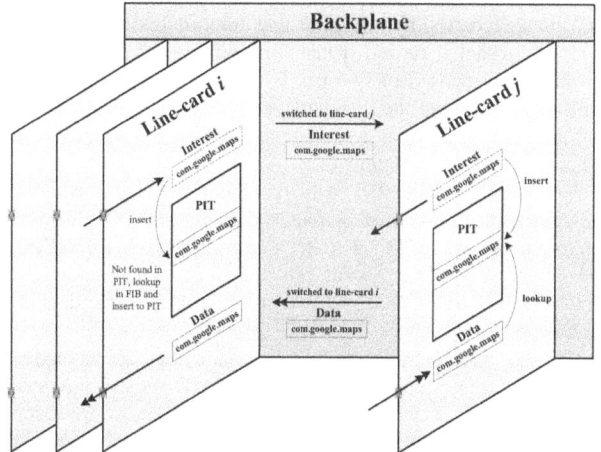

Figure 11: Place PIT on both incoming and outgoing line-cards.

encoded to 1 by invoking $f("maps", 3, 3)$. Assume that the arrival of another Interest name, say "com/google/maps", makes "google" be assigned another code 2 in CAS 2, the code 2 is appended to the appropriate hash table entry. Consequently, as a level-2 component, "google" has two corresponding codes – 2 and 3.

Next we take "cn/google/maps" as a Data name. Obviously, $f("cn", 1, 0)$ encodes "cn" to 2. $f("google", 2, 2)$ first hashes "google" to the appropriate entry of hash table and finds that "google" has two corresponding codes (2 and 3), then f recognizes that this "google" belongs to CAS 9 and returns code "3". (Of course, information about which CAS each code belongs to is also maintained.) At last, $f("maps", 3, 3)$ returns 1.

Though at level 2, "google" has two corresponding codes (2 and 3), but this will not bring any negative interference when selecting a correct code for "google". A potential drawback is, the code of the i-th component depends on the code of the $(i-1)$-th component, which makes the encoding process of each component in a name sequentially executed and may degrade the throughput. However, evaluation results show that, by a four-module accelerated hash-based encoding function f, this is not the bottleneck of the system.

5. WHERE PIT RESIDES?

Figure 1(a) and Figure 1(b) illustrate the Interest and Data packet forwarding process within a router, from which the PIT lookup and update process can be derived as well. NDN takes PIT as a global table and conceptually assume that all the Interest and Data packets can access that table, which is, however, impractical to implement in current router architecture. The router architecture is illustrated in Figure 10, with multiple line-cards plugged in the backplane.

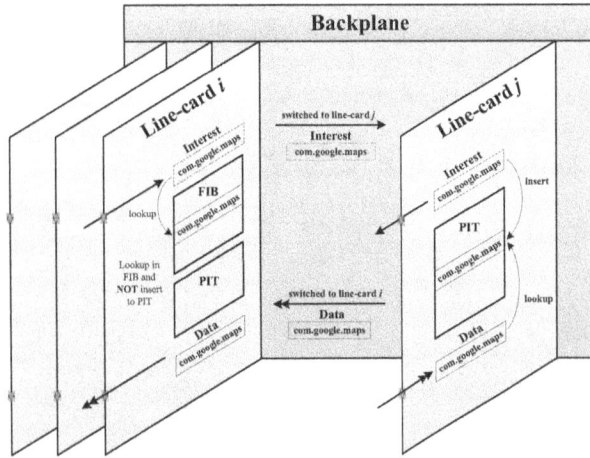

Figure 12: Place PIT on outgoing line-cards.

Packets are received by line-cards and are switched to another line-card via the backplane and switch fabric (not shown). Conceptually, a router is composed of two parts: data plane and control plane. Data plane is simply the line-cards, and the control plane is a CPU that draws network topology and computes FIB. For brevity, control plane is not shown in Figure 10.

When implementing PIT, a fundamental problem is where and how to reside PIT in current router architecture.[4] Obviously, it's impossible to place PIT on router's control plane since delivering every Interest and Data packet into control plane is not a wise idea and costs a fortune. Therefore, we place the PIT on the data plane (line-cards).

The most straightforward way is that each line-card maintains its own PIT – creates a corresponding PIT entry for every incoming Interest packet and removes the associative PIT entries for received Data packets, as shown by line-card i in Figure 10. However, this means will encounter serious problems. The first problem is, suppose that an incoming Interest packet I arrives at line-card i, and line-card i looks it up in its PIT first. If not found, an entry is inserted into i's PIT for this Interest. Then the Interest is switched to line-card j and further forwarded to the downstream routers. When the responding Data packet returns, due to the symmetric routing property, the Data packet will be received by line-card j. However, line-card j fails to find the Data packet in its PIT because the corresponding PIT entry is on line-card i, thus this Data packet does not know where to go and will be discarded. The second is problem, imagine that immediately after Interest I arrives at line-card i (I is not responded yet), an identical Interest I' arrives at line-card k for the first time, and line-card k looks it up in its PIT, and will certainly not find it since line-card k has never received an Interest identical to I' before. But according to the design of NDN, Interest I' should be found in PIT since an identical Interest I has been received by this router and is not responded yet. If we really want this, line-card i has to send I's corresponding PIT entry to all the other line-cards for synchronization after the entry is created, which will definitely consume a lot of resources and incur extra burden on the router.

A possible solution is to create entries for an Interest on both incoming and outgoing line-cards, as illustrated in Figure 11. When an Interest packet arrives at this router, line-card i and j both insert its carried name into their own PITs. When the responding Data packet comes back to line-card j, line-card j looks up the name of the Data packet in its PIT, obtains the proper destination interface and forwards it to line-card i. Thus this method solves the first

---

[4]Though NDN is a clean-slate network architecture, we believe that it will not make significant modifications to the router architecture.

aforementioned problem. But for the second problem above, Interest I' still cannot be found in line-card k's PIT, and will be forwarded to line-card j. However, if we lookup I' in line-card j's PIT, I' is there! This fact inspires us to lookup Interest names against the PITs on the outgoing line-cards. Therefore, the PIT entries on the incoming line-cards are over-provisioned. In Figure 11, the PIT entry for the incoming Interest "*com.google.maps*" on line-card i is redundant since it will never be looked up.

Consequently, we propose to only place the PIT entries on (egress channel of) the outgoing line-cards. This means each line-card only have to create PIT entries for outgoing Interests (in the egress channel) that are switched to itself from other line-cards, rather than the incoming Interests (in the ingress channel) from the outside, as depicted by line-card j in Figure 12. By this design, the Interest lookup and forwarding process slightly changes, as illustrated in Figure 1(c). (For brevity, the CS, which is a global buffer shared by all the line-cards, is not shown in Figure 12.) For a Interest packet that comes in through interface x of line-card i and goes out through interface y on line-card j:

1. line-card i checks if CS has cached the desired data chunk, if so, returns a copy by a Data packet,
2. otherwise, looks up the Interest against FIB for the outgoing interface y, and switches it to line-card j,
3. line-card j checks if PIT has an entry for this Interest, if so, appends x to this entry and discards this Interest,
4. otherwise, creates a PIT entry and fills it with Interest name and the arrival interface x. Then forwards the Interest to the interface y.

Data packet is looked by by the line-card that receives it, and the lookup and forwarding process does not change.

6. EVALUATION

6.1 Experimental Setup

We measure the performance of NCT and ENCT on the platform: Intel Xeon E5520, 2.27 GHz, 15.9 GB RAM.

The one-hour trace is captured from a 20 Gbps gateway link in CERNET, 17:00~17:59, Dec. 21st, 2011. The domain names are collected from ALEXA, DMOZ and our web crawler, and 9,834,747, about 10 million, domain names are collected in total. We also extracted 7,624,393, around 8 million, real URLs from the HTTP GET and HEAD requests in the trace. For brevity, we refer to these two name sets by 10M Name Set and 8M Name Set, respectively. The statistics of these two Name Sets are presented in Table 3.

The evaluation can be generally divided into two parts: 1) The size and access frequency of PIT, which have been previously illustrated by Table 1 and Table 2 in Section 3; 2) The performance of our proposed encoding-accelerated PIT access scheme (NCE and S^2TA), such as memory consumption, access frequency, and comparison with other methods.

6.2 Evaluation Results

6.2.1 Memory Usage

For the two Name Sets, we first measure their memory consumption of the: 1) original size, i.e., directly store the names as character strings in a table, 2) NPT, 3) ENPT (S^2TA) + hash table. (Hash table is required by function f.)

The overall results are shown in Table 3, more detailed results are given in Figure 13 and Figure 14. Some facts can be derived from these statistics. By comparing the *# of edges/components* in the NPT with the *# of total components* in Table 3, we find that 8M Name Set is more aggregatable than the 10M Name Set. Moreover, besides

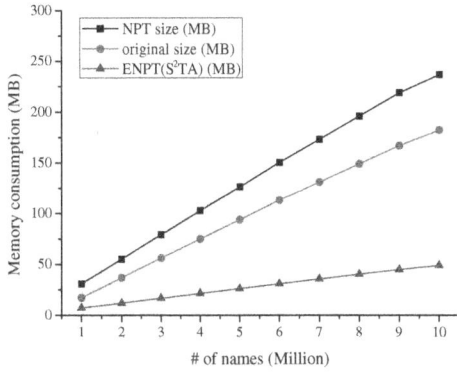

Figure 13: 10M Name Set memory consumption–original size, NPT size, ENPT size.

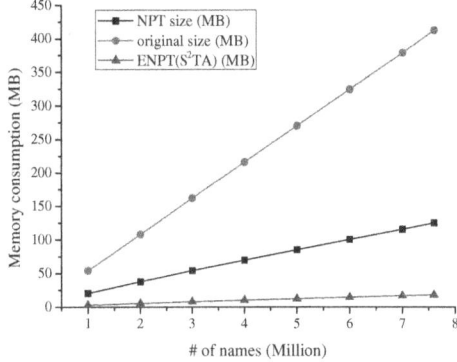

Figure 14: 8M Name Set memory consumption–original size, NPT size, ENPT size.

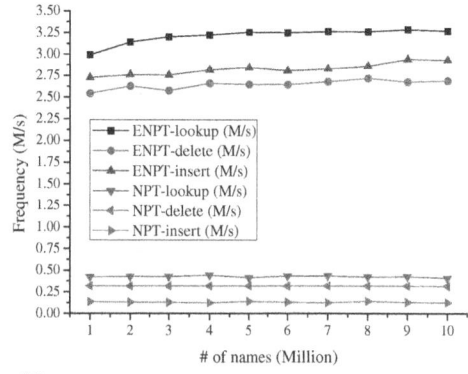

Figure 15: Lookup, insert and delete performance for the 10M Name Set.

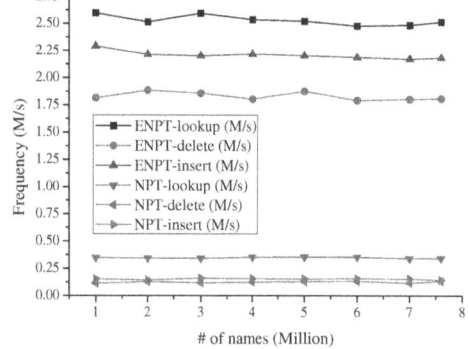

Figure 16: Lookup, insert and delete performance for the 8M Name Set.

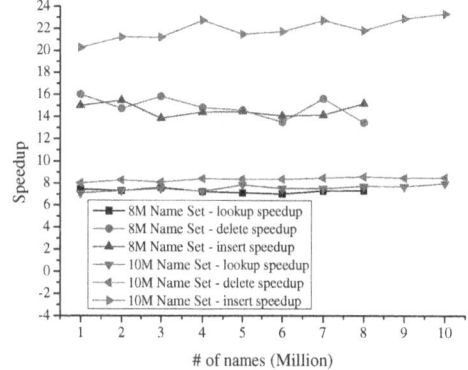

Figure 17: PIT access frequency speedup based on ENPT.

the name components, each NPT node stores additional information, such as state number, pointer to children, pointer to parent, etc. Therefore, the NPT of 10M Name Set is even larger than its original size. But these additional information makes NPT easier to manage than storing names directly.

As mentioned in section 4.6, we assign codes to components by invoking function f, which in turn resorts to a hash function. Thus a hash table is required, for the 10M Name Set, the size of hash table is approximately 67.11 MB, and around 33.55 MB for 8M Name Set. At last, the compression ratio of the 10M Name Set and 8M Name Set is 63.66% and 12.56%, respectively.

Please keep in mind that, ENPT is a logical data structure and we do not implement it directly. However, it is implemented by S^2TA. We can think of operations on the ENPT, but we actually operate the S^2TA to carry out those operations.

6.2.2 Lookup, insert and delete performance

Subsequently, we measure the PIT access (lookup, insert and delete) performance by ENPT. We do not measure the *update* performance, because updating PIT means appending interfaces to a specific PIT entry, whose performance it very close to that of lookup.

Deleting a name involves backtracking a path, which is easy to implement by NPT since NPT has parent pointers. The deleting process in ENPT is: first go straight to the leaf node, delete the leaf, then backtrack to see if the parent node still needs to be removed, so and so forth. (If a parent node's all the children are deleted, and it does not have a pointer to the PIT entry, it needs to be removed too.) Because there is no such pointer to parent node in S^2TA, we keep track of the node information along the path to a specific leaf when deleting a name. We do not actually delete the node and free the space of ENPT, in fact, deleting a node means the code of its preceding edge/component is freed, and that code can be reused within its CAS, as well as the corresponding space in S^2TA. (Refer to function f.)

The lookup, insert and delete performance for the two Name Sets are illustrated in Figure 15 and Figure 16, respectively. The lookup, insert and delete performance on the 10M Name Set can achieve 3.27 M/s, 2.93 M/s and 2.69 M/s, respectively, and that on the 8M Name Set achieve 2.51 M/s, 1.81 M/s and 2.18 M/s, respectively.

Obviously, the PIT access frequency based on ENPT is phenomenally promoted compared with that of the NPT, which is more clearly depicted by the speedups in Figure 17.

6.2.3 Code allocation function f

Each incoming name will be decomposed to multiple name components and each component will be assigned a code by invoking function $f(component, level, preceding\_code)$, which in turn calls a hash function. The hash function is borrowed from Python. The way how f works has been discussed in Section 4.6. Because this step of encoding components is just right before the step of lookup, insert or delete, the encoding performance should be no less than the lookup, insert or the delete performance. Otherwise, the encoding step will be the performance bottleneck. We utilize 4 parallel modules to improve the encoding throughput, and the performance

221

Table 3: Name Component Statistics of Two Name Sets.

URL set	# of names	# of total components	average component length (Byte)	average # of components per name	original size (MB)	# of components /edges in NPT	NPT size (MB)	ENPT size (MB)	ENPT+Hash Table size (MB)	compression ratio
10M Name Set	9,834,747	24,808,603	7.35	2.52	182.26	12,228,081	236.57	48.91	116.02	63.66%
8M Name Set	7,624,393	26,882,827	15.35	3.53	412.78	4,570,563	125.03	18.28	51.83	12.56%

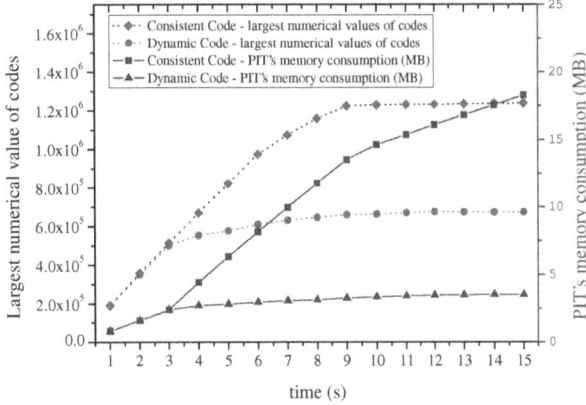

Figure 18: Largest numerical values of codes and PIT's memory consumption.

is 20.67 M components per second and 18.96 M components per second for the 10M Name Set and 8M Name Set, respectively. Divide them by the average number of components per name, we further compute the encoding performance: 8.20 M names per second and 5.37 M names per second for the 10M Name Set and 8M Name Set, respectively. Therefore, the encoding performance is better than the lookup performance and will not be the performance bottleneck.

6.2.4 Results of dynamic code

The drawback of assigning consistent codes to components has been discussed in Section 4.5. To demonstrate the effects of dynamic code, we replay the HTTP packets in the captured trace to mimic the packet (name) incoming and outgoing process, which will lead to a PIT of around 300 K (refer to Table 1) entries, and measure how large the numerical value of the codes will be. For comparison, both consistent and dynamic code method will be measured, as well as the PIT's actual memory consumption. The result is shown by Figure 18, the dotted curves represent the largest code of all the CASes, which show that as times goes on, names keep coming and going, the largest code increases. For consistent code, the largest keeps increasing at a high rate after the PIT reaches 300 K valid entries, thus the consumed memory of PIT increases as well. However, for dynamic code, after the PIT reaches 300 K valid entries, the largest code greatly slows down its increasing pace, making PIT's memory consumption remains stable (solid curves). In fact, the largest code by RULE 1 at each snapshot is the number of total components observed by a CAS until this snapshot, while the largest code by RULE 2 is the amount of components a CAS contains at each snapshot. The PIT's memory consumption exhibits similar growth law of the codes. The hash table size (not shown in Figure 18) is almost the same for both consistent and dynamic code methods, since the received names are the same, and thus the name components. The hash table size is 33.55 MB (for the 8M URL Name Set). Therefore, with NCE and dynamic code, PIT exhibits good scalability.

7. RELATED WORK

This section compares our NCE solution to our previous work in [14], and we name it Original NCE. In fact, this paper only continues the encoding idea, but the ways to assign codes, lookup, insert and delete are different. We conclude three major distinctions: 1) The data structure to implement the ENPT in Original NCE involves complicated memory management, such as data movement,

fragment management; 2) Original NCE allocates consistent codes to components and does not allow identical components be encoded to different, dynamic codes, which fundamentally contradicts with the code allocation function f in this paper; 3) Original NCE only achieves lookup speedup, but does not exhibit good support for insert and delete operations. However, in this paper, NCE not only significantly accelerates lookup, but also insert and delete operations.

8. CONCLUSION

NDN/CCN propose that PIT caches yet un-responded Interests, when the responding Data packets return, the names are removed from PIT. PIT brings significant features to NDN/CCN. However, none has conducted a measurement study to show the size and access requirements of PIT. Without these knowledge, we have no data to support the design of NDN routers or the actual deployment of NDN. In this paper, we are the first address three problems associated with the PIT: 1) the size and access (lookup, insert, delete) frequency of PIT; 2) how to address the large size and high access frequency problem with a scalable solution; 3) where does PIT reside within a router.

We emulate NDN's application-layer working paradigms by transferring the existing IP applications to the NDN platform. By mapping/translating a captured 20 Gbps gateway trace from IP to NDN scenario at the application perspective, we quantify the size and access frequency of PIT, which demands an efficient and scalable solution. Therefore, NCE is proposed to accelerate the access throughput of PIT, as well as to reduce its size. Moreover, the dynamic code allocation technique makes the NCE solution practical, and further keeps the actual memory consumption of PIT stable. At last, we propose to place PIT on the packets' outgoing line-cards (egress channel) when actually implementing PIT, which meets the PIT design in [15] and eliminates the cumbersome synchronization problem among multiple PITs on line-cards.

9. REFERENCES

[1] HTTP/1.1 RFC. http://www.ietf.org/rfc/rfc2616.txt.

[2] l7-filter. http://l7-filter.clearfoundation.com.

[3] TIE. http://tie.comics.unina.it.

[4] Tstat. http://tstat.tlc.polito.it.

[5] H. Balakrishnan, K. Lakshminarayanan, S. Ratnasamy, S. Shenker, I. Stoica, and M. Walfish. A layered naming architecture for the internet. In *Proc. of SIGCOMM*, 2004.

[6] A. Dainotti, W. de Donato, and A. Pescapé. TIE: A community-oriented traffic classification platform. In *TMA'09*, May 2009.

[7] C. Esteve, F. L. Verdi, and M. F. Magalhaes. Towards a new generation of information-oriented internetworking architectures. In *Proc. of ACM CoNEXT*, 2008.

[8] A. Finamore, M. Mellia, M. Meo, M. M. Munafò, P. di Torino, D. Rossi, and T. ParisTech. Experiences of internet traffic monitoring with tstat. *IEEE Network*, 25(3), May-June 2011.

[9] E. Fredkin. Trie memory. *Communications of the ACM*, 3(9):490–499, Sep 1960.

[10] V. Jacobson, D. K. Smetters, J. D. Thornton, M. Plass, N. Briggs, and R. Braynard. Networking named content. In *Proc. of ACM CoNEXT*, 2009.

[11] H. Jiang and S. Jin. Exploiting dynamic querying like flooding techniques in unstructured peer-to-peer networks. In *Proc. of IEEE ICNP*, 2005.

[12] A. Kumar, J. J. Xu, and E. W. Zegura. Efficient and scalable query routing for unstructured peer-to-peer networks. In *Proc. of IEEE INFOCOM*, 2005.

[13] S. Nilsson and G. Karlsson. IP-Address Lookup Using LC-tries. *IEEE Journal on Selected Areas in Communications*, 17(6):1083–1092, JUNE 1999.

[14] Y. Wang, K. He, H. Dai, W. Meng, J. Jiang, B. Liu, and Y. Chen. Scalable name lookup in ndn using effective name component encoding. In *Proc. of IEEE ICDCS*, 2012.

[15] L. Zhang, D. Estrin, V. Jacobson, and B. Zhang. Named Data Networking (NDN) Project. In *Technical Report, NDN-0001*, 2010.

Coexist: Integrating Content Oriented Publish/Subscribe Systems with IP

Jiachen Chen*, Mayutan Arumaithurai*†, Xiaoming Fu*, K.K. Ramakrishnan‡
* Institute of Computer Science, University of Goettingen, Germany,
Email: {jiachen, arumaithurai, fu}@cs.uni-goettingen.de
† NEC Laboratories Europe, Heidelberg, Germany, Email: arumaithurai@neclab.eu
‡ AT&T Labs-Research, Florham Park, NJ, U.S.A., Email: kkrama@research.att.com

ABSTRACT

Content-Centric Networking (CCN) seeks to meet the *content-centric* needs of users. In this paper, we propose hybrid-COPSS, a hybrid content-centric architecture. We build on the previously proposed Content-Oriented Publish/Subscribe System (COPSS) to address incremental deployment of CCN and elegantly combine the functionality of content-centric networks with the efficiency of IP-based forwarding including IP multicast. Furthermore, we propose an approach for incremental deployment of caches in generic query/response CCN environments that optimizes latency and network load. To overcome the lack of inter-domain IP multicast, hybrid-COPSS uses COPSS multicast with shortcuts in the CCN overlay. Our hybrid approach would also be applicable to the Named Data Networking framework.

To demonstrate the benefits of hybrid-COPSS, we use a multiplayer online gaming trace in our lab test-bed and microbenchmark the forwarding performance and queuing for both COPSS and hybrid-COPSS. A large scale trace-driven simulation (parameterized by the microbenchmark) on a representative ISP topology was used to evaluate the response latency and aggregate network load. Our results show that hybrid-COPSS performs better in terms of response latency in a single domain. In a multi-domain environment, hybrid-COPSS significantly reduces update latency and inter-domain traffic.

Categories and Subject Descriptors

C.2.1 [**Computer-communication Networks**]: Network Architecture and Design

General Terms

Design, Performance

Keywords

CCN, NDN, COPSS, Pub/Sub, Multicast, Incremental Deployment

1. INTRODUCTION

Content Centric Networking (CCN) [1–7] seeks to transform content as a first-class entity. Rather than the current *location-centric* network architecture, CCN presents a networking paradigm that enables users to issue a query for content instead of contacting a specific end host for that content. It allows the network to provide the content from any of the multiple sources where it is available (including the in-network cache), and meets the user's need to send and receive content, irrespective of where it is or who generated it. Named Data Networking (NDN) [4] is one of the more popular examples of CCN.

Publish/subscribe (pub/sub) systems are particularly suited for large scale information dissemination, and provide the flexibility for users to subscribe to information of interest, without being intimately tied to when that information is made available by publishers. This temporal separation between information generation and indication of interest is a highly desirable content centric feature. Our proposal, Content-Oriented Publish/Subscribe System (COPSS) [7,8] is built on top of NDN with a new CCN-oriented multicast capability, to meet the efficiency and scalability needs of large scale pub/sub systems. Examples we address include a content-focused Twitter-like pub/sub system [7] and applications with tight timeliness requirements such as online multiplayer gaming and content streaming [8].

A major component of the NDN design is the extensive use of caching at every hop of the network. Specifically, NDN requires every NDN router process the content request (rather than header-based forwarding) and store the named content to respond to subsequent requests from the cache, in order to achieve better performance than the case if the request for content was just forwarded by the network to ultimately be served from the publisher/source of the content. While the performance penalty of hop-by-hop processing of the content request and response is mitigated by caching and generating the response from the point where the first cache hit takes place, the NDN solution still requires every NDN router to do the complex parsing of the request to forward the request or respond to it. Having a cache at every NDN router is also expensive. As discussed in [7], we demonstrated that supporting content centricity entails significant additional processing in the forwarding engine. Thus, the excitement of the new content centric network architecture has to be tempered by the performance, cost and complexity consequences of the architecture. Furthermore, there is an important aspect on the incremental deployment and co-existence of CCN with the current IP-oriented network world.

In this paper, we examine how to evolve from an IP infrastructure to a CCN-oriented network by co-existing with the IP network. Hybrid-COPSS attempts to support all the *functionality* a COPSS-enhanced CCN network provides (both Query/Response and Publish/Subscribe) and provides users with name-oriented/content-oriented access to information. However, the network exploits the cheaper IP-like forwarding capability where appropriate. Cache hits from the key CCN nodes enable fast response to content requests, but this needs to be balanced against the cost of having a large number of complex CCN nodes. Therefore, additionally, by a judicious choice of placing a limited number of full-fledged CCN nodes that can also cache content at key points in combination with a larger number of hash-based forwarding (similar to IP forwarding), we address the problem of efficient migration to an Information Centric future. The NDN implementation treats CCN as an overlay using TCP/UDP between CCN overlay nodes. However, we believe a tightly integrated approach as proposed here provides the best of both worlds, with COPSS routers at the edge and at selected points, and the core routers in the network only performing IP forwarding. Additionally, we also propose a new FIB propagation mechanism and an incremental deployment strategy to increase the cache hit rate with the limited amount of memory (cache) size. Our contributions in this paper include:

- A detailed design of hybrid-COPSS with both pub/sub and query/response features in a hybrid (IP + CCN) scenario. We address the inter-working of COPSS with IP multicast to achieve both incremental deployment and forwarding efficiency of hash based multicast (similar to IP multicast). We also present how COPSS routers seamlessly integrate an IP network infrastructure and content-centric end hosts. Moreover, we present how CCN nodes with caches could be placed to optimize query/response functionality.

- Addressing the challenges of inter-domain multicast, through hybrid-COPSS's use of CCN overlay nodes at individual administrative domain edges. Moreover our hybrid-COPSS design provides maximum freedom to individual domains with the capability of distributing and managing their limited IP multicast space, while ensuring that global connectivity is maintained.

- An approach to map the very large (potentially unbounded) address space that a large scale CCN network may use, onto a limited IP multicast group address space. The choice of the address mapping has a direct impact on network efficiency. We provide an efficient mapping schema that reduces wasteful network traffic.

- A study of FIB size and cache hit rate in a pub/sub system using incremental CCN deployment. We propose an incremental deployment framework for ISPs and a FIB propagation mechanism to increase the cache hit rate and reduce the FIB size in such environment.

- Evaluating the performance of hybrid-COPSS against COPSS and IP multicast in an experimental test-bed.

We use a gaming trace to microbenchmark the forwarding performance and queueing in our lab test-bed and use a representative ISP topology in a simulation environment to compare the response latency, network traffic and rendezvous point throughput of the alternate approaches. The results show that hybrid-COPSS can be integrated into the current IP network architecture, and significantly improves latency compared to a CCN-only environment. While network traffic can be higher as a result of an extremely limited IP multicast group address space, we observe that even with a larger address space, an IP multicast-like approach can only perform as good as hybrid-COPSS. However, hybrid-COPSS significantly outperforms IP-multicast-like approaches in multi-domain environments, in terms of reducing inter-domain traffic.

Our paper is organized as follows. In §2 we discuss related work and background. Then, we give the basic pub/sub communication of hybrid-COPSS in §3. In §4, we describe our optimization on query/response in hybrid-COPSS. §5 solves the inter-domain multicast problem and §6 discusses scalability and management issues. Evaluation results are reported in §7 before concluding in §8.

2. RELATED WORK

Publish/subscribe systems (*e.g.* [6, 9–14]) are attractive because they relieve the consumers from the strict synchronization in time and location when the information is generated by publishers [15]. Although meant to be content-centric in nature, current systems require the users to know the location or identity of the publisher of the information of interest. In limited situations, information aggregators that collect, index and re-distribute the information in some form remove the burden from the users (and act as a rudimentary content-centric forwarder). Thus, most current pub/sub systems are built on a location-based architecture, which results in inefficiency both in data forwarding and information management at the user end.

NDN [4] is an instantiation of CCN. With NDN, a human readable *ContentName* is used to identify a data chunk. Request and response are represented by two basic kinds of packets *Interest* and *Data*. Information consumers send Interests (queries) to request for content. NDN routers forward Interests towards potential information providers according to the Forwarding Information Base (FIB) (we use $FIB(name)$ to denote the outgoing face(s) in FIB for Content-Name *name*). Interests not matching an entry in the Content Store (CS) are cached at the Pending Interest Table (PIT) of each router they traverse. Information providers subsequently send matched Data (response) packets along the reverse path of the pending Interests towards information consumers. Data packets will be cached at each router's Content Store for future use, and will consume each pending Interest when traversing the path. Proposals such as VoCCN [16, 17] try to support online live streaming and group communication over NDN architecture. However, such systems might incur significant overhead due to the inherent query/response communication model, when a simple pub/sub form of information dissemination is desired.

In [18], the authors motivate the need for channel-like (including push-based) communication in NDN. IP multicast solutions such as PIM-DM [19], PIM-SM [20] and SSM [21] provide multicast group communication to applications like IPTV. However, a lot of ISPs do not support IP multicast, especially for inter-domain traffic, due to complexity and charging concerns. Overlay multicast [22–24] is another alternative for multicast group communication, which allows data to be replicated at hosts along the dissemination structure (tree or mesh) thus saving some of the network traffic and publisher load. Data replication, multicast routing, group management and other functions are achieved at the application layer. Thus, it enables easier deployment without the need to change the current IP infrastructure. However, overlay multicast is agnostic to the underlying topology, likely resulting in forwarding inefficiencies.

COPSS [7, 8], created as an enhancement to NDN, provides an efficient pub/sub capability in the content-centric network architecture and effectively integrates pub/sub with query/response. COPSS supports content-centric features in the naming structure including the ability to include hierarchies. The fundamental component, a Content Descriptor (CD), is used in the CCN-based multicast framework. COPSS uses a Rendezvous Point (RP) based multicast structure, along with automatic RP balancing to avoid traffic concentration. However, the performance of COPSS may be improved in environments where incremental deployment or co-existence with IP networks is desired, by exploiting native IP multicast, to reduce latency. In this paper, we present a complete COPSS overlay based solution that is more efficient and scalable than the solution outlined in [7,8] that consisted of the 1st hop router simply mapping a Content Descriptor (CD) to an IP multicast group. Hybrid-COPSS facilitates the dynamic management of the CD to IP multicast and CD to RP mapping, which helps in reducing the unnecessary traffic being sent in the network. The solution we study in this paper also provides the means for a careful partitioning of the functionality at different nodes so as to allow the nodes to maintain the functionality (as we incrementally deploy) as much as possible when the network finally evolves into pure CCN/COPSS environment.

While we presented the basic philosphy for co-existence and graceful migration in a workshop paper [25], we present in this paper a detailed architecture that handles pub/sub, optimizes query/response with a 2-stage dissemination and RP balancing in a multi-domain scenario so that such RP balancing "affects only the first domain downstream". Furthermore, we provide a thorough evaluation to illustrate the cache hit rate and response latency with the optimized query/response solution. In this paper we look at the total update latency from microbenchmarking to illustrate that the use of an efficient IP forwarding, hybrid-CCN (hybrid-COPSS) can have better performance even in a simple topology. Our approach also offers increased scalability for the pub/sub topology with varying numbers of end-hosts.

3. Hybrid-COPSS PUB/SUB

In this section, we describe the multicast-based delivery model of hybrid-COPSS with our approach for incremental deployment, leveraging an IP network's efficient forwarding. We retain the content centric functionality from both the user's and the end-system's perspective. In [4, 6], the authors proposed that content centric network could be built as an overlay to achieve the CCN's functionality. NDN [4] proposes to use UDP packets to encapsulate NDN packets (Interest and Data) or TCP to transfer NDN protocol messages over an IP network. This links NDN forwarding engines via faces (address:port) and forwards packets on a hop-by-hop basis across the IP underlay. While the COPSS architecture can also be implemented as an overlay, we explore an integrated approach. We allow hybrid-COPSS to provide content oriented functionality that is integrated with the routing and forwarding functionality of an IP network.

To achieve forwarding efficiency for multicast (overlay or IP)-based information dissemination, we seek to reduce the time required for name resolution and complex protocol exchange at every hop in the overlay. Therefore, it is desirable for the heavy-weight COPSS forwarding function to be present only at critical positions and leave intermediate routers to focus only on forwarding. Note that the needs of a query/response system could be different from that of a multicast system. Therefore we do not place strict requirements on where the pure COPSS or pure IP routers need to be placed. In fact, we expect that the COPSS enabled nodes can be used either with their full CCN overlay functionality or with the more limited, but efficient, functionality (consisting of only multicast and IP like forwarding). This design allows a query/response application to utilize the CCN capability of intermediate nodes when and where needed. The overlay-underlay design of the nodes implies that where needed, there are CCN routers that provide query aggregation and caches (thereby the benefit of cache hits).

3.1 Packet Forwarding in COPSS

To provide a flexible publish and subscribe functionality, COPSS adopts a Content Descriptor (CD) based approach where a CD could refer to a keyword, tag or can be combined with a property (e.g., hierarchical structure) of the content. A CD is a human-readable, hierarchically structured string (similar to a ContentName in NDN). However, in NDN, a piece of data is identified by a globally unique ContentName. But, in COPSS a piece of data (e.g.,a document) can have multiple CDs and at the same time there may exist multiple data items (e.g., multiple documents) that are identified by a given CD.

COPSS is designed to use a Rendezvous Point (RP)-based multicast as the basic communication model. To reduce the (well-known) traffic concentration at an RP, several RPs are setup in the network based on the workload, and every RP serves a set of CDs. Considering the problems related to RP balancing and node failure, we do not constrain an RP to reside only at a given physical machine. It is a module identified by a ContentName that can be dynamically placed at a router. RN_{CD} is the ContentName that identifies the module implementing the RP serving the CD. CD to RP mapping is stored in the RP_Table on every edge router ($RN_{CD} = RP\_Table(CD)$). When an RP is setup (or moved), it will propagate the list of the CDs it is responsible for along with its own ContentName throughout the network. The COPSS routers would then have an FIB entry storing the ContentName and the outgoing face towards the RP. We use $FIB(RN_{CD})$ to denote the outgoing face towards the RP that serves CD.

COPSS routers are equipped with a Subscription Table (ST) to store the outgoing faces to reach subscribers downstream. The ST is a dictionary of {Face:Bloom-Filter(CDs)} such that if any CD in a Publish packet has a hit in the

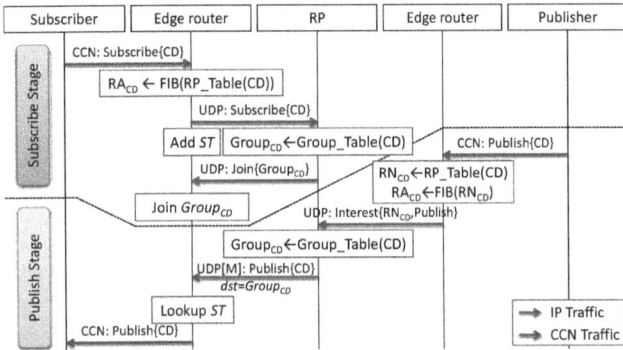

Figure 1: Basic Protocol Exchange

bloom filter for a face, the packet will be sent (and will only be sent once) through that face.

When an end host subscribes to a CD, it sends a Subscribe packet to its 1st hop router. The 1st hop router modifies its own ST and forwards it upstream, i.e., $FIB(RN_{CD})$, after checking if it has not already subscribed to this CD. The upstream routers change their local ST and forward the Subscribe packet until it reaches the RP or a router that has already subscribed to the CD (essentially an 'on-tree' node). On receiving a Publish packet (containing multiple related CDs and the content) from an end host, a 1st hop router encapsulates the Publish packet into an Interest packet with ContentName RN_{CD}. This packet will be forwarded to the RP that serves the CD. The RP recognizes the Interest packet and decapsulates it. The decapsulated Publish packet is forwarded downstream according to the routers' STs until it reaches all the subscribers.

3.2 Packet Forwarding in Hybrid-COPSS

In hybrid-COPSS, we seek to exploit the forwarding functionality of IP in the core of the network to achieve efficiency. The full-fledged functionality of COPSS is present on edge routers (routers directly linked to publishers and subscribers) and at the RPs. The edge routers are directly linked to the RPs on the overlay. The edge routers still maintain RP_Table. But since the FIB stores the {`address:port`} pair as an outgoing face, edge routers can find the address of the RPs seamlessly. Here, we denote RA_{CD} as the address of the RP module that serves CD. The RA_{CD} can be calculated by $FIB(RN_{CD})$ because the RPs and the edge routers are directly linked at the CCN overlay layer. Then, this packet will be forwarded to the RP through the underlay. CD to multicast group address ($Group_{CD}$) mapping is maintained in the Group_Table on the RP ($Group_{CD} = Group\_Table(CD)$). Note that in a multi-domain scenario, the routers at the borders of the domains could also have COPSS functionality. This will be briefly addressed in Section 5.

On receiving a Subscribe packet from an end host that seeks to subscribe to a CD, the edge router will first modify its ST and then forward it to the RP using the address RA_{CD}. The protocol exchange is shown as the "Subscribe Stage" in Fig. 1. When the COPSS RP receives this packet, it will assign an IP multicast group address to the CD if there is no group for the specified CD. It then sends a group join invitation to the edge router to which the edge router responds by joining the specified IP multicast group. In the

IP network, an RP-based tree or source-based tree will be formed according to the IP multicast protocol (e.g., PIM-SM).

To publish content, the publisher sends a Publish packet to the edge router (the user behavior is unchanged). The protocol exchange is shown as the "Publish Stage" in Fig. 1. The edge router encapsulates the packet using an Interest with RN_{CD} and sends it to RA_{CD}. When the RP receives this packet, it will decapsulate it and forward it based on the Group_Table, instead of the ST. We can also use the ST on the RP to maintain the Group_Table by replacing the face with the group address. This packet will be delivered to all the edge routers that subscribe to $Group_{CD}$. The edge routers check the CD in the Publish packet and forward it based on their own ST. Since CDs are used to represent content, the sheer volume of CDs could be an order (or multiple orders) of magnitudes greater than IP multicast group addresses. The mapping of the very large, hierarchical CCN namespace onto a bounded, flat IP multicast group address space will naturally result in wasteful traffic being sent on the network, which will have to be discarded by the edge router.

We believe that to a significant extent ISPs support IP multicast within their domain. The primary challenge is in supporting IP multicast across domains. However, in those cases where IP multicast is not supported within a domain, we rely on a pure COPSS overlay. We then exploit an overlay multicast tree to minimize the number of copies sent from the RP and on each overlay hop.

4. Hybrid-COPSS QUERY/RESPONSE

Query/response based dissemination is an essential part of content delivery, in addition to the pub/sub delivery mechanisms. It could be initiated by an end-host that requests a particular content or in response to receiving a snippet via the pub/sub delivery mechanism. The latter approach (called 2-stage dissemination) is performed when the end-host is interested in receiving the whole video after watching a trailer that was sent by the pub/sub mechanism. The 2-stage dissemination helps reduce the bandwidth used for data delivery since not every subscriber is interested in each piece of data published with the CD he subscribed to. At the same time, this strategy can help publishers with policy control in responding to subscriber requests. For example, the snippet can be seen as an advertisement and the publishers can have access control on the actual complete content. For the 1st stage, we use COPSS to minimize the announcement latency, and for the 2nd stage, we use query/response to get the best use of in-network cache, as is typically performed in NDN. This work focuses on adapting the query/response component to efficiently function in the envisioned hybrid scenario where it co-exists with IP as well as COPSS (multicast based dissemination) nodes.

As mentioned earlier, the needs of a query/response system could be different from that of a pub/sub system and therefore we do not place strict requirements on where CCN-aware or pure IP routers need to be in the network. Since hybrid-COPSS utilizes IP multicast, it is sufficient for edge routers and RPs to be CCN aware in the basic case. The simplest approach is to enable query/response functionality on the COPSS nodes. But for achieving improved overall efficiency, we believe that having more CCN-aware routers can help because of the in-network cache and FIB aggregation

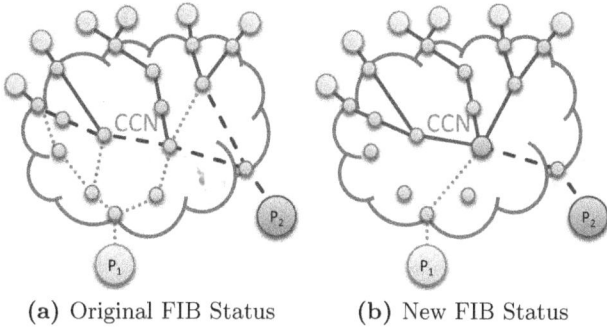

(a) Original FIB Status   (b) New FIB Status

Figure 2: FIB propagation for Query/Response

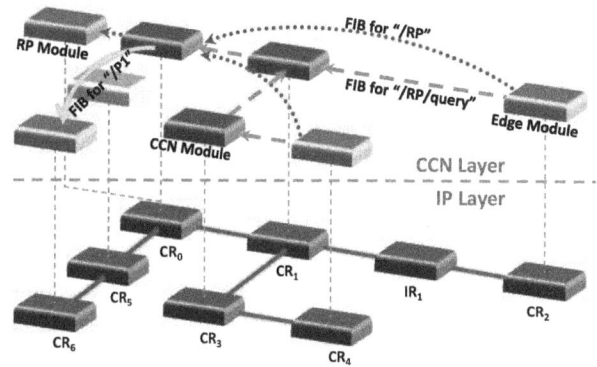

Figure 3: Overlay for Query/Response

capability. However, for the dissemination of the content, traversing a larger number of CCN-aware routers involves more processing at each of the hops and thus contributes to higher latency. Therefore, with the goal of achieving an incrementally deployable architecture and having an optimized delivery mechanism that reverts to hashing based forwarding whenever possible, we explore optimization strategies for placement of CCN caching nodes versus nodes that forward based only on IP.

4.1 RP-based Query/Response

To provide content-oriented query/response in an incremental way, we propose an optimized RP-based query/response architecture that allows for reducing the size of the FIB while increasing the cache hit rate. To allow the subscribers to obtain the content, the publishers need to send a globally unique ContentName along with the snippet and also propagate that he serves (the prefix of) the ContentName beforehand. The propagation results in an FIB entry being added either at the RP or in the CCN aware routers depending on the approach (see Fig. 2). In COPSS, the globally unique ContentName is divided into 2 parts: globally unique publisher name and the publisher's locally unique data ID. In the case of the NDN-based query/response, a globally unique name is used. NDN requires the network to know how to reach the publisher by creating a source-based tree rooted (i.e., starting from the edge of the network inwards) at the publisher's 1st hop router. Fig. 2a shows the FIB for /P1 (green dotted lines) and /P2 (red dashed lines). The blue solid lines are the shared part of the trees. Since there might be subscribers everywhere in the network, every router needs to know the outgoing face for all the publishers. The total FIB entry size in the network can be calculated as:

$$Size_{FIB} = (n_{rp} + n_{pub}) \times n_r, \qquad (1)$$

where n_{rp} is the number of RPs, n_r is the number of CCN-aware routers and n_{pub} is the number of publishers.

However, in a pub/sub system environment, the RPs are already well-known to the complete network (for multicast). If we aggregate the query/response tree at the RP, we can reduce the FIB entries stored on every router: only RPs need to know how to reach the publishers. In Fig. 2b, the blue solid lines from the subscribers to the RP are the FIB for /RP. In the hybrid world, FIB values are {IP:port} pairs, so we can point $FIB(P1)$ on RP directly to the 1st router

of P_1. The green dotted line in Fig. 2b is an overlay link that skipped an intermediate router in the underlay. The FIB entry size in the network can be calculated as:

$$Size'_{FIB} = n_{rp} \times n_r + n_{pub} \times n_{rp}. \qquad (2)$$

Since $n_{rp} \ll n_r$, the new FIB size can be much smaller than the original (in equation 1). Though this optimization is optimal for query/response in the 2 stage dissemination where the subscribers are aware of the RP, this model can also be applied to the NDN-like query/response model. In that case, RPs would be responsible for aggregating FIBs from data providers and would advertise themselves. The CCN routers would forward the query to the RPs, which in turn would forward it to one of the data providers in case a cache hit has not already occurred. The benefits of such an architecture for the NDN-like query response is that it limits the FIB sizes that are being propagated in the network. In the rest of the section, we detail the effect of placing CCN aware nodes on the pub/sub model.

FIB Propagation from Subscribers

In §3, we required the RP and all the edge routers be CCN-aware routers. In the overlay, we created FIB entries from the edge routers directly to the RP since there are no CCN-aware routers in between. When we have more CCN-aware routers deployed in the network to increase the cache-hit rate for query/response interactions, we slightly modify the "Subscribe Stage" to allow intermediate CCN-aware routers be part of query/response tree as well. But, in the "Publish Stage", the packet flow remains the same.

We use Fig. 3 as the example of our description. In Fig. 3, IR_1 is an IP router. CR_{0-6} are CCN-aware routers. $CR_{2,4,6}$ are edge routers. S_1, S_2 and P_1 are linked to CR_2, CR_4 and CR_6 respectively. In the overlay, we have the CCN module at CCN-aware routers. The edge routers have CCN modules with additional functionality (e.g., encapsulating, decapsulating and ST for end-hosts). Since a CCN-aware router can serve as 0, 1 or more RPs, we design the RP as a logically separate module for multicast and renaming. Thus, on router CR_0, there exists both CCN module and RP module. When an RP is setup on a router, the RP module registers a FIB entry on the router's CCN module: name=/RP, face=$R_{x:RP}$. We use $CR_{x:CCN}$ to denote the {address:port} pair of the CCN/edge module on CR_x and $CR_{x:RP}$ for RP module on CR_x.

On receiving a Subscribe packet from the subscriber (S_1), the edge router ($CR_{2:CCN}$) will add ST and forward it using UDP packet whose destination is $CR_{0:CCN}$ according to FIB(RP_Table(CD)). Here, we use a flag bit in the UDP packet to mark the special message type. When IR_1 receives the packet, since it is a normal IP router, it forwards the packet directly to CR_1. But CR_1 knows about the special flag bit and instead of forwarding this packet towards CR_0, it redirects the packet to $CR_{1:CCN}$. Its CCN module treats the Subscribe packet similar to an Interest packet. An entry (name=/RP/sub, face=$CR_{2:CCN}$) will be added into the Pending Interest Table (PIT) that contains the requests that are yet to be served. When $CR_{1:CCN}$ sends the UDP packet out, the *from* field of the UDP packet is changed from $CR_{2:CCN}$ to $CR_{1:CCN}$. When $CR_{0:CCN}$ receives the Subscribe packet, it does the same as $CR_{1:CCN}$, but the face of the PIT entry is $CR_{1:CCN}$. The packet is forwarded to $CR_{0:RP}$. $CR_{0:RP}$ responds with a Join packet which contains CD, RP and $Group_{CD}$ (the same behavior as described in §3.2). The CCN module treats this packet similar to Data packets with extra FIB add action. $CR_{0:CCN}$ adds $FIB(/RP/query) = CR_{0:RP}$, $CS(/RP/sub) = Join$. Subsequently, this Join packet will consume $PIT(/RP/sub)$ and be forwarded to $CR_{1:CCN}$. $FIB(/RP/query)$ will be created along the path to $CR_{2:CCN}$ and Content Store entry for /RP/sub will be stored in the intermediate routers. When edge module ($CR_{2:CCN}$) receives the Join packet, it checks if there is an end-host subscribing to CD. If so, it will join the IP multicast group specified by the Join packet (the same behavior as described in §3.2). When another subscriber S_2 also tries to join CD, $CR_{1:CCN}$ can respond directly instead of going all the way to $CR_{0:RP}$ since it already has $CS(/RP/sub) = Join$. If S_2 tries to subscribe to a sub entry of CD (e.g., CD/x), and the RP serves CD, $R_{4:CCN}$ will add $ST(CD/x) = S_2$ but subscribe to the CD upstream, so as to still get hit on $CR_{1:CCN}$. Fig. 3 shows the FIB for /RP (in red dotted lines, for multicast) and FIB for /RP/query (in green dashed lines, for query/response).

FIB Propagation from Publishers

Since every end-host in the network can be a possible publisher, and it is not necessary for some of the publishers to be known by the whole network (they only use single-stage pub/sub), setting up an FIB entry for every possible publisher on the RPs is not advisable in the pub/sub environment. We therefore require FIB creation information to be piggybacked with the first Publish packet. If a publisher needs to serve a prefix (he wants the subscribers to issue a query for the whole data), he will encapsulate the prefix in a Publish packet. In Fig. 3, P_1 will encapsulate /P1 in the Publish packet. On seeing the piggybacked information in the Publish packet, the $CR_{6:CCN}$ will setup $FIB(/P1) = P_1$ and forward the packet to $CR_{0:CCN}$ ($CR_{5:CCN}$ will not see the packet). $CR_{0:CCN}$ will add $FIB(/P1) = CR_{6:CCN}$. The FIB is shown in yellow solid line in Fig. 3.

Data Dissemination According to FIB

On receiving a snippet with CD and data ID /P1/Data1, S_1 queries data with /P1/Data1 and include CD as a reference. When $CR_{2:CCN}$ receives the packet, it adds $PIT(/RP/query/P1/Data1) = S_1$ and forwards it according to entry $FIB(/RP/query)$, which is $CR_{1:CCN}$. Then, $CR_{1:CCN}$ forwards it to $CR_{0:RP}$ through $CR_{0:CCN}$. $CR_{0:RP}$ removes

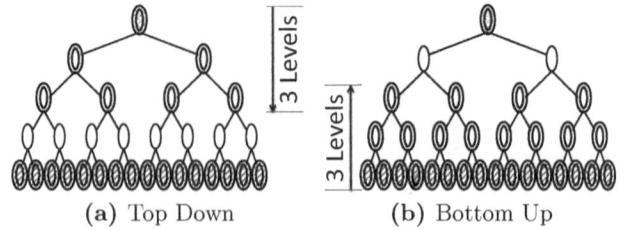

(a) Top Down   **(b)** Bottom Up

Figure 4: Possible Incremental Deployment Strategies

the /RP/query prefix and forwards the Interest (ContentName=/P1/Data1) to P_1 through $CR_{0:CCN}$. P_1 responds with the Data packet with name prefix /P1/Data1. It will be forwarded to $CR_{0:RP}$ and $CR_{0:RP}$ adds name prefix /RP/query to the packet. CCN modules on the other routers forward the Data packet and save it in the cache, just the case as defined in NDN. S_1 will receive data and the packet will be cached in CCN module CR_0, CR_1 and CR_2.

4.2 Strategy to Enable CCN-aware Routers

Although deploying a larger number of caches in the network can increase the cache hit rate, incremental deployment of the CCN nodes may suggest the need to examine the cost of deploying these caches. A higher cache hit reduces server/publisher load, network traffic and load on the nodes that have to process the content further upstream towards the source. However, with a larger number of such CCN enabled nodes, there are more nodes that have to do comprehensive processing of the packet, which increases cost as well as add latency at each of those hops. In this section we assume the ISPs have a limited maximum amount of cache that can be deployed across the various nodes in the network. This then raises the question of which routers should be replaced/enabled with the CCN (cache) functionality. As proposed earlier, in the case of the RP-based architecture, the query/response path follows the same tree as the one used for pub/sub multicast tree. Subscribers are at the leaves of the tree with the RP as the root node. In order to better understand the trade-off, we analyze 2 possible ways of deploying CCN-enabled routers in the network, considering the logical multicast topology: top down (from the RP down) and bottom up (starting from the end-hosts/subscribers). Fig. 4 shows a 5-level (binary) dissemination tree. The root node is the RP, the leaf nodes are the edge routers to the subscribers. According to the requirement of hybrid-COPSS, the RP and edge routers have to be CCN-enabled routers. The CCN enabled routers are marked as nodes with a double circle in the figure. The *top down* strategy deploys CCN enabled routers starting from the routers directly connected to the RP in the logical tree. Fig. 4a shows the structure with 3 levels of CCN-enabled routers according to top down strategy. The *bottom up* strategy deploys CCN enabled routers starting from the routers directly linked to edge (leaf) routers in the tree. Fig. 4b shows the structure with 3 levels of CCN-enabled routers based on the bottom up strategy. Note Fig. 4 is for illustration purposes (we realize here the number of CCN nodes in the two figures are different).

The advantage of a bottom up model is that since the cache nodes are deployed closer to the leaves (subscribers/querying nodes), a cache hit at the intermediate routers

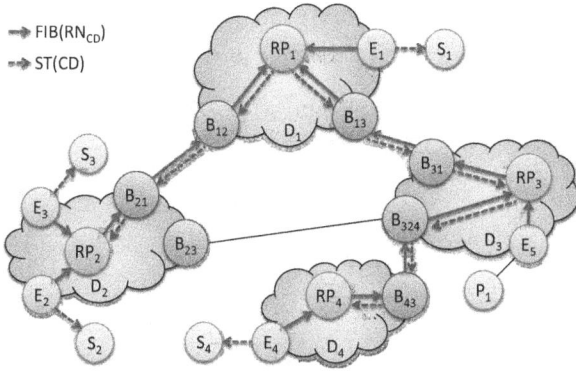

Figure 5: Inter-domain multicast

could result in lower latency as well as less network traffic. However, this strategy could suffer from the fact that the total cache is now divided across a larger number of CCN enabled routers. A smaller cache at each node may result in a lower cache hit rate. Alternately a smaller number of CCN enabled routers may be deployed with larger amount of cache on each router, but in the bottom-up case, this may result in only a subset of the end-nodes in the tree seeing the benefit of the cache. On the contrary, in the case of the top down model, the presence of a few cache nodes at the top levels of the tree allows for a larger cache and at the same time has the potential to serve a larger set of end hosts as well as take advantage of the aggregation of user requests, yielding a higher cache hit rate. However since they are farther away from the end hosts, it results in increased response latency and traffic. We will compare these two strategies in our evaluation.

5. Hybrid-COPSS INTER-DOMAIN

A problem with using IP multicast, to take advantage of forwarding efficiency, is the inability to go across domains (possibly due to business and deployment considerations). We combine overlay multicast at the COPSS layer and IP multicast in the underlay so that a global $Group_{CD}$ mapping is not required, i.e., all the IP multicast information is maintained within the individual domains. This allows us to have different CD to IP multicast mapping in each domain based on considerations such as the load and subscriber count for each CD. We take advantage of pure IP multicast, while making sure that the content-centric COPSS overlay recognizes the administrative boundaries at the IP layer.

As shown in Fig. 5, similar to the requirements of the single domain solution, the edge routers and the RP in each domain are COPSS aware routers. Additionally, in the multi-domain case, we require the boundary routers (marked B_x) to be COPSS aware. The overlay uses a COPSS multicast tree rooted at the first established RP (global RP) across all the domains. The individual domains have a local COPSS RP that subscribes to the global RP if there is at least one interested subscriber in its own domain, or a domain downstream.

5.1 RP Setup

The first subscription to a CD that is not yet served by any global RP initiates the process of setting up the RP within the originating domain. This RP serves as the root

of the global multicast tree (being the global RP). The RP disseminates the fact that it now serves the CD (for the mapping $RN_{CD} = RP\_Table(CD)$ identifying the RP) to all boundary routers (named outgoing boundaries) and the edge routers in its domain.

When an outgoing boundary receives information on the CDs that an RP (from its own domain) is serving, it will set up a forwarding table entry ($FIB(RN_{CD})$) for that particular RP and forwards the information to the other boundary routers (named incoming boundaries) in adjacent domains. When an incoming boundary receives the "CD serving" information, it first checks if there is already an RP in its domain (to avoid loops). If not, it sets up a FIB entry ($FIB(RN_{CD})$) pointing to the boundary from which it received the serving information and sets up a new local RP for this CD. The newly created local RP then sets up a $FIB(RN_{CD})$ pointing to the incoming boundary in its domain and propagates the serving information to all the edge routers and all the boundaries except the incoming boundary. To minimize the forwarding latency when going across domains, we suggest that the local RP be co-located on one of the incoming boundary nodes of a domain. E.g., If RP_3 is located on B_{31} in Fig. 5, the latency through D_3 will be $B_{31} \rightarrow B_{324}$ instead of $B_{31} \rightarrow RP_3 \rightarrow B_{324}$. The edge routers will set up $FIB(local\_RN_{CD})$ pointing to the RP in its own domain on receiving the serving information. The FIB information in Fig. 5 shows the result of RP setup started at domain D_1, triggered by S_1. Notice that B_{23} will not setup another RP in domain D_2 since there already exists an RP in D_2 when the new RP information was propagated to B_{23}. It will also not be considered an outgoing boundary or setup a FIB to RP_2. But B_{324} will serve as the boundary that serves D_4 (tree is routed through D_3), so it has a FIB entry pointing to RP_3.

5.2 Subscribe

The subscription procedure in the individual domains is similar to the subscriptions in a single domain case. The edge router forwards the subscription to the local RP, the local RP assigns an IP multicast group for the CD and then asks the edge router to join the local IP multicast group. However, the local RP will also forward the subscription upstream to the global RP (root) according to the forwarding table entry ($FIB(RN_{CD})$). The boundaries and RPs in between will setup their ST appropriately, but it is not necessary to assign an IP multicast group address within these domains. In the example shown, the ST information in Fig. 5 shows the result of a subscription by S_1 through S_4 (dashed lines showing the subscription tree). Since there is no subscriber in D_3, no IP multicast group is needed in D_3.

5.3 Publish

For the intra-domain multicast, the local RP multicasts the packet using local $Group_{CD}$. But on the overlay, we use a shortcut-enabled multicast tree to optimize forwarding performance and reduce inter-domain traffic. That is, on receiving a multicast packet (encapsulated into an Interest with the ContentName of the RP), the local RP decapsulates the packet and sends it downstream using $ST(CD)$, except on the incoming face. At the same time, it re-encapsulates this packet using its own RN_{CD} and forwards it according to the FIB. With this shortcut, the Multicast packet does not need to go all the way to the root of the tree and come back down. Instead, it is disseminated to subscribers while being forwarded upstream to the global RP.

For example, P_1 in Fig. 5 sends a Multicast packet to E_5. E_5 encapsulates the packet using the ContentName RP_3 and sends it to RP_3. With no subscriber in D_3 ($Group(CD) = null$), RP_3 will only send a COPSS Multicast (overlay) packet downstream (through B_{324} and B_{43} to RP_4) according to the ST. At the same time, RP_3 encapsulates the packet into an Interest using RN_{CD}, $i.e.$, RP_1, and forwards it according to the FIB. The Interest will be forwarded through B_{31} and B_{13} to RP_1. RP_1 decapsulates the packet and forwards it according to the ST to B_{12} (and then to B_{21}, RP_2). RP_1 will not forward it to B_{13} since it is the incoming face. Also, RP_1, RP_2 and RP_4 will send IP multicast with $Group_{CD}$ in D_1, D_2 and D_4 respectively. $Group_{CD}$ may differ in the different domains according to the subscription status in each domain. Edge routers receive the packet and forward it to subscribers.

5.4 Automatic Rendezvous Point Balancing

In G-COPSS [8], an automatic RP balancing method was proposed to relieve traffic concentration ("hot spots"). We adopt this solution for hybrid-COPSS to minimize the effect on inter-domain traffic by using different physical RPs ($i.e.$, $local\_RN_{CD}$) instead of just one global RN_{CD}. The introduction of a new RP and migration of CDs to it for load-balancing affects only the first domain downstream. $E.g.$, if the RP for a CD tries to move from RP_3 to RP_3', RP_3' will set $RN_{CD} = RP_1$, $FIB(RP_1) = B_{31}$ and a subscriber from it ($e.g..$, B_{31} will modify ST, but others like B_{13} will not be affected.) A new FIB entry $FIB(RP_3')$ will be created in RP_4, B_{43}, and B_{324} pointing upstream. RN_{CD} in RP_4 and edge routers in D_3 will be changed to RP_3'. But if there are other domains subscribing to D_4, they will not be affected.

6. MANAGING ADDRESS MAPPINGS

In this section, we examine efficiency and scalability related issues of hybrid-COPSS. First is the issue of management of the mapping table between the CDs and the Multicast groups on RPs ($i.e.$, $Group_{CD} = Group\_Table(CD)$). The mapping function controls the tradeoff between the IP multicast address space usage versus excess traffic carried over links when a group address is used for multiple CDs. The second issue is the management of the table containing the mapping of CDs to a Rendezvous Point ($i.e.$, $RN_{CD} = RP\_Table(CD)$). Although hybrid-COPSS requires that RN_{CD} is maintained in every COPSS edge router, we seek to limit the size of this table to enable the solution to scale better even when we have millions of CDs.

6.1 CD to Multicast Group Mapping

Since CDs are used to represent content, the sheer volume of CDs could be orders of magnitudes greater than the available space of IP multicast group addresses. Therefore there is a need to map multiple CDs to a particular IP multicast group id. The mapping of the unbounded, hierarchical CCN namespace onto a bounded, flat IP multicast group address space will naturally result in wasteful traffic being sent on links in the network. But different mapping functions can result in varying amounts of wastage. Imagine that there are 2 CDs: FireAlarm (subscribed to by almost everyone but does not have many updates) and CCNMailingList (subscribed to

by only a few people but with frequent updates). If we map them both onto one IP Multicast group, this group will be subscribed by almost everyone. This implies that the updates in the CCNMailingList will be received and discarded by almost every edge router, resulting in a large amount of wasted traffic carried by the network.

According to the example above, a mapping function that classifies CDs based on their subscribers and update frequency would be preferred. However, in the true sense of a Content-Centric Network, we assume that neither publishers nor RPs know who or where the subscribers are. Predicting the publication frequency of CDs is even more difficult since anyone in the network could be a publisher. We suggest that instead of predicting, it is better to dynamically adapt, by having a re-map function based on various criteria to ensure fair load distribution and reduction of wasteful traffic.

In our approach, every edge router calculates the wasted amount of traffic delivered over an IP multicast group, using a sliding window. Waste is defined as a packet that is received at a COPSS edge router, but for which there is no outgoing face according to its own ST. When the amount of waste packets in a group exceeds a certain threshold, the edge router reports the overhead (including the group address and the waste packets for every CD) to the local RP. Based on the total amount of wasteful traffic on every IP multicast group address used, the local RP splits the heterogeneous CDs and assigns them to a new IP multicast group. In some other cases, the RP may also try to combine several CDs into one multicast group when they have similar behavior, although we have not explored this in detail yet. When a new mapping is propagated to the whole network (within a domain), all the edge routers rejoin the new IP multicast group if there are subscriptions maintained by them that would be affected. Notice that, since IP multicast is used within a domain, such a re-mapping function will not affect the other domains. Each domain can have its own $Group_{CD}$ according to their subscription status. Our ongoing work is to examine the details of this mapping and its effect.

6.2 Management of the RP Table

To limit the size of table containing the mapping of a CD to an RP ($RN_{CD} = RP\_Table(CD)$) at every edge router and for ease of management, we use a broker to maintain the complete database in each domain. The smaller table for RN_{CD} at the edge routers are treated as a cache. When a router has a cache miss, it goes to the broker. Similar to the NDN design, intermediate routers can also respond to this request. The data structure of storing RN_{CD} on the router can be chosen from one of two options. The first is a traditional CD to RP mapping table. The index of CDs can be grouped into a tree structure to optimize search performance. The router would only have to map the CD once before it sends the packet. The second option is a bloom filter based RN_{CD} table. For every RP, there is a (counting) bloom filter storing the hash of the CDs it serves. This can compress the index greatly and reduce the cache miss rate. However, since we can have false positives, the router will have to test all the bloom filters before it can forward the packet. This could result in packets being sent to the wrong RPs (because of the false positives), thereby resulting in wasted network traffic and also computation overhead in the RPs to check if it indeed serves the CD in the packet.

Table 1: Avg. Forwarding Latency(95%CI)

(in µs)	COPSS	Hybrid-COPSS
1st Hop	2778.14(579.13)	2860.21(592.49)
Internal Unicast	**2679.05(575.13)**	**34.71(3.04)**
Rendezvous Point	2749.33(572.32)	2804.65(574.47)
Internal Multicast	**82.76(5.60)**	**33.18(2.90)**
Last Hop	83.26(6.10)	140.65(5.79)

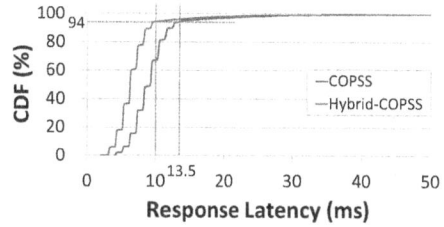

Figure 6: Test Bed Response Latency (CDF)

7. EVALUATION

In this section, we microbenchmark a hybrid-COPSS implementation on our test-bed compared with pure COPSS for the forwarding efficiency and queuing. We then use an online gaming trace and a Twitter trace to evaluate the performance of these architectures under load with a simulation parameterized by the microbenchmark results.

7.1 Microbenchmarking

We implemented hybrid-COPSS on our lab test-bed and compare it to the implementation described in [8]. Similar to [8], 62 players are used to load the test-bed implementation, but with a longer period ($2min$ warm up and $10min$ evaluation) from the gaming trace to get statistically significant results. We also traced every packet using Wireshark and calculate the average processing time for different actions at every router by tracking the arrival and departure time of that packet. Six different kinds of operations are defined here. For the 1st hop router, the last hop router and the RP, we do not breakdown the performance of individual encapsulation, decapsulation and forwarding functions. These routers are treated as black boxes. However, for internal routers, we separately measure the functionality of unicast (Interest in COPSS; UDP unicast in hybrid-COPSS) and multicast (Multicast in COPSS; UDP multicast in hybrid-COPSS) forwarding.

Microbenchmarking Results

We show the average forwarding latency on different routers along with the 95% confidence intervals in Table 1. The results confirmed the observation in [7] that CCN (especially NDN) forwarding is much more expensive than IP forwarding. The 1st hop router and the RP in both COPSS and hybrid-COPSS require FIB lookup functionality, as does COPSS unicast forwarding even at the internal routers. With hybrid-COPSS, the internal unicast is UDP/IP forwarding, which is far more efficient. The last hop router and the internal multicast take less time due to the simpler ST lookup in COPSS. With hybrid-COPSS, the internal multicast is IP multicast forwarding, which is quick. In hybrid-COPSS, although it incurs a slight overhead on the edge routers and the RP (around $70µs$), the internal routers even outperform COPSS multicast since no name resolution is required there.

The average update latency in hybrid-COPSS is $6.95ms$, compared to $9.54ms$ in COPSS. Fig. 6 shows the CDF of the total update latency. Observe that more than 94% (compared to only 67% for COPSS) of the new updates/publications in hybrid-COPSS incur a latency of less than $10ms$ while in COPSS the same 94% take $13.5ms$. Since we used a simple test-bed topology, with only 1-2 internal routers between the RP and edge routers, we expect higher performance gains in a typical network topology with more intermediate hops.

7.2 Large Scale Simulation

To further evaluate the performance of hybrid-COPSS in a large scale environment with realistic network topologies, we use trace-driven simulation, with two traces: one from a multiplayer online game, and the second from Twitter. The evaluations look at both single and multi-domain environments for the strategy for incremental deployment for pub/sub and query/response. Note that in the case of the Twitter trace evaluation, most subscribers are treated as pure receivers with only 50 of them being treated as publishers. We do this to emulate scenarios where the ratio of publishers to subscibers vary. On the other hand, for the gaming evaluation, all players send as well as receive updates. As the number of subscribers grows, additional load is generated per-player, making it more challenging to support in the network.

Data Traces

Game Trace: We first modified a Counter-Strike (CS) trace obtained on a busy CS server in a $7h05m25s$ period [26]. 414 unique players who published $10,686,950$ packets (average publish frequency is around $2.39ms/packet$) were in the trace. We also created several subsets of the trace (the # of players varied from 50 to 400) to evaluate the scalability of our architecture. All the players share a global hierarchically partitioned map divided into 5 regions. Each region is further divided into 5 zones. A player is able to see and modify objects based on his location in the game and the hierarchy of the area he belongs to. So the RP will actually be busier than the original server (where at most 22 players may share an instance of the game, based on the CS server configuration). In hybrid-COPSS, $Group_{CD}$ is manually assigned: CDs in one region share an IP multicast group (7 CDs in 1 group: $Group_{/i/*,/i} = 224.0.0.i$), and /0 uses a single group since

Twitter Trace: We used a Twitter trace on technical topics obtained from the public Internet during a one-week period in 2010. We identified and selected 25 popular keywords such as *iphone, ipad, blackberry, smartphone* as 1st level CDs and created a subset of the trace containing $41,613$ tweets from $22,987$ users that contain these keywords. These 25 key words are treated as the 1st level CDs. We further identify secondary popular keywords (up to 25 keywords) that are associated with these 1st level CD keywords. We build a tree structure off of all these CDs. There are a a total of 407 distinct leaf CDs. To have an adequate number of tweets originating from each publisher in our system, we assigned the $22,987$ users to 50 publishers. The individual users were hashed to a publisher using a power-law. Every 1st level CD uses an IP multicast address. To have a suffi-

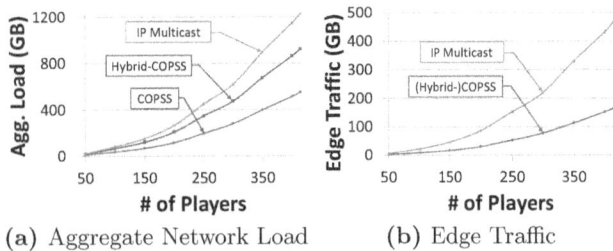

(a) Aggregate Network Load   **(b)** Edge Traffic

Figure 7: Single Domain Performance: Game Trace

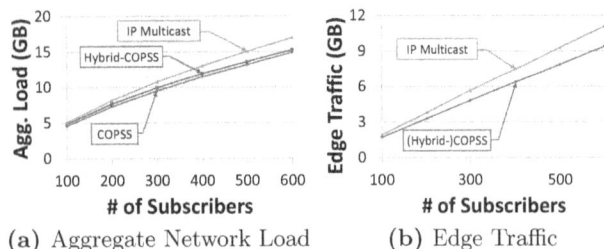

(a) Aggregate Network Load   **(b)** Edge Traffic

Figure 8: Single Domain Performance: Twitter Trace

ciently large message for the query/response case, we scaled the tweet size by a factor of 128. A publisher publishes the original message size as a snippet first and then the subscribers probabilistically query for the complete message on receiving the initial snippet (tweet).

Single Domain: Pub/Sub

In single domain case, we use RocketFuel 3697 [27] (79 routers) as the core topology. A total of 200 edge routers are randomly assigned to the core routers (1-3 edge router(s) per core router). For the evaluation of hybrid-COPSS in this case, we used both the Twitter and Gaming traces. The subscribers or players are evenly distributed on the edge routers. 3 routers with the minimum average shortest path distance to all the edge routers are chosen as the RPs (in [8], we showed that 3 RPs are sufficient to efficiently handle the gaming trace and we use the same environment for the Twitter trace). The IP multicast groups in hybrid-COPSS are distributed on the 3 chosen RPs (an RP can serve several multicast groups). Pure IP multicast is also compared using the same RP and $Group_{CD}$ settings.

In the Gaming trace, with cheaper IP forwarding, hybrid-COPSS achieves an average update latency of around $73.7ms$, compared to $84.6ms$ with COPSS. However, because of an insufficient # of IP multicast groups (we map 7 CDs onto 1 IP multicast group), hybrid-COPSS causes a larger load in the network compared to COPSS, but less than pure IP multicast, as shown in Fig. 7a when we vary the number of players. This demonstrates that our solution can be integrated into the current IP architecture without substantial performance degradation.

With the same # of multicast groups, IP multicast and hybrid-COPSS result in the same amount of traffic in the network core. However at the edge with IP multicast, since the last hop router does not do filtering, end hosts will have to receive all the unnecessary packets and discard them if they find them to be of no interest. The wastage on the edge

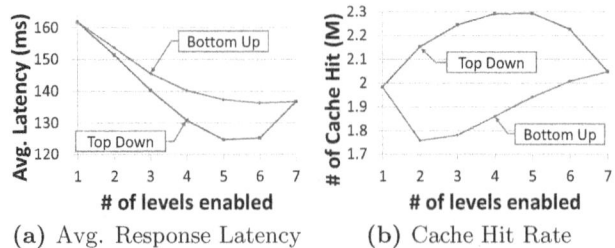

(a) Avg. Response Latency   **(b)** Cache Hit Rate

Figure 9: Performance of Different Incremental Deployment Strategies

is shown in Fig. 7b. It causes substantial computational and communication overhead on the end host as the number of players increases. For instance, this could be a substantial penalty on mobile devices with limited battery power. Since hybrid-COPSS does not change the user behavior, the edge traffic in hybrid-COPSS is the same as COPSS.

Fig. 8 illustrates the same trend for the Twitter trace, except that the network load(Fig. 8a) and edge traffic(Fig. 8b) grow sub-linearly. This is due to the fact that although there is an increase in the # of subscribers, the network is able to take advantage of multicast. The aggregation of subscribers due to their common interests results in increased efficiency for multicast.

Single Domain: Query/Response

We ran a simulation to compare the performance of the two strategies. To understand the results better, we view the topology in the abstract as a 8-level binary tree (composed of 255 routers). The publishers are at the root of the tree and the 256 subscribers are the leaves of the tree. The one-way latency on the links is set to $10ms$). We used the Twitter trace, where snippets are sent as an announcement and the subscribers then query for data they are interested in (*e.g.*, video clips) [7]. The subscribers have a 50% probability of querying for the data on receiving an announcement. The delay before issuing the query ranges from $5sec$ to $1hr$. The total cache size in the network is set to $2.55GB$, divided equally among all the CCN enabled routers in the network. The total published data size is around $4.5GB$.

The results are shown in Fig. 9, where the x-axis $level = n$ denotes the number of levels of routers that have their CCN cache enabled. Therefore $level = 2$ implies that 2 levels of routers have their cache enabled, both in the top down approach and bottom up approach. Note that the results (in terms of response latency, cache hit rate and network load) are the same for both top down and bottom up at $level = 1$ and 7. For $level = 1$, only the RP and the edge routers are CCN enabled, and for $level = 7$, all the routers in the network are CCN enabled.

There are multiple criteria interacting here that affect the tradeoff of where and how many CCN routers to deploy: *individual cache size*, *the cache locations* and the *strategy* (top-down or bottom-up). With the top-down strategy, when $level = 1$ (RP and 128 edge CCN routers), with a per node cache size of 20.24 MB, we see a lower overall cache hit rate than with $level = 2$ (RP, 128 edge and 2 extra CCN routers) where the per node cache size is smaller at 19.93 MB. The higher cache hit rate is because of the aggregation of requests across uesrs at the second level in the tree that the

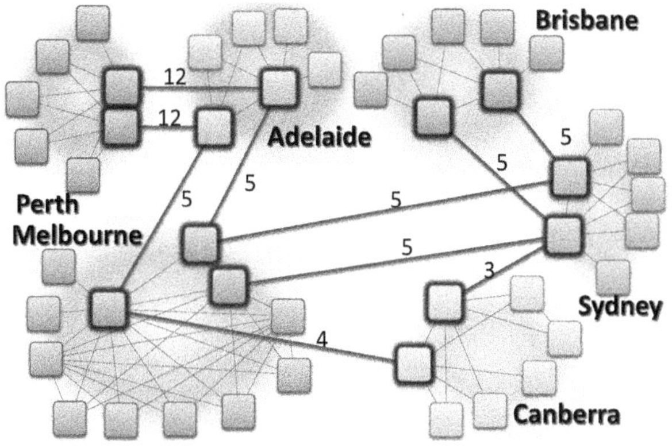

Figure 10: Multi-domain Topology: RocketFuel 1221

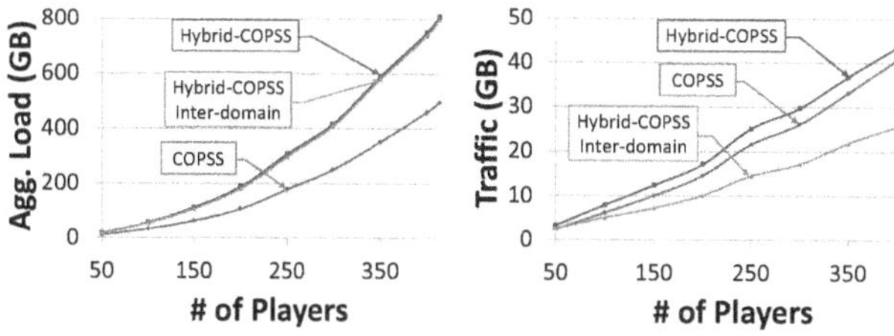

(a) Aggregate Network Load   (b) Inter-domain Traffic

Figure 11: Multi-Domain Performance: Game Trace

2 extra CCN routers can respond to from their caches. The latency also reduces as we increase the number of levels of CCN routers. This improvement with the top-down strategy continues up to a point (*level* = 5), reaching a peak in the overall cache hits and reduced response latency. When we go to *level* = 6 or *level* = 7, the cache hit rate reduces because the individual cache sizes are now smaller than the working set size. Correspondingly, the response latency goes up, even though there are more CCN routers further down the tree. When we look at the bottom-up strategy, we see that the cache hit rate behaves quite differently. Going from *level* = 1 to *level* = 2, the cache hit rate goes down, even though the number of CCN routers goes up. This is because the total cache is divided across more routers (RP, 128 edge and 64 extra CCN routers), because the extra CCN routers do not see a benefit of any aggregation of requests and do no yield an increase in overall cache hits. As we go up the hierarchy in the tree with the bottom-up strategy (going from *level* = 2 to *level* = 3 and higher), the cache hit rate increases, and also results in reducing the response latency. Finally, when we compare the top-down vs. bottom-up strategy, the cache hits for the top-down is consistently higher (better aggregation of requests at the top levels of the tree). The average response latency for the top-down strategy is also consistently better, even though fewer CCN routers may be used (*e.g.*, top-down *level* = 2 has only 3 extra CCN routers, bottom-up *level* = 2 has 64 extra CCN routers). This is because of the higher cache hit rate for the top-down strategy as a result of better aggregation of requests. The network load is also lower with the top-down strategy.

This result suggests that when an ISP has a fixed number of CCN nodes to deploy in the network (i.e., building an overlay network), it is better to take a top-down approach, enabling the CCN functionality starting from the nodes directly linked to the RP, but make sure every router having enough cache size compared to working set size. The top-down strategy can achieve lower response latency with fewer expensive CCN-enabled routers. This serves as an indication that the RP-based dissemination tree structure is better than a publisher focused, source-based tree (see Fig. 2a) solution. The source-based tree would have to inevitably employ a bottom-up approach from the edge to the core. Note that the strategy used here can be applied in the a multi-domain scenario as well.

Multi-Domain Simulation

We then investigate hybrid-COPSS in a multi-domain scenario using RocketFuel 1221 Australia, which has clear domain structure, as our core topology. According to [27], Telstra has hubs in major cities (Sydney, Melbourne, Perth) with spokes elsewhere. We consider every city as a domain and the routers at these major cities as the core routers. The topology (Fig. 10) also shows the weights of the inter-domain links between the 13 boundary routers (marked bold). 3 (global) RPs are selected on the boundary routers in Sydney, Canberra and Melbourne. We then add 207 edge routers onto the core routers, based on the proportions of the number of core routers in that city in the original topology. Every edge router can have 1 or 2 link(s) to a core router in the same city (but not to the boundary routers). The latency from an edge to core router is a random value between 2*ms* and 8*ms*. The subscribers or players are linked to the 207 edge routers evenly. Both the Gaming trace and the Twitter trace are used to compare the performance of pure COPSS solution, a simpler, basic hybrid-COPSS solution (that does not consider inter-domain properties) and the hybrid-COPSS inter-domain solution with varying number of players.

Similar to the single domain, for the multi-domain Gaming case, hybrid-COPSS achieves lower average update latency (around 52.09*ms* compared to 61.13*ms* in pure COPSS) but results in wasting network bandwidth because of the severe shortage of IP multicast group addresses. Inter-domain hybrid-COPSS, however, provides an alternative. Because multicast routing can take advantage of shortcuts across domains, a multicast packet doesn't have to go all the way to the global RP to be forwarded to subscribers. Our solution reduces the average update latency by about 2.46ms. Hybrid-COPSS also cuts the inter-domain traffic almost by half (Fig. 11b), and reduces the aggregate network load slightly compared to hybrid-COPSS (Fig. 11a). This means the inter-domain solution is much "cheaper" for ISPs even compared to a pure COPSS solution. Moreover, as described previously, this solution does not need routers in different domains to know each other's IP multicast group mappings. Thus, it becomes more practical than other solutions that depend on inter-domain multicast.

With the multi-domain Twitter trace, hybrid-COPSS has shorter response latency than pure COPSS (51.35*ms* vs. 59.05*ms*) and inter-domain solution reduces it by an additional 1*ms*. As for the inter-domain traffic, since subscribers do not publish (in the Twitter model) and there is at least one subscriber per domain in all the scenarios, the inter-domain traffic does not vary with an increasing number of

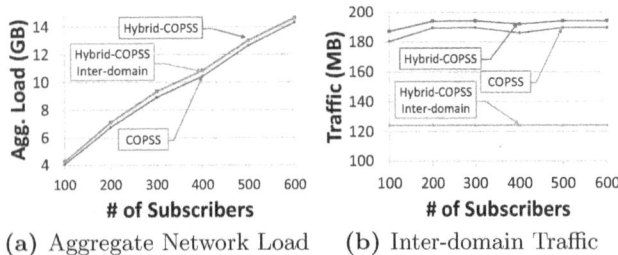

(a) Aggregate Network Load   (b) Inter-domain Traffic

Figure 12: Multi-Domain Performance: Twitter Trace

subscribers (Fig. 12a). But, we can observe that the inter-domain hybrid-COPSS solution reduces the traffic by almost 1/3 compared to a COPSS solution that is not aware of domain boundaries (Fig. 12b).

8. SUMMARY

In this paper, we present hybrid-COPSS, an architecture to integrate CCN functionality with the current IP architecture. We present a detailed solution to integrate both the pub/sub and the query/response based content-centric architecture in an IP network. Hybrid-COPSS is designed to be as generic as possible and is therefore applicable to the query/response based NDN solution as well. In this paper we have addressed the 3-way tradeoff that arises when considering the incremental deployment of CCN, in terms of *traffic*, *latency* and *cost*. There is a higher amount of packet traffic (both within a domain as well as inter-domain) as we go more towards a pure IP environment; there is increased latency in a pure CCN environment because of the additional per-hop processing; finally there is an increase in processing cost for each CCN hop because of the additional complexity. Our evaluations suggest that hybrid-COPSS strives to achieve a proper balance in this trade-off by putting CCN functionality at key points and hash-based forwarding at the other routers. Moreover, we optimize the query/response dimension of hybrid-COPSS and show that an RP based top down approach provides the best means for service providers to incrementally deploy caching-capable CCN nodes. The hybrid-COPSS inter-domain solution recognizes the current challenges in having inter-domain IP multicast and overcomes it with the use of CCN overlay nodes at the domain edges. The inter-domain hybrid-COPSS cuts inter-domain traffic almost by half even compared to our earlier CCN proposal, COPSS.

9. REFERENCES

[1] B. Segall, D. Arnold, J. Boot, M. Henderson, and T. Phelps, "Content Based Routing with Elvin," in *AUUG2K*, 2000.

[2] A. Carzaniga, M. Rutherford, and A. Wolf, "A routing scheme for content-based networking," in *INFOCOM*, 2004.

[3] T. Koponen, M. Chawla, B.-G. Chun, A. Ermolinskiy, K. H. Kim, S. Shenker, and I. Stoica, "A data-oriented (and beyond) network architecture," in *SIGCOMM*, 2007.

[4] L. Zhang, D. Estrin, J. Burke, V. Jacobson, and J. Thornton, "Named Data Networking (NDN) Project," PARC, Tech. Report NDN-0001, 2010.

[5] V. Jacobson, D. K. Smetters, J. D. Thornton, M. F. Plass, N. H. Briggs, and R. L. Braynard, "Networking Named Content," in *CoNEXT*, 2009.

[6] W. Fenner, D. Srivastava, K. K. Ramakrishnan, D. Srivastava, and Y. Zhang, "XTreeNet: Scalable Overlay Networks for XML Content Dissemination and Querying," in *WCW*, 2005.

[7] J. Chen, M. Arumaithurai, L. Jiao, X. Fu, and K. K. Ramakrishnan, "COPSS: An Efficient Content Oriented Publish/Subscribe System," in *ANCS*, 2011.

[8] J. Chen, M. Arumaithurai, X. Fu, and K. K. Ramakrishnan, "G-COPSS: A Content Centric Communication Infrastructure for Gaming," in *ICDCS*, 2012.

[9] V. Ramasubramanian, R. Peterson, and E. G. Sirer, "Corona: a high performance publish-subscribe system for the world wide web," in *NSDI*, 2006.

[10] A. R. Bharambe, S. Rao, and S. Seshan, "Mercury: a scalable publish-subscribe system for Internet games," in *NetGames*, 2002.

[11] C. Esteve, F. Verdi, and M. Magalhaes, "Towards a new generation of information-oriented Internetworking architectures," in *ReArch*, 2008.

[12] A. Carzaniga, D. S. Rosenblum, and A. L. Wolf, "Design and evaluation of a wide-area event notification service," *ACM TOCS*, pp. 332–383, 2001.

[13] G. Chockler, R. Melamed, Y. Tock, and R. Vitenberg, "SpiderCast: a scalable interest-aware overlay for topic-based pub/sub communication," in *DEBS*, 2007.

[14] S. Voulgaris, E. Riviére, A.-M. Kermarrec, and M. Van Steen, "Sub-2-Sub: Self-Organizing Content-Based Publish and Subscribe for Dynamic and Large Scale Collaborative Networks," INRIA, Research Report, December 2005.

[15] P. T. Eugster, P. A. Felber, R. Guerraoui, and A.-M. Kermarrec, "The many faces of publish/subscribe," *ACM Comput. Surv.*, vol. 35, no. 2, pp. 114–131, 2003.

[16] V. Jacobson, D. K. Smetters, N. H. Briggs, M. F. Plass, P. Stewar, J. D. Thornton, and R. L. Braynard, "VoCCN: voice-over content-centric networks," in *ReArch*, 2009.

[17] Z. Zhu, S. Wang, X. Yang, V. Jacobson, and L. Zhang, "ACT: audio conference tool over named data networking," in *ICN*, 2011.

[18] C. Tsilopoulos and G. Xylomenos, "Supporting diverse traffic types in information centric networks," in *ICN*, 2011.

[19] A. Adams and W. S. J. Nichols, "Protocol Independent Multicast - Dense Mode (PIM-DM): Protocol Specification (Revised)," RFC 3973, January 2005.

[20] B. Fenner, M. Handley, H. Holbrook, and I. Kouvelas, "Protocol Independent Multicast - Sparse Mode (PIM-SM): Protocol Specification (Revised)," RFC 4601, August 2006.

[21] H. Holbrook and B. Cain, "Source-specfic Multicast for IP," RFC 4607, August 2005.

[22] H. Eriksson, "Mbone: the multicast backbone," *Commun. ACM*, vol. 37, no. 8, pp. 54–60, 1994.

[23] Y. Cui, B. Li, and K. Nahrstedt, "ostream: asynchronous streaming multicast in application-layer overlay networks," *JSAC*, vol. 22, no. 1, pp. 91 – 106, 2004.

[24] J. Jannotti, D.-K. Gifford, and K.-L. Johnsonand, "Overcast: Reliable Multicasting with an Overlay Network," in *OSDI*, 2000.

[25] J. Chen, M. Arumaithurai, X. Fu, and K. K. Ramakrishnan, "Coexist: A Hybrid Approach for Content Oriented Publish/Subscribe Systems," in *ICN*, 2012.

[26] W. Feng, "On-line Games," http://www.thefengs.com/wuchang/work/cstrike/.

[27] R. Mahajan, N. Spring, D. Wetherall, and T. Anderson, "Inferring Link Weights using End-to-End Measurements," in *IMW*, 2002.

MCA²: Multi-Core Architecture for Mitigating Complexity Attacks

Yehuda Afek§, Anat Bremler-Barr†, Yotam Harchol‡, David Hay‡, Yaron Koral§

§Tel Aviv University
Tel Aviv, Israel

†The Interdisciplinary Center
Hertzelia, Israel

‡The Hebrew University
Jerusalem, Israel

{afek, yaronkor}@post.tau.ac.il, bremler@idc.ac.il, {yotamhc,dhay}@cs.huji.ac.il

ABSTRACT

This paper takes advantage of the emerging multi-core computer architecture to design a general framework for mitigating network-based complexity attacks. In complexity attacks, an attacker carefully crafts "*heavy*" messages (or packets) such that each heavy message consumes substantially more resources than a normal message. Then, it sends a sufficient number of heavy messages to bring the system to a crawl at best. In our architecture, called MCA²—Multi-Core Architecture for Mitigating Complexity Attacks—cores quickly identify such suspicious messages and divert them to a fraction of the cores that are dedicated to handle all the heavy messages. This keeps the rest of the cores relatively unaffected and free to provide the legitimate traffic the same quality of service as if no attack takes place.

We demonstrate the effectiveness of our architecture by examining cache-miss complexity attacks against Deep Packet Inspection (DPI) engines. For example, for Snort DPI engine, an attack in which 30% of the packets are malicious degrades the system throughput by over 50%, while with MCA² the throughput drops by either 20% when no packets are dropped or by 10% in case dropping of heavy packets is allowed. At 60% malicious packets, the corresponding numbers are 70%, 40% and 23%.

Categories and Subject Descriptors

C.2.0 [**Computer-Communication Networks**]: General—*Security and protection*; C.2.3 [**Computer-Communication Networks**]: Network Operations—*Network management, Network monitoring*

General Terms

Design, Reliability, Performance, Security

Keywords

Intrusion Detection, Multi-core, Complexity Attack, DDoS

1. INTRODUCTION

Security devices, such as Network Intrusion Detection/Prevention Systems (NIDS/NIPS), are the front defense line against cyber attacks over the Internet. Open source examples of such devices include Snort [27] and Bro [10]. In recent years, a trend of two-phase *combined attack* on security devices is becoming common: the attackers first neutralize the security device, for example, by overwhelming it with traffic, and then, when the security device has been knocked down, attack the assets it was protecting. For example, a recent attack on SONY, combined a DDoS attack with credit cards theft [29]. The combined attacks usually have different effect on NIDS and NIPS. In NIDS, where the stealth-mode device only monitors the traffic and issues alerts when it detects malicious activity, these DDoS attacks may force the device to stop inspecting part, or all, of the traffic and thereby allowing another attack to pass unnoticed. On the other hand, in-line NIPS, which inspects the packets on their critical path, might be forced to drop legitimate traffic and therefore practically causing a denial of service on the servers it protects. For example, Bro and Snort are both vulnerable to this kind of attacks [20].

This paper deals with *complexity attacks*, which are used for the first phase. These attacks exploit the gap between the amount of resources the system requires in processing normal packets and carefully crafted packets that consume drastically more resources (computing, memory, cache, or other). These crafted packets, which we call *heavy* packets, are, on one hand, easy to construct, while, on the other hand, they require very intensive processing from the system. This implies that a small effort on the attacker's side leads the target system to spend great effort, and therefore, it is bound to lose.

We present MCA²—a Multi-Core Architecture for Mitigating Complexity Attacks. MCA² essentially isolates the malicious traffic to a fraction of the cores and deals with legitimate traffic on the remaining cores, which are therefore not affected by the attack.

Our MCA² system can be configured to mitigate any complexity attack with the following properties:

1. There are heavy and normal packets, where heavy packets consume considerably more resources from the security device when being processed.

2. There is a method to identify heavy packets. This method requires very few resources.

3. Packets can be moved efficiently between system cores.

4. There is a special method that handles heavy packets more efficiently than the method used for normal packets.[1]

It turns out that there are quite a few complexity attacks that meet these criteria. However, we restrict our discussion to a central component of NIDS/NIPS, namely the Deep Packet Inspection (DPI) engine. DPI is the process in which the payload of the messages is inspected to detect predefined signatures of malicious activities. We consider three examples that have the above properties: *cache-miss attack* on Snort's signature detection engine; *active states explosion attack* on the Hybrid-FA [5] regular expression detection engine; and *force construction attack* on the Bro IDS regular expression detection engine.[2]

We focus on the first example and use it to explain our method and the above-mentioned list of properties. We then show that the active states explosion complexity attack fits our requirements as well. The third example is omitted due to space consideration. We back up all our findings with experimental results, showing the benefits of using MCA^2 in conjunction with the NIDS.

In general, any complexity attack that satisfies these four properties can be mitigated, given a proper heavy packet identification method. We discuss in detail two examples of such methods in this paper. Although each attack requires a different identification method, all methods share a common general technique of scanning the first few bytes and detecting malicious behavior as early as possible.

Specifically, considering cache-miss attacks, we target Snort's DPI engine, which uses some variant of the Aho-Corasick (AC) [1] algorithm for performing pattern matching. A complexity attack on the AC algorithm (in a stand-alone environment) is shown in [8]: AC uses a large deterministic finite automaton (DFA) that cannot fit entirely in the cache. The common traffic, however, uses only a very small part of it, resulting in fast memory references and few cache misses. An attacker can easily craft malicious packets that cause an exhaustive traversal over the DFA that pollutes the cache. In this paper, we show for the first time that Snort is indeed vulnerable to this attack: an attack on its DPI component degrades its *overall* performance by a factor of 4.2.

After establishing that the threat of this attack is real, we turn to investigate how MCA^2 mitigates such an attack. The key challenge is how to detect and isolate malicious traffic. This is done in two steps. First, training data is used to identify and mark the common states of the DFA. These are the states frequently visited while processing normal common traffic. Then, for each packet, we count the fraction of non-common states visited (out of the total number of states traversed by the packet). As soon as this fraction exceeds a certain threshold, the packet is marked *heavy*. When the fraction of heavy packets is above a certain threshold, we allocate one or more cores to deal with them exclusively, while the rest of the cores continue to process only normal traffic

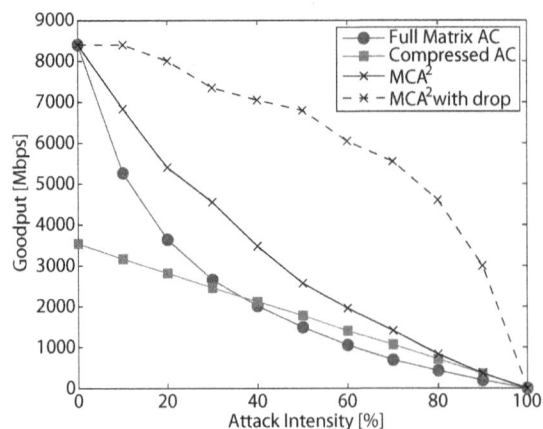

Figure 1: The goodput of MCA^2 for different attack intensities. MCA^2 with no drops maintains a balance between all cores.

(and to detect heavy packets); each subsequent *heavy packet* is moved to one of the dedicated cores. This process isolates the effect of heavy packets and protects the private caches of the non-dedicated cores from being polluted. MCA^2 can be further optimized by running on the dedicated cores an implementation that is optimized for heavy packets (albeit with penalty in the normal case).

The main performance measure we use is the *goodput* of the system, namely the volume of non-malicious packets that were processed. Our experimental results are summarized in Fig. 1, which shows the system's goodput under different attack intensities (namely, in 50% attack intensity, half of the incoming traffic is malicious). We compare the performance of MCA^2 with two implementations of the AC algorithm: the first, denoted "Full Matrix AC", is optimized for well-behaved normal traffic, and the second, denoted "Compressed AC", is optimized to work under cache-miss attacks (as described in Section 2.2).

When the system is not allowed to drop packets, MCA^2 uses the "Full Matrix AC" on the cores that process normal traffic and the "Compressed AC" on the dedicated cores. The number of cores of each type is dynamically determined as a function of the attack intensity. When there is no attack, MCA^2 is reduced to "Full Matrix AC".

We also consider the case when the NIDS/NIPS is allowed to drop packets. Dropping all heavy packets implies that no dedicated threads are required, freeing up all processing resources for the detection of heavy packets and processing of non-heavy (mostly legitimate) packets, thus increasing the goodput.

Our experiments show a significant goodput improvement: MCA^2 achieves up to twice the goodput of both implementations, even without dropping packets. Furthermore, it *always* outperforms a hybrid implementation that chooses the best of the previous implementations at any given time, with a goodput boost of up to 73%.

As for the second example, we use the regular expressions Hybrid-FA data structure to illustrate an *active states explosion attack*. Hybrid-FA uses a single "head-DFA" for commonly-used states while other parts of the automaton are kept as separate DFAs, which are activated simultaneously when required. Usually, only the "head-DFA" is ac-

---

[1] This special method usually handles normal packets poorly, otherwise it would have been used by the system in the first place.

[2] Bro takes a lazy approach to cope with the large DFA size. Namely, it constructs only the DFA parts it actually uses. Normal traffic uses only a small part of the DFA. Hence, a simple complexity attack forces Bro to construct a large portion of the DFA and, by that, degrades the performance significantly.

tivated. Our complexity attack causes the Hybrid-FA to activate many states in parallel, thus forcing the system to traverse several states per input byte; this degrades system throughput significantly. We show that MCA^2 in full-drop setup can mitigate such an attack: our experiments show that under a mild *active states explosion attack*, the goodput of the system is increased by a factor of 4.8.

This paper is organized as follows. In Section 2 we provide the necessary background on complexity attacks and DPI. Section 3 discusses related work. Section 4 presents the cache-miss attack and its impact on Snort. Section 5 describes the MCA^2 architecture. In Sections 6 and 7 we demonstrate how MCA^2 mitigates cache-miss attacks and active-states explosion attacks, respectively. Our experimental results appear in Section 8. Finally, we conclude in Section 9.

2. BACKGROUND

2.1 Complexity attack

In a complexity attack, the attacker exploits the system's worst-case performance, which differs from the average case that the system was designed for. Crosby and Wallach were among the first to demonstrate the phenomenon on the commonly-used *Open Hash* data structure [13]: an attacker designs an input that requires $O(n)$ elementary operations per insertion, instead of $O(1)$ operations that are required on the average.

Recent works show that many other systems and algorithms are vulnerable to complexity attack, including Quick-Sort [22], regular expression matcher [25], intrusion detection systems [8,15,26], the Linux route-table cache [33], SSL authentication algorithm [11], and the retransmission algorithm in wireless networks [7]. Complexity attacks on different components of NIDS/NIPS were suggested in the past. For example, Bro maintains a hash table with the IP header fields of packets as keys; thus, by tailoring the traffic with specific headers, one can cause the hash insert-operation to last significantly longer, resulting in Bro failure. While in some cases modifying the algorithm suffices to mitigate the problem (e.g., Crosby and Wallach's attack can be solved by using hash functions that are not known to the attacker), this does not hold in general. We believe that only a system approach like MCA^2, can alleviate the attack scenarios discussed in this paper.

2.2 Deep Packet Inspection (DPI) and Snort

DPI is a crucial component in contemporary security tools, which heavily relies on pattern matching to detect signatures of malicious traffic. We consider the following two classes of pattern matching: exact matching and regular expression matching. The former usually uses a Deterministic Finite Automaton (DFA), while the latter uses either a DFA or a Nondeterministic Finite Automaton (NFA) for the ongoing inspection of the input data [18].

In our main example, we focus mostly on the exact matching algorithms, which use DFA. A DFA is a five-tuple $\langle S, \Sigma, \delta, q_0, F \rangle$, where S is a finite set of states, Σ is a finite set of input symbols, $\delta : S \times \Sigma \to S$ is a transition function, returning the next state, given the current state and any symbol from the input, $s_0 \in S$ is the initial state, and $F \subseteq S$ is a set of accepting states. Aho-Corasick algorithm provides a method to build such an automaton (a.k.a. AC

DFA) from a set of patterns. Given the DFA, a packet is inspected by traversing the automaton symbol by symbol from s_0; a pattern is detected if a state in F is reached in this traversal. Fig. 2(a) depicts the AC DFA for the pattern set {E,BE,BD,BCD,CDBCAB,BCAA}.

In today's security tools, AC DFAs are huge—e.g., Snort's AC DFA has 77, 182 states for 31, 094 patterns—raising the question of how to store it efficiently in memory. The alternatives naturally trade memory space with execution time. Additionally, most security tools (including Snort) divide their patterns to several sets, according to the traffic type.

Snort uses a full-matrix encoding for its AC DFAs as presented in [1]. In this representation (see Fig. 2(b)), transitions are stored in a two-dimensional array with $|S|$ rows and $|\Sigma|$ columns. An entry at position (i, j) stores the value of $\delta(s_i, j)$, implying that the number of bits in each entry is at least $\log_2 |S|$. Typically, input inspection of one byte at a time results in an overall memory footprint of $256|S| \log_2 |S|$ ($|\Sigma| = 256$). For Snort's AC DFAs, this translates to a combined footprint of 75.15 MB. On the other hand, the main advantage of this encoding is that a transition consists of a *single* memory load operation, which reveals directly the next state.

Alternative encodings require more than one memory access, but offer significant memory reduction. Such encodings exist in the literature [4,8,30]. Fig. 2(d) depicts such encoding, as proposed in [8]; this encoding is based on a compressed automaton as depicted in Fig. 2(c).

3. RELATED WORK

The recent proliferation of multi-core general purpose processors motivated many researchers to reinvestigate well known problems in this new domain. Among these, there are several works that proposed multi-core solution for DPI processing. These papers' main focus is on different ways to load balance the system tasks between the available cores.

Current NIDS/NIPS systems such as Snort [27] and Bro [10] split the load to many *sequential* sub-tasks in a pipeline manner. Other works, such as [32], suggest fine-grained pipelining for parallelizing network applications on multi-core architectures. This partitioning is effective if the processing cost for each sub-task is similar, which is usually not the case for NIDS/NIPS.

A different line of research focuses on load balance the traffic flows equally between the different cores and performing the inspection in parallel [12,17,21,23,28]. The load balancing is based on both the packet header parameters and some layer-7 parameters. We note that such architectures are orthogonal to MCA^2 and can be applied to load balance the work between general threads that process the normal traffic. If MCA^2 is not used in conjunction with these architectures, they are all vulnerable to complexity attacks.

Becchi et al. [6] focus on DPI and present a performance evaluation scheme for multiprocessor systems. The proposed design also splits the traffic between several cores with the same DPI engine that supports regular expression matching. Their study identifies and evaluates algorithmic and architectural trade-offs and limitations, and highlights how the presence of caches affects the overall performance. However, it is geared at optimizing the normal case and is vulnerable to similar complexity attacks as we describe in the paper. Such attacks can be mitigated by incorporating MCA^2 to this scheme as well.

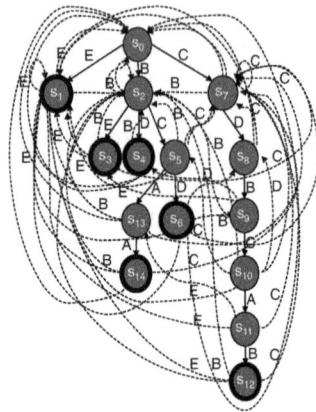

(a) The AC DFA for pattern-set {E, BE, BD, BCD, CDB-CAB, BCAA}

	A	B	C	D	E
S_0	0	2	7	0	1
S_1	0	2	7	0	1
S_2	0	2	5	4	3
S_3	0	2	7	0	1
S_4	0	2	7	0	1
S_5	13	2	7	6	1
S_6	0	9	7	0	1
S_7	0	2	7	8	1
S_8	0	9	7	0	1

(b) Full-matrix Encoding

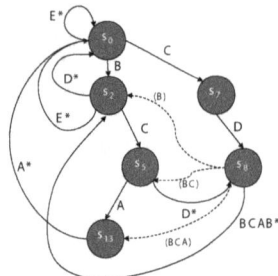

(c) Compressed Automaton

S_0	B: 2, C: 7, E: 0*, fail: 0
S_2	C: 5, D: 0*, E: 0*, fail: 0
S_5	A: 13, D: 8*, fail: 0
S_7	D: 8, fail: 0
S_8	BCAB: 2*, BCA: 13, BC: 5, B:2, fail:0
S_{13}	A: 0*, fail: 0

(d) Compressed Encoding

Figure 2: **Example of an AC DFA and two methods to store it in memory: non-compressed (full-matrix) encoding, and compressed encoding. The compressed encoding is derived from a compressed automaton, in which** *fail* **transitions are taken without consuming input symbols, and transitions marked with '*' indicate that a match was found.**

Another multi-core load-balancing approach is to partition the patterns among the cores (cf. [31, 34, 35]). Then different DPI algorithms, each specializing in different kinds of pattern sets, is run on each core. In some cases, the partitioning itself is done so as to balance the load between the algorithms. It is important to note that, unlike MCA^2, in this kind of architectures, each packet is examined by several cores (each performs only part of the inspection). In addition, it does not take into account the incoming traffic, and is vulnerable to an attack on each core separately.

Kumar et al. [19] present several methods to reduce regular-expressions-based DFA size. One of the mechanisms used in that paper is based on the assumption that normal flows rarely match more than the first few symbols of any signature. Thus, the most frequently visited portions of the automaton are used to build a *fast path* DFA, and the rest of the automaton is represented by a separated NFA, which is the *slow path*. The authors suggest a solution, which is similar to MCA^2 in that it handles heavy traffic with a different algorithm and applies a lightweight classification algorithm to distinguish between heavy and normal traffic. In addition, [19] proposed to protect against DoS attacks by attaching lower priority to flows with higher probability of being malicious. Nevertheless, that work analyzes the case of a single core, and therefore could not benefit from the multi-core properties as MCA^2 does. Furthermore, the pro-

posed protection in [19] fails under a continuous DoS attack because the heavy packets that receive lower priority eventually overload the system buffer. MCA^2 is also resilient to DoS attack of longer duration.

4. SNORT CACHE-MISS COMPLEXITY ATTACK

It has been shown that only a small number of states within the AC DFA is used, when scanning normal traffic [8]. Therefore, a very large fraction of the DPI memory accesses result in a cache hit. With this information, an adversary can launch a *Cache-Miss Attack*, consisting of input traffic that causes the DFA to traverse a large number of states, and therefore, having many cache misses. Such traffic can be constructed easily, since the signatures (and hence the AC DFA) are known publicly. These cache misses have two negative effects. First, a main-memory access is at least 10–20 times slower than a cache access, implying that it takes significantly more time to deal with this malicious input traffic. Second, and even more importantly, dealing with the malicious traffic causes significant cache pollution, which in turn slows down also the processing of well-behaved traffic. In the stand-alone setting considered in [8], the Cache-Miss Attack degrades the performance of the DPI routine by a factor of *four* and is considered an

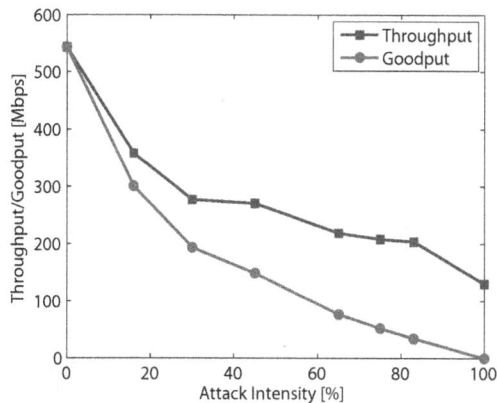

Figure 3: The effects of a cache-miss attack on the throughput and goodput of Snort, facing attacks of different intensities. All attacks do not cause any alert from Snort NIDS.

effective algorithmic complexity attack. The circles-curve in Fig. 1 shows the goodput reduction for different values of attack bandwidth.

While [8] demonstrates the attack on a stand-alone AC DFA, we show that the attack works on Snort, an entire NIDS, which is used in practice. Recall that Snort divides the pattern sets into classes according to traffic types. Among these, the largest DFA is the one that represents the HTTP traffic, with a memory footprint of about 32 MB. We devise a cache-miss attack in two steps. First, we collected the patterns for the automaton from Snort's publicly available signatures set. To prevent this attack from getting detected by Snort built-in mechanisms, we omit the last character of each pattern; this means that our attack packets, which contain these truncated patterns, go under the radar and do not activate any rule that alert the system. Then, we constructed a set of HTTP traffic traces that mix attack packets with normal HTTP traffic to the most-visited web sites [2]. We created eight different traces that differ in the proportion of attack traffic in them. The intensity of attack varies from 0% (only normal traffic) to 100% (only attack traffic).

Fig. 3 shows the overall goodput of Snort under these traces. The throughput of Snort drops by a factor of 1.5 when attack intensity is 16%, and *up to a factor of 4.2* as the cache miss attack becomes more intense. Namely, a Snort IDS with traffic bandwidth of 70% of its maximum capability would be knocked down or let packets go by uninspected under an attack that consumes only one sixth of this bandwidth. This proves the claim that the exact string matching engine is a bottleneck in Snort and shows the great impact that a cache-miss attack may have on such systems. We note that the exact matching in Snort is also an important building block for regular expression matchings: Snort breaks each regular expression into several exact patterns, and invokes a regular expression engine (for a single expression) upon matching all its exact patterns.

Next, we turn to discuss the solution for the complexity attacks that were presented in [8]. The gist of the solution is to use a compressed data-structure that fits mostly in cache (see Fig. 2(c)), and therefore is not prone to this kind of

attack. Recall, however, that this data structure requires more than a single memory access per input byte.

The compressed encoding of Snort patterns requires only 1.5 MB. Compressed AC implementation has almost the same throughput, regardless of the kind of input traffic (the squares-curve in Fig. 1 presents the goodput of this implementation; the linear goodput decrease is due to the increased bandwidth of the attack and not due to an overall throughput degradation). However, it is *two times slower* than that of the full-matrix encoding under normal traffic. This implies that the solution in [8] recommends to always cut the throughput by half in order to overcome cache-miss attacks.

In this paper, we show how a multi-core architecture can be used to break the barrier and enjoy both worlds: high throughput on normal traffic and resiliency to cache-miss attacks. It uses the two encodings as building blocks and provides an efficient way to use them simultaneously, such that each handles the kind of traffic it is best designed for.

5. THE MCA² SYSTEM DESCRIPTION

5.1 MCA² Design overview

MCA² operates over a multi-core platform as described in Fig. 4, where each core runs one or more hardware threads (typically two in the Intel machines). Each hardware thread receives references to packets for inspection via its incoming-packets queue. The Network Interface Card (NIC) receives incoming traffic-packets and places them in main memory. It also writes packet references to the cores' incoming-packets queues. We follow recent works [16, 17] to load balance the incoming traffic between the different hardware threads in the NIC. Note that each packet has a single copy in main memory (created by the NIC). Sending a packet into a queue (or moving it from one queue to another) is performed efficiently by passing a pointer between the cores' queues without a message copy.

The system works either in *routine-mode* or in *alert-mode*. In routine-mode, all threads operate the same: they receive packets from the NIC and process them with the same monitoring algorithm. However, upon switching to alert-mode, the dedicated threads' primary role is to handle heavy packets. Therefore, they might switch to an algorithm that is optimized for such traffic pattern (depending on the kind of the attack). From that point, the dedicated threads receive messages from other threads with references to heavy packets. Thus, the dedicated thread handles the packets references in the messages of its transfer queue, as well as the packets in its incoming packets queue. Furthermore, the following *stealing* policy is incorporated to prevent load imbalance and increased latency for non-heavy packets that were sent by the NIC to the dedicated threads: when a general thread sends a heavy packet to a dedicated thread, it "steals" one or more not-yet-processed packets from that dedicated thread's incoming packets queue and places them at the head of its own incoming packets queue. Our experiments show that the system becomes balanced when the number of packets traded for a single heavy packet is between two and four, depending on the algorithms in use.

The last component of MCA² is its *stress monitor*, whose role is to monitor the percentage of heavy packets in the system and to switch between system modes. Namely, when the percentage of heavy packets crosses a specific threshold, the

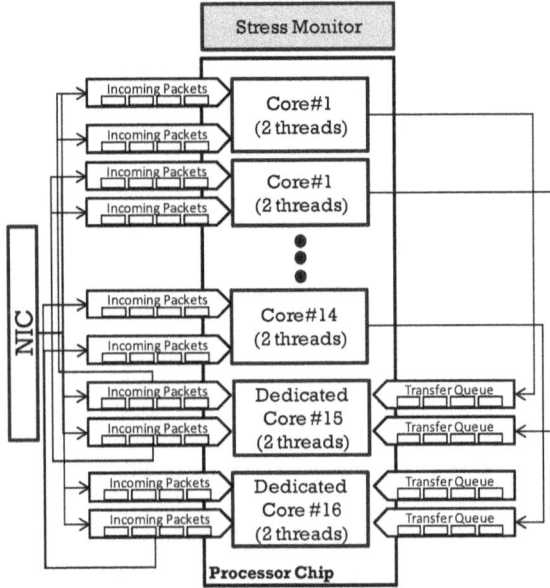

Figure 4: Illustration of MCA^2. Cores 1 through 14 are general, while Core 15 and 16 are dedicated. Each core has two threads. Threads of general cores transfer their heavy packet to a specific thread in a dedicated core through a transfer queue. Only the logical structure of the queues is presented.

system switches into alert-mode; conversely another threshold is used to determine when the system switches back into routine-mode. The thresholds are determined to maximize the system goodput.

We note that in some multi-core environments, the load balancing is done on the flow level (that is, all packets of the same flow are sent to the same core by the NIC) [23]. In such cases, MCA^2 should preserve this property; namely, after classifying a packet of some flow as heavy, all the consecutive packets of the same flow are treated as heavy packets.

5.2 Cross-Thread Communication Mechanism

Concurrency in multi-core systems usually suffers from cross-thread communication overhead, which might become significant in some constellations. The common cross-thread communication techniques require synchronization mechanisms that use expensive system calls and may cause blocking situations. In MCA^2, we use a non-blocking (that is, without any synchronization) mechanism with minimal overhead.

Notice that the most challenging stage of the cross-thread communication in MCA^2 is when writing references of heavy packets to the transfer queues: synchronization might be required since many general threads can transfer heavy packets to the same dedicated thread, resulting in simultaneous access to that queue. Therefore, we implement the transfer queue for each dedicated thread as a collection of queues, one for each general thread that transfers heavy packets to the corresponding dedicated thread. The dedicated thread, in turn, reads from these queues in a round-robin manner. Notice that each such queue is a single-writer single-reader queue.

In order to keep track over the state of records in the queue, the reader and writer threads use *phase* bits that al-

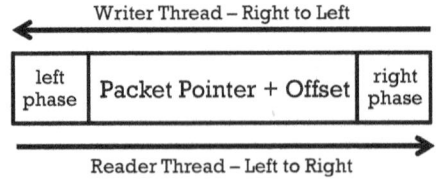

Figure 5: Sketch of a record in the bad packet queue

ternate every time a round of read/write from/to the queue is completed. Specifically, each queue is implemented as an array of records, where each record has a left phase bit, a right phase bit, and a content field. The content field contains a pointer to the location of the packet in memory, along with the *offset* in the payload, which indicates the last byte in which the AC scan was in the root state of the AC DFA (see Fig. 5). Moreover, each queue has two global bit-fields that track the phases of the threads. These fields are `writer_phase`, which keeps the phase of the (single) enqueuing thread, and `reader_phase`, which stores the phase of the (single) dequeuing thread. Finally, each queue has two global pointers: `head`, which points to the next array entry to write to, and `tail`, which points to the next array entry to read from. All these fields are accessible to both threads. However, `head` and `writer_phase` are only written by the enqueuing thread, while `tail` and `reader_phase` are only written by the dequeuing thread.

To write a packet, the enqueuing (general) thread first checks if it does not overwrite a packet that was not dequeued. This is done by checking whether `reader_phase` = `writer_phase` or `tail` > `head`. It can be easily proved that both cases of this condition imply that the dequeuing thread is at most Q packets away from the enqueuing thread, where Q is the length of the array.[3]

If the queue is full, then a packet is not enqueued; the general thread can either process this packet or stall and retry later. Otherwise, the enqueuing thread writes into the queue record *from right to left*. It first writes `writer_phase` in the `right_phase` field, then it writes the pointer and offset, and finally it writes `writer_phase` in the `left_phase` field. If the written entry is the last entry of the array, the thread flips the `writer_phase` bit and writes the next entry to the array beginning.

The dequeuing thread reads from the queue record *in the opposite direction*. It first reads the `left_phase` field, then it reads the pointer and offset, and finally the `right_phase` field. We distinguish between three possible cases when reading a record from the queue:

1. `left_phase` $\neq$ `right_phase`. This implies that the record is now being written. The dequeuing thread should stall shortly and retry reading the entry.

2. `left_phase` = `right_phase` $\neq$ `reader_phase`. This implies that the queue is empty (the dequeuing thread

---

[3]Proof outline: Assume that the condition does not hold. The absolute index of the first packet to overwrite another packet differs by exactly Q from the index of the next packet to read. This implies that `reader_phase` $\neq$ `writer_phase` and `tail` = `head` (namely, `tail` $\not>$ `head`), and hence a contradiction.

should do nothing, and try to dequeue a packet from the next queue).

3. `left_phase` = `right_phase` = `reader_phase`. This implies that the record is valid for reading and processing. The dequeuing thread starts processing the payload of the packet, after skipping its first *offset* bytes. [4]

When a record is read successfully from the last entry of the array, the dequeuing thread flips `reader_phase` and continues dequeuing from the first entry.

Similar mechanism is applied to all other queues in the system (except for the input packet queues of the dedicated threads, which use `test&set` locks to allow packet *stealing*, as discussed in Section 5.1). Our simulations show that even under worst-case traffic, the overhead of this communication mechanism does not exceed 0.98% degradation in system throughput.

5.3 Thread Allocation Scheme

The number of threads allocated to handle heavy packets depends on the exact setup in which the NIDS/NIPS system works. Specifically, we differentiate between two extreme cases: a *no-drop setup* in which no packets are dropped by the NIDS, and a *full drop setup* in which all heavy packets are considered malicious and are dropped immediately. In between, we also consider a *limited drop setup* that allows dropping heavy packets when their percentage exceeds a certain threshold. It is important to notice that in all setups non-heavy packets are not dropped.

The no-drop setup is adequate for an NIDS that only alerts upon an attack. On the other hand, the limited-drop and full-drop setups are used in NIPS; limited-drop is suitable when the security administrator wishes to invest only limited resources in the process, to monitor sporadic attacks over the network and to deal with false malicious traffic.

When deciding how many threads to allocate in each setup, our goal is to maximize *goodput* assuming that the system is balanced. Notice that this goal coincides with maximizing the overall *throughput* of the system. Determining the number of dedicated threads to allocate in the full-drop setup is trivial: no dedicated threads should be allocated and heavy packets are dropped immediately upon their identification. As for limited-drop setup, we need the minimal number of threads to handle only a small fraction of the heavy packets. This can be done either by using only a single thread or all hardware threads in a single core, depending on the attack's characteristic and the multi-core architecture.

A more challenging task is to determine the number of dedicated threads in the no-drop setup. This number depends on the parameters summarized in Table 1.

Naturally, the number of dedicated threads grows along with the fraction of the heavy packets. In addition, we take into account the performance of the two algorithms and consider how they perform while handling either only heavy packets or only normal packets. It is important to notice that the throughput of the algorithm usually depends on r—the fraction of heavy packets it handles. For brevity, our

---

[4]Since the AC DFA was in its root state, when scanning the byte in the offset position, it implies that patterns cannot begin before that byte and end afterwards. Hence, it is safe to skip the scanning up until this byte [9]

Parameter	Description
r	The fraction of heavy packets out of all traffic
$AlgG_h$	The throughput of the general threads' algorithm running solely on heavy packets
$AlgD_h$	The throughput of the dedicated threads' algorithm running solely on heavy packets
$AlgG_n$	The throughput of the general threads' algorithm running solely on normal packets
$AlgD_n$	The throughput of the dedicated threads' algorithm running solely on normal packets
N	The number of available threads

Table 1: **The parameters used to determine the number of dedicated threads (no-drop setting). All throughput values are given for a single thread.**

model does not consider these exact numbers and uses only the two extreme points.

Let β be the ratio between $AlgG_n$ and $AlgD_h$ (see Table 1), and let T be the system's throughput when the entire traffic is normal (that is, no heavy packets) and all threads are general. Thus, when the traffic has a fraction r of heavy packets, the best allocation scheme can achieve a throughput of

$$T \left((1-r) + \frac{r}{\beta} \right).$$

This throughput is achieved when the number of dedicated threads is

$$D_f = N \frac{r/\beta}{1 - r + r/\beta}.$$

Notice that D_f is not an integer, and therefore, it should be rounded to provide the required number of threads. According to the attack type and the multi-core architecture on which MCA[2] is running, one can choose to round D_f so that all hardware threads of the same core would be either dedicated or general. We denote this rounded number by D and it is the output of our model.

Note that we have presented a simplified model for deciding how many dedicated threads to allocate. A more accurate model supports additional aspects. For example, one can use the algorithm of the general threads by a dedicated thread for lower rates of r. This is beneficial when $AlgG_n$ is significantly larger than $AlgD_n$, and since r is too small, most of the packets handled by the dedicated threads are not heavy; second, one might consider the packets' detection overhead. If it is too large as compared to the gain in throughput, the system should not allocate any dedicated thread. Moreover, one can optimize the number of packets that are exchanged between dedicated and general threads, by taking into account the load balancing among them. Our experimental results with the thread allocation scheme are shown in Section 8.2.4. Finally, a useful practice is to limit the maximal value of D_f to preserve a share of general threads under any attack intensity.

5.4 Flow Affinity

NIDS/NIPS systems are required sometimes to preserve flow affinity; namely, all packets from the same flow should be processed in the same core (e.g., to communicate the results of different modules of the system, and to keep inter-packet context). In that case, MCA[2] marks *heavy flows* instead of heavy packets. We note that significant research effort has been devoted to flow affinity in multi-core environment (cf. [16, 17]). MCA[2] can be combined with any

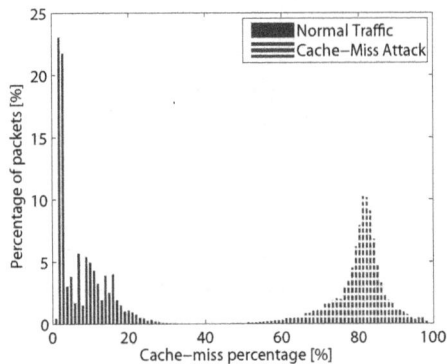

Figure 6: Distribution of cache-misses under normal traffic and under attack.

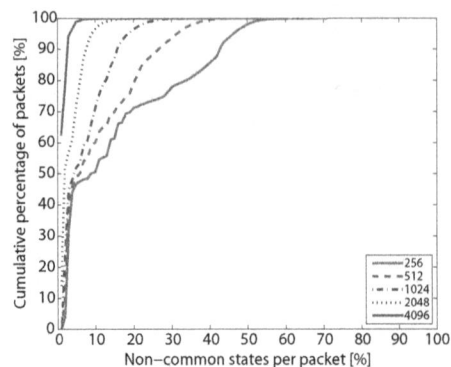

Figure 7: CDF of the percentage of normal traffic packets by their non-common states ratio for different numbers of common states, 256, 512,..., 4096.

Non-Heavy	Number of common states				
packets	256	512	1024	2048	4096
99.0%	**53**	**38**	**25**	**15**	**6**
97.5%	50	35	22	11	5
95.0%	47	32	19	10	4

Table 2: The non-common states ratio for different percentage of non-heavy packets and different numbers of common states.

method that provides flow affinity, yet to divert heavy flows to dedicated cores and thus reduce their effect on legitimate flows.

More specifically, given a system with a packet dispatcher that sets flow affinity, we add a *preliminary* data structure for fast determination of whether a flow was marked heavy or not. This data structure supports insertion of flows and deletion of flows when they either become inactive or when they recover (namely, when they stop behaving maliciously for a certain amount of time/packets). Due to their compact memory footprint and fast lookup time, we suggest using either a counting bloom filter [14] or a hash table with timestamps, so that outdated records can be easily removed.

6. MCA² FOR CACHE-MISS ATTACKS

In this section we present an algorithm for detecting heavy packets in AC DFA complexity attack. Cache-miss attacks are characterized by *a large number of different state machine traversals that cause cache-misses* (as compared with the normal system operations), as clearly illustrated in Fig. 6: On normal traffic,[5] the system has a very low average cache-miss per packet ratio of around 10%, where under a cache-miss attack it is around 80%, leaving an evident margin with a factor of more than 8.

A direct way to measure the value of this parameter is to actually monitor the system cache-misses through the hardware counters. However, this approach is processor-dependent and may not be applicable in our case (either due to lack of appropriate interface or due to the overhead that such monitoring introduces). A more efficient way that was used when implementing MCA² for these attacks, is to approximate the cache-miss upon each input symbol based on the underlying AC DFA itself. This is done by studying the set of states with training traces, ordering the states by the number of visits, and marking as *common states* the most visited states of the DFA when processing the normal packets as discussed below.

An important parameter that should be chosen is the number of common states (that is, in the ordered list of states, what is the rank above which a state is marked as common). Recall that upon normal traffic, DPI is performed with a full-matrix encoding, where each state is represented

[5]Namely, real-life web traffic, see Section 8 for discussion on this trace.

by a row in a matrix of size $256 \log |S| \approx 1KB$ (for Snort's AC DFA). One may suggest to keep the number of common states such that they all fit in the available cache bank. We state that $1KB$ is an overestimate, and in fact, many more states may fit in the cache without causing performance degradation. The reason is that only few outgoing transitions for each state are actually accessed, implying that only a small part of the state's row is actually loaded into the cache.

Prior to determining the number of common states, we explain the interplay between this number and the fraction of the packets that are eventually considered heavy. For each packet, let the *non-common states ratio* be the ratio between the number of non-common states visits and the overall number of DFA traversals per packet. A packet is marked heavy if its non-common states ratio exceeds a certain threshold. Our goal is that under well-behaved traffic the number of packets marked heavy would be very small, as it corresponds to false identifications. Naturally, as the number of common states increases, the number of potentially heavy packets decreases, and therefore the threshold may be increased.

Fig. 7 considers a normal traffic and depicts a CDF showing the percentage of packets by their non-common states ratio. As one can see, this percentage grows quickly as the number of common states increases. Table 2 shows the correlation between the non-common states ratio and the percentage of non-heavy packets. Since the normal traffic contains almost no heavy packets, we set the threshold so that only 1% of the normal packets are marked heavy. These thresholds, for each number of common states, are marked in bold in Table 2.

Using the above thresholds and the thread-allocation scheme (see Section 5.3), we ran experiments in which we mea-

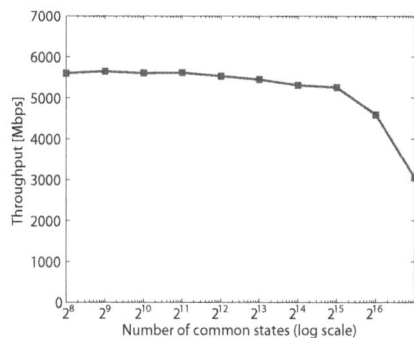

Figure 8: The total system throughput for a different number of common states, under an attack of intensity 33%.

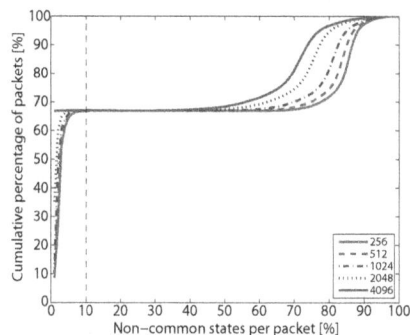

Figure 9: CDF of the percentage of mild attack (33%) traffic packets by their non-common states ratio.

sured the system throughput for different numbers of common states. Fig. 8 depicts these measurements under mild attack, in which 33% of the packets are malicious. Note that there is no significant difference between set sizes below 8 KB. We have repeated these experiments for various attacks scenarios and determined that the highest throughput is achieved when the number of common states is 1,024. This translates to a threshold of 25% traversals to non-common states to mark a packet as heavy. Finally, we note that under mild attack of 33% of the traffic, any threshold above 10% suffices; this is clearly evident by the CDF in Fig. 9.

7. MCA² FOR ACTIVE-STATES ATTACKS

In addition to exact-string matching, contemporary DPI engines usually support regular expression matching. However, unlike exact-string matching, a set of regular expressions is not represented in a DFA, due to the infamous *state blow-up* phenomenon, which implies that these DFAs have prohibitively large memory footprint. A common approach is to replace the DFA with a non-deterministic finite automaton (NFA) [24].[6] When using an NFA, the matching algorithm keeps a vector of active states and for each input

---

[6]Another common practice, used by Snort, is to extract exact-string anchors from the regular expressions and use a DFA to match these anchors. If an anchor is matched, the regular expression engine is applied on the packet for matching only the relevant expression. This reduces the problem

symbol, it computes the next state according to all active states. Naturally, this makes NFA significantly less efficient (namely, when k states are active at the same time on average, an NFA performs k times slower than a DFA).

Becchi et al. [5] proposed a hybrid approach that combines NFA with DFA. Therefore, they have noticed that in the process of transforming an NFA to a corresponding DFA, the states that cause a space blow-up can be easily determined. They interrupt the transformation of these specific states by keeping them non-deterministic, such that they connect two deterministic automatons. This process produces a hybrid finite automaton (*Hybrid-FA*) that consists of a *head DFA*, which is a regular DFA, though some of its leaves are "border states"—states that are non-deterministic and lead to another DFA, named *tail DFA*. As border states are non-deterministic, reaching such a state during traversal requires keeping more than one active state at a time. Thus, this data structure trades space for time by letting more than one active state at a time, but doing so only when space blow-up is actually prevented.

As discussed also in [5], on certain inputs, the average number of active states may be potentially higher by a factor of 30 than on an average case input. This gap reveals a potential complexity DoS attack on a system that uses the algorithm. To illustrate the attack we used the Hybrid-FA code [3] (provided by the authors of [5]), along with a set of regular expressions taken from the Bro NIDS (which was also provided in [3]). We carefully crafted one malicious packet that causes activation of at most eight active states simultaneously.[7] To simulate an attack, we used a trace with 90% legitimate web traffic and only 10% malicious traffic. Our experiment on this trace shows a slowdown of 83% in goodput, implying that the system is very vulnerable even under very mild attack.

We have replaced the pattern matching module, described in Section 6, with the Hybrid-FA pattern matching code [3] to combine MCA² with Hybrid-FA. To identify heavy packets in this case we used a window of 40 bytes in which we examined the average number of active states (Hybrid-FA code keeps a vector of active states, therefore it is simple to poll its size at any time). If during packet processing, the average number of active states in a 40 bytes window exceeds a certain threshold, then the packet is marked as heavy. If the MCA² system is in alert mode it can either drop the packet or send it to a dedicated core, according to the selected configuration.

Fig. 10 shows the distribution of the highest average number of active states per 40 bytes window, per packet, under traffic that contains 90% real-life web packets and 10% attack packets: while normal packets do not exceed 3.1 active states per window on average, our attack packets have maximal average of 7.1. Thus, one can easily differentiate between legitimate and malicious traffic. In Section 8.3 we show the results of our experiments with Hybrid-FA and MCA².

---

to the exact-string matching problem, which was discussed in Section 6.

[7]To create the malicious packet we selected prefixes of regular expressions that contain a "dot-star" in them. We only got eight active states as this is the limit of the specific pattern-set we used, and also since it is enough to illustrate the DoS attack. Many different such packets can be crafted, for convenience we use one example.

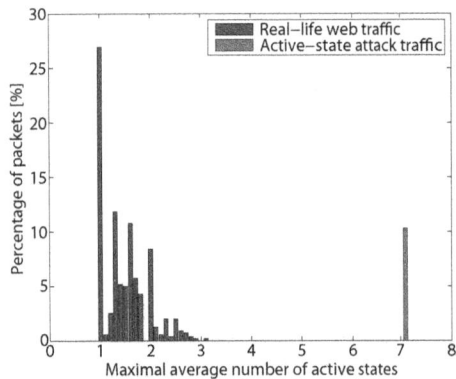

Figure 10: Distribution of maximal average number of active states per 40 bytes window, per packet, under real-life web traffic and under attack.

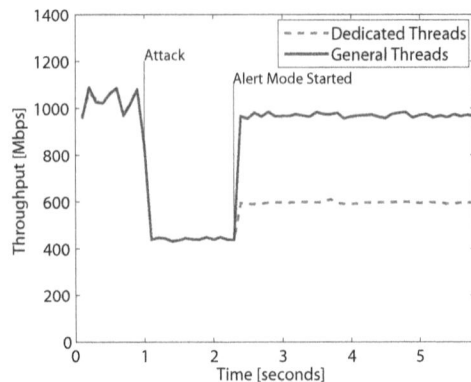

Figure 11: Average throughput per thread over time, when a sudden *cache-miss attack* happens. The system uses eight threads, and when alert mode starts, two of them become dedicated threads.

8. EXPERIMENTAL RESULTS

8.1 Experimental Environment

We use a system with Intel Sandybridge Core i7 2600 CPU, quad-core, each core has two hardware threads, 32 KB L1 data cache (per core), 256 KB L2 cache (per core), and 8 MB L3 cache (shared among cores). The system runs Linux Ubuntu 11.10. Since hardware threads of the same core share the L1 and L2 caches, we have treated them together, as illustrated in Fig. 4. Thread affinity was used to associate threads to cores. In such a way, dedicated threads share the same core, and do not mix with the general threads.

Two web traffic traces are used with size of 145 MB each. These traces contain traffic from randomly selected URLs taken from Alexa top web-sites list [2]. One of these traces is used as our *real-life web traffic* trace and the other trace is used as a training set for determining the common states set for cache-miss attack, as described in Section 6. To simulate a cache-miss attack, we created several *cache-miss attack traffic* traces. These traces contain both normal packets, which cause few cache misses, and flows of specific malicious packets. The latter contains a concatenated list of all patterns from the pattern-set, in order to make as many cache misses as possible. These traces contain different volume of such malicious packets, corresponding to the intensity of the attack. We use cache-miss attack traffic traces with a growing rate of malicious packets, from 0% to 100% (that is, the 20% attack intensity trace contains 20% malicious packets and 80% normal packets). Note that these traces were also used for Fig. 1. To simulate an active-state attack we created another set of traces. These traces also mix normal packets with malicious traffic, with a growing rate of malicious traffic. Adversarial packets in these traces are clones of the malicious packet described in Section 7. The intensity of attack also varies from 0% to 100%.

8.2 Cache-Miss Attack Simulation Results

8.2.1 Goodput

Fig. 1 depicts the *goodput* of MCA^2 when processing the different traffic traces. Note that, as the attack intensity increases, the goodput decreases, since the the non-malicious

traffic occupies smaller portion of the entire traffic. In addition, the penalty of identifying heavy packets (including initial inspection, packet loading, counters initialization, etc.) becomes more significant. Upon an attack we gain a goodput improvement of 67%–102%, as compared to the full-matrix implementation, which do not take cache-miss attack into consideration.

Fig. 1 also depicts the *full-drop setup*, as described in Section 5.3, and also shows a significant improvements in goodput.

Finally, we ran MCA^2 on our web traffic traces. As expected, alert-mode was never activated, and a throughput of 8219 Mbps was obtained on average. This is statistically the same as a light (yet vulnerable) implementation with no MCA^2 at all.

8.2.2 Accuracy

To analyze the *heavy packet isolation*, we first examine the isolation results of our mechanism on the *real-life web traffic* and on the *cache-miss attack* traces. According to the analysis in Section 6, the system would ideally identify 99% of the packets in the first trace as non-heavy. However, as the mechanism estimates of the non-common states rate based on the packet's prefix, it is accurate. Our tests show that the actual rate of packets in the *real-life web traffic* trace that are identified as non-heavy is 96%. These 4% packets would be transferred to a dedicated thread, although being legitimate, and suffer some slowdown. Nevertheless, without MCA^2, these packets would suffer even lower throughput under such attacks. In the *cache-miss attack* traces we know exactly which packet is heavy and can measure precise values for false identification rate. Neither one of the configurations from Section 6 falsely classified malicious packet as normal in more than 0.001% of the trace. We see that our detection mechanism provides an accurate isolation.

8.2.3 Identifying Cache-Miss Attacks

In order to experiment system behavior upon a sudden *cache-miss attack*, we have created a trace that consists of the web traffic trace in which, at some point of time, 33% of the traffic is a cache-miss attack traffic. We set the time interval for checking the rate of heavy packets to *one second*

r	D_F	Optimal Thread Allocation
0	0	0
0.2	0.75	2
0.33	1.26	2
0.5	2.35	4
1	8	8

Table 3: Validation of the thread allocation model of Section 5.3.

for this experiment. We measure the approximate throughput of each thread per intervals of 100ms each. Then, we average the timing per interval for all general threads and dedicated threads. Fig. 11 depicts the result of this experiment. The system starts when all its eight threads are 'general threads'. At the beginning, from time 0 to time 1, input traffic is regular web traffic. Then, at time 1, attack packets begin to arrive, lowering threads throughput by a ratio of about 68%. After a second (time 2), the system identifies the attack and switches to *alert mode*. It sets a pair of threads that belong to a single core as 'dedicated threads' with the compressed matching algorithm, optimized for handling heavy packets. General threads now transfer heavy packets to dedicated threads and therefore are much less affected by them, preserving high *goodput* (general threads still have to scan the first few bytes of heavy packets in order to classify them as heavy. This causes the slight relative slowdown in their throughput as compared to their performance before the attack has started).

8.2.4 Thread Allocation

We validate our thread allocation model as described in Section 5.3. First, we determine the value of β based on our experiments, where all the threads are running either the full-matrix or the compressed implementation. Specifically, we got that $AlgG_n = 1040.6$ Mbps (per thread) and $AlgD_h = 444.3$ Mbps (per thread), therefore $\beta = 2.34$. Table 3 presents the fractional number D_f obtained by our model, and compares it with the allocation that achieves the highest throughput in our experiments (that is, for each value of r we tried all possible thread allocations and picked the optimal ones). Since, in our system, threads are allocated in pairs, all the optimal experimental results coincide with the model's calculations.

8.3 Active-State Attack Simulation Results

Fig. 12 depicts the goodput of Hybrid-FA when processing the different traffic traces, and the goodput of Hybrid-FA when combined with MCA^2 full-drop setup. The MCA^2 full-drop setup provides significant improvements in goodput (as in Section 8.2.1, goodput decreases as attack intensity increases as the the non-malicious traffic occupies smaller portion of the entire traffic).

Considering the other possible configurations of MCA^2, unlike for cache-miss attack, we do not have an off-the-shelf algorithm that can be used on the dedicated cores to boost performance on heavy packets in an active-state attack. The designing of such an algorithm is left for future research.

In terms of accuracy of isolation, MCA^2 isolates our attack traffic from the legitimate traffic. Nevertheless, attacker can create lighter packets that might go under the radar, however such packets are bound to induce much smaller slow-

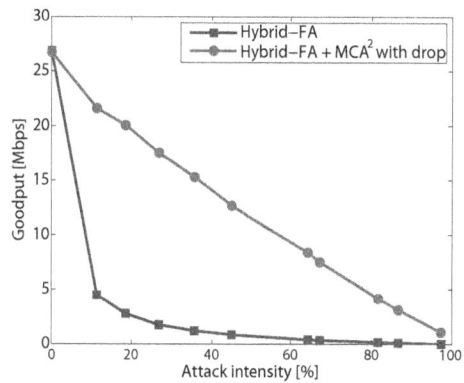

Figure 12: Goodput of Hybrid-FA and of Hybrid-FA with MCA^2 full-drop setup, facing different intensity of active-state attack.

down, if any. We also note that in different legitimate traffic, some packets may be identified as heavy (if they use more active states), but we did not find such traffic in our traces.

9. CONCLUSION

In this paper, we expose a known security hole, the complexity attack, demonstrate its effectiveness, and provide a system solution to mitigate the attack. In the demonstrated complexity attack, negligible effort on the attacker side results in a substantial effort (namely, resource consumption) on the target system. This is a security hole calling for a DDoS attack.

A simple method to mitigate a complexity attack is to throw more computing resources into the system. Obviously, often this is a prohibitively expensive and wasteful approach. An alternative approach is to design algorithms that are efficient in processing malicious packets (e.g., compressing states in the Aho Corasick DPI algorithm). Unfortunately, in many cases an algorithm that works well on malicious packets performs worse on normal packets, thus again requiring more computing resources at normal times. Our MCA^2 architecture provides a method to enjoy from both, special treatment is given to suspicious (a.k.a. heavy) packets in dedicated cores with an optimized algorithm designed for the heavy, while treating the rest of the traffic in the other cores with the best algorithm for the average traffic. This architecture provides several advantages, first the overall system throughput is increased; second, treating heavy packets on the side with dedicated cores isolates the normal traffic from the suspicious traffic; third, we can choose different treatments for heavy packets, without affecting the normal packets; and finally the system may shift gears and decide how many resources to allocate for the processing of heavy packets.

MCA^2 architecture is a general framework to deal with different kinds of complexity attacks. While in this paper we have demonstrated it on one domain—Deep Packet Inspection in NIDS—we are looking to apply the framework for the mitigation of other attacks.

Acknowledgments

This work was supported by the European Research Council under the European Union's Seventh Framework Programme (FP7/2007-2013)/ERC Grant agreement n° 259085.

10. REFERENCES

[1] A. V. Aho and M. J. Corasick. Efficient string matching: an aid to bibliographic search. *Commun. ACM*, 18:333–340, June 1975.

[2] Alexa: The web information company, Dec 2011. http://www.alexa.com/topsites.

[3] M. Becchi. Regular expression processor. http://regex.wustl.edu.

[4] M. Becchi and P. Crowley. An improved algorithm to accelerate regular expression evaluation. In *ACM/IEEE ANCS*, pages 145–154, 2007.

[5] M. Becchi and P. Crowley. A hybrid finite automaton for practical deep packet inspection. In *ACM CoNEXT*, pages 1:1–1:12, December 2007.

[6] M. Becchi, C. Wiseman, and P. Crowley. Evaluating regular expression matching engines on network and general purpose processors. In *ACM/IEEE ANCS*, pages 30–39, 2009.

[7] U. Ben-Porat, A. Bremler-Barr, H. Levy, and B. Plattner. On the Vulnerability of the Proportional Fairness Scheduler to Retransmission Attacks. In *IEEE INFOCOM*, pages 1431–1439, Apr. 2011.

[8] A. Bremler-Barr, Y. Harchol, and D. Hay. Space-time tradeoffs in software-based deep packet inspection. In *IEEE HPSR*, 2011.

[9] A. Bremler-Barr, D. Hay, and Y. Koral. CompactDFA: Generic state machine compression for scalable pattern matching. In *INFOCOM*, pages 659–667, 2010.

[10] The Bro Network Security Monitor. http://bro-ids.org.

[11] C. Castelluccia, E. Mykletun, and G. Tsudik. Improving Secure Server Performance by Re-balancing SSL/TLS Handshakes. In *USENIX Security Symposium*, Apr. 2005.

[12] W. Cong, J. Morris, and W. Xiaojun. High performance deep packet inspection on multi-core platform. In *IEEE BNMT*, pages 619 –622, Oct. 2009.

[13] S. A. Crosby and D. S. Wallach. Denial of service via algorithmic complexity attacks. In *USENIX Security Symposium*, pages 29–44, 2003.

[14] L. Fan, P. Cao, J. Almeida, and A. Z. Broder. Summary cache: a scalable wide-area web cache sharing protocol. *IEEE/ACM Trans. Netw.*, 8(3):281–293, June 2000.

[15] M. Fisk and G. Varghese. Fast Content-Based Packet Handling for Intrusion Detection. Technical report, University of California at San Diego, CA, USA, 2001.

[16] F. Fusco and L. Deri. High speed network traffic analysis with commodity multi-core systems. In *ACM IMC*, pages 218–224, 2010.

[17] D. Guo, G. Liao, L. N. Bhuyan, B. Liu, and J. J. Ding. A scalable multithreaded l7-filter design for multi-core servers. In *ACM/IEEE ANCS*, pages 60–68, 2008.

[18] J. E. Hopcroft, R. Motwani, and J. D. Ullman. *Introduction to Automata Theory, Languages, and Computation*. Pearson/Addison Wesley, 3rd edition, 2007.

[19] S. Kumar, B. Chandrasekaran, J. Turner, and G. Varghese. Curing regular expressions matching algorithms from insomnia, amnesia, and acalculia. In *ACM/IEEE ANCS*, pages 155–164, 2007.

[20] W. Lee, J. a. B. D. Cabrera, A. Thomas, N. Balwalli, S. Saluja, and Y. Zhang. Performance adaptation in real-time intrusion detection systems. In *RAID*, pages 252–273, Octover 2002.

[21] T. Liu, Y. Sun, and L. Guo. Fast and memory-efficient traffic classification with deep packet inspection in CMP architecture. In *IEEE NAS*, pages 208–217, 2010.

[22] M. D. McIlroy. A Killer Adversary for Quicksort. *Software–Practice and Experience*, pages 341–344, 1999.

[23] T. Nelms and M. Ahamad. Packet scheduling for deep packet inspection on multi-core architectures. In *ACM/IEEE ANCS*, pages 21:1–21:11, 2010.

[24] PCRE - perl compatible regular expressions. http://www.pcre.org.

[25] T. Peters. Algorithmic Complexity Attack on Python, May 2003. http://mail.python.org/pipermail/python-dev/2003-May/035916.html.

[26] R. Smith, C. Estan, and S. Jha. Backtracking Algorithmic Complexity Attacks Against a NIDS. In *ACM ACSAC*, Dec. 2006.

[27] Snort: The Open Source Network Intrusion Detection System. http://www.snort.org.

[28] R. Sommer, V. Paxson, and N. Weaver. An architecture for exploiting multi-core processors to parallelize network intrusion prevention. *Concurr. Comput. : Pract. Exper.*, 21:1255–1279, July 2009.

[29] Sony Ericsson Latest Victim of SQL Injection Attack, 2011. http://www.eweek.com/c/a/Security/Sony-Data-Breach-Was-Camouflaged-by-Anonymous-DDoS-Attack-807651.

[30] N. Tuck, T. Sherwood, B. Calder, and G. Varghese. Deterministic memory efficient string matching algorithms for intrusion detection. In *IEEE INFOCOM*, 2004.

[31] O. Villa, D. P. Scarpazza, and F. Petrini. Accelerating real-time string searching with multicore processors. *IEEE Computer*, 41(4):42–50, April 2008.

[32] J. Wang, H. Cheng, B. Hua, and X. Tang. Practice of parallelizing network applications on multi-core architectures. In *ICS*, pages 204–213, 2009.

[33] F. Weimer. Algorithmic Complexity Attacks and the Linux Networking Code. http://www.enyo.de/fw/security/notes/linux-dst-cache-dos.html.

[34] B. Xu, K. Zheng, Y. Xue, and J. Li. Scalable string matching framework enhanced by pattern clustering. *Ubiquitous Computing and Communication Journal*, 5(2):16–26, June 2010.

[35] K. Zheng, H. Lu, and E. Nahum. Scalable pattern matching on multicore platform via dynamic differentiated distributed detection. *IEEE Trans. on Comput.*, 60:346–359, 2011.

Malacoda: Towards High-Level Compilation of Network Security Applications on Reconfigurable Hardware

Sascha Muehlbach
Center for Advanced Security Research
Darmstadt (CASED)
Secure Things Group
64293 Darmstadt, Germany
sascha.muehlbach@cased.de

Andreas Koch
Technische Universitaet Darmstadt
Department of Computer Science
Embedded Systems and Applications Group
64289 Darmstadt, Germany
koch@esa.cs.tu-darmstadt.de

ABSTRACT

While the use of reconfigurable computing for tasks such as packet header processing or deep packet-inspection in high-speed networks has been widely studied, efforts to extend the technology to application-level processing have only recently been made. One issue that has prevented wider use of reconfigurable platforms in that context is the unfamiliar programming environment: Such systems commonly require expertise in computer architecture and digital logic design generally foreign to networking experts. To make the technology more accessible to potential users, we present the high-level domain-specific language Malacoda for application-level network processing and an associated compiler that automatically translates Malacoda descriptions into high-performance hardware blocks for insertion into an FPGA-based processing platform. We evaluate our approach on the use-case of a hardware-accelerated secure honeypot-in-a-box, programmed in Malacoda, and implemented on the NetFPGA 10G board. Results from a live-test of the system connected to a 10G Internet uplink complete the evaluation.

Categories and Subject Descriptors

B.5.2 [**Hardware**]: Design Aids—*Automatic Synthesis*; C.2.0 [**Computer Communication Networks**]: General—*Security*

Keywords

10G, High-Level Languages, FPGA, Networking, Honeypot

1. INTRODUCTION

Modern high-speed networks tax the capabilities of conventional software-based solutions to provide the required performance. Reconfigurable technology has long been used for packet-header processing (e.g., switching, routing, firewalls etc.) and has also been successfully applied to Deep Packet-Inspection (DPI), e.g., to accelerate the Snort Network Intrusion Detection System (NIDS) [21] using dedicated hardware [7].

But high performance is not the only advantage of using reconfigurable logic for network processing. Especially in the security domain, the *absence* of a general-purpose software-programmable microprocessor, which could be compromised by an appropriate attack injecting malicious code, can be exploited to security-harden front-line systems. An application that perfectly fits here are honeypot systems. Honeypots emulate vulnerable applications to attract potential attackers and can be used for various purposes: gathering information about new attacks, collecting statistics about propagation of Malware, or even for active network protection (e.g., [16]). Due to their exposed placement, a common risk for software-based honeypots is their possible subversion into attacking other hosts, if the systems are not carefully monitored. This is even more crucial if the honeypot is participating in an active defense scenario.

We exploit both the high processing performance as well as the improved security of dedicated hardware to showcase the implementation of MalCoBox, a hardware honeypot-in-a-box [13]: Following the well-known low-interaction honeypot approach [6, 4], the MalCoBox emulates an attack surface of hundreds of thousands of vulnerable hosts executing applications having security flaws and collects the malicious attack packets, which can then be studied by security researchers to derive anti-virus signatures or other defenses. Since the system processes data at wire speed, it cannot be overwhelmed, e.g., by a distributed denial of service (DDoS) attack. Furthermore, due to the lack of a compromisable processor, the honeypot cannot be abused by attackers for malicious activities.

While a prototype based on our underlying high-speed network processing platform NetStage has already been presented previously [15], its practical use was limited due to the programming requiring experience in hardware design and the associated tool flows, in addition to networking expertise. Even with these skills, the low-level implementation of accelerated protocol handlers in hardware description languages (HDL) such as VHDL or Verilog still takes significant effort. This is doubly detrimental for our honeypot use-case. On one hand, network security researchers will generally lack the hardware design experience. On the other, with the dynamic attack landscape of the Internet, new vulnerability emulations must be created quickly in order to keep up.

As a solution, the major new contributions of this work are:

- A specialized high-level language (Malacoda) for concisely describing service emulations and application-level (ISO / OSI Layer 7) vulnerabilities for the honeypot.

- An associated compiler for creating fast hardware units executing on NetStage.

- A fully functional implementation of the described honeypot system on the NetFPGA 10G card, programmed in Malacoda, and stress-tested in a real data center environment.

A brief overview of the NetStage platform is given in Section 3. Section 4 covers the Malacoda language for programming in the honeypot domain, followed by a description of the current prototype compiler in Section 5. We evaluate our approach using it for an implementation of the hardware honeypot-in-a-box on the NetFPGA 10G board and discuss the results of a long-term live evaluation run in Section 6, before we conclude and look forward to further research.

2. RELATED WORK

Making reconfigurable network processing platforms more accessible for non-hardware designers has been the subject of considerable research. This has ranged from focused (but very effective) approaches such as compiling the Snort payload signature ruleset (regular expressions) into hardware accelerators [7] to more general-purpose solutions such as G [3] and Chimpp [22].

G is a general-purpose language, but specialized for packet-header processing. It allows users to flexibly specify the packet format (fields and positions) and conditional rules for modifying these fields depending on packet contents. The programs are then compiled into hardware units to be integrated into a larger system (not described in [3]). While capable of payload processing, G lacks regular expression handling and extended support for protocols above the level of processing individual incoming packets. The focus on header processing in G is also emphasized by the example applications, which deal with switching or MPLS routing [3].

Chimpp is more general framework in that it relies on an XML description for the composition of arbitrary packet-handling hardware blocks. These blocks can be of various granularities (e.g., ARP lookup or simple TTL decrement), but must be implemented manually in a synthesizable HDL. Chimpp only supplies the interfacing / composition capabilities. The authors propose a basic library of modules with focus on routing applications for the NetFPGA platform [11], which they use to build an IP router and NAT gateway as sample application. However, modules for higher-level payload processing are not provided. The packet header-processing roots of Chimpp are apparent when considering that it takes it inspiration from Click [8], a popular software framework for the description of routing operations.

NetThreads [9] uses an alternative approach to improve the programmability of the reconfigurable network processing system: Instead of generating custom logic, NetThreads defines specialized 4-way multi-threaded processors on the NetFPGA platform, which are then software-programmable

in languages such as ANSI C. While this offers networking experts a familiar programming environment, for complex tasks the performance of the system does not reach the performance of dedicated hardware accelerators. E.g., for a sample application that does regular expression matching to classify HTTP packets, a performance of roughly 2000 Packets/s is given in [9], which is comparable to a throughput of approx. 16 Mb/s for 1024B packets. Without dedicated hardware accelerators for regular expression or protocol processing, it appears questionable that the approach has performance benefits exceeding those of existing hardwired network processor ASICs [24, 18].

While it would be possible to support a C programming environment for custom-generated reconfigurable network processing units by using one of the commonly available C-to-Gates compilers (e.g., [12, 23]), Brebner [3] shows a productivity gain of more than 6x for using a domain-specific language such as G versus a similar implementation in hardware-synthesizable C (which also has to take the language idiosyncrasies of the specific C-to-Gates compiler into account).

In summary, we are aiming for a system with the flexibility of NetThreads (supporting full protocol interaction processing) and the conciseness of a domain-specific language such as G. To this end, we require not only header, but more advanced payload processing capabilities such as regular expression matching and state tracking. The descriptions should be compiled into multi-threaded hardware units tightly integrated into our high-performance 10G network processing architecture, which provides the underlying general-purpose Internet communication functionality. None of the prior solutions match these requirements.

In terms of the honeypot application, MalCoBox represents the first attempt (to our knowledge) to implement such a system entirely on dedicated hardware. In contrast to the work of Pejovic et al. [20], where memory table-based state machine are interpreted to describe the client-server interaction, we rely on dedicated hardware accelerators for this task. Pejovic et al. implemented a prototype on a Virtex-4 FPGA, but unfortunately did not publish any performance benchmarks. While their table-driven approach is easier to implement in hardware, we expect memory bandwidth to become a bottleneck when network speeds of 10+ Gb/s are considered. Also, they propose the use of a PowerPC CPU to implement parts of the higher-level protocols, which we strictly avoid on our architecture for security reasons.

3. PLATFORM ARCHITECTURE

Figure 1 shows the current architecture of NetStage [15] for the NetFPGA 10G card, which meets the requirements of a basic network server application. Core features include a specialized implementation of the basic Internet protocols (IP, ARP, ICMP, UDP, TCP) as well as facilities for routing packets, scheduling time-based events (e.g., for packet retransmissions), and per-thread state (context) storage. An external management interface allows the monitoring and control of the system independently of the production network.

The network communication core (Fig. 1.a) implements the low- and mid-level Internet protocols. All of the modules have separate 128b wide transmit/receive datapaths, allowing full-duplex operation and reach 20 Gb/s throughput at the nominal 156.25 MHz clock frequency of the network in-

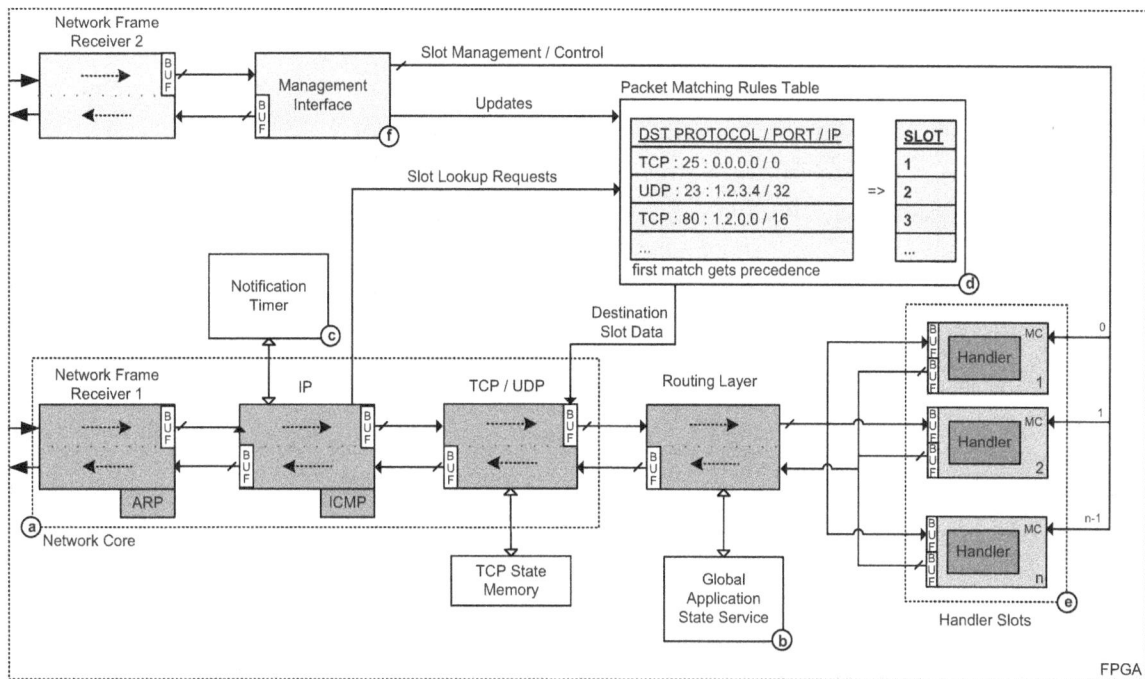

Figure 1: Core architecture of NetStage

terfaces (note that higher speeds of the core are possible, see Section 6). All of the processing stages are decoupled using buffer queues to limit the impact of throughput variations (e.g., during dynamic partial reconfiguration of new Handlers into Handler Slots, Fig 1.e).

While the core provides all communication facilities, it does not deal with the actual application-level protocol processing, which is provided in the form of dedicated hardware *Handlers*. These are attached to the packet routing layer in pre-defined *Slots*, allowing them to be easily replaced (including by partial reconfiguration). The automatic creation of these Handlers as part of the honeypot application forms the main subject of this work. While some of the platform features relevant for that discussion will also be presented here, the platform architecture and capabilities are described in greater detail in [15].

The base NetStage architecture is highly portable: Initially, it has been evaluated on the BEE3 hardware platform [2], fully exploiting the four Virtex 5 devices. For the current research, it was ported to the NetFPGA 10G board [17], trading reconfigurable area for access to fast FPGA-external QDRII SRAMs to provide more context storage and timed events. Additionally, remote management of the complete system is simplified since the NetFPGA 10G card is easily plugged into a standard 2U rackmount server.

3.1 Reconfigurable Protocol Handlers

Protocol Handlers (Fig 1.e) are responsible for the actual application-level processing of network data. In the honeypot scenario, each protocol Handler emulates a certain service and/or one or more application vulnerabilities. Specifically, the Handlers react to incoming packets and generate response packets according to predefined rules that can also track per-session state. However, the handler hardware units

themselves do not have a direct access to a long-term memory storing application-level session information for multiple connections. Instead, all context data is stored externally in the Global State Memory (see Section 3.3) and provided to the Handler along with the session packets. This allows multi-threaded processing in each Handler, where packets of different sessions are processed on the same hardware in an interleaved manner.

Figure 2 shows the architecture of a Handler. It consists of the actual protocol state machine, an (optional) regular expression matching engine, and an (optional) set of response packets described as stored templates. These three components need to be customized for each application, which previously required writing RTL HDL, but is now automatically performed using our new compiler (Section 5). A Slot "wrapper" acts as standardized interface, which provides buffering (implemented as ring buffer) of incoming and outgoing messages and simplifies the attachment of Handlers to the core system.

3.2 Message-based Communication

Core stages and Handlers use a message-based communication scheme for data exchange. Generally, messages encapsulate packet payloads by prefixing them with an Internal Control Header (ICH, see Figure 3). Additionally, the system uses ICH-only messages to transport control data independent of network traffic.

The routing of messages between the core and the Handlers is determined by a Packet Matching Rules Table (Fig. 1.d) that selects a target Slot considering the protocol, port, and IP address of an incoming packet. The use of netmask-based prefix matching allows to bind a Handler to an entire subnet of addresses, which is essential for the honeypot to span large address ranges. Slot lookup is efficiently im-

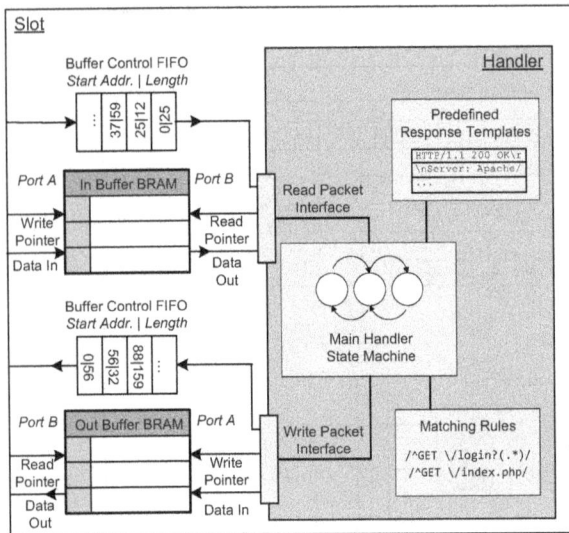

Figure 2: Slot wrapper and Handler architecture

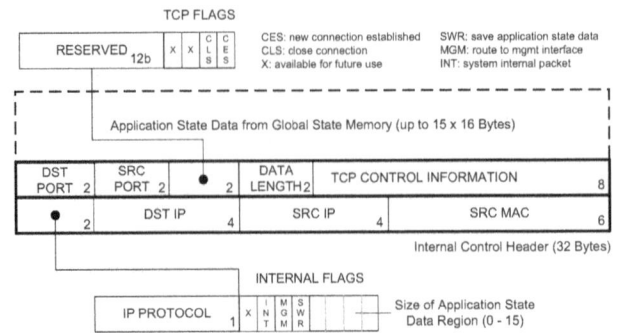

Figure 3: Internal Control Header (ICH) preceding each message

plemented using a content-addressable memory (CAM) and fully pipelined. The rules are configured using the management interface. In contrast to [15], the routing layer has been separated from the core in the NetFPGA implementation to allow for efficient handling of more application state data without burdening the core.

3.3 Global State Memory

Instead of storing per-session state locally in the Handler, this is done centrally in the Global State Memory (Fig. 1.b). This simplifies Handler design as well as swapping in and out Handlers in a dynamic partial reconfiguration (DPR) scenario [15].

The state data is transported as part of the ICH to and from the Handlers. The routing layer manages reading and writing state data from an external QDRII SRAM. Memory addresses are given by the packet source-address / protocol / port combination (hashed together). The SWR control flag in the ICH is used to request that the ICH Application Data Region is written back to the Global State Memory. The size of the state data required for each Handler is defined at design time, stored together with the routing rules, retrieved when performing a Slot lookup, and entered in an ICH field.

Access to the state data itself is pipelined between the core and the routing layer. To further improve platform efficiency, the routing layer contains two queues: one for packets requiring state data, a second one for packets without state data. The latter can then be processed without waiting for state data to become available.

The NetFPGA 10G implementation of the platform supports up to fifteen 128b data words of per-session state (240 bytes), which is generally sufficient for our honeypot use-case to hold passwords, session IDs, or session states.

3.4 Notification Timer

Each Handler can produce response packets at least at line speed. While this is advantageous for high-speed environments, scenarios are conceivable where a client would be overwhelmed with packets at these rates, leading to packet loss or failing communication. Therefore, the platform can

throttle the data transfer by letting the producer thread on the Handler sleep before sending the next packet.

This mode of operation is supported by the Notification Timer (Fig 1.c), which allows a connection (thread) on a Handler to sleep, freeing the Handler for the next connection, and waking up the sleeping thread after the required time (selectable from two globally configured time intervals) has passed. The thread desiring to sleep sends an appropriate control message as internal message (with the ITP flag set) holding both its internal state as well as a selector for the desired timer to the core for requesting a later wake-up. On wake-up, the Notification Timer sends an internal wake-up message to the Handler, restoring the state (context) of the original thread and allowing it to continue execution.

The same functionality is also used for implementing TCP retransmission of unacknowledged packets. Instead of storing the previously crafted response packet each time it is sent out (which would waste external memory), the Handler creates a notification packet. This allows it to rebuild the packet after a certain time period, if the packet has not been acknowledged in the meantime. For the honeypot application, a useful sleep time is $50\mu s$ for throttling and the retransmission timer is set to 200ms.

4. PROGRAMMING IN MALACODA

Many of the solutions to simplify FPGA programming (see Section 2) are achieving good results when focusing on particular problems, e.g., by introducing a Domain Specific Language (DSL). Such a DSL has advantages both for the programmer as well as the compiler. A DSL allows the programmer to describe a specific problem in his domain, while the compiler can generate highly efficient hardware circuits due to its more precise knowledge about language use and the target architecture. Furthermore, by using a DSL, characteristics of the target hardware platform (e.g., multi-threading etc.) can be reflected already in the language specification, and need not be retrofitted as pragmas or library calls.

Together, these aspects make a very strong point for using a DSL to allow network engineers to describe new Handlers in their traditional application domain. An automatic compiler can then generate high-performance hardware blocks matching the execution model of the NetStage architecture (see Section 3.1). This approach has initially been presented at the conceptual level in [14]. The resulting feedback from

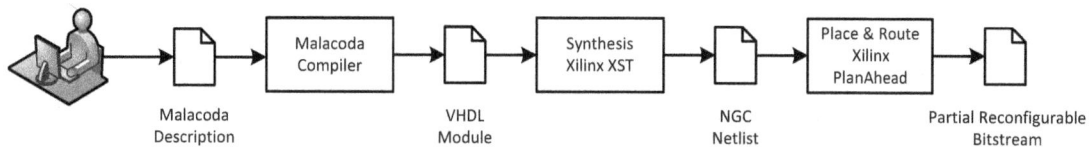

Figure 4: Tool flow for generating hardware from Malacoda programs

domain experts has since led to the creation of the Malacoda DSL and its compiler, both presented here for the first time.

4.1 Background

Service emulations in current software honeypots [6, 4] are generally described in the implementation language of the core honeypot system (commonly, script languages such as Tcl/Tk or Perl), there is no unified way of describing such emulations. Furthermore, such general-purpose programming languages are difficult to efficiently compile into hardware. On the other hand, to lower the barrier of entry for Malacoda, its syntax was inspired by Perl, which should be familiar to most network engineers and security researchers.

In contrast to a general-purpose processor, the complexity and performance of dedicated hardware is highly dependent on the current task to be executed. There are limiting factors, e.g., available FPGA resources, that often require a trade-off between resource use and performance. Based on an analysis of existing honeypot scripts and service emulations, we have designed Malacoda with a balancing of these trade-offs firmly in mind. The analysis has shown the following major operations to be essential for modeling a wide range of service Handlers for the honeypot scenario:

- Describe states and transitions that reflect the communication session.

- Evaluate the incoming request packet and craft a proper response packet by filling-in static template data and inserting parts from the original request packet, based on certain rules.

- Notify an administration station about certain protocol stages of a client conversation.

We will focus both language specification as well as compiler construction on these crucial base functionalities.

4.2 Syntax

Malacoda describes the sequential operations processed in a single NetStage thread. Parallel operations are executed by the automatic multi-threading performed by the NetStage core. Listing 1 shows a sample Malacoda description emulating a Telnet login into a shell.

A Malacoda description starts with a name, followed by an optional section to define state variables. The basic template of a Malacoda description is a protocol dialogue that contains multiple states and (conditional) transitions. A dialogue models the communication session required for emulating a certain service or vulnerability. Each state is identified by an assigned name, with **DEFAULT** indicating the initial state. That state is entered on a newly opened TCP connection or for any arriving UDP packet.

The body of a state description consists of actions (commands and assignments). An assignment to the reserved variable **state** indicates a transition to the indicated state after processing the current packet. State actions include the sending and receiving of packets, while conditional execution is expressed as **if**/**elsif**/**else** constructs. Additionally, the language supports regular expression matching using a subset of the Perl operators.

Listing 1: Sample Malacoda program

```
// Emulate login to a root shell
TELNET {
  // define variables
  dynamic username[14];
  // main fsm
  dialogue {
    // default and initial state for a new
        connection
    DEFAULT:
      addresponse("login:");
      $state = LOGIN;
    // next state
    LOGIN:
      // extract user name
      $username = chomp($INPKG);
      addresponse("password:");
      log("TELNET: Login attempt detected");
    ...
    SHELL:
      // emulate Unix uname command
      if ($INPKG =~ /^uname -a/) {
        // send the system identification
        addresponse("Linux myhost 2.6.35.6 ...");
        addresponse("\n");
        addresponse("[localhost]# ");
      }
    ...
  }
}
```

4.3 Malacoda Commands

Malacoda allows the following commands in state actions currently. *SOURCE* can be either a string (of ASCII characters or byte values, expressed by prefixing two hex digits with \\), or a variable name (reserved or user-defined). Note that a response packet may be incrementally constructed with multiple commands. It will only be sent once all actions have been processed for a state. Furthermore, all commands implicitly operate on the output buffer.

- **addresponse**(*SOURCE*): Append a byte sequence to the response packet buffer.

- **addresponse**(*SOURCE*, s, n): Copy n bytes starting at index s from *SOURCE* to the response packet.

- **addresponse**(**file**:*STRING*): Send a given byte sequence defined at compile-time in an external file (useful for larger responses that would make the Malacoda program hard to read if embedded into the Malacoda source).

251

- **log** (*SOURCE*): Send log packet with the given byte sequence to management interface.
- **if/elsif/else** (*expression*): Conditionally execute commands depending on the value of *expression*.
- **replace**(*s*, *SOURCE*): Replace a single byte or a byte sequence of the response packet with the value given by *SOURCE* starting at index *s*.
- **close**: Send a close connection notification with this response packet to the client (only available for TCP connections).

Beyond the special commands, the Perl command **chomp** is supported to remove any newline character from a byte string.

4.4 Expressions

Malacoda supports arithmetic, regular expression, and comparison operators in expressions. The current version of the language uses unsigned byte sequences as the fundamental data type. The reserved variable **$INPKG** indicates the entire payload of the current input packet.

Sub-ranges of a variable may be selected by the [ ] operator: $VARIABLE[n]$ selects an individual byte of a variable, while $VARIABLE[a,b]$ selects the given byte sub-sequence of a variable (from index a to index b, inclusive). Individual bits of a byte (e.g., required to set a flag in a custom application protocol), are accessed with $VARIABLE[n][p]$, where p is the index of the bit.

4.5 User-Defined Variables

For advanced emulations, Malacoda allows the explicit storage of per-session state in user-defined variables. These are held in the Global State Memory (see Section 3.3) for the duration of the entire client session. Variables store unstructured byte sequences that are interpreted in context of their current operator. However, they can be declared differently depending on whether they have a variable length (up to a static upper limit ≤ 255) or a fixed length (Listing 2). In the first case, an additional byte of storage is used to track the length of the variable. Longer values will simply be truncated to the maximum variable at the fixed length.

Listing 2: User-defined Variables

```
// variable with dynamic length (in bytes)
dynamic variable1[8];
// variable with fixed length (in bytes)
fixed variable2[4];
```

5. COMPILING MALACODA

The Malacoda compiler has the following design goals:

- Make the MalCoBox hardware honeypot accessible to security and network engineers without hardware design expertise.

- Enable hardware-experienced engineers to quickly generate template code for Handlers that can be later manually optimized for more complex under-the-hood operations.

The resulting compile flow is organized as shown in Figure 4 and produces synthesizable VHDL descriptions. Each

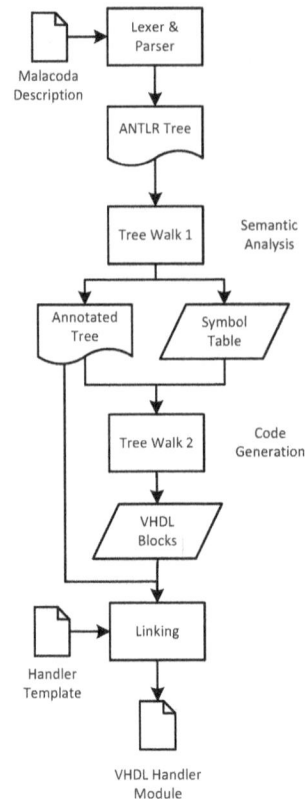

Figure 5: Malacoda compiler organization

VHDL module implements a single Handler and can be compiled into a bitstream using the standard FPGA development tools (e.g., Xilinx ISE [27]). Experienced hardware engineers can easily add custom code to the VHDL handler representation (e.g., for handling special-cases such as handler-local calculations), while non-hardware developers can rely on automatic scripts to generate the FPGA bitstream. Note that our support for dynamic partial reconfiguration on the Xilinx platform requires the use of the PlanAhead [26] floorplanning tool in addition to the usual logic synthesis and FPGA mapping steps.

5.1 Compiler Design

The construction of the Malacoda compiler is shown in Figure 5. Due to the highly specialized nature of Malacoda, much of the complexity of general-purpose high-level language compilers can be avoided. Most of the basic compiler operations are performed using Java code automatically generated by the ANTLR v3 compiler-construction tool [19]. It not only generated the lexer and parser from a formal representation of Malacoda, but also the creation of the Abstract Syntax Tree (AST) used as intermediate representation.

During the semantic analysis pass, the AST is traversed to build a symbol table of states, variables, and regular expressions. Furthermore, the pass discovers the basic blocks and their control predicates, storing this data by annotating the AST. That information is exploited in the code generation pass, which expands a pre-defined Handler template in VHDL by replacing placeholders with the actual signal declarations, output assignments etc. The template already contains the buffered interface to the NetStage core and a skeleton FSM for receiving and sending messages, which is then extended with the Handler-specific processing.

5.2 Handler Execution Model

The execution model of a compiled Handler is split into two phases: reading an input packet and writing an output packet, word by word. At run-time, the generated hardware initially evaluates all conditions in the handler in parallel by reading the entire incoming packet. This is possible, since Handler state is only updated atomically on a state transition. Malacoda assignment semantics are thus similar to the non-blocking assignments in VHDL and Verilog. After reading the entire packet, the condition results are evaluated in program order, selecting a state and predicating the execution of the state actions, which are then executed sequentially in the second phase. To reduce Handler complexity, the compiler does not yet optimize to stop reading packet data if no conditions remain that could potentially match. E.g., if a Handler checks a single condition in the first 128b of the packet, the remaining words of the packet could be skipped (if no other command requires reading them).

However, the current lack of this optimization will not lead to major slowdowns, since the majority of the processing time is often spent constructing output packets by copying the data from the Handler-internal template storage to the output queue in the second phase. In this phase, the generation of the output packet defines the state sequence. Actions are reordered to execute in the order their output occurs in the response packet. This avoids idle cycles by continually streaming data to the packet-under-construction.

If the packet-under-construction has grown to the MTU size, it is transmitted (performing a segmentation-like operation). The building of the response then continues in a new packet. Depending on the user-defined policy, this next packet is either constructed and sent out immediately, or explicitly delayed using an internal timer notification request message. For TCP connections, an internal timer notification request message is always generated to schedule a possible retransmission of the constructed packet after the appropriate time-interval. Finally, if any log packet has been assembled, it is now output to the management interface.

5.3 Regular Expression Matching

The compiler generates a dedicated matching engine for each regular expression in a Malacoda program. Currently, these engines are implemented as simple FSMs with maximally parallel comparators to ensure that a result is available immediately after the last word of a packet has been read. While this approach is feasible for the current prototype which focuses on basic character and string matching (e.g., the compiler currently does not support character classes), it would be worthwhile to integrate more refined matching architectures that are both smaller and support a larger set of regular expression operators (e.g., [5, 25]).

5.4 Packet Construction

Response packets can be generated by copying data from stored templates which are modified on-the-fly using Malacoda commands (e.g., **replace**) at run-time. The compiler can implement these template ROMs (which are also 128b wide) either as LUTs (for small templates) or BRAMs (for larger ones). Depending on the amount of template data,

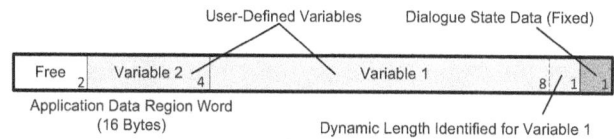

Figure 7: Variable allocation in Global State Memory

the compiler decides which model fits better. Based on the synthesis results (see Section 6.3), the upper limit for a LUT-based implementation has been set to 1 KB of static data.

For some protocols (e.g., DNS), the response packet contains much of the data received in the request packet. The compiler optimizes this special case by detecting if an **addresponse($INPKG)** command (that copies from the input buffer to the output buffer) occurs in the Malacoda program. In that case, dedicated wiring is generated to perform this copying of data in parallel hardware while the packet is read during the condition evaluation phase. Note that this operation is speculative: If a control condition would actually select a different execution path at run-time, the copied packet is removed from the output buffer simply by resetting its write pointer.

Similarly, the **replace** operations in the program are not mapped to byte-wise copy-and-select steps when reading a template. Instead, they too are turned into a dedicated wiring/logic network that modifies all bytes of a 128b template ROM data word in parallel (e.g., by permutation, insertion, deletion, replacement) to achieve a high throughput. To this end, the compiler needs to differentiate between fixed and variable length output packets, shown in Figure 6: If the length of the output packet is fixed (e.g., when always copying a fixed number of bytes from the input packet or by sending only data from a response template), the compiler can create the logic required to route data into the output packet to address fixed write offsets.

This task is more difficult for variable packet sizes (see Fig. 6-b). If data from a template should be appended to a response packet that already contains a variable-length variable, the byte offset of 128b word of template data in the output buffer depends on the number of bytes previously written to the buffer. If this case is detected at compile-time, the compiler generates a wide barrel-shifter that can move the template data to the appropriate offset within the output buffer within a single cycle. Since the barrel-shifter requires many FPGA resources, it is only created if required by the current Malacoda program.

Log packets to the management console are generated in a similar fashion, but they will always be tagged with the IP source and destination address/port information of the original packet, as well as a system-wide 128b time-stamp for better correlation of log messages with traffic dumps.

5.5 Global State Memory Allocation

The compiler also allocates proper space in the Global State Memory (see Section 3.3), both for the reserved internal (e.g., **state**), as well as for the user-defined variables (Figure 7). Dynamic variables occupy their maximum length plus a byte tracking their current length, static variables always have a fixed size. After allocating all variables, the compiler reports the number of state words required for this

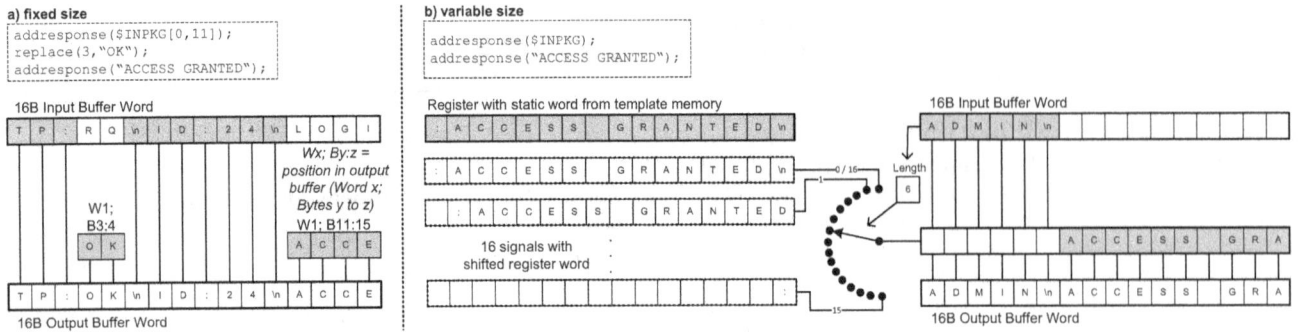

Figure 6: Construction of output packet for fixed and variable-sized response

particular Handler. This information is later required when configuring the routing rules (see Section 3.3).

The state variable is always used automatically if more than the DEFAULT state exists within a dialogue. Symbolic state names are binary-encoded into integers.

6. EXPERIMENTAL RESULTS

6.1 Hardware and Environment

The system has been implemented on the NetFPGA 10G board [17]. It uses two 10G Ethernet interfaces, a Xilinx Virtex-5 TX240 FPGA, and three 72 Mb QDRII SRAMs (holding the Global State Memory, the Notification Timer event queues, and the TCP connection state). A MicroBlaze CPU is present in the system, but only configures the Ethernet SFP+ link parameters, it does not have any contact with network data.

For the honeypot live-test, the NetFPGA card has been placed in a 2U Linux server, equipped with an additional 10G network card that connects to the management interface. For analysis, we implemented a monitoring option for NetStage that copies every packet received from the public interface to the management port. These packets are then captured on the Linux host using the standard tcpdump command. Separation between actual management traffic and this mirroring traffic is achieved by considering the different MAC addresses.

The entire project has been synthesized with Xilinx ISE / EDK 13.3 [27] and mapped with PlanAhead 13.3 [26]. We have configured our platform with six Handler Slots, each with a maximum of 6080 LUTs and 8 BRAMs of FPGA resources. The network core runs at a target frequency of 175 MHz, giving a raw internal throughput of 22.4 Gb/s.

6.2 Synthesis Results

Table 1 shows area results for the portable NetStage core and the board-specific interface and control logic. We then list the FPGA areas required when building the system for dynamic reconfiguration using six Handler Slots, or alternatively, when statically compiling our six sample Handlers (see next Section) into the design.

Note that the platform itself, including the message routing network for six Handlers, requires only 22% of the LUTs in the TX240 FPGA. However, the heavy use of buffers imposes a high demand for BRAMs. Even when reserving space for dynamic reconfiguration in the form of six Slots, the total design requires less than 50% of the TX240 LUTs,

Table 1: Synthesis results for system and components

Module	LUT	FF	BRAM
NetStage core (w/o Handler)	23,278	29,512	156
Infrastructure	9,504	11,029	28
Platform w/o Handlers	**32,782**	**40,541**	**184**
+ 6 Slots	**69,262**	**77,021**	**232**
+ 6 Handlers	**46,372**	**46,066**	**182**

leaving ample space for more or larger Handler Slots, or additional core functionality. The statically configured version does not suffer from Slot-internal fragmentation and is even smaller, but no longer has the self-adaptation capability using DPR.

6.3 Handlers

For our evaluation, we have selected six different Handlers for the emulation of typical network services or actual vulnerabilities:

- Web server: Imitates a webmail service running on a vulnerable web server (identified by a corresponding version header), collecting information about web server attacks.

- Telnet: Emulates a faux system administration CLI accepting any login / password to gather data about what combinations attackers try and commands being executed after login.

- Mail server: Pretends to be an open relay simply accepting every mail (SMTP protocol) to gather information about spam attempts.

- MSSQL Slammer detection: Responds to MSSQL Ping and detects a malicious packet as sent by the Slammer [10] worm.

- SMB login detection: Emulates the first steps of the protocol until client login. Used to gather information about attack attempts on the SMB service.

- DNS server: Emulates a DNS server that resolves a single domain. Used to collect information about DNS attacks.

Table 2: Synthesis results for compiled Handler modules

Handler	Opt.	LUT	FF	BRAM	Max Freq. MHz
SMB	LUT	3,383	1,624	0	185
DNS	LUT	2,864	1,447	0	223
MSSQL	LUT	1,894	1,288	0	212
Telnet	LUT	3,921	1,643	0	175
Mail	LUT	2,460	1,543	0	193
Web	BRAM	2,285	1,355	4	203
Mail	BRAM	2,432	1,584	4	183
Web	LUT	5,796	1,346	0	193

Table 3: Latency for selected operations

Handler	Operation	Latency [Cycles]
Web	120 B Request	$8 + \lceil 120\,B/16\,B \rceil$
	1024 B Response	$11 + \lceil 1024\,B/16\,B \rceil$
Mail	14 B Request	$9 + \lceil 14\,B/16\,B \rceil$
	16 B Response	$8 + \lceil 16\,B/16\,B \rceil$
DNS	33 B Request	$8 + \lceil 33\,B/16\,B \rceil$
	99 B Response	$5 + \lceil 99\,B/16\,B \rceil$

Each Handler has been programmed in Malacoda and compiled into hardware using the developed compiler. Table 2 shows the required resources for each of them. For Mail and Web, we also list alternate results when manually choosing a different LUT / BRAM implementation option for the Template ROMs (see Section 5.4).

Compared to the other Handlers, the MSSQL emulation requires only few LUTs. The size of response templates has a major impact on the Handler size. E.g., the Telnet Handler has 570 B of replies in 18 templates, while the MSSQL Handler has only one, since it just detects an attack and logs its occurrence. Furthermore, the Web server Handler demonstrates that implementing large portions of response templates (in that case, 7 KB) in BRAM instead of LUTs has a significant advantage in terms of resource usage. In terms of code complexity, the Malacoda Handler descriptions have 16 ... 80 lines of code, while the resulting VHDL modules have 625 ... 2220.

To evaluate the efficiency of the compiler, we compare an automatically compiled Handler to a manually developed one. We cannot use the previous Handlers originally developed for [15], since the new features introduced with the NetFPGA port cause changes in the Handler-internal structure. Thus, we created a special Malacoda description for the Web (HTTP) Handler of [15] and stripped from the compiled VHDL code the functionality specifically required for the NetFPGA version of NetStage, thus leaving a version comparable to the original one (which targeted the BEE3 platform). This compiled version of the Web Handler is slightly larger (1,570 LUTs, 665 FFs) compared to the original manual implementation (1,026 LUTs and 586 FFs), but both achieve nearly the same performance when creating response packets (the compiled version needs two additional cycles). The overhead in LUTs is due to a more complex regular expression matching implementation and additional logic in the generic implementation of output packet generation. Since the compiler has been optimized for generating high-performance hardware, the increased use of resources is acceptable here, since the Handlers are still relatively small compared to the overall FPGA capacity.

6.3.1 Performance

The latency of an individual Handler consists of a fixed number of clock cycles for administrative functions (processing header data, register notifications), and a variable number of clock cycles depending on the size of the packet for content-related activities (as the compiler generates hardware that processes an entire 16 B input / output word in one clock cycle). Table 3 shows performance data for example packets processed by the Web, Mail and DNS Handler. Note that these latency limits are maintained up to the throughput limit, as the Handlers are implemented using dedicated (non-shared) resources that do not depend on the system load. The maximum throughput can be calculated from the current core target frequency of 175 MHz.

Here, the Web Handler achieves a raw data throughput of 14.5 Gb/s including administrative messages (but excluding packet header processed by the core), which would be sufficient to saturate the 10G link. Administrative messages (e.g., TCP retransmission notification) are transmitted using the same channel as network data, thus reducing the core network throughput for outgoing packets. Handlers which generate a smaller volume of output data are affected more by the administrative overhead: At 175 MHz, the Mail Handler could reply to SMTP HELO messages with an external throughput of only 5 Gb/s (including the 54 B of protocol headers added by the core). This could be improved by an additional pipeline stage inside the Handler (leading to a throughput of 11 Gb/s in the Mail example). Note that the Mail Handler requires an extra cycle of latency for the administrative operation of accessing the global application state memory, but completely avoids TCP throttling notifications, as current response packets always fit in one network packet.

The UDP-based DNS service does not need any notification messages at all. Therefore, the number of fixed clock cycles for response generation is further reduced. In this example, the DNS Handler achieves an external throughput (including the 42 B of protocol headers) of 10.6 Gb/s.

6.4 Live Test

For the live test, the NetFPGA 10G card has been connected to a 10G data center uplink at a major German university, with two dedicated /25 subnets (= 256 IPs) assigned to the honeypot. The public network traffic for the honeypot was dumped for later analysis on the management server. The Handlers were configured to listen on all IP addresses and the test was run for one month. During that time, 1.74 Million connection requests were reaching the honeypot. Table 4 lists the Top-10 services requested, as well as numbers for the remaining services for which the honeypot has active Handlers (active Handlers are shaded gray).

With a connection rate of more than 50%, the Microsoft SMB protocol is leading the list. This is unsurprising, since due to its widespread use and various known vulnerabilities, SMB is a promising target for attackers. In total, our four

Table 4: Number of connections by service

Nr.	# Conn.	Port	Service
1.	977,549	445/TCP	MS-DS (SMB)
2.	167,430	80/TCP	HTTP
3.	82,882	139/TCP	NETBIOS Session
4.	36,167	3389/TCP	MS WBT Server
5.	31,093	1433/TCP	MS SQL Server
6.	30,966	8080/TCP	HTTP Alternate
7.	27,063	22/TCP	SSH
8.	20,118	23/TCP	Telnet
9.	15,618	210/TCP	Z39.50
10.	13,627	25/TCP	SMTP
44.	1838	1434/UDP	MS SQL Monitor
189.	243	53/UDP	DNS

Table 5: Number of monitoring events

Event	# Occurrences
Webserver: GET URL	118,384
MSSQL: Slammer Worm	1,588
SMB: Login Attempt	24,566
Mailserver: Mail Queued	3,778
Telnet: Login	11,438

TCP-based Handlers are among the Top-10, such that the honeypot had a good coverage of network traffic.

In addition to counting the raw connections, we also implemented monitoring points inside the Handlers, using the Malacoda **log** command to log when a certain step has been reached. The occurrence of these events is given in Table 5. While some portion of the connections from Table 4 was coming from simple portscans, many of the clients actually interacted with our honeypot. We discuss these results in the following subsections.

6.4.1   Web Server

Around 70% of the clients were requesting a particular URL. The majority of these were attempts to reach a vulnerable service (e.g., phpMyAdmin). On the other hand, we did not observe clients trying to log into the webmail facade served by the Handler. It appears that automatic attack tools do not try to take advantage of a login form, and that human attackers did not interact with this part of the honeypot.

In addition to the HTTP web service, we also observed many TCP SYN ACK requests (362,381) hitting our system. These packets were not initiated by our honeypot (since we never sent out SYN request), but instead originate from SYN flooding attacks to real web servers by attackers using spoofed IP addresses belonging to our darknet. This is a well known procedure; commercial service providers exist that detect DDoS attacks by monitoring such darknets using distributed sensors [1].

6.4.2   Mail Server

Around 25% of the connecting clients actually tried to send a mail. The contents of these mails appear to be initial identification messages from spam engines, checking whether a server that looks like an open relay actually does deliver the mails. The mails were addressed to cryptic recipient addresses hosted at public webmail services. However, we refrained from actually delivering these messages to avoid impacting the owner institution of our darknet IP range. In a later test, these probe mails could be selectively forwarded to induce the attackers to send real spam to the honeypot, and allow it to collect any attached malware.

6.4.3   DNS Server

Attackers only had limited interest in the DNS server emulation. All requests indicate coming from automatic vulnerability scanners and simply request arbitrary domain names. However, we observed spoofed response packets similar to those hitting the web server. We received 13,804 DNS responses from real DNS servers, despite our system never requesting a lookup.

6.4.4   MSSQL Slammer Detection

Even nine years after the large outbreak and the massive effort to remove the worm from infected systems, we count 20-40 Slammer requests per day trying to infect the honeypot. They originate from other infected systems or automatic scripts all over the world (60% from China, 19% from India and 5% from the United States). This shows the difficulty of eradicating a worm such as Slammer once it has been released on a large scale.

6.4.5   Telnet Shell Emulation

More than 50% of all connecting clients tried to log in. The majority used the username / password combination root / admin or simply root with no password. The commands that were executed after the client has logged in were, e.g., echo test or echo connectioncheck, most likely coming from automatic scripts looking for open Telnet servers.

6.4.6   SMB Login Detection

This Handler was the most frequently accessed service of our honeypot setup. While the majority of requests were simple port scans, some of the clients were actually following the protocol interaction and tried an anonymous login without password.

We did not perform more detailed analysis, as the SMB protocol with its many variable field lengths and partial encryption is not handled very efficiently by the current Malacoda compiler prototype. These limitations will be addressed in future revision of the system.

6.5   Summary

Summing up, the results from the compiler and the live evaluation clearly show the feasibility of the described architecture. The hardware honeypot can be operated just as any low-interaction software honeypot, collecting data unattended for long time periods, but without the risk of the system becoming compromised. The implementation on the NetFPGA 10G has proven its stability and the Internet protocol implementation of the NetStage communication core demonstrated its ability to establish communication sessions with many different clients on the Internet. Due to the simplified programming interface offered by Malacoda, the system has a high potential in research, education, and production environments. The compiled Handlers have a similar performance to manually optimized ones, but a significantly reduced development effort.

7. CONCLUSION AND FUTURE WORK

With NetStage, we have already demonstrated the high potential of reconfigurable computing beyond the commonly used switching, routing, and deep packet-inspection applications. Using MalCoBox, our honeypot-in-a-box, as a demonstrator, we exploit hardware-accelerated operations not only for higher performance, but also for hardened security.

This work has begun to address a common problem limiting the use of reconfigurable technology for data processing purposes, namely the lack of high-level design tools. We have approached this issue by defining Malacoda, a domain-specific language focused on network processing, specifically for active security applications such as honeypots. It allows networking experts not proficient in hardware design to easily and concisely describe the protocol interactions typical for the emulated network services of a honeypot.

The associated compiler, even though it is just a prototype, has already succeeded in compiling Malacoda programs into high-performance Handlers for execution on the platform, fully exploiting its capabilities for multi-threading, parallel execution, and deep pipelining. The evaluation of the actual implementation on the NetFPGA 10G platform and the long-term live test not only demonstrate the practical feasibility of the developed architecture, but also show directions for future research.

The focus will be on improving the compiler, e.g., by integrating more efficient regular expression matching hardware, better optimization during condition evaluation, and optional optimization for single connection throughput by deeper pipelining. Beyond these issues, support for cryptographic operations, more arithmetic and logic computations, as well as the ability to seamlessly access manually designed IP blocks from Malacoda code, are already planned.

8. ACKNOWLEDGMENTS

This work was support by CASED (www.cased.de) and Xilinx, Inc.

9. REFERENCES

[1] ARBOR Networks. Active Threat Level Analysis System (ATLAS). Available online at: atlas.arbor.net.

[2] BEEcube Inc. *BEE3 Hardware User Manual*, 2008.

[3] G. Brebner. Packets everywhere: The great opportunity for field programmable technology. *Proc. Intl. Conf. on Field Programmable Technology*, pages 1–10, 2009.

[4] Dionaea. Available online at: dionaea.carnivore.it.

[5] T. Ganegedara, Y.-H. E. Yang, and V. K. Prasanna. Automation Framework for Large-Scale Regular Expression Matching on FPGA. In *Proc. Intl Conf. Field Programmable Logic and Applications*, pages 50–55, 2010.

[6] Honeyd. Available online at: www.honeyd.org.

[7] T. Katashita, Y. Yamaguchi, A. Maeda, and K. Toda. FPGA-Based Intrusion Detection System for 10 Gigabit Ethernet. *IEICE Trans. Information and Systems*, E90-D:1923–1931, 2007.

[8] E. Kohler, R. Morris, B. Chen, et al. The Click modular router. *ACM Trans. Computer Systems*, 18:263–297, 2000.

[9] M. Labrecque, J. G. Steffan, G. Salmon, et al. NetThreads: Programming NetFPGA with Threaded Software. In *Proc. NetFPGA Developers Workshop*, 2009.

[10] D. Litchfield. Microsoft SQL Server 2000 Unauthenticated System Compromise. Available online at: http://marc.info/?l=bugtraq&m=102760196931518.

[11] J. Lockwood, N. McKeown, G. Watson, et al. NetFPGA - An Open Platform for Gigabit-Rate Network Switching and Routing. In *Proc. Intl. Conf. Microelectronic Systems Education*, pages 160 –161, 2007.

[12] Mentor Graphics. Catapult C. Available online at: www.mentor.com.

[13] S. Mühlbach, M. Brunner, C. Roblee, and A. Koch. MalCoBox: Designing a 10 Gb/s Malware Collection Honeypot using Reconfigurable Technology. In *Proc. 20th Intl. Conf. on Field Programmable Logic and Applications*, pages 592–595, 2010.

[14] S. Mühlbach and A. Koch. A novel network platform for secure and efficient malware collection based on reconfigurable hardware logic. In *Proc. 2011 World Congress on Internet Security*, pages 9–14, 2011.

[15] S. Mühlbach and A. Koch. NetStage/DPR: A Self-adaptable FPGA Platform for Application-Level Network Security. In *Proc. 7th Intl. Symposium on Reconfigurable Computing: Architectures, Tools and Applications*, pages 328–339, 2011.

[16] Mykonos. Mykonos Web Security. Available online at: www.mykonossoftware.com.

[17] NetFPGA 10G. Available online at: www.netfpga.org.

[18] NetLogic Microsystems. NETL7 Layer 7 knowledge-based processor. Available online at: www.netlogicmicro.com.

[19] T. Parr. ANTLR Parser Generator v3, 2008. Available online at: http://www.antlr.org/.

[20] V. Pejovic, I. Kovacevic, S. Bojanic, et al. Migrating a Honeypot to Hardware. In *Proc. Intl. Conf. on Emerging Security Information, Systems, and Technologies*, pages 151–156, 2007.

[21] M. Roesch. Snort - Lightweight Intrusion Detection for Networks. In *Proc. 13th USENIX Conf. System Administration*, LISA '99, pages 229–238, 1999.

[22] E. Rubow, R. McGeer, J. Mogul, and A. Vahdat. Chimpp: a click-based programming and simulation environment for reconfigurable networking hardware. In *Proc. 6th ACM/IEEE Symposium on Architectures for Networking and Communications Systems*, pages 36:1–36:10, 2010.

[23] Synopsys. Synphony C. Available online at: synopsys.com.

[24] Tilera. TILE64 processor. Available online at: tilera.com.

[25] H. Wang, S. Pu, et al. A modular NFA architecture for regular expression matching. *Proc. Intl. Symposium on Field Programmable Gate Arrays*, pages 209–218, 2010.

[26] Xilinx, Inc. *UG632: PlanAhead User Guide v. 13.3*, 2011.

[27] Xilinx, Inc. *UG681: ISE Design Suite Software Manuals and Help v. 13.3*, 2011.

Author Index

Afek, Yehuda 235

Agarwal, Apoorv 137

Ahuja, Vishal 39

Akella, Aditya 199

Al-Fares, Mohammad 61

Alim, Abdul 143

Anderson, James W. 49

Arumaithurai, Mayutan 223

Balakrishnan, Hari 151

Bezahaf, Mehdi 143

Borkmann, Daniel 75

Braud, Ryan 49

Brebner, Gordon 1

Bremler-Barr, Anat 235

Burley, Brett 145

Calvert, Ken 147

Cao, Zizhong 111

Chasaki, Danai 149

Chen, Fuyu 73

Chen, Jiachen 223

Chen, Xinming 141

Chen, Yan 211

Dai, Huichen 211

Das, Sambit 61

De Carli, Lorenzo 175

Dong, Lijun 15

Dragga, Christopher 199

Estan, Cristian 175

Farrens, Matthew 39

Fleming, Kermin Elliott 151

Fu, Wenliang 139

Fu, Xiaoming 223

Ganapathy, Vinod 163

Gember, Aaron 199

Ghosal, Dipak 39

Griffioen, James 147

Harchol, Yotam 235

Harris, Eric N. 175

Hay, David 235

Horne, William 163

Huang, Shufeng 147

Iannucci, Peter A. 151

Iqbal, Muhammad Faisal 123

John, Lizy K. 123

Kapoor, Rishi 49, 61

Kekely, Lukáš 77

Keller, Ariane 75

Keslassy, Isaac 99

Kirovski, Darko 3

Koch, Andreas 247

Kolodny, Avinoam 99

Koponen, Teemu 135

Koral, Yaron 235

Korcek, Pavol 81

Kořenek, Jan 77, 81

Kosar, Vlastimil 81

Lee, Ki Suh 27

Li, Fu 73

Li, Jun 15, 83

Lin, Cheng-Hung 79

Liu, Bin 15

Liu, Bin 211

Liu, Jyh-Charn 79

Liu, Zhi 83

Lu, Jianyuan 15

Manadhata, Pratyusa 163

Marian, Tudor 27

Mathy, Laurent 143

Matsuoka, Naoki 87

Muehlbach, Sascha 247

Murai, Jun 187

Narayanan, Ashok 85

Neuhaus, Stephan 75

Ng, Tze Sing Eugene 137

Onoue, Koichi 87

Oran, Dave 85

Panwar, Shivendra S. 111

Perry, Jonathan 151

Porter, George 49

Porter, George 61

Prabhakar, Balaji 61

Puš, Viktor 77

Qi, Yaxuan 83

Ramakrishnan, K. K. 223

Rao, Prasad 163

Sankaralingam, Karthikeyan 175

Shah, Devavrat 151

Shigechika, Noriyuki 187

Shin, Ji-Yong 3

Sirer, Emin Gün 3

So, Won 85

Song, Tian 139

Sun, Xiaoye 137

Tanaka, Jun 87

Tazaki, Hajime 187

Uehara, Keisuke 187

Vahdat, Amin 49, 61

Van Meter, Rodney 187

Viktorin, Jan 81

Wakikawa, Ryuji 187

Wang, Shian 139

Wang, Xiang 83

Wang, Xiaojun 139

Wang, Xin 15

Wang, Yaogong 85

Wang, Yi 15, 211

Wasmundt, Samuel L. 175

Weatherspoon, Hakim 3, 27, 61

Wolf, Tilman 141

Wu, Hao 15

Wu, Jianming 73

Xie, Haiyong 73

Yang, Liu 163

Yang, Qimin 145

Zadnik, Martin 81

Zahavi, Eitan 99

Zhang, Yanyong 15

www.ingramcontent.com/pod-product-compliance
Lightning Source LLC
Chambersburg PA
CBHW061356210326
41598CB00035B/6003